Advanced Structural Mechanics

Advanced Structural Mechanics

Alberto Carpinteri

CRC Press
Taylor & Francis Group
Boca Raton London New York

CRC Press is an imprint of the
Taylor & Francis Group, an **informa** business

A SPON PRESS BOOK

CRC Press
Taylor & Francis Group
6000 Broken Sound Parkway NW, Suite 300
Boca Raton, FL 33487-2742

First issued in paperback 2019

ISBN-13: 978-0-415-58037-3 (hbk)
ISBN-13: 978-0-367-86473-6 (pbk)

Library of Congress Cataloging-in-Publication Data

Names: Carpinteri, Alberto, 1952- author.
Title: Advanced structural mechanics / Alberto Carpinteri.
Description: Boca Raton : CRC Press, Taylor & Francis Group, 2017. | Includes bibliographical references and index.
Identifiers: LCCN 2016026415 | ISBN 9780415580373
Subjects: LCSH: Structural engineering. | Structural analysis (Engineering)
Classification: LCC TA633 .C35389 2017 | DDC 624.1/7--dc23
LC record available at https://lccn.loc.gov/2016026415

Visit the Taylor & Francis Web site at
http://www.taylorandfrancis.com

and the CRC Press Web site at
http://www.crcpress.com

Contents

Preface   *xi*
Author   *xiii*

1 Plane frames   **1**

   1.1 Introduction 1
   1.2 Beam systems with axial symmetry 4
   1.3 Beam systems with axial skew-symmetry 9
   1.4 Beam systems with polar symmetry 13
   1.5 Beam systems with polar skew-symmetry 14
   1.6 Rotating-node frames 17
   1.7 Translating-node frames 27
   1.8 Thermal loads and imposed displacements 35
   1.9 Frames with nonorthogonal beams 40
   1.10 Frames loaded out of their own plane 44

2 Statically indeterminate beam systems: method of displacements   **49**

   2.1 Introduction 49
   2.2 Parallel-arranged bar systems 49
   2.3 Parallel-arranged beam systems 54
   2.4 Automatic computation of beam systems with
       multiple degrees of indeterminacy 58
   2.5 Plane trusses 65
   2.6 Plane frames 67
   2.7 Plane grids 69
   2.8 Space trusses and frames 71

3 Plates and shells   **75**

   3.1 Introduction 75
   3.2 Plates in flexure 75
   3.3 Sophie Germain's equation 82
   3.4 Shells with double curvature 84
   3.5 Nonsymmetrically loaded shells of revolution 88
   3.6 Symmetrically loaded shells of revolution 91

3.7 Membranes and thin shells of revolution 93
3.8 Circular plates 97
3.9 Cylindrical shells 104
3.10 Cylindrical pressurized vessels with bottoms 106
3.11 Three-dimensional bodies of revolution 110

4 Finite element method 113

4.1 Introduction 113
4.2 Single-degree-of-freedom system 113
4.3 Principle of minimum total potential energy 116
4.4 Ritz–Galerkin method 118
4.5 Application of the principle of virtual work 121
4.6 Kinematic boundary conditions 127

5 Dynamics of discrete systems 129

5.1 Introduction 129
5.2 Free vibrations 129
  5.2.1 Undamped free vibrations ($c = 0$) 131
  5.2.2 Damped free vibrations ($c > 0$) 132
5.3 Harmonic loading and resonance 137
  5.3.1 Undamped systems 137
  5.3.2 Systems with viscous damping 138
5.4 Periodic loading 142
5.5 Impulsive loading 143
5.6 General dynamic loading 146
5.7 Nonlinear elastic systems 149
5.8 Elastic–perfectly plastic spring 152
5.9 Linear elastic systems with two or more degrees of freedom 154
5.10 Rayleigh ratio 156
5.11 Stodola–Vianello method 158

6 Dynamics of continuous elastic systems 161

6.1 Introduction 161
6.2 Modal analysis of deflected beams 162
6.3 Different boundary conditions for the single beam 165
  6.3.1 Simply supported beam 165
  6.3.2 Cantilever beam 167
  6.3.3 Rope in tension 168
  6.3.4 Unconstrained beam 169
  6.3.5 Double clamped beam 170
  6.3.6 Clamped–hinged beam 172
6.4 Continuous beam on three or more supports 173
6.5 Method of approximation of rayleigh–ritz 175
6.6 Dynamics of beam systems 180
6.7 Forced oscillations of shear-type multistory frames 182

6.8   *Vibrating membranes 189*
6.9   *Vibrating plates 193*
6.10   *Dynamics of shells and three-dimensional elastic solids 195*
6.11   *Dynamics of elastic solids with linear viscous damping 200*

7 Buckling instability in slender, thin, and shallow structures   203

7.1   *Introduction 203*
7.2   *Discrete mechanical systems with one degree of freedom 204*
7.3   *Discrete mechanical systems with two or more degrees of freedom 206*
7.4   *Rectilinear elastic beams with different constraint conditions 214*
7.5   *Framed beam systems 223*
7.6   *Rings and cylindrical shells subjected to external pressure 227*
7.7   *Lateral torsional buckling 231*
7.8   *Plates subjected to compression 234*
7.9   *Shallow arches and shells subjected to vertical loading:*
       interaction between buckling and snap-through 239
7.10   *Trussed vaults and domes: the case of progressive snap-through 244*

8 Long-span structures: dynamics and buckling   253

8.1   *Introduction 253*
8.2   *Influence of dead loads on natural frequencies 254*
8.3   *Discrete systems with one or two degrees of freedom 254*
8.4   *Flexural oscillations of beams subjected to compression axial loads 261*
8.5   *Oscillations and lateral torsional buckling of deep beams 264*
8.6   *Finite element formulation for beams, plates, and shells 266*
8.7   *Nonconservative loading and flutter 270*
8.8   *Wind effects on long-span suspension or cable-stayed bridges 277*
8.9   *Torsional divergence 278*
8.10   *Galloping 280*
8.11   *Flutter 283*

9 High-rise structures: statics and dynamics   293

9.1   *Introduction 293*
9.2   *Parallel-arranged system of vertical cantilevers: general algorithm 294*
9.3   *Vlasov's theory of thin-walled open-section beams in torsion 301*
9.4   *Capurso's method: lateral loading distribution between the*
       thin-walled open-section vertical cantilevers of a tall building 309
9.5   *Diagonalization of Vlasov's equations 314*
9.6   *Dynamic analysis of tall buildings 316*
9.7   *Numerical example 319*

10 Theory of plasticity   329

10.1   *Introduction 329*
10.2   *Elastic–plastic flexure 332*

10.3   *Incremental plastic analysis of beam systems 338*

10.4   *Law of normality of incremental plastic deformation
and of convexity of plastic limit surface 351*

10.5   *Theorems of plastic limit analysis 354*

  10.5.1   *Theorem of maximum dissipated energy 354*

  10.5.2   *Static theorem (upper bound theorem) 355*

  10.5.3   *Kinematic theorem (lower bound theorem) 356*

  10.5.4   *Mixed theorem 356*

  10.5.5   *Theorem of addition of material 356*

10.6   *Beam systems loaded proportionally by concentrated forces 357*

10.7   *Beam systems loaded proportionally by distributed forces 362*

10.8   *Nonproportionally loaded beam systems 371*

10.9   *Cyclic loading and shake-down 375*

10.10 *Deflected circular plates 379*

10.11 *Deflected rectangular plates 383*

11 Plane stress and plane strain conditions   389

11.1   *Introduction 389*

11.2   *Plane stress condition 389*

11.3   *Plane strain condition 392*

11.4   *Deep beam 394*

11.5   *Thick-walled cylinder 399*

11.6   *Circular hole in a plate subjected to tension 404*

11.7   *Concentrated force acting on the edge of an elastic half-plane 408*

11.8   *Analytical functions 411*

11.9   *Kolosoff–Muskhelishvili method 415*

11.10 *Elliptical hole in a plate subjected to tension 420*

12 Mechanics of fracture   427

12.1   *Introduction 427*

12.2   *Griffith's energy criterion 429*

12.3   *Westergaard's method 433*

12.4   *Mode II and mixed modes 442*

12.5   *Williams' method 446*

12.6   *Relation between energy and stress treatments: Irwin's theorem 451*

12.7   *Crack branching criterion in mixed mode condition 458*

12.8   *Plastic zone at the crack tip 462*

12.9   *Size effects and ductile–brittle transition 466*

12.10 *Cohesive crack model and snap-back instability 472*

12.11 *Eccentric compression on a cracked beam:
opening versus closing of the crack 482*

12.12 *Stability of fracturing process in reinforced
concrete beams: the bridged crack model 484*

References	*493*
Appendix I	*497*
Appendix II	*501*
Appendix III	*505*
Appendix IV	*507*
Appendix V	*511*
Appendix VI	*521*
Appendix VII	*525*
Index	*527*

Preface

Advanced Structural Mechanics intends to provide a complete and uniform treatment of the more advanced and modern themes of structural mechanics, whereas the more fundamental and traditional have been already treated in my previous volume, *Structural Mechanics Fundamentals*, published in 2014 by CRC Press, Taylor & Francis Group. The present text represents the second edition of a substantial part (8 chapters over 20) of my original book *Structural Mechanics: A Unified Approach*, published in 1997 by E & FN SPON, an imprint of Chapman & Hall. Four new chapters have been added, while the old ones have been revised and improved.

The theory of statically indeterminate beam systems is initially presented, with the solution of numerous examples and the plotting of the corresponding diagrams of axial force, shearing force, and bending moment. For the framed structures, approached on the basis of the method of displacements, automatic computational procedures, normally involving the use of computers, are introduced in both the static and the dynamic regimes.

The mechanics of linear elastic plates and shells is then studied adopting a matrix formulation, which is particularly useful for numerical applications. The kinematic, static, and constitutive equations, once composed, provide an operator equation which has as its unknown the generalized displacement vector. Moreover, constant reference is made to duality, i.e., to the strict correspondence between statics and kinematics that emerges as soon as the corresponding operators are rendered explicit, and it is at once seen how each of them is the adjoint of the other. In this context, the finite element method is illustrated as a method of discretization and interpolation for the approximate solution of elastic problems, in the static as well as in the dynamic regimes.

In the second part of the volume, the more frequently occurring phenomena of global structural collapse are studied: instability of elastic equilibrium (buckling), plastic collapse, and brittle fracture. The unifying aspects, such as those regarding postcritical states, and the discontinuous phenomena of snap-back and snap-through are emphasized. Numerous examples regarding frames previously examined in the elastic regime are once more taken up and analyzed incrementally in the plastic regime. Furthermore, comparison with the results based on the two theorems of plastic limit analysis (the static and kinematic theorems) is made. Regarding fracture mechanics, the conceptual distinction between "concentration" and "intensification" of stresses is highlighted, and the stress treatment and energy treatment are set in direct correlation. Size scale effects, as well as the associated ductile–brittle transition, are discussed. Analogous transition from plastic collapse to buckling by increasing the structural slenderness is considered, as well as the interaction between buckling and resonance, with a particular focus on suspension bridges.

The computational analysis of frames is treated in Chapters 1 and 2; plate and shell theory in Chapter 3, as well as the finite element method in Chapter 4; Chapters 5 and 6 develop the dynamics of discrete and continuous systems. Chapters 7–9 treat the stability and dynamics of long-span and high-rise structures, whereas Chapters 10–12 treat plastic collapse and fracture mechanics.

The book has been written to be used as a text by graduate students of either architecture or engineering, as well as to serve as a useful reference for researchers and practicing engineers. A suitable selection of various chapters may constitute a convenient support for different types of courses, from short monographic seminars to courses covering an entire academic year.

This text is the fruit of many years of lecturing at international courses and conferences, as well as of teaching in Italian universities, formerly at the University of Bologna and currently at the Politecnico di Torino, where I have been professor of structural mechanics since 1986. Chapters 1–6 represent the contents of my course, "Advanced Structural Mechanics" (80 hours); Chapters 7–9 represent the contents of my course, "Static and Dynamic Instability of Structures" (60 hours), whereas Chapters 10–12 cover the major part of my course, "Fracture and Plasticity" (80 hours).

A constant reference and source of inspiration for me in writing this book has been the tradition of the Italian School, to which I am sincerely indebted. At the same time, it has been my endeavor to update and modernize a basic, and in some respects dated, discipline by merging classical topics with ones that have taken shape in more recent times. The logical sequence of the subjects dealt with makes it possible in fact to introduce, with a minimum effort, even topics that are by no means elementary and that are of differing nature.

Finally, I wish to express my most sincere gratitude to all those colleagues, collaborators, and students, who, having attended my lectures or having read the original manuscript, have, with their suggestions and comments, contributed to the text as it appears in its definitive form. In particolar, I wish to thank: Pietro Cornetti, Giuseppe Lacidogna, Amedeo Manuello, Gianfranco Piana, Federico Accornero. I further wish to thank my PhD student, Giuseppe Nitti, for his very effective help in the proofs correction, as well as in the figure quality improvement.

Alberto Carpinteri

Author

Alberto Carpinteri received his doctoral degrees in nuclear engineering cum laude (1976) and in mathematics cum laude (1981) from the University of Bologna (Italy). After two years working for the Consiglio Nazionale delle Ricerche, he was appointed assistant professor at the University of Bologna in 1980.

He moved to the Politecnico di Torino in 1986 as professor, and became the chair of Solid and Structural Mechanics, and the director of the Fracture Mechanics Laboratory. During this period, he has held different positions of responsibility: head of the Department of Structural Engineering (1989–1995), and founding director of the Postgraduate School of Structural Engineering (1989–2014).

Prof. Carpinteri was a visiting scientist at Lehigh University, Pennsylvania, USA (1982–1983), and was appointed as a fellow of several academies and professional societies, namely the European Academy of Sciences (2009), the International Academy of Engineering (2010), the Turin Academy of Sciences (2005), the American Society of Civil Engineers (1996). He is presently the head of the Engineering Division in the European Academy of Sciences (2016).

Prof. Carpinteri was the president of different scientific associations and research institutions: the International Congress on Fracture, ICF (2009–2013), the European Structural Integrity Society, ESIS (2002–2006), the International Association of Fracture Mechanics for Concrete and Concrete Structures, lA-FraMCoS (2004–2007), the Italian Group of Fracture, IGF (1998–2005), the National Research Institute of Metrology, INRIM (2011–2013).

Prof. Carpinteri was appointed as a member of the Congress Committee of the International Union of Theoretical and Applied Mechanics, IUTAM (2004–2012), a member of the executive board of the Society for Experimental Mechanics, SEM (2012–2014), a member of the editorial board of fourteen international journals, and the editor-in-chief of the journal, *Meccanica* (Springer, IF=1.949). He is the author or editor of over 900 publications, of which more than 400 are papers in refereed international journals and 54 are books or journal special issues.

Prof. Carpinteri received numerous honors and awards: the Robert L'Hermite Medal from RILEM (1982), the Griffith Medal from ESIS (2008), the Swedlow Memorial Lecture Award from ASTM (2011), the Inaugural Paul Paris Gold Medal from ICF (2013), and the Frocht Award from SEM (2017), among others.

Chapter 1

Plane frames

1.1 INTRODUCTION

A frame is a system of beams having many degrees of indeterminacy. In this chapter, we shall refer expressly to plane frames loaded in their own plane, with the exception of a brief look at the case of a portal frame loaded by forces perpendicular to its own plane.

In reference to the deformed configuration of their own fixed-joint nodes, plane frames can be subdivided into

1. **Rotating-node frames** (Figure 1.1)
2. **Translating-node frames** (Figure 1.2)

In the former, the fixed-joint nodes rotate elastically but do not undergo translation, provided that it is possible to neglect the axial deformability of the individual beams. In the latter, the fixed-joint nodes not only rotate but also undergo considerable translation. In some cases, as in that of the portal frame of Figure 1.1a, the symmetry of the load condition implies the annihilation of the translations of the fixed-joint nodes. The same portal frame loaded in a nonsymmetrical fashion (Figure 1.2a) proves to be a translating-node frame. More precisely, the two upper nodes undergo a translation in the horizontal direction. In general, it is possible to state that the beam systems sufficiently constrained externally prove to be rotating-node frames (Figure 1.1b). By suppressing some of the external constraints, the same frames can transform into translating-node systems (Figure 1.2b). Notice, however, how symmetry does not always imply nontranslation of nodes. For example, the lack of a central column in the frame of Figure 1.2c renders the two central nodes vertically translating.

When, in the next chapter, we deal with the automatic computation of plane frames, the abovementioned distinction between rotating-node frames and translating-node frames will not be made. In that case, the method of resolution will be not susceptible to such a distinction. In the present chapter, we shall propose a method of solution in which the translations of the fixed-joint nodes are among the unknowns of the problem, together with the redundant moments which develop at the fixed-joint nodes themselves. It is, in other words, a hybrid method, halfway between the method of forces and the method of displacements (see next chapter), in which the equations that resolve the problem consist partly of relations of congruence and partly of relations of equilibrium.

The method outlined consists in disconnecting, with respect to rotation, all the external built-in supports and the internal fixed-joint nodes, putting hinges in them, and applying the corresponding redundant moments (Figures 1.3a and b). On account of the equilibrium

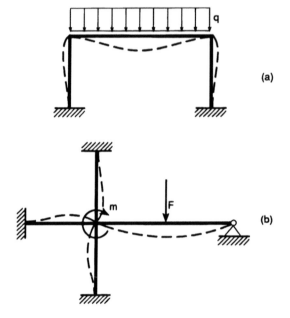

Figure 1.1

with regard to rotation of the fixed-joint node, the redundant moments are linked together by the following relation (Figures 1.3c and d):

$$\sum_i X_i = 0$$

It is important to distinguish between the moments X_i exerted by the fixed-joint node on the convergent beams and assumed as being counterclockwise in the schemes of Figures 1.3b and c, and the moments X_i exerted by the beams on the fixed-joint node, which are the opposite of the previous ones (Figure 1.3d). In the case where the fixed-joint node is loaded by a concentrated moment m (Figure 1.3e), the relation that expresses the equilibrium of the node becomes (Figure 1.3f)

$$\sum_i X_i = m$$

Once hinges have been inserted in all built-in supports and fixed-joint nodes, we obtain a beam system, called the **associated truss structure**, which may be either redundant, isostatic, or hypostatic. In the former two cases, the original frame is a rotating-node frame (Figure 1.4), with the unknowns consisting of the redundant moments alone, and the equations for resolving the problem consisting of the relations of angular congruence alone. In the latter case, the original frame is a translating-node frame (Figure 1.5), with the supplementary unknowns consisting of the displacements of the nodes and the supplementary equations consisting of as many relations of equilibrium, in general expressible via the application of the principle of virtual work. In the case, for instance, of the structure of Figure 1.2a, the associated truss (Figure 1.5a) consists of an articulated parallelogram whose only degree of

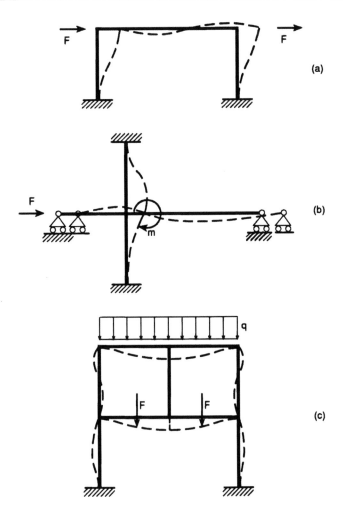

Figure 1.2

freedom can be described by the horizontal translation of the upper nodes. The same thing can be said for the structure of Figure 1.2b, whose associated truss (Figure 1.5b) presents the possibility of translations of the horizontal cross member. However, the truss associated with the frame of Figure 1.2c presents three degrees of freedom, corresponding, respectively, to the horizontal translations of the two cross members and to the vertical translation of the central column (Figures 1.5c through e).

At the beginning of the chapter, we shall consider beam systems with properties of geometrical, constraint, and loading symmetry. These are structures that present axes or centers of symmetry about which the distribution of matter, the constraints, and the loads applied are symmetrical. As we shall be able to investigate more fully in the following, the most notable consequence deriving from the properties of symmetry of a statically indeterminate structure is the reduction of its effective degree of indeterminacy. The increase in regularity with respect to structures devoid of symmetry is, in general, the cause of a decrease in static indeterminacy. Both axial symmetry and polar symmetry will be considered, as well as the corresponding skew symmetries.

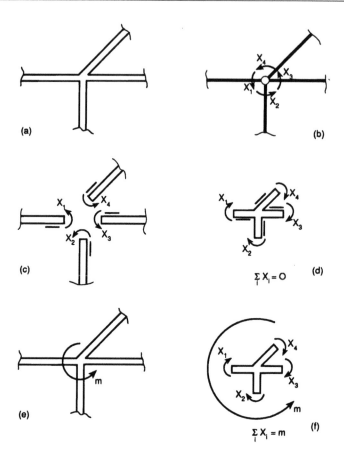

Figure 1.3

1.2 BEAM SYSTEMS WITH AXIAL SYMMETRY

A beam system is said to be symmetrical with respect to an axis when one of the two halves into which the structure is subdivided by the axis comes to superpose itself on the other, if it is made to rotate by 180° about the axis itself (Figure 1.6a). A beam system with axial symmetry is said to be symmetrically loaded if, in the abovementioned rotation, the loads that act on one half also come to superpose themselves on those acting on the other half. In addition to the beams and the loads, the constraints, both external and internal, must, of course, also respect the condition of symmetry so that the structural behavior should be specularly symmetrical.

In a beam system with axial symmetry, the structural response, whether static or kinematic, must logically prove symmetrical. This means that the characteristics, both static and deformation, must be specular. Whereas, then, the axial force and the bending moment are equal and have the same sign in the pairs of symmetrical points, the shearing force is equal but has an opposite sign, with a skew-symmetrical diagram. Likewise, the elastic rotations and the elastic displacements perpendicular to the axis of symmetry are equal and opposite, while the elastic displacements in the direction of the axis are equal and of the same sign.

If the conditions described are to be maintained on the axis of symmetry, it will be necessary for the shearing force, as well as the rotation and the component of the displacement orthogonal to the axis of symmetry, to vanish. Thus, whereas the shearing force T vanishes

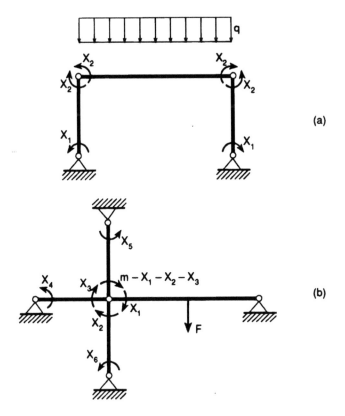

Figure 1.4

to satisfy symmetry and equilibrium simultaneously, the displacement u and rotation φ vanish to satisfy symmetry and congruence simultaneously (Figure 1.6a). These conditions, both static and kinematic, at the points where the axis of symmetry encounters the structure, are realized by a double rod perpendicular to the axis itself (Figure 1.6b). It is, therefore, possible to reduce the study of the entire structure to that of one-half, constrained at a point corresponding to its axis of symmetry by a double rod. Consider, then, that the axial force and bending moment diagrams are symmetrical, whereas the shearing force diagram is skew-symmetrical.

The reduced structure of Figure 1.6b has two degrees of indeterminacy, while the original structure apparently has three (Figure 1.6a). This means that the structure of Figure 1.6a actually has two degrees of indeterminacy for reasons of symmetry. Whereas, in fact, the vertical reactions are each equal to one-half of the vertical load, the fixed-end moments and the horizontal reactions are represented by equal and symmetrical loadings, which remain, however, statically indeterminate (Figure 1.6c).

In the case where, instead of the internal fixed-joint, there is a weaker constraint at the axis of symmetry, it is possible to apply once again what has already been said, but excluding *a priori* from the conditions of symmetry the characteristics not transmitted by the constraint itself, and at the same time including the relative displacements permitted by it. If, for example, the two symmetrical parts of a structure are connected by a hinge (Figure 1.7a), the moment in the center will vanish by definition of the hinge constraint, while the shear will vanish by virtue of symmetry. Hence, the only remaining static characteristic transmitted by the hinge will be the axial force. The existence of the hinge allows, on the other

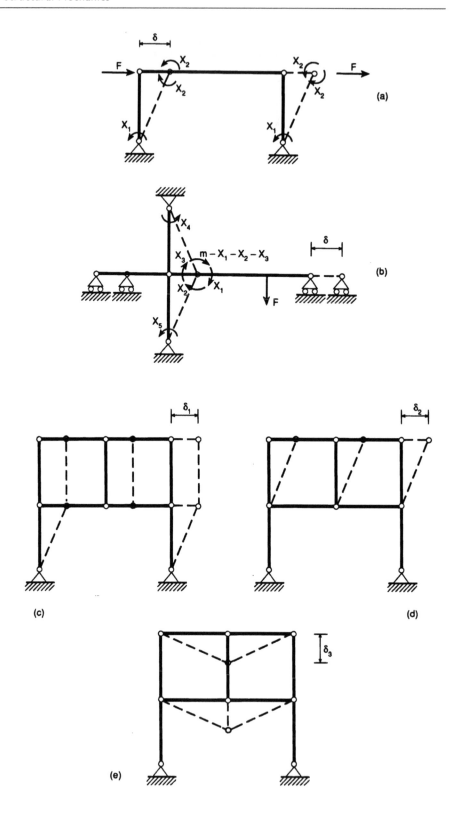

Figure 1.5

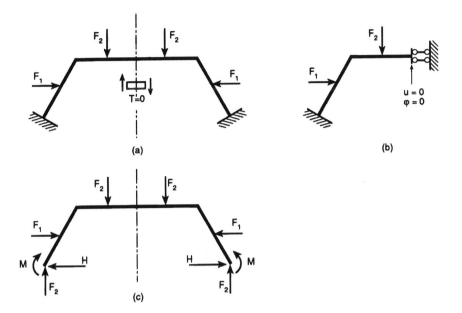

Figure 1.6

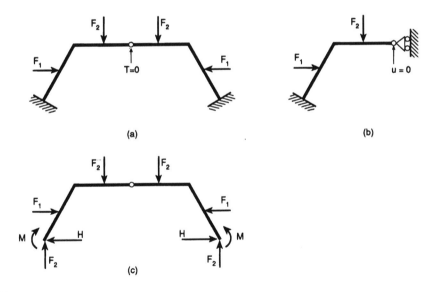

Figure 1.7

hand, relative rotations between the two parts, just as symmetry allows displacements of the center in the direction of the axis. However, displacements of the center are not possible in either direction perpendicular to the axis for reasons of symmetry, nor are detachment or overlapping possible for reasons of congruence.

These conditions, both static and kinematic, are realized by a vertically moving rolling support or by a horizontal connecting rod. The equivalent structure of Figure 1.7b has one

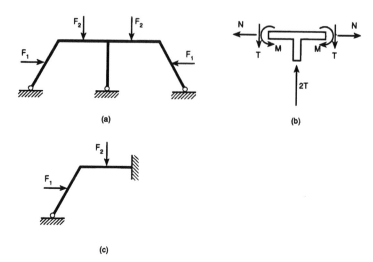

Figure 1.8

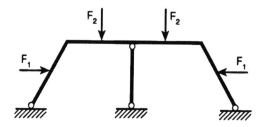

Figure 1.9

degree of indeterminacy, while the original structure of Figure 1.7a apparently has two. As before, the vertical reactions are statically determinate and each equal to one-half of the vertical load, while the horizontal reactions and the fixed-end moments are linked together by the equation of equilibrium to rotation of each part about the hinge (Figure 1.7c).

In the case where there are columns or uprights on the axis of symmetry (Figure 1.8a), it is necessary to consider, in addition to the conditions of symmetry, the conditions of equilibrium of the central fixing-node (Figure 1.8b). It is simple to conclude that the upright is loaded by an axial force that, in absolute value, is twice the shear transmitted by each of the two horizontal beams, while the characteristics of shear and bending moment are zero on the upright for reasons of symmetry. If the upright is considered as axially undeformable, the equivalent structure is reduced to that of Figure 1.8c, where the center is constrained with a perfect fixed-joint. This structure is, thus, indeterminate to the second degree, whereas the original structure is apparently indeterminate to the third degree (Figure 1.8a). If, instead, we wish to take into account the axial compliance of the central upright, it is necessary to consider a fixed-joint elastically compliant to the vertical translation, having a stiffness of $EA/2h$, where h denotes the height of the upright.

Nothing changes with respect to the previous case, if the central upright, instead of being fixed, is only hinged to the horizontal beam, and thus consists of a simple vertical connecting rod (Figure 1.9). As before, it will transmit only a vertical force to the overlying beam. The equivalent scheme is thus yet again that of Figure 1.8c.

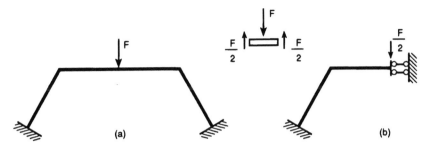

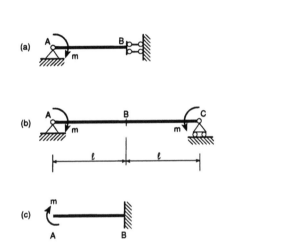

Figure 1.10

Figure 1.11

Finally, let us take the case where a concentrated force is applied in the center (Figure 1.10a). By reason of the equilibrium of the central beam element and from symmetry, it is possible to refer to the equivalent scheme of Figure 1.10b, where the end constrained by the double rod is also loaded by a force equal to one-half of the total.

Applying in inverse manner the considerations so far made, it is possible to calculate the elastic rotation at the hinged end of a beam, constrained at the opposite end by a double rod (Figure 1.11a). From symmetry, this scheme is, in fact, equivalent to a beam of twice the length, hinged at the ends and loaded by two specular loads (Figure 1.11b). In the specific case of the moment applied to the hinged end (Figure 1.11a), the scheme of Figure 1.11b yields the rotation

$$\varphi_A = -\varphi_C = -\frac{m(2l)}{3EI} - \frac{m(2l)}{6EI} = -\frac{ml}{EI} \tag{1.1}$$

The elastic line of the beam of Figure 1.11a is, on the other hand, equal, but for an additional constant, to that of the cantilever of Figure 1.11c.

1.3 BEAM SYSTEMS WITH AXIAL SKEW-SYMMETRY

A symmetrical beam system is said to be loaded in a skew-symmetrical way when the loads acting on one of the halves are the opposite of, and symmetrical to, the loads acting on the remaining half (Figure 1.12a).

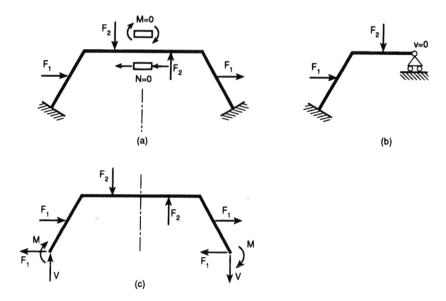

Figure 1.12

In a beam system with axial skew-symmetry, the structural response, both static and kinematic, must logically be skew-symmetrical. This means that the characteristics, both static and deformation, must be the opposite of those produced at the symmetrical points.

Whereas, then, the so-called symmetrical characteristics—axial force and bending moment—will present a skew-symmetrical diagram, the skew-symmetrical characteristic—the shearing force—will present a symmetrical diagram. Likewise, the elastic rotations and the elastic displacements orthogonal to the axis of symmetry are equal and have the same sign, while the elastic displacements in the direction of the axis are equal and opposite.

If these conditions are also to be respected on the axis of symmetry, both the axial force and the bending moment must vanish, as must the component of the displacement in the direction of the axis of symmetry. Thus, whereas the axial force N and the bending moment M vanish to satisfy skew-symmetry and equilibrium simultaneously, the displacement v vanishes to satisfy skew-symmetry and congruence simultaneously (Figure 1.12a). These conditions, which are static and kinematic, at the points where the axis of symmetry encounters the structure, are realized by a roller support moving orthogonally to the axis itself (Figure 1.12b). The study of the original structure is thus reduced to that of one of its halves, constrained in the center with a roller support. The reduced structure of Figure 1.12b thus has one degree of redundancy, whereas the original structure apparently has three (Figure 1.12a). While, in fact, the horizontal reactions are each equal to one-half of the horizontal load, the fixed-end moments and the vertical reactions are represented by equal and skew-symmetrical loads, which, respecting the condition of equilibrium to rotation, remain only once indeterminate (Figure 1.12c).

In the case where, instead of the internal fixed-joint, there is a weaker constraint at the axis of symmetry, these considerations may be repeated, taking into account, however, only the reactions transmitted by the constraint and adding the relative displacements allowed by this. If, for instance, the two symmetrical parts of a structure loaded skew-symmetrically are connected by a hinge (Figure 1.13a), the shear, as before, is the only characteristic transmitted, and hence the reduced scheme is again that of Figure 1.12b. This scheme is indeterminate to the first degree, while the original structure of Figure 1.13a is apparently

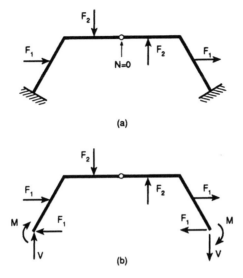

(a)

(b)

Figure 1.13

indeterminate to the second degree. The vertical reactions V and the fixed-end moments M (Figure 1.13b) are, in fact, linked by a condition of equilibrium to rotation, and are thus once indeterminate.

If, instead, the two symmetrical parts are connected by a double rod (Figure 1.14a), the constraint cannot transmit either of the two symmetrical characteristics and the equivalent scheme will be represented by the cantilever beam of Figure 1.14b. The original structure of Figure 1.14a is therefore substantially statically determinate, since the horizontal reactions

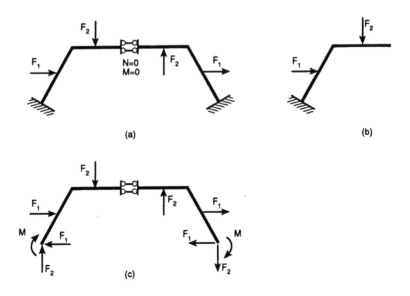

(a)

(b)

(c)

Figure 1.14

are each equal to one-half of the horizontal load, just as the vertical reactions are each equal to the vertical load acting on the corresponding part (Figure 1.14c). The fixed-end moments may, on the other hand, be determined via an equation of global equilibrium with regard to rotation.

In the case where there are columns or uprights on the axis of symmetry (Figure 1.15a), it is necessary to consider, in addition to the conditions of skew-symmetry, the conditions of equilibrium of the central fixing-node (Figure 1.15b). Unlike the case of symmetry, the upright is subjected to moment and shear, while the axial force exerted on it is zero. The bending moment and shearing force in the immediate vicinity of the node, are equal, respectively, to twice the moment and twice the axial force acting on the horizontal beams. It is thus possible to consider the reduced scheme of Figure 1.15c, where the material of the upright is considered with its elastic modulus halved.

In the case where, in place of the upright, there is a simple connecting rod (Figure 1.16), it is obvious that the reduced scheme that must be referred to remains that of Figure 1.12b, with a hinge instead of the built-in support.

When a concentrated moment m is applied in the center (Figure 1.17a), it is possible to consider this as a skew-symmetrical load consisting of two moments equal to $m/2$ and having the same sign. The equivalent scheme is that of Figure 1.17b. In the case, therefore, of

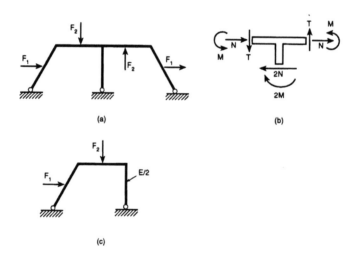

Figure 1.15

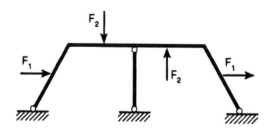

Figure 1.16

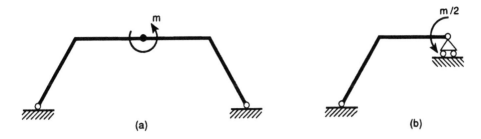

Figure 1.17

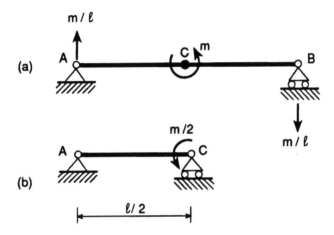

Figure 1.18

a simply supported beam (Figure 1.18a), we revert to the supported beam of halved length (Figure 1.18b). The elastic rotation of the ends thus equals

$$\varphi_A = \varphi_B = -\frac{\left(\dfrac{m}{2}\right)\left(\dfrac{l}{2}\right)}{6EI} = -\frac{ml}{24EI} \tag{1.2}$$

while the elastic rotation of the center section is

$$\varphi_C = \frac{\left(\dfrac{m}{2}\right)\left(\dfrac{l}{2}\right)}{3EI} = \frac{ml}{12EI} \tag{1.3}$$

1.4 BEAM SYSTEMS WITH POLAR SYMMETRY

A beam system is said to be symmetrical with respect to a pole when one of the two halves into which the structure is subdivided by the pole may be superposed on the other half, if made to rotate by 180° about the pole itself (Figure 1.19a). A system of beams with polar

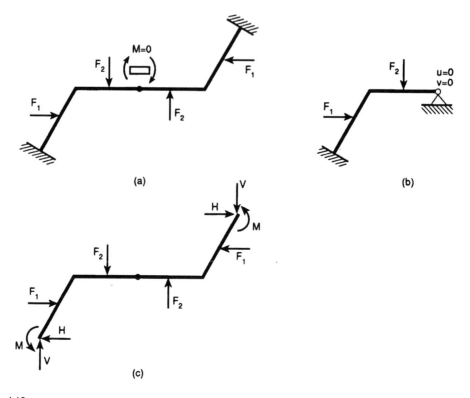

Figure 1.19

symmetry is said to be symmetrically loaded if, in this rotation, the loads acting on one half are also superposed on those acting on the remaining half.

If the structure presents polar symmetry, it is logical that, from polar-symmetrical causes, there follow polar-symmetrical effects that are both static and kinematic. In particular, at the pole, the moment vanishes to satisfy polar symmetry and equilibrium simultaneously, just as the displacement vanishes to satisfy polar symmetry and congruence simultaneously. These conditions are realized by a hinge, so that the equivalent reduced structure appears as in Figure 1.19b. This scheme thus proves to have two degrees of indeterminacy. The original structure also has two degrees of indeterminacy (Figure 1.19c). The reactions H, V, M are, in fact, linked only by the global equation of equilibrium with regard to rotation, the equations of equilibrium with regard to translation already being identically satisfied.

Of course, in the case where there is originally a hinge at the pole, the considerations outlined again all apply.

Finally, in the case where a concentrated moment m is applied at the pole (Figure 1.20a), this load can be considered as polar-symmetrical and as consisting of two moments equal to $m/2$ and having the same sign. The reduced scheme is thus that of Figure 1.20b.

1.5 BEAM SYSTEMS WITH POLAR SKEW-SYMMETRY

A polar-symmetrical beam system is said to be loaded skew-symmetrically when the loads that act on one of the halves are the opposite of and symmetrical to those acting on the remaining half (Figure 1.21a).

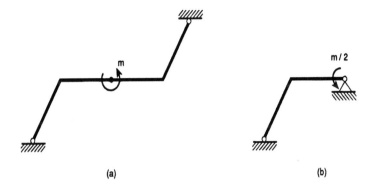

(a)   (b)

Figure 1.20

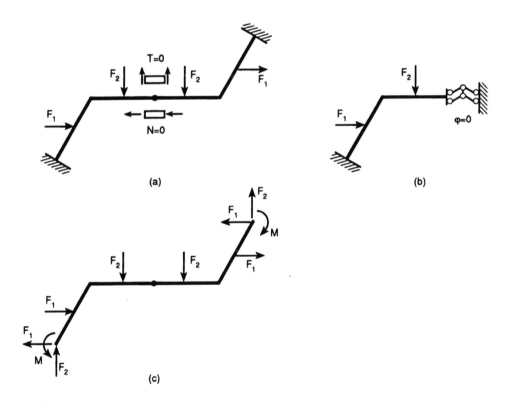

Figure 1.21

At the pole, the axial force and the shearing force vanish to satisfy polar skew-symmetry and equilibrium simultaneously, just as the elastic rotation vanishes to satisfy polar skew-symmetry and congruence simultaneously. These conditions are realized by a double articulated parallelogram (Figure 1.21b). The equivalent scheme appears to be indeterminate to the first degree, whereas the original structure is apparently indeterminate to the third degree. The degree of residual redundancy is due to the indeterminacy of the fixed-end moment M (Figure 1.21c).

In the case where there is a hinge at the pole (Figure 1.22a), the equivalent scheme reduces to the cantilever beam of Figure 1.22b. The structure is therefore substantially statically determinate. The fixed-end moment M in this case is, in fact, determined via a condition of partial equilibrium to rotation about the hinge (Figure 1.22c).

Finally, in the case where a concentrated force F is applied at the pole (Figure 1.23a), this load can be considered polar skew-symmetrical and consisting of two forces equal to $F/2$, having the same sign. The reduced scheme is thus that shown in Figure 1.23b.

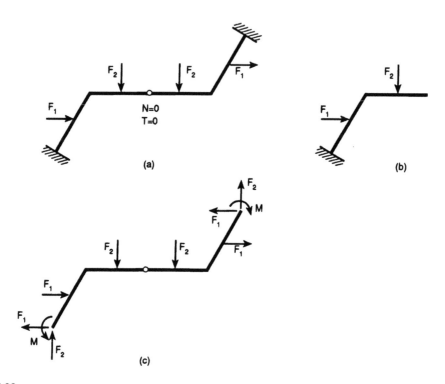

Figure 1.22

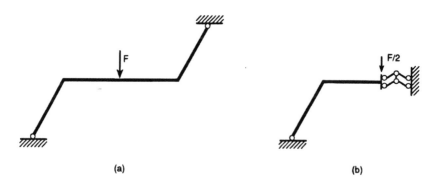

Figure 1.23

1.6 ROTATING-NODE FRAMES

The solution of **rotating-node frames** is altogether analogous to the solution of continuous beams and consists of writing down a number of equations of angular congruence equal to the number of unknown redundant moments. Once these moments are known (they turn out to be end moments for the individual beams), it is then easy to draw the moment diagrams, superposing the diagrams corresponding to external loads on the linear functions that correspond to the redundant moments.

Consider the frame of Figure 1.24a, consisting of a horizontal continuous beam and a vertical upright built-in, at the top, in the center of the beam and, at the bottom, to the foundation. Let two hinges be inserted, respectively, in the fixed-joint node B and in the external built-in support D, applying at the same time the corresponding redundant moments (Figure 1.24b). Note that, owing to the equilibrium of the node B, the redundant moment acting on the end B of the upright DB is a function of the remaining two redundant moments acting on the same node. More precisely, if we designate by X_1 the moment transmitted between the fixed-joint node and the horizontal beam AB, and by X_2 the moment transmitted between the fixed-joint node and the horizontal beam CB, the moment exchanged between the fixed-joint node and the upright is equal to (X_1+X_2) and has a sense opposite to the previous moments. Clearly, the senses assumed are altogether conventional and may not be confirmed by calculation, should negative redundant moments be obtained.

There are, thus, basically three redundant unknowns, X_1, X_2, X_3 (Figure 1.24b), just as there are three equations of angular congruence:

$$\varphi_{BA} = \varphi_{BC} \tag{1.4a}$$

$$\varphi_{BA} = \varphi_{BD} \tag{1.4b}$$

$$\varphi_{DB} = 0 \tag{1.4c}$$

Equations 1.4a and b express the fact that all three branches of the fixed-joint node of Figure 1.24b rotate by the same amount, and that, hence, the fixed-joint node as a whole rotates rigidly without relative rotations between the various branches. Equation 1.4c expresses the existence in D of a built-in support, which prevents any displacement and hence also the absolute rotation of the base section D of the upright. It is then possible to show how the associated truss structure of Figure 1.24b constitutes a statically determinate system of beams, and hence the frame under examination may be considered a rotating-node type.

The terms of Equations 1.4 can be rendered explicit by taking into account exclusively elastic rotations, which are well known for the elementary schemes of a supported beam:

$$-\frac{X_1 l}{3EI} = -\frac{X_2 l}{3EI} + \frac{ml}{6EI} \tag{1.5a}$$

$$-\frac{X_1 l}{3EI} = \frac{(X_1+X_2)l}{3EI} - \frac{X_3 l}{6EI} \tag{1.5b}$$

$$\frac{X_3 l}{3EI} - \frac{(X_1+X_2)l}{6EI} = 0 \tag{1.5c}$$

Multiplying all three equations by $6EI/l$, we obtain

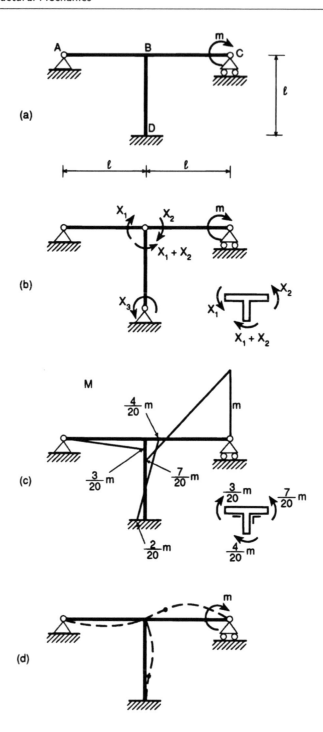

Figure 1.24

$$-2X_1 = -2X_2 + m \tag{1.6a}$$

$$-2X_1 = 2X_1 + 2X_2 - X_3 \tag{1.6b}$$

$$2X_3 - X_1 - X_2 = 0 \tag{1.6c}$$

The first expresses X_2 as a function of X_1,

$$X_2 = X_1 + \frac{m}{2} \tag{1.7}$$

and, substituting this expression into Equations 1.6b and c, we obtain a system of two linear algebraic equations in the unknowns X_1 and X_3:

$$6X_1 + m = X_3 \tag{1.8a}$$

$$2X_3 - 2X_1 - \frac{m}{2} = 0 \tag{1.8b}$$

Substituting Equation 1.8a into Equation 1.8b, we have

$$X_1 = -\frac{3}{20}m \tag{1.9a}$$

so that Equations 1.7 and 1.8a yield, respectively,

$$X_2 = \frac{7}{20}m \tag{1.9b}$$

$$X_3 = \frac{m}{10} \tag{1.9c}$$

The fixed-joint node B is thus in equilibrium under the action of the moments of Figure 1.24c. The moment diagram is obtained by laying the segments out to scale on the side of the stretched fibers, and joining the ends of these with rectilinear segments. The moment diagram is, in fact, linear, as no transverse distributed loads are present. The deformed configuration of the frame must be consistent with the external constraints and with the moment diagram (Figure 1.24d). Notice the two points of inflection corresponding to the sections in which the bending moment vanishes. The fixed-joint node rotates in a counterclockwise direction by the amount $ml/20EI$.

As regards the shear diagram, this is obtainable from the schemes of equilibrium of the individual supported beams to which we are brought back (Figure 1.25a), loaded by the external loads and by the redundant moments. This is obviously constant on each individual beam and considerably higher on the beam loaded externally (Figure 1.25b). The sign of the shear results from the usual convention. It is possible, however, to deduce the shear as a derivative of the moment function, referring the latter to a right-handed system YZ, with an origin at one of the two ends of the individual beams.

Finally, the values of the axial force are also constant in each individual beam. The beam BC is not subjected to axial force, as it is constrained by a horizontally moving roller support

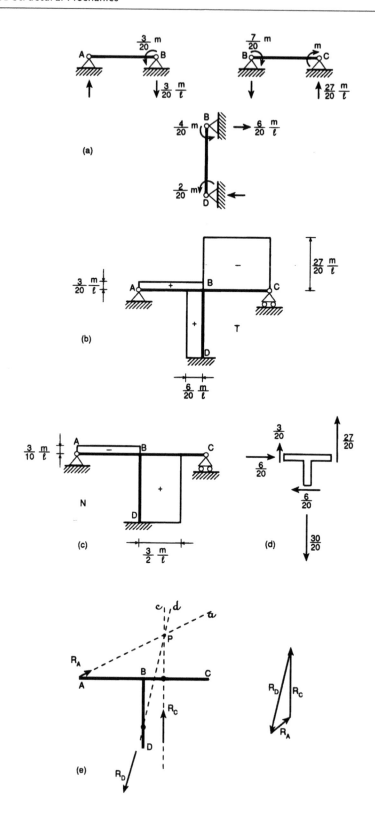

Figure 1.25

in C, while the beam AB absorbs an axial force of compression equal in absolute value to the shear of the upright DB (Figure 1.25c). In turn, the upright DB absorbs axially the shears of the two horizontal beams AB and CB, being loaded in tension by a total force equal to 3/2 m/l. An effective check at this point consists of considering the equilibrium also with regard to the translation of the fixed-joint node B (Figure 1.25d).

Many of the results just obtained can find further confirmation from the pressure line. The structure, considered as being completely disconnected externally, is basically subjected to three forces (Figure 1.25e). The first is the reaction R_A of the hinge A, which has as its components the axial force 6/20 m/l and the shearing force 3/20 m/l. The second is the resultant R_C of the applied moment m and the reaction V_C of the roller support. The third force is the reaction R_D of the built-in support D, which has as its components the axial force 3/2 m/l and the shearing force 3/10 m/l, and passes at a distance X_3/R_D from point D. It is possible to verify how these three forces pass through the same pole P and constitute a system equivalent to zero.

The pressure line may be defined portion by portion in the following manner:

- Portion AB: straight line a
- Portion CB: straight line c
- Portion DB: straight line d

It may thus be noted that the pressure line passes through those sections in which the bending moment vanishes (Figure 1.24c) and in which the elastic deformed configuration presents points of inflection (Figure 1.24d).

As a second example, let us consider the square closed configuration of beams shown in Figure 1.26a, subjected to two equal and opposite forces F. The double axial symmetry allows the structure to be considered as a rotating-node frame, and hence there is a single unknown redundant moment X (Figure 1.26b). The equation that resolves the problem will then be furnished by the condition of angular congruence

$$\varphi_{AB} = \varphi_{AC} \tag{1.10}$$

which is identical to the other three conditions for the vertices B, C, D. Rendering the condition of Equation 1.10 explicit, we have

$$\frac{Xl}{3EI} + \frac{Xl}{6EI} - \frac{Fl^2}{16EI} = -\frac{Xl}{3EI} - \frac{Xl}{6EI} \tag{1.11}$$

whence there results

$$X = \frac{Fl}{16} \tag{1.12}$$

The moment diagram is thus given in Figure 1.26c. It shows four points of annihilation, symmetrical with respect to both axes of symmetry. The elastic deformed configuration of Figure 1.26d presents, of course, points of inflection corresponding to the abovementioned sections. While the vertical beams are entirely convex outward, the horizontal beams are convex outward only in the end regions. The four fixed-joint nodes are not displaced but all rotate by the same amount $Fl^2/32EI$; A and D rotate clockwise, B and C counterclockwise.

The shear diagram for the vertical beams is zero and, for the horizontal beams, reproduces that of the scheme of the supported beam (Figure 1.26e). On the other hand, the axial force diagram for the horizontal beams is zero and, for the vertical beams, is compressive and equal to F/2 (Figure 1.26f).

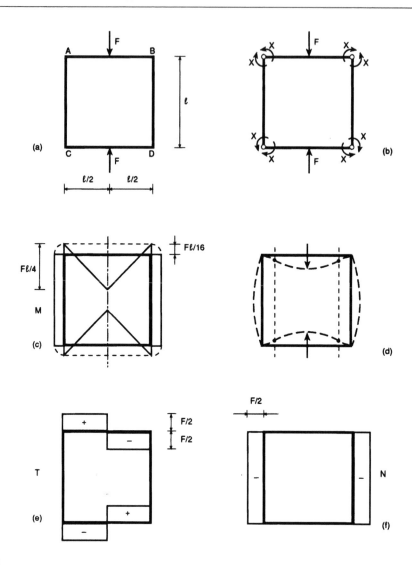

Figure 1.26

At this point, it is simple to verify that the pressure line is formed by the two vertical straight lines passing through the points of inflection (Figure 1.26d).

Let the symmetrical portal frame of Figure 1.27a be loaded by two symmetrical distributions of horizontal forces acting on the upper half of the two uprights. This frame is a rotating-node type by symmetry, and hence an equivalent statically determinate structure can be obtained by disconnecting with regard to rotation the fixed-joint nodes B and C and eliminating the connecting rod FG (Figure 1.27b).

In this case, the two equations of congruence will have to express the angular congruence of the node B (or C) as well as the annihilation of the horizontal displacement of the point F (or G):

$$\varphi_{BA} = \varphi_{BC} \tag{1.13a}$$

$$u_F = 0 \tag{1.13b}$$

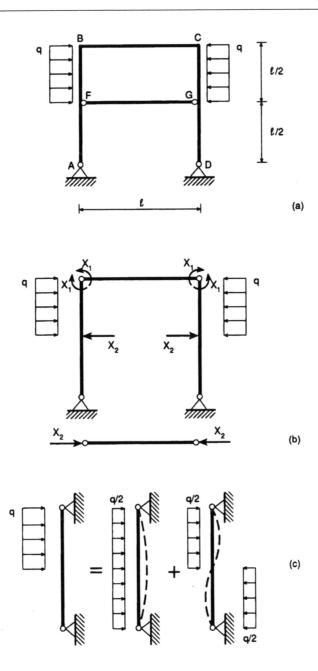

Figure 1.27

Rendering the foregoing relations explicit as functions of the redundant unknowns X_1 and X_2 (Figure 1.27b), we obtain the following two equations that resolve the problem:

$$-\frac{X_1 l}{3EI} - \frac{X_2 l^2}{16EI} + \frac{9}{384}\frac{q l^3}{EI} = \frac{X_1 l}{3EI} + \frac{X_1 l}{6EI} \qquad (1.14a)$$

$$\frac{X_1 l^2}{16EI} + \frac{X_2 l^3}{48EI} - \frac{5}{384}\frac{q l^4}{2EI} = 0 \qquad (1.14b)$$

The contributions corresponding to the load q acting on the half-beam can be obtained by considering the principle of superposition and splitting this load into two components, one symmetrical and the other skew-symmetrical (Figure 1.27c). The rotation of the end is

$$\varphi = \frac{\frac{q}{2}l^3}{24EI} + \frac{\frac{q}{2}\left(\frac{l}{2}\right)^3}{24EI} = \frac{9}{384}\frac{ql^3}{EI} \tag{1.15}$$

while the displacement in the center:

$$\upsilon = \frac{5}{384}\frac{\frac{q}{2}l^4}{EI} + 0 = \frac{5}{384}\frac{ql^4}{2EI} \tag{1.16}$$

is given by the symmetrical contribution alone.

Reordering Equations 1.14, we obtain

$$320X_1 + 24X_2 l = 9ql^2 \tag{1.17a}$$

$$48X_1 + 16X_2 l = 5ql^2 \tag{1.17b}$$

Multiplying the former by 2 and the latter by −3, and then adding them together, the unknown X_2 is eliminated:

$$X_1 = \frac{3}{496}ql^2 \tag{1.18a}$$

whence we obtain

$$X_2 = \frac{73}{248}ql \tag{1.18b}$$

Considering the scheme of Figure 1.28a, rotated by 90° with respect to its actual orientation, and writing the relations of equilibrium with regard to translation and to rotation with respect to the center:

$$V_A + V_B + X_2 = q\frac{l}{2} \tag{1.19a}$$

$$-V_A\frac{l}{2} + V_B\frac{l}{2} - X_1 - q\frac{l}{2}\frac{l}{4} = 0 \tag{1.19b}$$

we at once obtain the transverse reactions V_A and V_B. Substituting Equations 1.18 into Equations 1.19, we have

$$V_A + V_B = \frac{51}{248}ql \tag{1.20a}$$

$$V_A - V_B = -\frac{65}{248}ql \tag{1.20b}$$

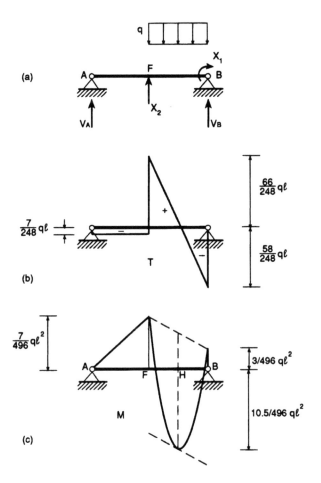

Figure 1.28

from which there follows:

$$V_A = -\frac{7}{248}ql \qquad (1.21\text{a})$$

$$V_B = \frac{58}{248}ql \qquad (1.21\text{b})$$

The shear diagram on the upright AB is depicted in Figure 1.28b. The extreme values are equal in absolute value to the reactions V_A and V_B, while in the center there is a positive jump in the function equal to the reaction X_2 of the connecting rod.

As regards bending moment, this presents the following three notable values (Figure 1.28c):

$$M(A)=0$$

$$M(F) = \frac{7}{248}ql \times \frac{l}{2} = \frac{7}{496}ql^2$$

$$M(B) = X_1 = \frac{3}{496}ql^2$$

whereas in the portion AF, the diagram is simply linear (Figure 1.28c); in the portion FB to the linear diagram there should be added graphically the parabolic diagram corresponding to the distributed load:

$$M(H) = \frac{M(F) + M(B)}{2} - \frac{1}{8}q\left(\frac{l}{2}\right)^2 = -\frac{21}{992}ql^2$$

A faster way of resolving the twice statically indeterminate structure of Figure 1.27a is that of interrupting the continuity of the uprights and then inserting two hinges: one at F (where beams FA, FB, and FG converge), and the other at G (where beams GD, GC, and GF converge) (Figure 1.29). In this way, we have to deal purely with notable schemes of supported beams, with moments applied at the ends and loads uniformly distributed over the entire span. The two conditions of congruence are, in this case, both rotational:

$$\varphi_{BC} = \varphi_{BF} \tag{1.22a}$$

$$\varphi_{FB} = \varphi_{FA} \tag{1.22b}$$

Equations 1.22 can be expressed as functions of the two unknown redundant moments X_1, X_2:

$$\frac{X_1 l}{3EI} + \frac{X_1 l}{6EI} = -\frac{X_1\left(\frac{l}{2}\right)}{3EI} - \frac{X_2\left(\frac{l}{2}\right)}{6EI} + \frac{q\left(\frac{l}{2}\right)^3}{24EI} \tag{1.23a}$$

$$\frac{X_2\left(\frac{l}{2}\right)}{3EI} + \frac{X_1\left(\frac{l}{2}\right)}{6EI} - \frac{q\left(\frac{l}{2}\right)^3}{24EI} = -\frac{X_2\left(\frac{l}{2}\right)}{3EI} \tag{1.23b}$$

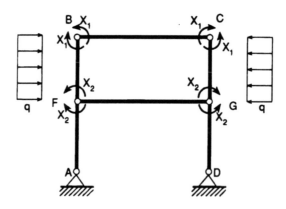

Figure 1.29

Multiplying by $6EI/l$, we obtain

$$2X_1 + X_1 = -X_1 - \frac{1}{2}X_2 + \frac{1}{32}ql^2 \tag{1.24a}$$

$$X_2 + \frac{1}{2}X_1 - \frac{1}{32}ql^2 = -X_2 \tag{1.24b}$$

Cross addition of the two sides of Equations 1.24 eliminates the term q:

$$X_2 = \frac{7}{3}X_1 \tag{1.25}$$

from which we obtain as a verification of the results previously reached otherwise:

$$X_1 = \frac{3}{496}ql^2 \tag{1.26a}$$

$$X_2 = \frac{7}{496}ql^2 \tag{1.26b}$$

The moment diagram will thus be constant over the cross member BC, and on the two uprights will present the same pattern already defined in Figure 1.28c.

1.7 TRANSLATING-NODE FRAMES

The solution for **translating-node frames** differs substantially from that for rotating-node frames, starting from the very way in which it is set out. The truss structure associated with the frame is, in fact, hypostatic, which means that the insertion of the hinges in all the fixed-joint nodes is an excessive operation with respect to the degree of redundancy of the structure. In other words, the associated truss, subjected to the action of external loads and redundant moments, must be in equilibrium for the particular loading condition. In addition to the redundant moments, the generalized coordinates that define the rigid deformed configuration of the associated truss will also be unknown. On the other hand, together with the relations of angular congruence, a number of equilibrium equations equal to the degrees of freedom of the associated truss will also combine to make up the system of equations that provide the solution.

Consider, for instance, the asymmetrical portal frame of Figure 1.30a, loaded by a horizontal force F. The associated truss is made up of the articulated parallelogram of Figure 1.30b, from which the components of rigid rotation of the two uprights are immediately drawn, while the cross member undergoes a horizontal translation. From the scheme of Figure 1.30c, obtained by restraining the displacements of the associated truss, the components of elastic rotation at the ends of the individual beams are derived instead.

The unknowns of the problem are, thus, the two redundant moments X_1, X_2 and the rigid rotation φ of the left-hand upright (Figure 1.30b). There must, therefore, be three equations that resolve the problem, made up of two equations of angular congruence (taking into account also the rigid rotations) and of an application of the principle of virtual work:

$$\varphi_{BA} = \varphi_{BC} \tag{1.27a}$$

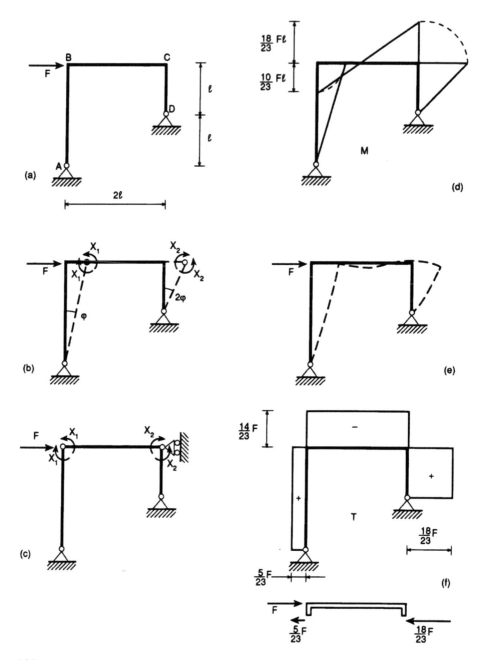

Figure 1.30

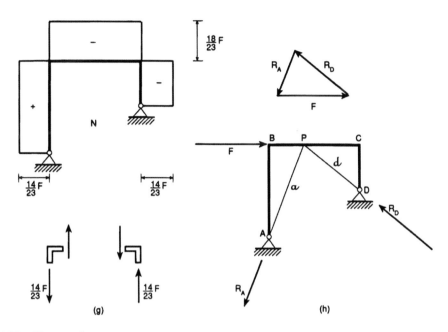

Figure 1.30   (Continued)

$$\varphi_{CB} = \varphi_{CD} \tag{1.27b}$$

principle of virtual work   (1.27c)

Rendering Equations 1.27 explicit, we obtain

$$-\frac{X_1(2l)}{3EI} - \varphi = \frac{X_1(2l)}{3EI} + \frac{X_2(2l)}{6EI} \tag{1.28a}$$

$$-\frac{X_2(2l)}{3EI} - \frac{X_1(2l)}{6EI} = \frac{X_2 l}{3EI} - 2\varphi \tag{1.28b}$$

$$F(2l\varphi) + X_1\varphi - X_2(2\varphi) = 0 \tag{1.28c}$$

Multiplying the first two equations by $3EI/l$ and dividing the third by φ, the system of Equations 1.28 transforms as follows:

$$-2X_1 - \frac{3EI}{l}\varphi = 2X_1 + X_2 \tag{1.29a}$$

$$-2X_2 - X_1 = X_2 - \frac{3EI}{l}(2\varphi) \tag{1.29b}$$

$$X_1 = 2X_2 - 2Fl \tag{1.29c}$$

Substituting Equation 1.29c into the first two, we obtain a system in X_2 and φ:

$$X_2 = \frac{8}{9}Fl - \frac{EI}{3l}\varphi \tag{1.30a}$$

$$X_2 = \frac{2}{5}Fl + \frac{6EI}{5l}\varphi \tag{1.30b}$$

Equating the right-hand sides of Equations 1.30, we obtain

$$\varphi = \frac{22}{69}\frac{Fl^2}{EI} \tag{1.31}$$

from which then

$$X_2 = \frac{18}{23}Fl \tag{1.32a}$$

$$X_1 = -\frac{10}{23}Fl \tag{1.32b}$$

The redundant moment X_1 turns out to be negative, and this means that, contrary to our initial assumption, at the fixed-joint node B the internal fibers are in fact stretched (Figure 1.30b). The bending moments at the two fixed-joint nodes thus being known, it is simple to draw the corresponding diagram, taking into account that the moment vanishes at the hinges, which are at the feet of the two uprights (Figure 1.30d).

The elastic deformed configuration of the translating-node frames must be drawn coherently with the external and internal constraints (angular congruence), with the displacements undergone by the fixed-joint nodes, and with the bending moment diagram. The deformed configuration of the asymmetrical portal frame is represented in Figure 1.30e. Note that the two fixed-joint nodes of the horizontal beam have been translated horizontally by the same amount and rotated clockwise, while the hinges at the foundation allow rotations but not translations of the base sections of the uprights. An inflection appears at the point where the moment vanishes on the cross member.

The fibers of the beams are thus stretched internally in the left-hand part and externally in the right-hand part of the portal frame.

The shear is constant on each member and equal, in absolute value, to the slope of the moment diagram (Figure 1.30f). The axial force also is constant on each member, being compressive on the horizontal beam and on the right-hand upright, and tensile on the left-hand upright (Figure 1.30g). Whereas, on the uprights, the axial force is equal, in absolute value, to the shear acting on the horizontal beam, the axial force in the horizontal beam is equal, in absolute value, to the shear acting on the right-hand upright. The schemes of equilibrium to the horizontal translation of the beam and the vertical translation of the nodes B and C are represented in Figures 1.30f and g, respectively.

The pressure line is made up of two straight lines (Figure 1.30h):

- Portion AB: straight line a
- Portion BD: straight line d

These two straight lines meet at the pole P, which is coincident both with the point of annihilation of the bending moment (Figure 1.30d) and with the point of inflection of the elastic deformed configuration (Figure 1.30e).

Also, the frame of Figure 1.31a is a translating-node frame. This consists of a horizontal beam loaded on the overhang and of two externally hinged uprights. The abovementioned frame is equivalent to the scheme of Figure 1.31b, in which the overhang has been eliminated and replaced by the reactions transmitted to the rest of the structure: the vertical force $F = ql$ and the moment $m = ql^2/2$.

The truss structure associated with the statically indeterminate scheme of Figure 1.31b has one degree of freedom and, as in the previous case, the scheme of Figure 1.32a provides the components of rigid rotation of the two uprights, with the cross member

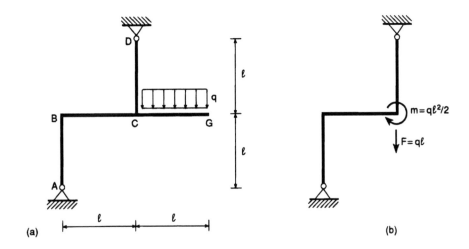

(a)   (b)

Figure 1.31

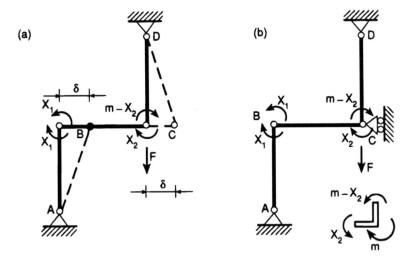

Figure 1.32

undergoing a horizontal translation. The restrained scheme of Figure 1.32b, in turn, provides the components of elastic rotation at the ends of the beams. Note how the moment m, acting on the fixed-joint node C (Figure 1.31b), has been split into its two components, $(m-X_2)$ and X_2, the former acting at the end of the beam CD, and the latter at the end of the beam CB. In this way, the equilibrium with regard to rotation of the node C is automatically satisfied (Figure 1.32b). The force F, unlike the moment m, does not generate bending moments in the scheme of Figure 1.31b, but only tensile axial force on the upright CD.

The equations of congruence and the equilibrium relation, which in implicit form appear as follows:

$$\varphi_{BA} = \varphi_{BC} \tag{1.33a}$$

$$\varphi_{CB} = \varphi_{CD} \tag{1.33b}$$

principle of virtual work $\tag{1.33c}$

may be expressed in explicit form, taking into account the rigid translation of the cross member δ:

$$-\frac{X_1 l}{3EI} - \frac{\delta}{l} = \frac{X_1 l}{3EI} + \frac{X_2 l}{6EI} \tag{1.34a}$$

$$-\frac{X_2 l}{3EI} - \frac{X_1 l}{6EI} = -\frac{(m-X_2)l}{3EI} + \frac{\delta}{l} \tag{1.34b}$$

$$X_1 \frac{\delta}{l} - (m-X_2)\frac{\delta}{l} = 0 \tag{1.34c}$$

Performing the calculations, we obtain

$$X_1 = \frac{m}{6} = \frac{1}{12}ql^2 \tag{1.35a}$$

$$X_2 = \frac{5}{6}m = \frac{5}{12}ql^2 \tag{1.35b}$$

$$\delta = -\frac{ml^2}{4EI} = -\frac{ql^4}{8EI} \tag{1.35c}$$

The positive signs of the redundant moments indicate that the real senses are those assumed, while the negative sign of the displacement δ points to a leftward translation of the cross member. The moment diagram is given in Figure 1.33a, complete with the part that regards the overhang CG. The equilibrium of the node C is guaranteed by the moments that have been determined (Figure 1.33b), while the lack of points of annihilation in the moment

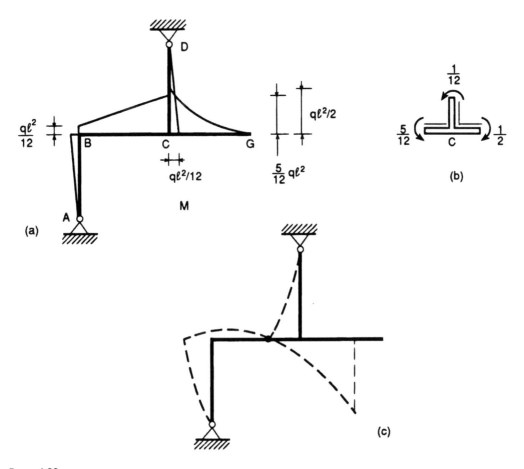

Figure 1.33

function (except for the external hinges A and D and the end G of the overhang) implies an elastic deformed configuration without inflections (Figure 1.33c).

The shear diagram may be obtained very simply from the schemes of equilibrium of the individual supported beams into which the frame has been subdivided (Figure 1.34a). The shear is constant on the two uprights and the cross member BC, whereas it is obviously linear on the cantilever CG (Figure 1.34b).

The axial force also is constant on all the beams of the frame (Figure 1.34c). Whereas it is zero on the overhang, on the uprights it is equal, in absolute value, to the shear of the cross member and *vice versa*. Finally, we have to remember the additional contribution to the axial force on the upright CD made by the force $F=ql$ (Figure 1.31b).

The pressure line is represented in Figure 1.35, and basically consists of three straight lines that meet at a common point P. In fact, the structure, considered as a single body and completely disconnected externally, is in equilibrium under the action of the external reactions R_A and R_D, as well as the resultant F of the active forces. As may be seen, none of the three straight lines encounters the axis of the beam of which it represents the state of loading. This is consistent with the absence of points of annihilation in the moment diagram (Figure 1.33a) and of points of inflection in the elastic deformed configuration (Figure 1.33c). The only virtual point at which the moment vanishes (apart from the two

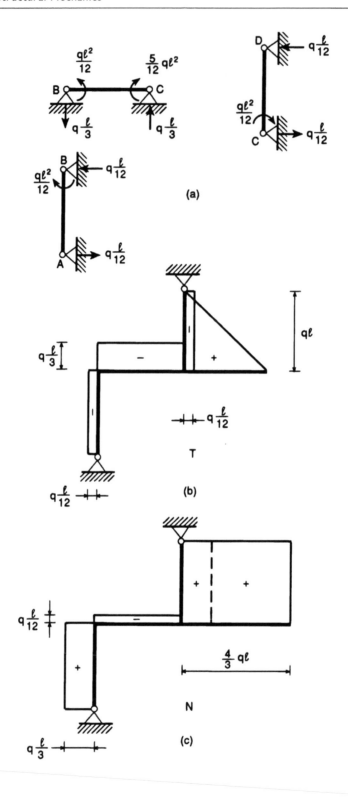

Figure 1.34

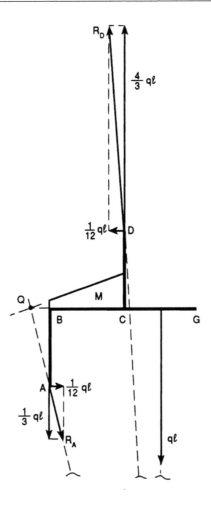

Figure 1.35

hinges) is the point Q of Figure 1.35, which represents the intersection of the axis of the cross member with the line of action of the reaction R_A.

1.8 THERMAL LOADS AND IMPOSED DISPLACEMENTS

In the case where a translating-node frame is subjected to thermal dilations or imposed displacements, the components of rigid rotation deriving directly from these anomalous loads must be found from the restrained truss scheme, which, being statically determinate, does not oppose any resistance to such movements.

As an example, let us take again the asymmetrical portal frame of Figure 1.36a, loaded by a uniform thermal variation on the cross member. The associated truss has one degree of freedom, and the generalized coordinate characterizing its rigid deformed configuration is the angle of rotation φ of the upright AB or the translation δ of the cross member (Figure 1.36b). From the restrained truss (Figure 1.36c), it is possible to find, as well as the

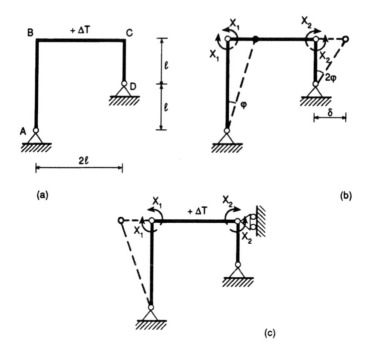

Figure 1.36

elastic rotations, the rigid rotation of the upright AB, produced directly by the thermal dilation of the cross member. In this case, the three Equations 1.27 are expressed as follows:

$$-\frac{X_1(2l)}{3EI} - \varphi + \frac{\alpha\Delta T(2l)}{2l} = \frac{X_1(2l)}{3EI} + \frac{X_2(2l)}{6EI} \tag{1.36a}$$

$$-\frac{X_1(2l)}{6EI} - \frac{X_2(2l)}{3EI} = \frac{X_2 l}{3EI} - 2\varphi \tag{1.36b}$$

$$X_1\varphi - X_2(2\varphi) = 0 \tag{1.36c}$$

Multiplying the first two equations by $3EI/l$, and dividing the third by φ, we obtain

$$4X_1 + X_2 = \frac{3\alpha\Delta TEI}{l} - \varphi\frac{3EI}{l} \tag{1.37a}$$

$$X_1 + 3X_2 = 2\varphi\frac{3EI}{l} \tag{1.37b}$$

$$X_1 = 2X_2 \tag{1.37c}$$

Substituting Equation 1.37c into the two previous ones, we have

$$X_2 = \frac{\alpha \Delta T E I}{3l} - \varphi \frac{EI}{3l} \tag{1.38a}$$

$$X_2 = \frac{6}{5} \varphi \frac{EI}{l} \tag{1.38b}$$

from which, by the transitive law, we find

$$\varphi = \frac{5}{23} \alpha \Delta T \tag{1.39a}$$

$$X_1 = \frac{12}{23} \alpha \Delta T \frac{EI}{l} \tag{1.39b}$$

$$X_2 = \frac{6}{23} \alpha \Delta T \frac{EI}{l} \tag{1.39c}$$

The moment diagram (Figure 1.37a) envisages the fibers of the beams, as always, stretched outward, with the absence of points of inflection in the elastic deformed configuration (Figure 1.37b). The translation of the cross member toward the right is

$$\delta = 2\varphi l = \frac{10}{23} \alpha \Delta T l \tag{1.40}$$

while the thermal elongation of the cross member, neglecting its axial elastic deformability, is

$$\Delta(2l) = 2\alpha \Delta T l \tag{1.41}$$

In actual fact, the node C shifts rightward by δ, while the node B is displaced leftward by $[\Delta(2l)-\delta]$.

The shear diagram is constant on each beam (Figure 1.37c), as is also the axial force diagram (Figure 1.37d), which is compressive on the cross member and on the left-hand upright, and tensile on the right-hand upright.

Finally, the pressure line is represented by the straight line AD, joining the two external hinges (Figure 1.37e). The frame is, in fact, in equilibrium under the action of two equal and opposite forces, the components of which are given by the shearing force diagram and the axial force diagram. The pressure line does not intersect the axis of the frame at any point, and this is consistent with the absence of points of moment annihilation (Figure 1.37a) and with the absence of inflection points in the elastic deformed configuration (Figure 1.37b).

As a second example of a translating-node frame subjected to distortions, let us look at the symmetrical portal frame of Figure 1.38a, where a vertical displacement is imposed on the hinge D. The associated truss structure (Figure 1.38b) has one degree of freedom and furnishes the rigid rotations, while the restrained truss scheme (Figure 1.38c) provides

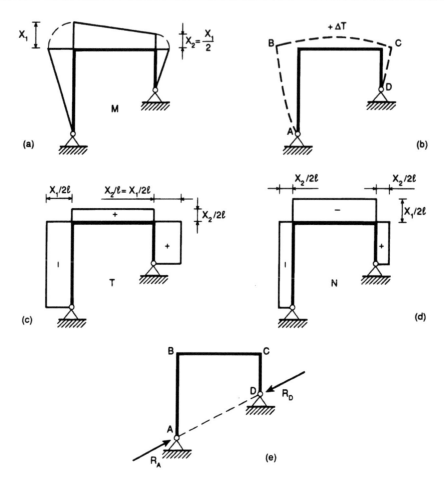

Figure 1.37

the elastic rotations as well as the rigid rotations deriving directly from the settlement (Figure 1.38d). The two equations of angular congruence and the principle of virtual work provide the three equations for resolving the problem:

$$-\frac{X_1 l}{3EI} - \frac{\delta}{l} = \frac{X_1(2l)}{3EI} + \frac{X_2(2l)}{6EI} - \frac{\eta_0}{2l} \tag{1.42a}$$

$$-\frac{X_2(2l)}{3EI} - \frac{X_1(2l)}{6EI} - \frac{\eta_0}{2l} = \frac{X_2 l}{3EI} - \frac{\delta}{l} \tag{1.42b}$$

$$X_1\frac{\delta}{l} - X_2\frac{\delta}{l} = 0 \tag{1.42c}$$

Note how the principle of virtual work is always to be applied to the associated truss system (Figure 1.38b), neglecting the imposed displacements. From Equation 1.42c, we have

$$X_1 = X_2 \tag{1.43}$$

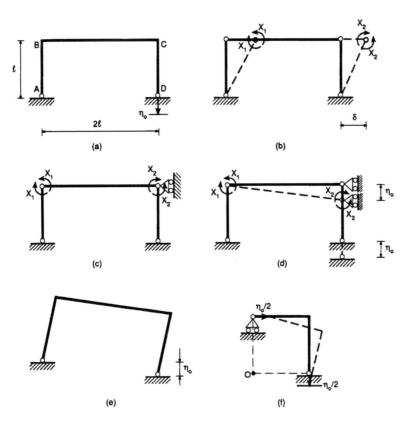

Figure 1.38

and multiplying Equations 1.42a and b by $6EI$:

$$8X_1 l = -\frac{6EI}{l}\left(\delta - \frac{\eta_0}{2}\right) \tag{1.44a}$$

$$8X_1 l = -\frac{6EI}{l}\left(\frac{\eta_0}{2} - \delta\right) \tag{1.44b}$$

Finally, we obtain

$$\delta = \eta_0/2 \tag{1.45a}$$

$$X_1 = X_2 = 0 \tag{1.45b}$$

The resolution of the frame leads us to note that the static characteristics, as well as the external reactions, are zero. This is simply because the imposed displacement η_0 is actually compatible with the constraints of the frame. The hinge A can, in fact, function as the center of rotation and allow an infinitesimal rigid rotation of the whole structure in a clockwise direction (Figure 1.38e). The trajectory of the point D, as it must be orthogonal to the radius vector AD, is vertical. If we consider alone the skew-symmetrical component of the loading (the symmetrical one producing a trivial vertical translation downward, equal to $\eta_0/2$), the kinematic scheme of Figure 1.38f provides an immediate justification for Equation 1.45a.

1.9 FRAMES WITH NONORTHOGONAL BEAMS

So far, we have considered only frames where the individual beams are mutually orthogonal. This is the case that usually concerns the frameworks of buildings, where the columns are vertical and the floors, with their joists, are horizontal. It is not, however, out of place to also consider the case of frames where the individual beams are not mutually orthogonal.

The resolution of frames made up of nonorthogonal beams is accomplished in the same way as that already seen in the foregoing sections. The only differences are represented by an associated truss system that presents more complicated kinematics since, in this case, there is no simple translation of the cross members, and by shearing force diagrams and axial force diagrams that are no longer directly derivable from one another by exchanging their components.

Consider, for instance, the portal frame of Figure 1.39a, which presents the right-hand stanchion inclined at an angle of 45° to the horizontal. The associated truss has one degree of freedom, it being a mechanism whose diagrams of horizontal and vertical displacements

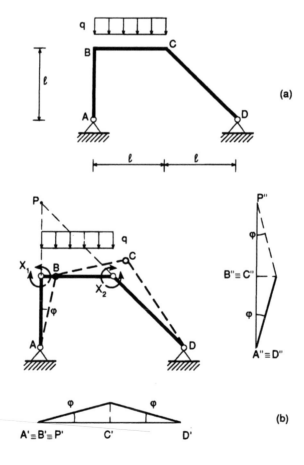

Figure 1.39

are given in Figure 1.39b. The three unknowns of the problem are the redundant moments X_1, X_2, and the rigid rotation φ of the upright stanchion AB, while the three equations that resolve the problem are the two equations of angular congruence plus the application of the principle of virtual work:

$$-\frac{X_1 l}{3EI} - \varphi = \frac{X_1 l}{3EI} + \frac{X_2 l}{6EI} + \varphi - \frac{q l^3}{24EI} \tag{1.46a}$$

$$-\frac{X_2 l}{3EI} - \frac{X_1 l}{6EI} + \varphi + \frac{q l^3}{24EI} = \frac{X_2 l \sqrt{2}}{3EI} - \varphi \tag{1.46b}$$

$$X_1 \varphi + X_1 \varphi - X_2 \varphi - X_2 \varphi - q l \left(\frac{l}{2} \varphi\right) = 0 \tag{1.46c}$$

Multiplying the first two equations by $6EI/l$ and dividing the third by φ, we have

$$4X_1 + X_2 = -12\varphi \frac{EI}{l} + \frac{q l^2}{4} \tag{1.47a}$$

$$X_1 + 2X_2(1 + \sqrt{2}) = 12\varphi \frac{EI}{l} + \frac{q l^2}{4} \tag{1.47b}$$

$$2(X_1 - X_2) = \frac{1}{2} q l^2 \tag{1.47c}$$

Resolving, we obtain

$$X_1 = \frac{16 + 3\sqrt{2}}{112} q l^2 \tag{1.48a}$$

$$X_2 = -\frac{3(4 - \sqrt{2})}{112} q l^2 \tag{1.48b}$$

$$\varphi = -\frac{8 + 5\sqrt{2}}{448} \frac{q l^3}{EI} \tag{1.48c}$$

Passing from these irrational expressions to decimal expressions, we have

$$X_1 \cong 0.18 q l^2 \tag{1.49a}$$

$$X_2 \cong -0.07 q l^2 \tag{1.49b}$$

$$\varphi \cong -0.03 \frac{q l^3}{EI} \tag{1.49c}$$

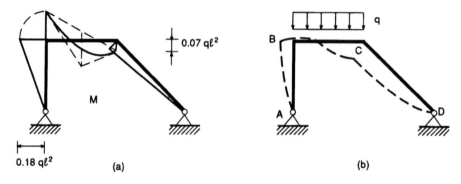

Figure 1.40

The unknowns X_2 and φ turn out to be negative and, hence, it follows that, at the node C, the internal fibers are stretched, and that the rigid rotation of the upright stanchion AB is counterclockwise.

The bending moment diagram (Figure 1.40a) is obtained by the graphical summation of the parabolic diagram for the distributed load q and the linear diagram for the redundant moments.

The deformed configuration of the frame (Figure 1.40b) is constructed by displacing the two fixed-joint nodes B and C on the basis of the mechanism of Figure 1.39b. Both the nodes translate leftward by the amount φl, while the node C alone translates downward by the same quantity. The deformed configuration, moreover, respects the external constraint conditions, the angular congruence at the nodes B and C, and the moment diagram (Figure 1.40a). At the point where the moment vanishes, there corresponds the inflection of the deformed configuration, which shows the fibers stretched externally in the left-hand part and internally in the right-hand part.

The schemes of equilibrium of the individual beams, loaded by the external load q and by the redundant moments (Figure 1.41a), immediately furnish the shear diagram (Figure 1.41b). The shear is constant on the two stanchions, where the moment is linear, and linear on the cross member, where the moment is parabolic. The point of zero shear corresponds to the section subjected to the maximum moment.

Only on the upright stanchion AB and the cross member BC do the shearing force and axial force exchange roles. They are both subject to a compression, equal to $0.75ql$ and $0.18ql$, respectively. As regards the stanchion CD, since it is inclined at an angle of $135°$ with respect to the cross member, its axial force can be determined as the sum of the axial components of the horizontal force ($0.18ql$) and the vertical force ($0.25ql$) that the cross member transmits to it (Figure 1.42a). The same result may be arrived at by considering the equilibrium with regard to translation of the fixed-joint node C (Figure 1.42a):

$$N\frac{\sqrt{2}}{2} - T\frac{\sqrt{2}}{2} + 0.18ql = 0 \tag{1.50a}$$

$$N\frac{\sqrt{2}}{2} + T\frac{\sqrt{2}}{2} + 0.25ql = 0 \tag{1.50b}$$

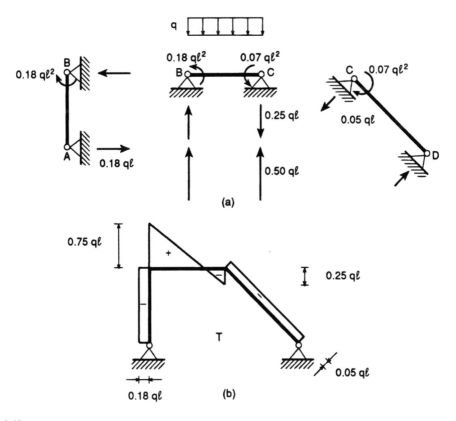

Figure 1.41

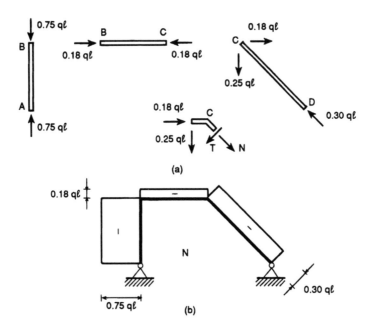

Figure 1.42

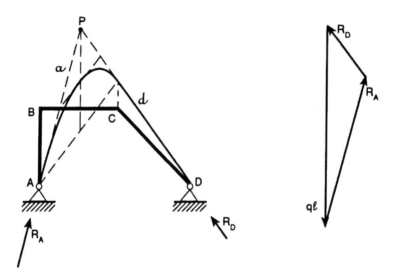

Figure 1.43

Resolving Equations 1.50, we obtain

$$N \cong -0.30ql \tag{1.51a}$$

$$T \cong -0.05ql \tag{1.51b}$$

Equation 1.51b reconfirms the result already obtained following another approach.

The structure, totally disconnected from the foundation and considered as a single body, is in equilibrium under the action of three forces (Figure 1.43). The reactions of the two hinges and the resultant of the distributed load pass through the pole P. The pressure line consists of the line of action of the external reaction, for each stanchion, and of the arc of parabola, which has as its extreme tangents the abovementioned lines of action, for the cross member. Notice how the arc of parabola corresponds, but for a negative scale factor, to the moment diagram of Figure 1.40a. Both the curves pass through the section of the cross member that is the site of the deformative inflection point (Figure 1.40b).

1.10 FRAMES LOADED OUT OF THEIR OWN PLANE

If we deal with frames loaded out of their own plane, the situation is notably complicated with respect to the case of the plane frames, because potentially all six static or kinematic characteristics are involved.

If we wish to proceed along the same lines as those followed in this chapter, we must disconnect the fixed-joint nodes with **spherical hinges**, instead of with the normal **cylindrical hinges**, and apply both bending and twisting redundant moments at the ends of the beams. If the associated truss has n degrees of freedom, it will be necessary to consider n kinematic parameters among the unknowns of the problem, as well as n applications of the principle of virtual work among the equations for resolving the problem. For each spherical hinge, we have, on the other hand, three unknown redundant moments and three equations of angular congruence.

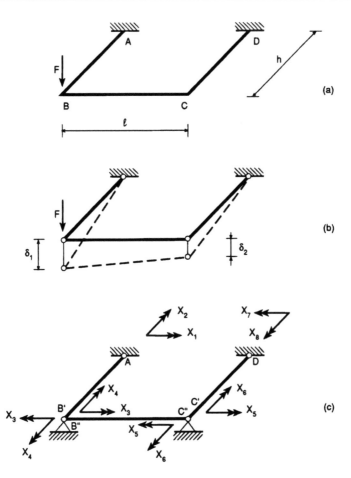

Figure 1.44

It may prove convenient to apply this method only in the simplest cases, as, for example, in those of portal frames or balconies. For the solution of the structure of Figure 1.44a, loaded by a force F on one of the two internal nodes, already as many as 10 equations in 10 unknowns are required. Two unknowns are represented by the deflections δ_1 and δ_2 of the two internal nodes (Figure 1.44b), while the other eight unknowns are represented by the bending and twisting moments that the beams exchange with one another or that the beams exchange with the built-in constraints in the external wall (Figure 1.44c). As regards the equations, there will be two applications of the principle of virtual work and eight conditions of angular congruence:

$$\varphi_A = 0 \tag{1.52a}$$

$$\vartheta_A = 0 \tag{1.52b}$$

$$\varphi_{B'} = \vartheta_{B''} \tag{1.52c}$$

$$\vartheta_{B'} = \varphi_{B''} \tag{1.52d}$$

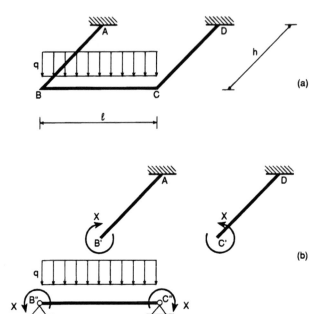

Figure 1.45

$$\varphi_{C'} = \vartheta_{C''} \tag{1.52e}$$

$$\vartheta_{C'} = \varphi_{C''} \tag{1.52f}$$

$$\varphi_D = 0 \tag{1.52g}$$

$$\vartheta_D = 0 \tag{1.52h}$$

where φ denotes the angles of deflection and ϑ those of torsion.

In the case where the previous structure is loaded by a uniform distributed load q on the cross member (Figure 1.45a), there is, by symmetry, a single redundant unknown represented by the moment X, which is a bending moment in the case of the cross member and a twisting moment in the case of the two cantilevers (Figure 1.45b). The condition of congruence imposes equality of the angle of deflection of the cross member with that of torsion of the cantilevers:

$$\varphi_{B''} = \vartheta_{B'} \tag{1.52d}$$

and in explicit form:

$$\frac{Xl}{3EI} + \frac{Xl}{6EI} - \frac{ql^3}{24EI} = -\frac{Xh}{GI_p} \tag{1.53}$$

where $I_p = 2I$, if the cross section of the beams is assumed to be circular. From Equation 1.53, we obtain

$$X = \frac{ql^3}{12[l + 2h(1 + \nu)]} \tag{1.54}$$

Whereas then the cross member is subject to bending moment and shear, the cantilever beams are subject to the constant shear $ql/2$, to the linear bending moment, which ranges from zero to a maximum of $qlh/2$ at the built-in support, as well as to the constant twisting moment X.

Chapter 2

Statically indeterminate beam systems

Method of displacements

2.1 INTRODUCTION

The **method of displacements** is the dual of the method of forces. Thus, it is a process of identifying a single set of kinematic parameters that, in addition to congruence, also implies equilibrium.

From the operative standpoint, the **method of displacements** consists of imposing certain displacements or rotations, characteristic of the system, in such a way that the redundant reactions satisfy an equal number of relations of equilibrium. In the following, this method will be applied to beam systems of various types. In illustrating the method, we shall start from simple cases of parallel-arranged elements and close the chapter by presenting fundamental concepts on which the **automatic computation** of beam systems with multiple degrees of indeterminacy (trusses, plane frames, plane grids, space frames) is based.

2.2 PARALLEL-ARRANGED BAR SYSTEMS

Consider a rigid cross member of symmetrical shape constrained with a symmetrical system of parallel connecting rods, which may present different lengths and different cross-sectional areas, and may consist of materials with different elastic moduli (Figure 2.1a). If this symmetrical system is loaded symmetrically, for instance by a vertical force F acting in the center of the cross member, the resulting deformation will be symmetrical and can be globally described by a single datum: the vertical translation δ of the cross member. Each connecting rod will, in fact, undergo the same elongation δ, since it is hinged to the cross member. For the congruence of the system, it is therefore possible to set

$$\delta_i = \frac{X_i l_i}{E_i A_i} = \delta \tag{2.1}$$

where:
   δ_i is the elongation of the ith connecting rod
   X_i is the corresponding axial tensile force
   l_i, E_i, A_i are, respectively, the length, the elastic modulus, and the cross-sectional area of
       the connecting rod

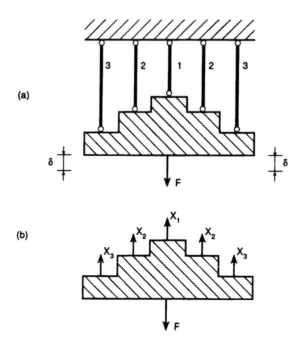

Figure 2.1

Congruence is thus implicitly assumed, while the kinematic unknown δ is determined by considering the equilibrium of the cross member with regard to vertical translation (Figure 2.1b):

$$F = \sum_i X_i \tag{2.2}$$

From Equation 2.1, we obtain, in fact, the value of each redundant unknown X_i as a function of the primary unknown δ:

$$X_i = \frac{E_i A_i}{l_i} \delta \tag{2.3}$$

and thus Equation 2.2 yields the displacement that is sought:

$$\delta = \frac{F}{\sum_i \dfrac{E_i A_i}{l_i}} \tag{2.4}$$

The foregoing expression may also be cast in the form:

$$\delta = \frac{F}{K} \tag{2.5}$$

where

$$K = \sum_i K_i = \sum_i \frac{E_i A_i}{l_i} \qquad (2.6)$$

denotes the total stiffness of the system. This turns out to be equal to the summation of the partial stiffnesses K_i of the parallel elements.

Finally, Equation 2.3 furnishes the reactions of the individual connecting rods:

$$X_i = F \left(\frac{E_i A_i}{l_i} \right) \Bigg/ \left(\sum_i \frac{E_i A_i}{l_i} \right) = F \frac{K_i}{K} \qquad (2.7)$$

where the ratio K_i/K, between partial stiffness and total stiffness, is called the **coefficient of distribution** and indicates the fraction of the total load supported by the ith element. This fraction is proportional to the partial stiffness of the element itself.

Hence, in this particular case of parallel elements, the convenience of the method of displacements, compared with the method of forces, emerges clearly. In fact, if we were to apply the latter method, we should have to resolve $(n-1)$ equations of congruence, n being the number of unknown redundant reactions (in the example of Figure 2.1, we have $n = 3$).

As a second example of parallel-arranged elements, consider a rigid cross member of asymmetrical shape, constrained by a system of parallel connecting rods of varying lengths and cross sections, and of different constitutive materials (Figure 2.2a). Let this system then be loaded in an altogether generic manner, on condition that no components of horizontal force are present, with respect to which the system is free. Consider, for instance, a generic vertical force applied to the cross member at a distance d from the center (Figure 2.2a). With respect to this force, the system presents $(n-2)$ degrees of indeterminacy, n being the total number of connecting rods.

Once deformation has taken place, the cross member will be rotated by the angle φ with respect to the undeformed configuration, so that there will be two kinematic unknowns; for example, the vertical translation δ of the midpoint of the cross member together with the angle of rotation φ. The congruence of the system thus imposes

$$\delta_i = \frac{X_i l_i}{E_i A_i} = \delta + \varphi x_i \qquad (2.8)$$

where x_i denotes the abscissa of the ith connecting rod.

Just as there are two primary unknowns, there are also two equations for resolving them, which are represented by the equations of equilibrium of the cross member with regard to vertical translation and to rotation about the center:

$$F = \sum_i X_i \qquad (2.9a)$$

$$Fd = \sum_i X_i x_i \qquad (2.9b)$$

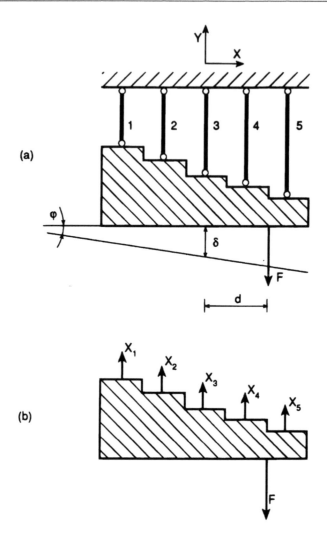

Figure 2.2

Since each individual redundant reaction may be expressed as a function of the two kinematic unknowns, via Equation 2.8:

$$X_i = \frac{E_i A_i}{l_i}(\delta + \varphi x_i)$$

(2.10)

Equations 2.9 are rendered explicit as follows:

$$\delta\left(\sum_i \frac{E_i A_i}{l_i}\right) + \varphi\left(\sum_i \frac{E_i A_i x_i}{l_i}\right) = F$$

(2.11a)

$$\delta\left(\sum_i \frac{E_i A_i x_i}{l_i}\right) + \varphi\left(\sum_i \frac{E_i A_i x_i^2}{l_i}\right) = Fd$$

(2.11b)

from which it emerges clearly how these represent a linear algebraic system of two equations in the two kinematic unknowns δ, φ. The ordered set of four coefficients represents the stiffness matrix of the system, which is symmetrical by virtue of Betti's reciprocal theorem.

Finally, consider a set of connecting rods concurrent at a point belonging to the plane (instead of at infinity as in the cases of Figures 2.1 and 2.2). Let the point of concurrence be represented by a hinge-node (Figure 2.3a), and let that node be loaded by a generic force F, inclined at the angle β with respect to the horizontal. Once deformation has occurred, the hinge-node will be displaced with respect to its original position by an amount u in the horizontal direction and by an amount v in the vertical direction (Figure 2.3a). Since all the connecting rods are hinged at the end, all sharing the same node, and since the principle of superposition holds, the elongation of each individual connecting rod will be given by the two contributions, one corresponding to the displacement u and the other to the displacement v:

$$\Delta l_i = \frac{X_i l_i}{E_i A_i} = u \cos \alpha_i + v \sin \alpha_i \tag{2.12}$$

The equations of equilibrium with regard to horizontal and vertical translation of the hinge-node constitute the equations that resolve the problem (Figure 2.3b):

$$F \cos \beta = \sum_i X_i \cos \alpha_i \tag{2.13a}$$

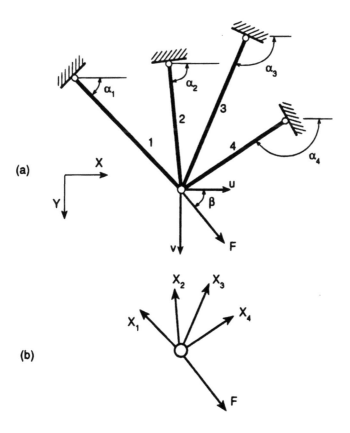

Figure 2.3

$$F\sin\beta = \sum_i X_i \sin\alpha_i \tag{2.13b}$$

Expressing, according to Equation 2.12, each redundant reaction as a function of the two kinematic unknowns u and υ:

$$X_i = \frac{E_i A_i}{l_i}\left(u\cos\alpha_i + \upsilon\sin\alpha_i\right) \tag{2.14}$$

Equations 2.13 transform as follows:

$$u\left(\sum_i \frac{E_i A_i}{l_i}\cos^2\alpha_i\right) + \upsilon\left(\sum_i \frac{E_i A_i}{l_i}\sin\alpha_i\cos\alpha_i\right) = F\cos\beta \tag{2.15a}$$

$$u\left(\sum_i \frac{E_i A_i}{l_i}\cos\alpha_i\sin\alpha_i\right) + \upsilon\left(\sum_i \frac{E_i A_i}{l_i}\sin^2\alpha_i\right) = F\sin\beta \tag{2.15b}$$

Also in this case, the stiffness matrix of the system is symmetric, in accordance with Betti's reciprocal theorem.

2.3 PARALLEL-ARRANGED BEAM SYSTEMS

Consider n beams connected together by a single fixed-joint node and each constrained at its other end in any manner whatsoever to the foundation (Figure 2.4a). Let these beams present different lengths, cross sections, and elastic moduli, and let the fixed-joint node be loaded by a counterclockwise moment m. When deformation has taken place, the fixed-joint node will be rotated by an angle φ, as will the end sections of the individual beams that converge at the fixed-joint node itself (Figure 2.4a):

$$\varphi_i = \frac{X_i l_i}{c_i E_i I_i} = \varphi \tag{2.16}$$

where:

    φ_i denotes the angle of rotation of the nodal section of the ith beam
    X_i is the reactive bending moment due to the abovementioned distortion
    l_i, I_i, E_i are, respectively, the length, the moment of inertia, and the elastic modulus of the beam
    c_i is a numerical coefficient dependent on the remaining constraint

Considering the imposed displacements, we can deduce, for instance, the coefficients c_i, corresponding to the beams of the scheme of Figure 2.4a. For Beam 1, we have $c_1=4$; for Beams 2 and 4, we have $c_2=c_4=3$; while for Beam 3, we have $c_3=1$.

An alternative and equivalent way of expressing the proportionality relation between the kinematic unknown φ_i and the redundant moment X_i is that of transforming the fixed-joint

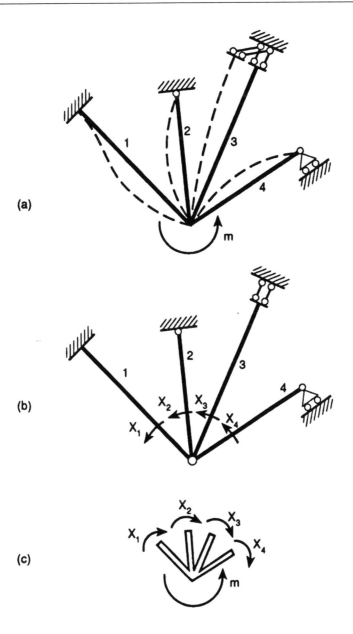

Figure 2.4

node into a hinge-node (Figure 2.4b) and to apply all the redundant unknowns X_i. It will thus be a question of determining the elastic rotations φ_i on all the various statically indeterminate or statically determinate schemes thus obtained. This time, the coefficients c_i are thus derived by assigning to the moments X_i the role of cause, and to the rotations φ_i that of effect, as opposed to the previously adopted procedure. The equilibrium of the fixed-joint node, on the other hand, imposes (Figure 2.4c)

$$m = \sum_i X_i \tag{2.17}$$

whereby, using Equation 2.16, we obtain

$$m = \varphi \sum_i \frac{c_i E_i I_i}{l_i} \qquad (2.18)$$

and hence the primary unknown of the problem:

$$\varphi = \frac{m}{\sum_i \dfrac{c_i E_i I_i}{l_i}} = \frac{m}{K} \qquad (2.19)$$

where K designates the total stiffness of the beam system, which yet again is equal to the summation of the partial stiffnesses of the individual elements.

From Equations 2.16 and 2.19, it is thus possible to derive the individual redundant unknowns:

$$X_i = m \left(\frac{c_i E_i I_i}{l_i} \right) \bigg/ \left(\sum_i \frac{c_i E_i I_i}{l_i} \right) = m \frac{K_i}{K} \qquad (2.20)$$

where the ratio K_i/K is the **flexural coefficient of distribution** and represents the fraction of external loading supported by the ith beam.

Consider, as a second fundamental case, a horizontal rigid cross member, constrained to the foundation by a series of uprights of various lengths, having different moments of inertia and consisting of different materials (Figure 2.5a). Let the horizontal cross member be loaded by a horizontal force F. When deformation has occurred, the cross member will be translated horizontally by the amount δ, as will the ends of the uprights. From congruence, we therefore have

$$\delta_i = \frac{T_i l_i^3}{c_i E_i I_i} = \delta \qquad (2.21)$$

where:

T_i is the shear transmitted to the cross member by the ith upright

l_i, I_i, and E_i are, as before, the characteristics of the upright

c_i is a numerical coefficient depending on the constraint holding the upright to the foundation

We can easily deduce the coefficients c_i corresponding to the exemplary scheme of Figure 2.5a. For Upright 1, we have $c_1 = 12$; for Uprights 2 and 4, we have $c_2 = c_4 = 3$; while for Upright 3, we have $c_3 = 0$, since the horizontal translations of the upright are permitted by the double rod.

The displacement δ_i having been imposed, the redundant shear reaction T_i is thus obtained. Inverting the roles, as has already been seen in the previous fundamental scheme, it is possible to apply the force T_i and find the elastic displacement δ_i. The equation that resolves the

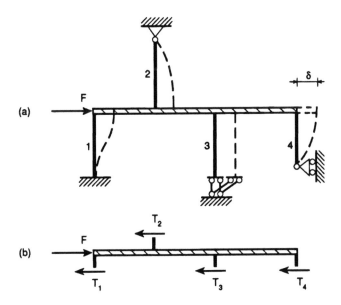

Figure 2.5

problem is thus of equilibrium, with regard to horizontal translation of the cross member (Figure 2.5b):

$$F = \sum_i T_i \tag{2.22}$$

from which, taking into account Equation 2.21, we obtain

$$\delta = \frac{F}{\sum_i \dfrac{c_i E_i I_i}{l_i^3}} = \frac{F}{K} \tag{2.23a}$$

the total stiffness K being provided by the summation of the partial stiffnesses.

The redundant reactions T_i are, instead, expressible as follows:

$$T_i = F \left(\frac{c_i E_i I_i}{l_i^3} \right) \bigg/ \left(\sum_i \frac{c_i E_i I_i}{l_i^3} \right) = F \frac{K_i}{K} \tag{2.23b}$$

where the ratio K_i/K is the **shear coefficient of distribution.**

Notice how the uprights, in addition to the shear T_i, generally also transmit to the cross member a moment M_i and an axial force N_i. Equilibrium with regard to vertical translation and to rotation of the cross member will thus be ensured by the combination of these loadings. Whereas, however, the moments M_i are, by now, known and deducible from schemes already referred to, the axial forces N_i constitute a set of redundant unknowns that may be calculated on the basis of a scheme of parallel connecting rods similar to that of Figure 2.2a, in which the external load consists of the moments M_i.

In the same way, the shearing forces and axial forces transmitted to the fixed-joint node of Figure 2.4c must allow equilibrium with regard to translation of the latter. Of course, in this case the shearing forces are known and their summation constitutes the total force acting on the node, while the axial forces are again unknown but deducible from a scheme of converging connecting rods such as that of Figure 2.3a.

2.4 AUTOMATIC COMPUTATION OF BEAM SYSTEMS WITH MULTIPLE DEGREES OF INDETERMINACY

Consider a system of rectilinear beams lying in the plane, constrained at the ends, both mutually and externally, by fixed-joint nodes or hinge-nodes (Figure 2.6). Let the fixed-joint nodes and the hinge-nodes be numbered, starting from the internal nodes and ending with the external ones. In the case of fixed-joint nodes, there are three kinematic parameters that characterize their elastic configuration: two orthogonal translations and the rotation (Figure 2.7a). In the case of hinge-nodes, the kinematic parameters that characterize their elastic configuration are equal to the number of the beams that converge in the node, augmented by two; it will, in fact, be necessary to take into account an independent rotation for each end section and the two orthogonal translations (Figure 2.7b). Finally, in the case of connections of a mixed type, fixed-joint and hinge (Figure 2.7c), there will be three kinematic parameters, plus the number of beams that converge at the hinge.

Further, let the beams be numbered and let each be disposed within a local reference system Y^*Z^*, which will, in general, be rototranslated with respect to the global reference system YZ (Figure 2.6). Let us then consider a generic loading of the system, consisting of concentrated and distributed loads on the individual beams and of concentrated loads (forces and couples) acting on the internal nodes.

At this point, let each beam with ends ij be isolated, and let it be considered as constrained by built-in supports in the end sections (Figure 2.8a). The procedure will thus be to impose, at each end, the three generalized displacements, and find the redundant reactions at the

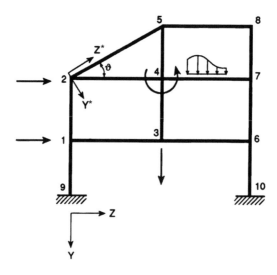

Figure 2.6

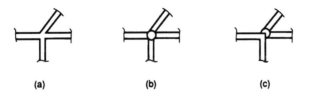

Figure 2.7

built-in ends. If we designate the reactions at the built-in constraints as M, T, and N, and the rotation and the two imposed displacements as φ, υ, and w, respectively, and if we assume the conventional positive senses indicated in Figure 2.8b, which are the same at either end, we have the following matrix relation:

$$
\begin{bmatrix} M_i \\ T_i \\ N_i \\ M_j \\ T_j \\ N_j \end{bmatrix} = EI \begin{bmatrix} \frac{4}{l} & -\frac{6}{l^2} & 0 & \frac{2}{l} & \frac{6}{l^2} & 0 \\ -\frac{6}{l^2} & \frac{12}{l^3} & 0 & -\frac{6}{l^2} & -\frac{12}{l^3} & 0 \\ 0 & 0 & \frac{A}{ll} & 0 & 0 & -\frac{A}{ll} \\ \frac{2}{l} & -\frac{6}{l^2} & 0 & \frac{4}{l} & \frac{6}{l^2} & 0 \\ \frac{6}{l^2} & -\frac{12}{l^3} & 0 & \frac{6}{l^2} & \frac{12}{l^3} & 0 \\ 0 & 0 & -\frac{A}{ll} & 0 & 0 & \frac{A}{ll} \end{bmatrix} \begin{bmatrix} \varphi_i \\ \upsilon_i \\ w_i \\ \varphi_j \\ \upsilon_j \\ w_j \end{bmatrix} - \begin{bmatrix} M_i^0 \\ T_i^0 \\ N_i^0 \\ M_j^0 \\ T_j^0 \\ N_j^0 \end{bmatrix}
\tag{2.24}
$$

This relation expresses the **constraint reaction vector** as the sum of two contributions; the first deriving from the imposed displacements and the second balancing the external loads acting on the beam. The symmetric (6×6) matrix that multiplies the **displacement vector** is referred to as the **stiffness matrix of the element**. Each column of the matrix is obtained by imposing the relative displacement or rotation and by calculating the redundant reactions at the ends. In place of the vector of the forces balancing the external load acting on the beam, there appears in Equation 2.24 the **vector of the forces equivalent to the external load**, with

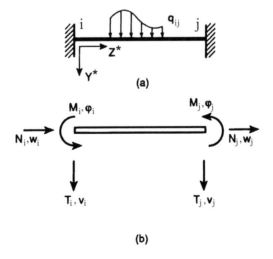

Figure 2.8

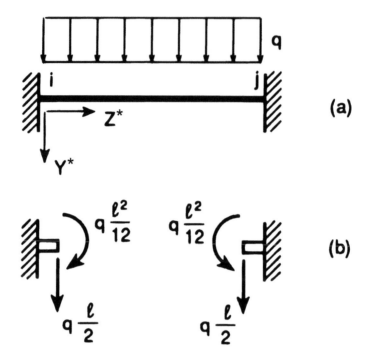

Figure 2.9

a negative algebraic sign. In the case of uniform distributed load (Figure 2.9), we have the vector:

$$\left[-\frac{ql^2}{12},\frac{ql}{2},0,\frac{ql^2}{12},\frac{ql}{2},0\right]^{\mathrm{T}}$$

(2.25)

while, with the butterfly-shaped thermal distortion (Figure 2.10) and the uniform thermal distortion (Figure 2.11), it is necessary to consider the following forces transmitted by the beam to the built-in constraint nodes:

$$\left[2EI\alpha\frac{\Delta T}{h},0,0,-2EI\alpha\frac{\Delta T}{h},0,0\right]^{\mathrm{T}}$$

(2.26)

$$\left[0,0,-EA\alpha\Delta T,0,0,EA\alpha\Delta T\right]^{\mathrm{T}}$$

(2.27)

In Equation 2.24, it is possible to group together the terms corresponding to the two ends i and j:

$$\begin{bmatrix}Q_i^* \\ Q_j^*\end{bmatrix}=\begin{bmatrix}K_{ii} & K_{ij} \\ K_{ji} & K_{jj}\end{bmatrix}\begin{bmatrix}\delta_i^* \\ \delta_j^*\end{bmatrix}-\begin{bmatrix}F_i^* \\ F_j^*\end{bmatrix}$$

(2.28)

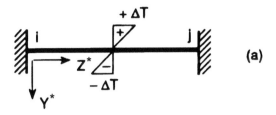

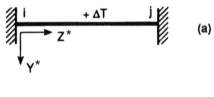

Figure 2.10

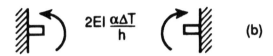

Figure 2.11

where the quantities marked by an asterisk are to be understood as referring to the local coordinate axes Y^*Z^*, corresponding to the beam ij (Figure 2.6). In compact form:

$$\underset{(6\times6)}{[K_e]}\underset{(6\times1)}{\{\delta_e^*\}} = \underset{(6\times1)}{\{Q_e^*\}} + \underset{(6\times1)}{\{F_e^*\}} \tag{2.29}$$

The quantities referred to the local system are expressible as functions of the same quantities referred to the global system:

$$\{\delta_e^*\} = [N]\{\delta_e\} \tag{2.30a}$$

$$\{Q_e^*\} = [N]\{Q_e\} \tag{2.30b}$$

$$\{F_e^*\} = [N]\{F_e\} \tag{2.30c}$$

[N] being the matrix of rotation that transforms the global reference system YZ into the local reference system Y*Z*:

$$[N] = \begin{bmatrix} 1 & 0 & 0 & 0 & 0 & 0 \\ 0 & \cos\vartheta & \sin\vartheta & 0 & 0 & 0 \\ 0 & -\sin\vartheta & \cos\vartheta & 0 & 0 & 0 \\ 0 & 0 & 0 & 1 & 0 & 0 \\ 0 & 0 & 0 & 0 & \cos\vartheta & \sin\vartheta \\ 0 & 0 & 0 & 0 & -\sin\vartheta & \cos\vartheta \end{bmatrix} \tag{2.31}$$

Substituting Equations 2.30 into Equation 2.29, we obtain

$$[K_e][N]\{\delta_e\} = [N](\{Q_e\} + \{F_e\}) \tag{2.32}$$

Finally, premultiplying both sides by $[N]^T$, we obtain

$$([N]^T[K_e][N])\{\delta_e\} = \{Q_e\} + \{F_e\} \tag{2.33}$$

The two vectors on the right-hand side are, respectively, the vector of the constraint reactions and the vector of the equivalent nodal forces, while the matrix in parentheses is the stiffness matrix of the element, reduced to the global reference system.

The **assemblage** operation consists of an **expansion** of the vectors $\{\delta_e\}$, $\{Q_e\}$, $\{F_e\}$ from the local dimension ($3\times2 = 6$) to the global dimension n, where n is the total number of kinematic parameters identifying the deformed configuration of the beam system. Thus, the procedure will be to order all the kinematic parameters of the system in one vector, in such a way as to be able to insert the end displacements of the generic element e into the positions that they should occupy. This can be achieved by premultiplying the vector of the local displacements $\{\delta_e\}$ by a suitable assemblage matrix $[A_e]^T$ of dimensions ($n\times6$), where all the elements are zero, with the exception of six elements having a value of unity arranged in the six different rows to be filled and corresponding to the six associated columns:

$$\{\delta^e\} = [A_e]^T\{\delta_e\} \tag{2.34a}$$

$$\{Q^e\} = [A_e]^T\{Q_e\} \tag{2.34b}$$

$$\{F^e\} = [A_e]^T\{F_e\} \tag{2.34c}$$

In the case of Beams 2 through 5 of the frame of Figure 2.6, for example, the matrix $[A_e]^T$ is of dimensions (30×6), where the first element of the fourth row, the second element of the fifth row, the third element of the sixth row, the fourth element of the thirteenth row, the fifth element of the fourteenth row, and the sixth element of the fifteenth row are different from zero and have a value of unity:

$$
[A_{2-5}]^{T} =
\begin{bmatrix}
\\
\end{bmatrix}
\left.\begin{array}{l} 1 \\ 1 \\ 1 \end{array}\right\} \text{node 2}
\qquad
\left.\begin{array}{l} 1 \\ 1 \\ 1 \end{array}\right\} \text{node 5}
$$

$$(30 \times 6)$$

(2.35)

Substituting the inverse relations of Equations 2.34 into Equation 2.33, we obtain

$$\left([N]^{T}[K_{e}][N] \right)[A_{e}]\{\delta^{e}\} = [A_{e}]\left(\{Q^{e}\} + \{F^{e}\} \right)$$

(2.36)

Premultiplying by $[A_{e}]^{T}$, we have

$$\left([A_{e}]^{T}[N]^{T}[K_{e}][N][A_{e}] \right)\{\delta^{e}\} = \{Q^{e}\} + \{F^{e}\}$$

(2.37)

Equation 2.37 remains valid even if the expanded vector of local displacements $\{\delta^e\}$ is replaced by the **global vector of nodal displacements** $\{\delta\}$:

$$\underset{(n\times n)}{\left[K^e\right]}\underset{(n\times 1)}{\{\delta\}} = \underset{(n\times 1)}{\{Q^e\}} + \underset{(n\times 1)}{\{F^e\}} \tag{2.38}$$

where $[K^e]$ denotes the matrix of local stiffness, reduced to the global reference system and expanded to the dimension n.

The local relation in expanded Equation 2.38 may be summed together with the similar relations corresponding to the other beams:

$$\left(\sum_e \left[K^e\right]\right)\{\delta\} = \sum_e \left(\{Q^e\} + \{F^e\}\right) \tag{2.39}$$

Noting that, for the equilibrium of nodes, we must have (Figure 2.12)

$$-\sum_e \{Q^e\} + \{F\} = \{0\} \tag{2.40}$$

Equation 2.39, which resolves the problem, may be cast in the form:

$$[K]\{\delta\} = \{F\} + \{F_{eq}\} \tag{2.41}$$

where:
  $[K]$ is the **matrix of global stiffness**
  $\{F\}$ is the **vector of actual nodal forces**
  $\{F_{eq}\}$ is the **vector of equivalent nodal forces**

The vectors of the actual and equivalent nodal forces summarize the external loads acting on the beam system and are to be considered as known terms or data of the problem. The stiffness matrix of the system is also known, once the geometry and the elastic properties

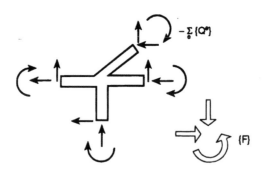

Figure 2.12

of the beams are known. The unknown of the problem is represented by the vector of nodal displacements $\{\delta\}$. It is evident, then, how the calculation of a beam system with multiple degrees of statical indeterminacy can be accommodated within the method of displacements and can be performed automatically via computer by means of a preordered succession of elementary operations.

So far, we have not taken into account the external constraint conditions. To do this, let us partition the vectors and the matrix that appear in Equation 2.41, so as to isolate the free displacements from the constrained displacements:

$$\begin{bmatrix} K_{LL} & K_{LV} \\ K_{VL} & K_{VV} \end{bmatrix} \begin{bmatrix} \delta_L \\ \delta_V \end{bmatrix} = \begin{bmatrix} F_L \\ F_V \end{bmatrix} \tag{2.42}$$

Whereas the constrained displacements $\{\delta_V\}$ are zero (or, at most, predetermined in the case of inelastic settlement), the free displacements $\{\delta_L\}$ represent the true unknowns of the problem:

$$[K_{LL}]\{\delta_L\} = \{F_L\} - [K_{LV}]\{\delta_V\} \tag{2.43}$$

from which we obtain

$$\{\delta_L\} = [K_{LL}]^{-1} \left(\{F_L\} - [K_{LV}]\{\delta_V\} \right) \tag{2.44}$$

On the other hand, from Equation 2.42, we have

$$\{F_V\} = [K_{VL}]\{\delta_L\} + [K_{VV}]\{\delta_V\} \tag{2.45}$$

and hence, by virtue of Equations 2.44 and 2.45, we obtain the external constraint reactions:

$$\underset{(v\times 1)}{\{Q_V\}} = \underset{(v\times l)}{[K_{VL}]} \underset{(l\times l)}{[K_{LL}]^{-1}} \left(\underset{(l\times 1)}{\{F_L\}} - \underset{(l\times v)}{[K_{LV}]} \underset{(v\times 1)}{\{\delta_V\}} \right) + \underset{(v\times v)}{[K_{VV}]} \underset{(v\times 1)}{\{\delta_V\}} - \underset{(v\times 1)}{\{F_V^0\}} \tag{2.46}$$

where $\{F_V^0\}$ is the vector of the equivalent nodal forces corresponding to the constrained nodes. Notice that, beneath the elements of Equation 2.46, the dimensions of the vectors and matrices that are involved are given.

Once all the kinematic parameters $\{\delta_L\}$ are known, it is then simple, by applying for each beam the initial Equation 2.24, to determine the internal reactions and hence the diagrams of the static characteristics.

2.5 PLANE TRUSSES

In the case of plane trusses (Figure 2.13), the kinematic parameters of the system reduce to the hinge-node displacements. The local stiffness matrix in this case reduces to a (2×2) matrix:

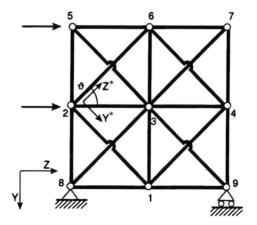

Figure 2.13

$$\begin{bmatrix} N_i \\ N_j \end{bmatrix} = EA \begin{bmatrix} \frac{1}{l} & -\frac{1}{l} \\ -\frac{1}{l} & \frac{1}{l} \end{bmatrix} \begin{bmatrix} w_i \\ w_j \end{bmatrix} \tag{2.47}$$

where only the axial force and the axial displacement are present (Figure 2.14). The vector of the forces balancing the external load acting on the individual bar is zero, since, in general, trusses are assumed as being loaded only at the hinge-nodes.

Equation 2.47 can be cast in the compact form:

$$\underset{(2\times2)\ (2\times1)}{[K_e]\{\delta_e^*\}} = \underset{(2\times1)}{\{Q_e^*\}} \tag{2.48}$$

where the quantities referred to the local system have been marked with an asterisk. On the other hand, both the axial displacement and the axial force can be projected on the axes of the global reference system YZ, giving rise to the four-component vectors $\{\delta_e\}$ and $\{Q_e\}$:

$$\{\delta_e^*\} = [N]\{\delta_e\} \tag{2.49a}$$

$$\{Q_e^*\} = [N]\{Q_e\} \tag{2.49b}$$

where $[N]$ denotes the orthogonal (2×4) matrix:

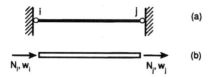

Figure 2.14

$$[N] = \begin{bmatrix} -\sin\vartheta & \cos\vartheta & 0 & 0 \\ 0 & 0 & -\sin\vartheta & \cos\vartheta \end{bmatrix}$$

(2.50)

The assemblage operation is based on the relation:

$$\underset{(n\times n)\,(n\times 1)}{[K^e]\{\delta\}} = \underset{(n\times 1)}{\{Q^e\}}$$

(2.51)

where:

$$\underset{(n\times n)}{[K^e]} = \underset{(n\times 4)}{[A_e]^T} \underset{(4\times 2)}{[N]^T} \underset{(2\times 2)}{[K_e]} \underset{(2\times 4)}{[N]} \underset{(4\times n)}{[A_e]}$$

(2.52)

is the local stiffness matrix, reduced to the global reference system and expanded to the dimension n

$\{\delta\}$ is the vector of nodal displacements, which presents two components for each node with a total of n components

$\{Q^e\}$ is the vector of nodal reactions, also reduced to the global reference system and expanded to the dimension n

Summing up all the Equations 2.51, as the index e characterizing each individual bar varies, we obtain

$$[K]\{\delta\} = \{F\}$$

(2.53)

which is the equation that resolves the problem.

Once all the kinematic parameters $\{\delta\}$ are known, and hence the displacements in the individual local reference systems:

$$\underset{(2\times 1)}{\{\delta_e^*\}} = \underset{(2\times 4)}{[N]} \underset{(4\times n)}{[A_e]} \underset{(n\times 1)}{\{\delta\}}$$

(2.54)

it is then possible to determine the axial force acting in each single bar, by applying the initial Equation 2.47.

2.6 PLANE FRAMES

Plane frames have already been dealt with extensively in Section 2.4. Here, we shall consider two important particular cases:

1. Rotating-node frames
2. Translating-node frames with rigid horizontal cross members (also called **shear-type frames**)

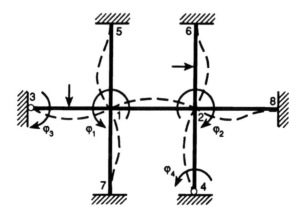

Figure 2.15

In **rotating-node frames**, the nodes are not displaced but simply rotate, except for the contributions due to axial deformability of the beams. Figure 2.15 provides an example of a rotating-node frame with four nodes effectively free to rotate (two of these are internal and two external). In these cases, if we neglect the axial deformability of the beams, the local stiffness matrix is, as in the case of trusses, a (2×2) matrix:

$$\begin{bmatrix} M_i \\ M_j \end{bmatrix} = EI \begin{bmatrix} \dfrac{4}{l} & \dfrac{2}{l} \\ \dfrac{2}{l} & \dfrac{4}{l} \end{bmatrix} \begin{bmatrix} \varphi_i \\ \varphi_j \end{bmatrix} - \begin{bmatrix} M_i^0 \\ M_j^0 \end{bmatrix} \tag{2.55}$$

In this case, the **vector of rigid joint moments** $[M_i^0, M_j^0]^T$ is also present.

Equation 2.55 can be put in the synthetic form of Equation 2.29. This is one of those cases in which the local and global kinematic parameters coincide, so that Equations 2.30 remain valid, even though the orthogonal matrix $[N]$ is the identity matrix of dimensions (2×2). The assemblage matrix $[A_e]$ will therefore have the dimensions $(2 \times n)$, where n is the number of the nodes of the frame.

In **translating-node frames with rigid horizontal cross members**, we have the situation complementary to the preceding one; the nodes do not rotate but are displaced only horizontally, neglecting the contribution due to the axial deformability of the columns. Figure 2.16 gives an example with three cross members effectively free to translate. Only the vertical beams are thus considered, hence the local stiffness matrix has, as in the foregoing cases (trusses and rotating-node frames), dimensions (2×2):

$$\begin{bmatrix} T_i \\ T_j \end{bmatrix} = EI \begin{bmatrix} \dfrac{12}{l^3} & -\dfrac{12}{l^3} \\ -\dfrac{12}{l^3} & \dfrac{12}{l^3} \end{bmatrix} \begin{bmatrix} \upsilon_i \\ \upsilon_j \end{bmatrix} - \begin{bmatrix} T_i^0 \\ T_j^0 \end{bmatrix} \tag{2.56}$$

Also in this case, the local kinematic parameters coincide with the global ones, which are precisely the horizontal translations of the cross members. The assemblage matrix $[A_e]$ will,

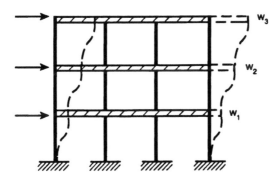

Figure 2.16

therefore, have the dimensions $(2 \times n)$, where n is the number of horizontal cross members plus one.

2.7 PLANE GRIDS

Consider a system of rectilinear beams lying in a plane, mutually constrained via fixed-joint nodes and externally via built-in ends or supports orthogonal to the foundation (Figure 2.17). The kinematic parameters characterizing the deformed configuration of each node are three: two rotations with orthogonal axes lying in the plane and the displacement orthogonal to the plane.

Let each beam be disposed in a local reference system X^*Z^*, which will, in general, be rototranslated with respect to the global reference system XZ (Figure 2.17), and let a generic loading of the system be considered, consisting of distributed and concentrated loads perpendicular to the plane XZ, acting on the individual beams or on the individual nodes.

Let each beam be isolated, it being considered as built in at the end sections i and j (Figure 2.18a). The procedure will be to impose on each end the three generalized displacements and to find the redundant reactions at the built-in constraints. Designating as M, T, M_t the reactions at the built-in constraints, and as φ, υ, ϑ, respectively, the rotation about

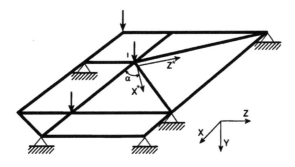

Figure 2.17

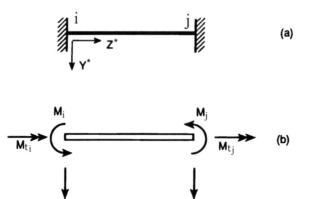

(a)

(b)

Figure 2.18

the axis X^*, the transverse displacement, and the rotation about the axis Z^* (Figure 2.18b), we have the following matrix relation:

$$
\begin{bmatrix} M_i \\ T_i \\ M_{ti} \\ M_j \\ T_j \\ M_{tj} \end{bmatrix} = EI \begin{bmatrix} \frac{4}{l} & -\frac{6}{l^2} & 0 & \frac{2}{l} & \frac{6}{l^2} & 0 \\ -\frac{6}{l^2} & \frac{12}{l^3} & 0 & -\frac{6}{l^2} & -\frac{12}{l^3} & 0 \\ 0 & 0 & \frac{GI_t}{EIl} & 0 & 0 & -\frac{GI_t}{EIl} \\ \frac{2}{l} & -\frac{6}{l^2} & 0 & \frac{4}{l} & \frac{6}{l^2} & 0 \\ \frac{6}{l^2} & -\frac{12}{l^3} & 0 & \frac{6}{l^2} & \frac{12}{l^3} & 0 \\ 0 & 0 & -\frac{GI_t}{EIl} & 0 & 0 & \frac{GI_t}{EIl} \end{bmatrix} \begin{bmatrix} \varphi_i \\ \upsilon_i \\ \vartheta_i \\ \varphi_j \\ \upsilon_j \\ \vartheta_j \end{bmatrix} - \begin{bmatrix} M_i^0 \\ T_i^0 \\ M_{ti}^0 \\ M_j^0 \\ T_j^0 \\ M_{tj}^0 \end{bmatrix}
\tag{2.57}
$$

where the local stiffness matrix is identical to the one present in Equation 2.24, except for the four terms corresponding to torsion, which replace those corresponding to axial force.

The matrix of Equation 2.57 can be cast in the compact form of Equation 2.29. The quantities referred to the local reference system X^*Z^* are expressible as functions of the same quantities referred to the global system XZ. Equations 2.30 therefore continue to hold, $[N]$ being the following rotation matrix:

$$
[N] = \begin{bmatrix} \cos\alpha & 0 & \sin\alpha & 0 & 0 & 0 \\ 0 & 1 & 0 & 0 & 0 & 0 \\ -\sin\alpha & 0 & \cos\alpha & 0 & 0 & 0 \\ 0 & 0 & 0 & \cos\alpha & 0 & \sin\alpha \\ 0 & 0 & 0 & 0 & 1 & 0 \\ 0 & 0 & 0 & -\sin\alpha & 0 & \cos\alpha \end{bmatrix}
\tag{2.58}
$$

Just as Equation 2.31 reduces shearing force and axial force to the global reference system, so Equation 2.58 likewise reduces bending moment and twisting moment to the global reference system (Figure 2.19).

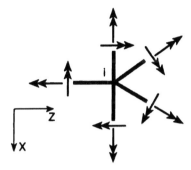

Figure 2.19

As regards calculation of the external constraint reactions, the procedure to follow is yet again that formally outlined in Section 2.4.

2.8 SPACE TRUSSES AND FRAMES

To conclude, let us consider the most general case, which comprises, as particular cases, both plane frames and plane grids. This is the case of **space frames**, which are three-dimensional systems of rectilinear beams mutually constrained and externally constrained by fixed-joints, cylindrical hinges, and spherical hinges (Figure 2.20). In the case, for instance, of fixed-joint constraint nodes, there are six kinematic parameters characterizing the deformed configuration; three mutually orthogonal translations and three rotations about three mutually

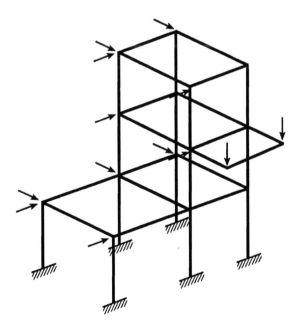

Figure 2.20

orthogonal axes. In the case of spherical hinges (e.g., three-dimensional trusses), the kinematic parameters are equal to three times the number of beams that converge at the node, augmented by three. In the case of mixed-type connections, the kinematic parameters will be of a lower number. Let each beam be disposed within a local reference system $X^*Y^*Z^*$, which will, in general, be rototranslated with respect to the global reference system XYZ, and let each be isolated by being considered as built-in at the end sections ij. Following the same procedure as in the previously considered cases, let six generalized displacements be imposed at either end, so as to find the 12 redundant reactions at the built-in supports. Designating as T_x, T_y, N, M_x, M_y, and M_z the reactions at the built-in supports, and u, v, w, φ_x, φ_y, and φ_z the imposed displacements, we have the following matrix relation:

$$
\begin{bmatrix} T_{xi} \\ T_{yi} \\ N_i \\ M_{xi} \\ M_{yi} \\ M_{zi} \\ T_{xj} \\ T_{yj} \\ N_j \\ M_{xj} \\ M_{yj} \\ M_{zj} \end{bmatrix}
=
\begin{bmatrix}
\frac{12EI_y}{l^3} & 0 & 0 & 0 & -\frac{6EI_y}{l^2} & 0 & -\frac{12EI_y}{l^3} & 0 & 0 & 0 & -\frac{6EI_y}{l^2} & 0 \\
0 & \frac{12EI_x}{l^3} & 0 & -\frac{6EI_x}{l^2} & 0 & 0 & 0 & -\frac{12EI_x}{l^3} & 0 & -\frac{6EI_x}{l^2} & 0 & 0 \\
0 & 0 & \frac{EA}{l} & 0 & 0 & 0 & 0 & 0 & -\frac{EA}{l} & 0 & 0 & 0 \\
0 & -\frac{6EI_x}{l^2} & 0 & \frac{4EI_x}{l} & 0 & 0 & 0 & \frac{6EI_x}{l^2} & 0 & \frac{2EI_x}{l} & 0 & 0 \\
-\frac{6EI_y}{l^2} & 0 & 0 & 0 & \frac{4EI_y}{l} & 0 & \frac{6EI_y}{l^2} & 0 & 0 & 0 & \frac{2EI_y}{l} & 0 \\
0 & 0 & 0 & 0 & 0 & \frac{GI_t}{l} & 0 & 0 & 0 & 0 & 0 & -\frac{GI_t}{l} \\
-\frac{12EI_y}{l^3} & 0 & 0 & 0 & \frac{6EI_y}{l^2} & 0 & \frac{12EI_y}{l^3} & 0 & 0 & 0 & \frac{6EI_y}{l^2} & 0 \\
0 & -\frac{12EI_x}{l^3} & 0 & \frac{6EI_x}{l^2} & 0 & 0 & 0 & \frac{12EI_x}{l^3} & 0 & \frac{6EI_x}{l^2} & 0 & 0 \\
0 & 0 & -\frac{EA}{l} & 0 & 0 & 0 & 0 & 0 & \frac{EA}{l} & 0 & 0 & 0 \\
0 & -\frac{6EI_x}{l^2} & 0 & \frac{2EI_x}{l} & 0 & 0 & 0 & \frac{6EI_x}{l^2} & 0 & \frac{4EI_x}{l} & 0 & 0 \\
-\frac{6EI_y}{l^2} & 0 & 0 & 0 & \frac{2EI_y}{l} & 0 & \frac{6EI_y}{l^2} & 0 & 0 & 0 & \frac{4EI_y}{l} & 0 \\
0 & 0 & 0 & 0 & 0 & -\frac{GI_t}{l} & 0 & 0 & 0 & 0 & 0 & \frac{GI_t}{l}
\end{bmatrix}
\begin{bmatrix} u_i \\ v_i \\ w_i \\ \varphi_{xi} \\ \varphi_{yi} \\ \varphi_{zi} \\ u_j \\ v_j \\ w_j \\ \varphi_{xj} \\ \varphi_{yj} \\ \varphi_{zj} \end{bmatrix}
-
\begin{bmatrix} T_{xi}^0 \\ T_{yi}^0 \\ N_i^0 \\ M_{xi}^0 \\ M_{yi}^0 \\ M_{zi}^0 \\ T_{xj}^0 \\ T_{yj}^0 \\ N_j^0 \\ M_{xj}^0 \\ M_{yj}^0 \\ M_{zj}^0 \end{bmatrix}
$$

$$(2.59)$$

The local stiffness matrix in this case has dimensions (12×12).

The rotation matrix $[N]$ also has dimensions (12×12) and can be partitioned into 16 submatrices of dimensions (3×3), of which 12 are zero and the four diagonal ones are mutually identical:

$$
[N] = \begin{bmatrix} N_0 & 0 & 0 & 0 \\ 0 & N_0 & 0 & 0 \\ 0 & 0 & N_0 & 0 \\ 0 & 0 & 0 & N_0 \end{bmatrix}
\tag{2.60}
$$

$$[N_0] = \begin{bmatrix} \cos \widehat{X^*X} & \cos \widehat{X^*Y} & \cos \widehat{X^*Z} \\ \cos \widehat{Y^*X} & \cos \widehat{Y^*Y} & \cos \widehat{Y^*Z} \\ \cos \widehat{Z^*X} & \cos \widehat{Z^*Y} & \cos \widehat{Z^*Z} \end{bmatrix} \tag{2.61}$$

The assemblage operation and the determination of the external constraint reactions are formally identical to those corresponding to plane systems, described in Section 2.4.

Chapter 3

Plates and shells

3.1 INTRODUCTION

In this chapter, the elastic problem of deflected plates and shells will be presented. The study of shells having double curvature also includes the case of shells of revolution, loaded both symmetrically or otherwise with respect to the axis of symmetry. This study will then be particularized to the specific, but, from the technical standpoint, highly significant cases of thin shells of revolution, circular plates, and cylindrical shells. In this context, reference will also be made to the problem of pressurized vessels having a cylindrical shape and flat or spherical caps. Finally, the problem of axisymmetrical three-dimensional bodies, loaded either symmetrically or otherwise with respect to the axis of symmetry, will be dealt with.

3.2 PLATES IN FLEXURE

Plates are structural elements where one dimension is negligible in comparison with the other two. This dimension is termed **thickness**. **Plane plates**, in particular, are cylindrical bodies whose generators are at least one order of magnitude smaller than the dimensions of the faces (the reverse of the situation we have in the case of the Saint-Venant solid).

Let us consider a plate of thickness h, loaded by distributed forces orthogonal to the faces and constrained at the edge (Figure 3.1). Let XY be the middle plane of the plate and Z the orthogonal axis. The so-called **Kirchhoff kinematic hypothesis** assumes that the segments orthogonal to the middle plane, after deformation has occurred, remain orthogonal to the deformed middle plane (Figure 3.2). Denoting then as φ_x the angle of rotation about the Y axis and as φ_y the angle of rotation about the negative direction of the X axis, the displacement of a generic point P of coordinates x, y, z will present the following three components:

$$u = \varphi_x z = -\frac{\partial w}{\partial x} z \tag{3.1a}$$

$$\upsilon = \varphi_y z = -\frac{\partial w}{\partial y} z \tag{3.1b}$$

$$w = w(x, y) \tag{3.1c}$$

Equation 3.1c indicates that all the points belonging to one and the same segment, orthogonal to the middle plane, are displaced in that direction by the same quantity.

From the kinematic hypothesis of Equations 3.1, it follows by simple derivation that the strain field is

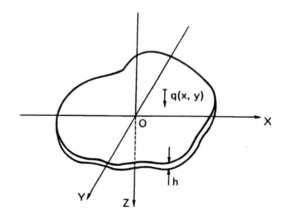

Figure 3.1

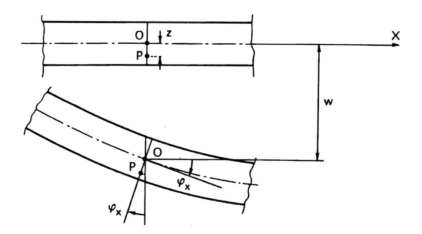

Figure 3.2

$$\varepsilon_x = \frac{\partial u}{\partial x} = \frac{\partial \varphi_x}{\partial x} z = -\frac{\partial^2 w}{\partial x^2} z \tag{3.2a}$$

$$\varepsilon_y = \frac{\partial v}{\partial y} = \frac{\partial \varphi_y}{\partial y} z = -\frac{\partial^2 w}{\partial y^2} z \tag{3.2b}$$

$$\varepsilon_z = \frac{\partial w}{\partial z} = 0 \tag{3.2c}$$

$$\gamma_{xy} = \frac{\partial u}{\partial y} + \frac{\partial v}{\partial x} = \left(\frac{\partial \varphi_x}{\partial y} + \frac{\partial \varphi_y}{\partial x} \right) z = -2 \frac{\partial^2 w}{\partial x \partial y} z \tag{3.2d}$$

$$\gamma_{xz} = \frac{\partial u}{\partial z} + \frac{\partial w}{\partial x} = 0 \tag{3.2e}$$

$$\gamma_{yz} = \frac{\partial v}{\partial z} + \frac{\partial w}{\partial y} = 0 \qquad (3.2f)$$

Kirchhoff's kinematic hypothesis, therefore, generates a condition of plane strain. The three significant components of strain may be expressed as follows:

$$\varepsilon_x = \chi_x z \qquad (3.3a)$$

$$\varepsilon_y = \chi_y z \qquad (3.3b)$$

$$\gamma_{xy} = \chi_{xy} z \qquad (3.3c)$$

where χ_x and χ_y are the flexural curvatures of the middle plane in the respective directions, and χ_{xy} is twice the unit angle of torsion of the middle plane in the X and Y directions.

For the condition of plane stress, the constitutive relations become

$$\varepsilon_x = \frac{1}{E}(\sigma_x - \nu\sigma_y) \qquad (3.4a)$$

$$\varepsilon_y = \frac{1}{E}(\sigma_y - \nu\sigma_x) \qquad (3.4b)$$

$$\gamma_{xy} = \frac{1}{G}\tau_{xy} \qquad (3.4c)$$

It is important to note, however, that a condition cannot, at the same time, be both one of plane strain and one of plane stress. The thickness h is, on the other hand, assumed to be so small as to enable very low, and consequently negligible, stresses σ_z to develop. This is an assumption that we shall take up and discuss in greater depth in Chapter 11. From Equations 3.4a and b, we find

$$\sigma_x - \nu\sigma_y = E\varepsilon_x \qquad (3.5a)$$

$$\nu\sigma_y - \nu^2\sigma_x = E\nu\varepsilon_y \qquad (3.5b)$$

whence, by simple addition, we obtain the expressions:

$$\sigma_x = \frac{E}{1-\nu^2}(\varepsilon_x + \nu\varepsilon_y) \qquad (3.6a)$$

$$\sigma_y = \frac{E}{1-\nu^2}(\varepsilon_y + \nu\varepsilon_x) \qquad (3.6b)$$

$$\tau_{xy} = \frac{E}{2(1+\nu)}\gamma_{xy} \qquad (3.6c)$$

From Equations 3.3, there thus follows the stress field of the plate:

$$\sigma_x = \frac{E}{1-v^2}(\chi_x + v\chi_y)z \qquad (3.7a)$$

$$\sigma_y = \frac{E}{1-v^2}(\chi_y + v\chi_x)z \qquad (3.7b)$$

$$\tau_{xy} = \frac{E}{2(1+v)}\chi_{xy}z \qquad (3.7c)$$

Integrating, over the thickness, the stresses expressed by Equations 3.7, we obtain the **characteristics of the internal reaction,** which are bending and twisting moments per unit length (Figure 3.3):

$$M_x = \int_{-h/2}^{h/2} \sigma_x z\,dz \qquad (3.8a)$$

$$M_y = \int_{-h/2}^{h/2} \sigma_y z\,dz \qquad (3.8b)$$

$$M_{xy} = M_{yx} = \int_{-h/2}^{h/2} \tau_{xy}z\,dz \qquad (3.8c)$$

Substituting Equations 3.7 into Equations 3.8, we obtain finally the **constitutive equations** of the plate:

$$M_x = D(\chi_x + v\chi_y) \qquad (3.9a)$$

$$M_y = D(\chi_y + v\chi_x) \qquad (3.9b)$$

$$M_{xy} = M_{yx} = \frac{1-v}{2}D\chi_{xy} \qquad (3.9c)$$

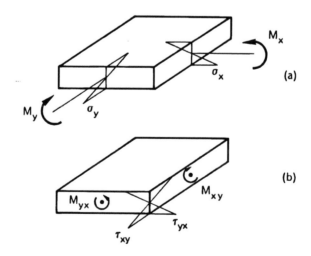

Figure 3.3

where

$$D = \frac{Eh^3}{12(1-\nu^2)} \tag{3.10}$$

is the flexural rigidity of the plate.

Let us then determine the **indefinite equations of equilibrium**, considering an infinitesimal element of the plate subjected to the external load and to the static characteristics. The condition of equilibrium with regard to rotation about the Y axis (Figure 3.4a) yields

$$\left(\frac{\partial M_x}{\partial x} dx \right) dy + \left(\frac{\partial M_{xy}}{\partial y} dy \right) dx - (T_x dy) dx = 0 \tag{3.11}$$

from which we deduce

$$\frac{\partial M_x}{\partial x} + \frac{\partial M_{xy}}{\partial y} - T_x = 0 \tag{3.12a}$$

Likewise, we have

$$\frac{\partial M_{xy}}{\partial x} + \frac{\partial M_y}{\partial y} - T_y = 0 \tag{3.12b}$$

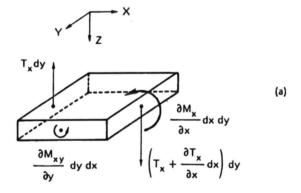

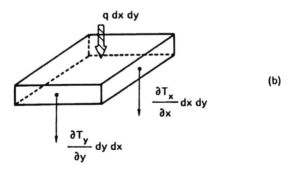

Figure 3.4

The condition of equilibrium with regard to translation in the direction of the Z axis (Figure 3.4b) gives

$$\left(\frac{\partial T_x}{\partial x}dx\right)dy+\left(\frac{\partial T_y}{\partial y}dy\right)dx+q\,dx\,dy=0 \qquad (3.13)$$

from which we deduce

$$\frac{\partial T_x}{\partial x}+\frac{\partial T_y}{\partial y}+q=0 \qquad (3.14)$$

The remaining three conditions of equilibrium—that with regard to rotation about the Z axis and those with regard to translation in the X and Y directions—are identically satisfied, in that the plate has been assumed to be loaded by forces not exerted on the middle plane.

If γ_x and γ_y denote the shearing strains due to the shearing forces, T_x and T_y, respectively, the kinematic equations define the characteristics of deformation as functions of the generalized displacements, in the following way:

$$\begin{bmatrix} \gamma_x \\ \gamma_y \\ \chi_x \\ \chi_y \\ \chi_{xy} \end{bmatrix} = \begin{bmatrix} \dfrac{\partial}{\partial x} & +1 & 0 \\[2mm] \dfrac{\partial}{\partial y} & 0 & +1 \\[2mm] 0 & \dfrac{\partial}{\partial x} & 0 \\[2mm] 0 & 0 & \dfrac{\partial}{\partial y} \\[2mm] 0 & \dfrac{\partial}{\partial y} & \dfrac{\partial}{\partial x} \end{bmatrix} \begin{bmatrix} w \\ \varphi_x \\ \varphi_y \end{bmatrix} \qquad (3.15)$$

It is to be noted that the shearing strains γ_x and γ_y have so far been neglected, starting from Equations 3.1a and b, as also in Equations 3.2e and f.

The **static Equations** 3.12 and 3.14, on the other hand, are presented in matrix form as follows:

$$\begin{bmatrix} \dfrac{\partial}{\partial x} & \dfrac{\partial}{\partial y} & 0 & 0 & 0 \\[2mm] -1 & 0 & \dfrac{\partial}{\partial x} & 0 & \dfrac{\partial}{\partial y} \\[2mm] 0 & -1 & 0 & \dfrac{\partial}{\partial y} & \dfrac{\partial}{\partial x} \end{bmatrix} \begin{bmatrix} T_x \\ T_y \\ M_x \\ M_y \\ M_{xy} \end{bmatrix} + \begin{bmatrix} q \\ 0 \\ 0 \end{bmatrix} = \begin{bmatrix} 0 \\ 0 \\ 0 \end{bmatrix} \qquad (3.16)$$

Also in the case of plates, **static–kinematic duality** is expressed by the fact that the static matrix, neglecting the algebraic sign of the unity terms, is the transpose of the kinematic matrix and *vice versa*.

Finally, the **constitutive Equations** 3.9 can also be cast in matrix form:

$$
\begin{bmatrix} T_x \\ T_y \\ M_x \\ M_y \\ M_{xy} \end{bmatrix} =
\begin{bmatrix}
\dfrac{5}{6}Gh & 0 & 0 & 0 & 0 \\
0 & \dfrac{5}{6}Gh & 0 & 0 & 0 \\
0 & 0 & D & \nu D & 0 \\
0 & 0 & \nu D & D & 0 \\
0 & 0 & 0 & 0 & \dfrac{1-\nu}{2}D
\end{bmatrix}
\begin{bmatrix} \gamma_x \\ \gamma_y \\ \chi_x \\ \chi_y \\ \chi_{xy} \end{bmatrix}
\tag{3.17}
$$

where the factor 5/6 is the inverse of the shear factor corresponding to a rectangular cross section of unit base and height h. Note that, while the height h appears in the first two rows raised to the first power, in the remaining rows it appears raised to the third power, in agreement with Equation 3.10. The shearing stiffness appears therefore more important than the flexural stiffness, by as much as two orders of magnitude, and this explains why the shearing strains are often neglected.

The kinematic Equation 3.15 and the static Equation 3.16, as well as the constitutive Equation 3.17, may be cast in compact form:

$$
\{q\} = [\partial]\{\eta\}
\tag{3.18a}
$$

$$
[\partial]^*\{Q\} + \{\mathcal{F}\} = \{0\}
\tag{3.18b}
$$

$$
\{Q\} = [H]\{q\}
\tag{3.18c}
$$

so that if, as is possible to do in the cases of the three-dimensional body and of the beam, we denote by

$$
\underset{(3\times3)}{[\mathcal{L}]} = \underset{(3\times5)}{[\partial]^*} \underset{(5\times5)}{[H]} \underset{(5\times3)}{[\partial]}
\tag{3.19}
$$

the Lamé matrix operator, the elastic problem of the deflected plate is represented by the following operator equation furnished with the corresponding boundary conditions:

$$
[\mathcal{L}]\{\eta\} = -\{\mathcal{F}\}, \quad \forall P \in S
\tag{3.20a}
$$

$$
[\mathcal{N}]^{\mathrm{T}}\{Q\} = \{p\}, \quad \forall P \in \mathcal{C}_p
\tag{3.20b}
$$

$$
\{\eta\} = \{\eta_0\}, \quad \forall P \in \mathcal{C}_\eta
\tag{3.20c}
$$

In these equations, $\mathcal{C}_p$ denotes the portion of the edge $\mathcal{C}$ on which the static conditions are assigned, while $\mathcal{C}_\eta$ denotes the complementary portion on which the kinematic (or constraint) conditions are assigned.

Designating as $\{n\}$ the unit vector normal to the boundary portion $\mathcal{C}_p$ (Figure 3.5), it is simple to render explicit the boundary condition of equivalence in Equation 3.20b:

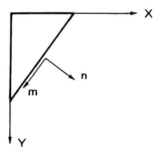

Figure 3.5

$$\begin{bmatrix} n_x & n_y & 0 & 0 & 0 \\ 0 & 0 & n_x & 0 & n_y \\ 0 & 0 & 0 & n_y & n_x \end{bmatrix} \begin{bmatrix} T_x \\ T_y \\ M_x \\ M_y \\ M_{xy} \end{bmatrix} = \begin{bmatrix} T_n \\ M_{nx} \\ M_{ny} \end{bmatrix} \tag{3.21}$$

M_{nx} and M_{ny} being the components of the moment vector acting on the section of normal n. As in the case of the three-dimensional body, also in the case of the plate, the matrix $[\mathcal{N}]^T$ is linked to the matrix operator $[\partial]^*$ of the static Equation 3.16, matching each partial derivative with the corresponding direction cosine of the normal to the boundary.

Finally, the **elastic problem of the deflected plate** can be summarized as follows:

$$[\mathcal{L}]\{\eta\} = -\{\mathcal{F}\}, \quad \forall P \in S \tag{3.22a}$$

$$\left([\mathcal{N}]^T[H][\partial]\right)\{\eta\} = \{p\}, \quad \forall P \in \mathcal{C}_p \tag{3.22b}$$

$$\{\eta\} = \{\eta_0\}, \quad \forall P \in \mathcal{C}_\eta \tag{3.22c}$$

Equations 3.22 are formally identical to those of the three-dimensional elastic problem, once the three-dimensional domain V has been replaced by the two-dimensional domain S of the plate, and the external surface S by the boundary $\mathcal{C}$ of the plate.

3.3 SOPHIE GERMAIN'S EQUATION

Neglecting the shearing deformability of the plate, it is possible to arrive at a differential equation in the single kinematic unknown w. Deriving Equations 3.12, we find

$$\frac{\partial T_x}{\partial x} = \frac{\partial^2 M_x}{\partial x^2} + \frac{\partial^2 M_{xy}}{\partial x \partial y} \tag{3.23a}$$

$$\frac{\partial T_y}{\partial y} = \frac{\partial^2 M_{xy}}{\partial x \partial y} + \frac{\partial^2 M_y}{\partial y^2} \tag{3.23b}$$

and substituting Equations 3.23 into Equation 3.14, we obtain

$$\frac{\partial^2 M_x}{\partial x^2} + 2\frac{\partial^2 M_{xy}}{\partial x \partial y} + \frac{\partial^2 M_y}{\partial y^2} + q = 0 \qquad (3.24)$$

The constitutive Equations 3.9, if we neglect the shearing strains, become

$$M_x = -D\left(\frac{\partial^2 w}{\partial x^2} + v\frac{\partial^2 w}{\partial y^2}\right) \qquad (3.25a)$$

$$M_y = -D\left(\frac{\partial^2 w}{\partial y^2} + v\frac{\partial^2 w}{\partial x^2}\right) \qquad (3.25b)$$

$$M_{xy} = -D(1-v)\frac{\partial^2 w}{\partial x \partial y} \qquad (3.25c)$$

Substituting Equations 3.25 into Equation 3.24, we deduce, finally, **Sophie Germain's equation:**

$$\frac{\partial^4 w}{\partial x^4} + 2\frac{\partial^4 w}{\partial x^2 \partial y^2} + \frac{\partial^4 w}{\partial y^4} = \frac{q}{D} \qquad (3.26)$$

which is the fourth-order differential equation governing the **elastic plate.** If we indicate by

$$\nabla^2 = \frac{\partial^2}{\partial x^2} + \frac{\partial^2}{\partial y^2} \qquad (3.27)$$

the Laplacian operator, Equation 3.26 can also be written as

$$\nabla^2(\nabla^2 w) = \frac{q}{D} \qquad (3.28)$$

or, even more synthetically:

$$\nabla^4 w = \frac{q}{D} \qquad (3.29)$$

Note the formal analogy between Equation 3.29 and the equation of the elastic line.

The term **principal directions of moment** relative to a point of the deflected plate is given to the two orthogonal directions along which the twisting moment M_{xy} vanishes, and consequently the shearing stresses τ_{xy} likewise vanish. These directions thus coincide with the principal ones of stress. The term **principal directions of curvature** is applied to the two orthogonal directions along which the unit angle of torsion $\chi_{xy}/2$ vanishes. In the case where the material is assumed to be isotropic, the constitutive Equation 3.9c shows how the principal directions of moment and the principal directions of curvature must coincide.

Deflected plates having polar symmetry are dealt with in Section 3.8.

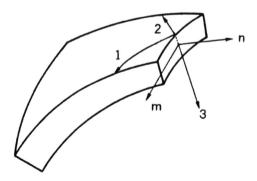

Figure 3.6

3.4 SHELLS WITH DOUBLE CURVATURE

Let us consider a shell of thickness h, the middle surface of which has a double curvature. On this surface, there exists a system of principal curvilinear coordinates s_1, s_2 (Figure 3.6), in correspondence with which the middle surface presents the minimum and maximum curvature. The **membrane regime** consists of the normal forces N_1, N_2, and the shearing force N_{12} contained in the plane tangential to the middle surface (Figure 3.7a), as well as the dilations ε_1, ε_2, and the shearing strain ε_{12} between the principal directions of curvature. The **flexural regime** consists of the shearing forces T_1, T_2, perpendicular to the tangent plane, of the bending moments M_1, M_2 and the twisting moment M_{12} (Figure 3.7b), as well as the shearing strains γ_1, γ_2 between each principal direction of curvature and the direction normal to the tangent plane, the flexural curvatures χ_1, χ_2, and twice the unit angle of torsion χ_{12}.

The **kinematic equations,** which define the characteristics of deformation as functions of the generalized displacements, may be put in matrix form:

$$
\begin{bmatrix} \varepsilon_1 \\ \varepsilon_2 \\ \varepsilon_{12} \\ \gamma_1 \\ \gamma_2 \\ \chi_1 \\ \chi_2 \\ \chi_{12} \end{bmatrix} =
\begin{bmatrix}
\dfrac{\partial}{\partial s_1} & +\dfrac{R_2}{R_1(R_2-R_1)}\dfrac{\partial R_1}{\partial s_2} & +\dfrac{1}{R_1} \\[2ex]
+\dfrac{R_1}{R_2(R_1-R_2)}\dfrac{\partial R_2}{\partial s_1} & \dfrac{\partial}{\partial s_2} & +\dfrac{1}{R_2} \\[2ex]
\dfrac{\partial}{\partial s_2}-\dfrac{R_2}{R_1(R_2-R_1)}\dfrac{\partial R_1}{\partial s_2} & \dfrac{\partial}{\partial s_1}-\dfrac{R_1}{R_2(R_1-R_2)}\dfrac{\partial R_2}{\partial s_1} & 0 \\[2ex]
-\dfrac{1}{R_1} & 0 & \dfrac{\partial}{\partial s_1} \\[2ex]
0 & -\dfrac{1}{R_2} & \dfrac{\partial}{\partial s_2} \\[2ex]
0 & 0 & 0 \\[2ex]
0 & 0 & 0 \\[2ex]
0 & 0 & 0
\end{bmatrix}
$$

$$
\begin{bmatrix}
0 & 0 \\
0 & 0 \\
0 & 0 \\
+1 & 0 \\
0 & +1 \\
\dfrac{\partial}{\partial s_1} & +\dfrac{R_2}{R_1(R_2-R_1)}\dfrac{\partial R_1}{\partial s_2} \\
+\dfrac{R_1}{R_2(R_1-R_2)}\dfrac{\partial R_2}{\partial s_1} & \dfrac{\partial}{\partial s_2} \\
\dfrac{\partial}{\partial s_2}-\dfrac{R_2}{R_1(R_2-R_1)}\dfrac{\partial R_1}{\partial s_2} & \dfrac{\partial}{\partial s_1}-\dfrac{R_1}{R_2(R_1-R_2)}\dfrac{\partial R_2}{\partial s_1}
\end{bmatrix}
\begin{bmatrix} u_1 \\ u_2 \\ u_3 \\ \varphi_1 \\ \varphi_2 \end{bmatrix}
\tag{3.30}
$$

where:

u_1, u_2, u_3 are the components of the displacement on the two principal axes of curvature and on the normal to the tangent plane

φ_1, φ_2 are the rotations about the principal directions of curvature 2 and 1, respectively

R_1, R_2 are the two principal radii of curvature

Equation 3.30 may be rewritten in compact form:

$$\{q\}=[\partial]\{\eta^*\} \tag{3.31}$$

where $\{\eta^*\}$ is the displacement vector in the rotated reference system 123.

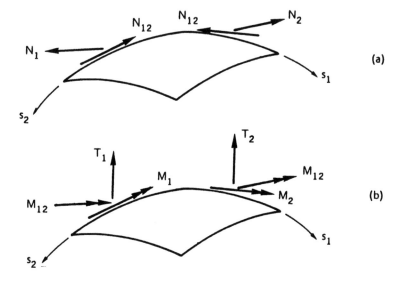

Figure 3.7

The **indefinite equations of equilibrium,** on the other hand, are five and express the equilibrium with regard to translation in the three directions 1, 2, 3, and the equilibrium with regard to rotation about the tangential axes 1 and 2:

$$
\left[
\begin{array}{cccccccc}
\dfrac{\partial}{\partial s_1}+\dfrac{R_1}{R_2(R_1-R_2)}\dfrac{\partial R_2}{\partial s_1} & -\dfrac{R_1}{R_2(R_1-R_2)}\dfrac{\partial R_2}{\partial s_1} & \dfrac{\partial}{\partial s_2}+\dfrac{2R_2}{R_1(R_2-R_1)}\dfrac{\partial R_1}{\partial s_2} & +\dfrac{1}{R_1} & 0 & 0 & 0 & 0 \\[12pt]
-\dfrac{R_2}{R_1(R_2-R_1)}\dfrac{\partial R_1}{\partial s_2} & \dfrac{\partial}{\partial s_2}+\dfrac{R_2}{R_1(R_2-R_1)}\dfrac{\partial R_1}{\partial s_2} & \dfrac{\partial}{\partial s_1}+\dfrac{2R_1}{R_2(R_1-R_2)}\dfrac{\partial R_2}{\partial s_1} & 0 & +\dfrac{1}{R_2} & 0 & 0 & 0 \\[12pt]
-\dfrac{1}{R_1} & -\dfrac{1}{R_2} & 0 & \dfrac{\partial}{\partial s_1}+\dfrac{R_1}{R_2(R_1-R_2)}\dfrac{\partial R_2}{\partial s_1} & \dfrac{\partial}{\partial s_2}+\dfrac{R_2}{R_1(R_2-R_1)}\dfrac{\partial R_1}{\partial s_2} & 0 & 0 & 0 \\[12pt]
0 & 0 & 0 & -1 & 0 & \dfrac{\partial}{\partial s_1}+\dfrac{R_1}{R_2(R_1-R_2)}\dfrac{\partial R_2}{\partial s_1} & -\dfrac{R_1}{R_2(R_1-R_2)}\dfrac{\partial R_2}{\partial s_1} & \dfrac{\partial}{\partial s_2}+\dfrac{2R_2}{R_1(R_2-R_1)}\dfrac{\partial R_1}{\partial s_2} \\[12pt]
0 & 0 & 0 & 0 & -1 & -\dfrac{R_2}{R_1(R_2-R_1)}\dfrac{\partial R_1}{\partial s_2} & \dfrac{\partial}{\partial s_2}+\dfrac{R_2}{R_1(R_2-R_1)}\dfrac{\partial R_1}{\partial s_2} & \dfrac{\partial}{\partial s_1}+\dfrac{2R_1}{R_2(R_1-R_2)}\dfrac{\partial R_2}{\partial s_1}
\end{array}
\right]
\begin{bmatrix} N_1 \\ N_2 \\ N_{12} \\ T_1 \\ T_2 \\ M_1 \\ M_2 \\ M_{12} \end{bmatrix}
+
\begin{bmatrix} p_1 \\ p_2 \\ q \\ m_1 \\ m_2 \end{bmatrix}
=
\begin{bmatrix} 0 \\ 0 \\ 0 \\ 0 \\ 0 \end{bmatrix}
\tag{3.32}
$$

Equation 3.32 may be written synthetically:

$$
[\partial]^{*}\{Q\}+\{\mathcal{F}^{*}\}=\{0\}
\tag{3.33}
$$

where $\{\mathcal{F}^{*}\}$ is the vector of the external forces in the rotated reference system 123.

It should be noted that the kinematic equations represent a more complete version than those proposed by Novozhilov. On the other hand, they have been obtained heuristically, that is, considering the adjoint of the static operator.

The constitutive equations are the following:

$$
\begin{bmatrix} N_1 \\ N_2 \\ N_{12} \\ T_1 \\ T_2 \\ M_1 \\ M_2 \\ M_{12} \end{bmatrix} =
\begin{bmatrix}
\dfrac{12D}{h^2} & v\dfrac{12D}{h^2} & 0 & 0 & 0 & 0 & 0 & 0 \\[2mm]
v\dfrac{12D}{h^2} & \dfrac{12D}{h^2} & 0 & 0 & 0 & 0 & 0 & 0 \\[2mm]
0 & 0 & \dfrac{1-v}{2}\dfrac{12D}{h^2} & 0 & 0 & 0 & 0 & 0 \\[2mm]
0 & 0 & 0 & (1-v)\dfrac{5D}{h^2} & 0 & 0 & 0 & 0 \\[2mm]
0 & 0 & 0 & 0 & (1-v)\dfrac{5D}{h^2} & 0 & 0 & 0 \\[2mm]
0 & 0 & 0 & 0 & 0 & D & vD & 0 \\[2mm]
0 & 0 & 0 & 0 & 0 & vD & D & 0 \\[2mm]
0 & 0 & 0 & 0 & 0 & 0 & 0 & \dfrac{1-v}{2}D
\end{bmatrix}
\begin{bmatrix} \varepsilon_1 \\ \varepsilon_2 \\ \varepsilon_{12} \\ \gamma_1 \\ \gamma_2 \\ \chi_1 \\ \chi_2 \\ \chi_{12} \end{bmatrix}
$$

(3.34)

or, in compact form:

$$\{Q\} = [H]\{q\}$$

(3.35)

The vectors of the external forces and of the displacements in the local system 123 may be obtained by premultiplying the corresponding vectors in the global reference system XYZ by the orthogonal matrix of rotation $[N]$:

$$\{\mathcal{F}^*\} = [N]\{\mathcal{F}\}$$

(3.36a)

$$\{\eta^*\} = [N]\{\eta\}$$

(3.36b)

where $[N]$ is the following (5×6) matrix:

$$
[N] = \begin{bmatrix}
\cos \widehat{1X} & \cos \widehat{1Y} & \cos \widehat{1Z} & 0 & 0 & 0 \\
\cos \widehat{2X} & \cos \widehat{2Y} & \cos \widehat{2Z} & 0 & 0 & 0 \\
\cos \widehat{3X} & \cos \widehat{3Y} & \cos \widehat{3Z} & 0 & 0 & 0 \\
0 & 0 & 0 & \cos \widehat{2X} & \cos \widehat{2Y} & \cos \widehat{2Z} \\
0 & 0 & 0 & -\cos \widehat{1X} & -\cos \widehat{1Y} & -\cos \widehat{1Z}
\end{bmatrix}
$$

(3.37)

and $\{\mathcal{F}\}, \{\eta\}$ are the following six-component vectors:

$$\{\mathcal{F}\}^{\mathrm{T}} = [q_x, q_y, q_z, m_x, m_y, m_z]$$

(3.38a)

$$\{\eta\}^{\mathrm{T}} = [u, v, w, \varphi_x, \varphi_y, \varphi_z]$$

(3.38b)

The matrix $[N]$ is not square, since the moment and rotation vectors, which are always contained in the tangent plane, in the global XYZ system generally possess three components.

Substituting Equations 3.36 into Equations 3.31 and 3.33, we obtain

$$[\mathcal{L}]\{\eta\} = -\{\mathcal{F}\} \tag{3.39}$$

the Lamé operator being given by the following matrix product:

$$\underset{(6\times6)}{[\mathcal{L}]} = \underset{(6\times5)}{[N]^{\mathrm{T}}} \underset{(5\times8)}{[\partial]^{*}} \underset{(8\times8)}{[H]} \underset{(8\times5)}{[\partial]} \underset{(5\times6)}{[N]} \tag{3.40}$$

On the other hand, the boundary condition of equivalence takes the form:

$$\underset{(5\times8)}{[\mathcal{N}]}\underset{(8\times1)}{\{Q\}} = \underset{(5\times6)}{[N]} \underset{(6\times1)}{\{p\}} \tag{3.41}$$

where:
  $\{p\}$ is the vector of the forces and of the moments applied to the unit length of the boundary and referred to the global system XYZ
  $[\mathcal{N}]^{\mathrm{T}}$ is the matrix that transforms the static characteristics into the aforesaid loadings, referred to the local system 123

Designating as n the axis belonging to the tangent plane and orthogonal to the boundary (Figure 3.6), and as m the axis tangential to the boundary, in such a way that the reference system $nm3$ is right-handed, we have

$$[\mathcal{N}]^{\mathrm{T}} = \begin{bmatrix} n_1 & 0 & n_2 & 0 & 0 & 0 & 0 & 0 \\ 0 & n_2 & n_1 & 0 & 0 & 0 & 0 & 0 \\ 0 & 0 & 0 & n_1 & n_2 & 0 & 0 & 0 \\ 0 & 0 & 0 & 0 & 0 & n_1 & 0 & n_2 \\ 0 & 0 & 0 & 0 & 0 & 0 & n_2 & n_1 \end{bmatrix} \tag{3.42}$$

where the submatrix formed by the last three rows and the last five columns replicates Equation 3.21, obtained for the plate. The submatrix obtained from the first two rows and the first three columns reproduces the equations for the plane stress condition. Therefore, it is apparent, also in the framework of the boundary conditions, how the membrane and flexural regimes remain separate and are not interacting. The matrix $[\mathcal{N}]^{\mathrm{T}}$ is directly correlated to the operator $[\partial]^{*}$.

The elastic problem of the shell with double curvature is thus summarized in the following equations:

$$[\mathcal{L}]\{\eta\} = -\{\mathcal{F}\}, \quad \forall P \in S \tag{3.43a}$$

$$([N]^{\mathrm{T}}[\mathcal{N}]^{\mathrm{T}}[H][\partial][N])\{\eta\} = \{p\}, \quad \forall P \in C_p \tag{3.43b}$$

$$\{\eta\} = \{\eta_0\}, \quad \forall P \in C_\eta \tag{3.43c}$$

3.5 NONSYMMETRICALLY LOADED SHELLS OF REVOLUTION

The term **shell of revolution** refers to a shell, generally of double curvature, generated by the complete rotation of a plane curve $r(z)$ about the axis of symmetry Z (Figure 3.8).

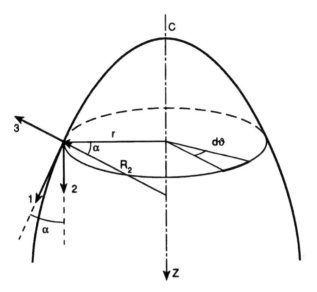

Figure 3.8

The set of infinite configurations that the generating curve $r(z)$ assumes in its rotation are called **meridians**. The set of infinite circular trajectories described by the individual points of the curve are called **parallels**. The meridians and the parallels represent the so-called lines of curvature, on which a system of principal curvilinear coordinates s_1, s_2 can be defined.

Denoting by s_1 the curvilinear coordinate along the meridians and by s_2 the curvilinear coordinate along the parallels, we have (Figure 3.8)

$$ds_1 = ds = \frac{dz}{\cos\alpha} \tag{3.44a}$$

$$ds_2 = r\,d\vartheta \tag{3.44b}$$

where α is the angle that the tangent to the meridian forms with the axis of symmetry, which is also equal to the angle that the normal to the surface forms with the radius r (Figure 3.8), and where ϑ represents the longitude. On the basis of **Meusnier's theorem**, the radius r and the principal radius of curvature R_2 are linked by the following relation:

$$r = R_2 \cos\alpha \tag{3.45}$$

Whereas the variations of the radius of curvature R_1 are equal to zero with respect to the coordinate s_2 (i.e., along the parallels), the variations of the radius of curvature R_2 with respect to the coordinate s_1 (i.e., along the meridians) are

$$\frac{\partial R_2}{\partial s_1} = \frac{\partial}{\partial s}\left(\frac{r}{\cos\alpha}\right)$$

$$= \frac{\partial r}{\partial s}\left(\frac{1}{\cos\alpha}\right) + r\frac{\sin\alpha}{\cos^2\alpha}\frac{\partial\alpha}{\partial s} \tag{3.46}$$

By the definition of curvature, $\partial\alpha/\partial s = -1/R_1$ and thus from Equation 3.46 we have

$$\frac{\partial R_2}{\partial s_1} = \frac{\sin\alpha}{\cos\alpha} - \frac{R_2}{R_1}\frac{\sin\alpha}{\cos\alpha}$$

$$= \tan\alpha\left(1 - \frac{R_2}{R_1}\right) \tag{3.47}$$

The term $1/\rho_2$, recurring in the kinematic Equation 3.30 and the static Equation 3.32, can be expressed as follows:

$$\frac{1}{\rho_2} = \frac{R_1}{R_2(R_1 - R_2)}\frac{\partial R_2}{\partial s_1} = \frac{\tan\alpha}{R_2} = \frac{\sin\alpha}{r} \tag{3.48}$$

The kinematic equations for the shells of revolution nonsymmetrically loaded are thus

$$
\begin{bmatrix}
\varepsilon_s \\
\varepsilon_\vartheta \\
\gamma_{s\vartheta} \\
\gamma_s \\
\gamma_\vartheta \\
\chi_s \\
\chi_\vartheta \\
\chi_{s\vartheta}
\end{bmatrix}
=
\begin{bmatrix}
\frac{\partial}{\partial s} & 0 & +\frac{1}{R_1} & 0 & 0 \\
+\frac{\sin\alpha}{r} & \frac{1}{r}\frac{\partial}{\partial\vartheta} & +\frac{1}{R_2} & 0 & 0 \\
\frac{1}{r}\frac{\partial}{\partial\vartheta} & \left(\frac{\partial}{\partial s} - \frac{\sin\alpha}{r}\right) & 0 & 0 & 0 \\
-\frac{1}{R_1} & 0 & \frac{\partial}{\partial s} & +1 & 0 \\
0 & -\frac{1}{R_2} & \frac{1}{r}\frac{\partial}{\partial\vartheta} & 0 & +1 \\
0 & 0 & 0 & \frac{\partial}{\partial s} & 0 \\
0 & 0 & 0 & +\frac{\sin\alpha}{r} & \frac{1}{r}\frac{\partial}{\partial\vartheta} \\
0 & 0 & 0 & \frac{1}{r}\frac{\partial}{\partial\vartheta} & \left(\frac{\partial}{\partial s} - \frac{\sin\alpha}{r}\right)
\end{bmatrix}
\begin{bmatrix}
u \\
v \\
w \\
\varphi_s \\
\varphi_\vartheta
\end{bmatrix}
\tag{3.49}
$$

The static equations are dual with respect to Equation 3.49:

$$
\begin{bmatrix}
\left(\frac{\partial}{\partial s} + \frac{\sin\alpha}{r}\right) & -\frac{\sin\alpha}{r} & \frac{1}{r}\frac{\partial}{\partial\vartheta} & +\frac{1}{R_1} & 0 & 0 & 0 & 0 \\
0 & \frac{1}{r}\frac{\partial}{\partial\vartheta} & \left(\frac{\partial}{\partial s} + \frac{2\sin\alpha}{r}\right) & 0 & +\frac{1}{R_2} & 0 & 0 & 0 \\
-\frac{1}{R_1} & -\frac{1}{R_2} & 0 & \left(\frac{\partial}{\partial s} + \frac{\sin\alpha}{r}\right) & \frac{1}{r}\frac{\partial}{\partial\vartheta} & 0 & 0 & 0 \\
0 & 0 & 0 & -1 & 0 & \left(\frac{\partial}{\partial s} + \frac{\sin\alpha}{r}\right) & -\frac{\sin\alpha}{r} & \frac{1}{r}\frac{\partial}{\partial\vartheta} \\
0 & 0 & 0 & 0 & -1 & 0 & \frac{1}{r}\frac{\partial}{\partial\vartheta} & \left(\frac{\partial}{\partial s} + \frac{2\sin\alpha}{r}\right)
\end{bmatrix}
$$

$$
\begin{bmatrix}
N_s \\
N_\vartheta \\
N_{s\vartheta} \\
T_s \\
T_\vartheta \\
M_s \\
M_\vartheta \\
M_{s\vartheta}
\end{bmatrix}
+
\begin{bmatrix}
p_s \\
p_\vartheta \\
q \\
m_s \\
m_\vartheta
\end{bmatrix}
=
\begin{bmatrix}
0 \\
0 \\
0 \\
0 \\
0
\end{bmatrix}
\tag{3.50}
$$

3.6 SYMMETRICALLY LOADED SHELLS OF REVOLUTION

When a shell of revolution is loaded symmetrically with respect to axis Z, Equations 3.49 and 3.50 simplify, since only the curvilinear coordinate s is present as an independent variable, while the displacement υ along the parallels and the rotation φ_ϑ about the meridians vanish, as well as the deformations $\gamma_{s\vartheta}$, γ_ϑ, $\chi_{s\vartheta}$ and the corresponding internal reactions $N_{s\vartheta}$, T_ϑ, $M_{s\vartheta}$:

$$
\begin{bmatrix} \varepsilon_s \\ \varepsilon_\vartheta \\ \gamma_s \\ \chi_s \\ \chi_\vartheta \end{bmatrix} = \begin{bmatrix} \frac{d}{ds} & \frac{1}{R_1} & 0 \\ +\frac{\sin\alpha}{r} & \frac{1}{R_2} & 0 \\ -\frac{1}{R_1} & \frac{d}{ds} & +1 \\ 0 & 0 & \frac{d}{ds} \\ 0 & 0 & +\frac{\sin\alpha}{r} \end{bmatrix} \begin{bmatrix} u \\ w \\ \varphi_s \end{bmatrix}
\tag{3.51}
$$

$$
\begin{bmatrix} \left(\frac{d}{ds}+\frac{\sin\alpha}{r}\right) & -\frac{\sin\alpha}{r} & \frac{1}{R_1} & 0 & 0 \\ -\frac{1}{R_1} & -\frac{1}{R_2} & \left(\frac{d}{ds}+\frac{\sin\alpha}{r}\right) & 0 & 0 \\ 0 & 0 & -1 & \left(\frac{d}{ds}+\frac{\sin\alpha}{r}\right) & -\frac{\sin\alpha}{r} \end{bmatrix} \begin{bmatrix} N_s \\ N_\vartheta \\ T_s \\ M_s \\ M_\vartheta \end{bmatrix} + \begin{bmatrix} p_s \\ q \\ m_s \end{bmatrix} = \begin{bmatrix} 0 \\ 0 \\ 0 \end{bmatrix}
\tag{3.52}
$$

Observe that, again for reasons of symmetry, the conditions of equilibrium to translation along the parallels and to rotation around the meridians are identically satisfied and thus do not appear in Equation 3.52. Finally, we have three equations of equilibrium (respectively, with regard to translation along the meridians, to translation along the normal n, and to rotation about the parallels) in the five static unknowns N_s, N_ϑ, T_s, M_s, M_ϑ (Figure 3.9). The elastic problem for shells of revolution thus has two degrees of internal redundancy, while the more general problem of shells with double curvature appears to have three degrees of redundancy. Just as for beam systems, then, symmetry reduces the degree of statical inde-terminacy of the elastic problem also for shells.

Equation 3.52 is verified by imposing these three conditions of equilibrium on an infinites-imal shell element, bounded by two meridians located at an infinitesimal distance $ds_2 = rd\vartheta$ and by two parallels located at an infinitesimal distance $ds_1 = ds$ (Figure 3.9).

The condition of equilibrium with regard to translation along the meridians yields the equation (Figures 3.9a, c, and e):

$$
dN_s r d\vartheta + N_s dr d\vartheta - N_\vartheta \sin\alpha\, ds\, d\vartheta + T_s \frac{ds}{R_1} r\, d\vartheta + p_s\, r\, ds\, d\vartheta = 0
\tag{3.53a}
$$

which, divided by $rdsd\vartheta$, coincides with the first of Equations 3.52.

The condition of equilibrium with regard to translation along the normal n furnishes the equation (Figures 3.9c through e):

$$
-N_s \frac{ds}{R_1} r d\vartheta - N_\vartheta ds\, d\vartheta \cos\alpha + dT_s r d\vartheta + T_s dr\, d\vartheta + q\, r\, ds\, d\vartheta = 0
\tag{3.53b}
$$

which, divided by $rdsd\vartheta$, coincides with the second of Equations 3.52.

Finally, the condition of equilibrium with regard to rotation about the parallels furnishes the equation (Figures 3.9b and c):

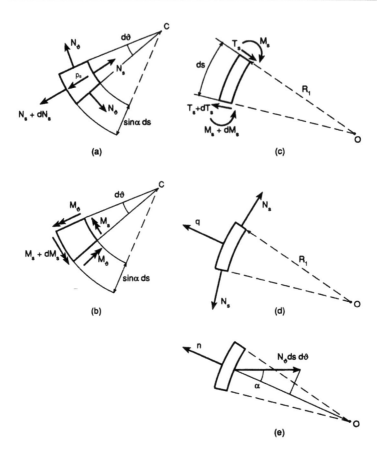

Figure 3.9

$$-T_s r \, d\vartheta \, ds + dM_s r \, d\vartheta + M_s \, dr \, d\vartheta - M_\vartheta \sin\alpha \, ds \, d\vartheta + m_s r \, d\vartheta \, ds = 0 \qquad (3.53c)$$

which, divided by $r ds d\vartheta$, coincides with the third of Equations 3.52.

Notice that, in the indefinite equilibrium Equation 3.52, the terms $\sin\alpha/r$, which are not negligible, have been enclosed. These contributions are because the parallel curvilinear sides of the shell element of Figures 3.9a and b differ by the amount $dr d\vartheta$, as well as because the different action lines of the forces acting on the remaining two meridian sides are convergent. In the static matrix of Equation 3.52, they appear in the first element of the first row (N_s), in the second element of the first row (N_ϑ), in the third element of the second row (T_s), in the fourth element of the third row (M_s), and in the fifth element of the third row (M_ϑ). It is remarkable that N_s, T_s, M_s are related to the parallel sides, whereas N_ϑ, M_ϑ are related to the meridian sides.

While the static Equation 3.52 can be obtained straightforwardly, the derivation of the kinematic Equation 3.51, where the terms $\sin\alpha/r$ are partially absent (only the terms related to ε_ϑ, χ_ϑ, and the meridian actions N_ϑ, M_ϑ are maintained), is more troublesome. From the principle of virtual work and the static Equation 3.52, it is possible to derive the kinematic Equation 3.51. Such a demonstration is reported in Appendix I.

3.7 MEMBRANES AND THIN SHELLS OF REVOLUTION

Membranes are two-dimensional structural elements without flexural rigidity. These elements can sustain only tensile forces contained in the tangent plane. A similar but opposite case is provided by **thin shells**, which are shells of such small thickness that they present an altogether negligible flexural rigidity. These elements can sustain only compressive forces contained in the tangent plane. In the case of membranes, therefore, a zero compressive stiffness is assumed, whereas in the case of thin shells, a zero tensile stiffness is assumed. Both hypotheses imply a zero flexural rigidity.

As regards membranes and thin shells of revolution (thin domes), the kinematic and static equations simplify notably compared with Equations 3.51 and 3.52, since only forces along the meridians and the parallels, N_s and N_ϑ, are present, as well as the displacements along the meridians and those perpendicular to the middle surface, u and w, respectively:

$$\begin{bmatrix} \varepsilon_s \\ \varepsilon_\vartheta \end{bmatrix} = \begin{bmatrix} \dfrac{d}{ds} & \dfrac{1}{R_1} \\ \dfrac{\sin\alpha}{r} & \dfrac{1}{R_2} \end{bmatrix} \begin{bmatrix} u \\ w \end{bmatrix} \tag{3.54a}$$

$$\begin{bmatrix} \left(\dfrac{d}{ds}+\dfrac{\sin\alpha}{r}\right) & -\dfrac{\sin\alpha}{r} \\ -\dfrac{1}{R_1} & -\dfrac{1}{R_2} \end{bmatrix} \begin{bmatrix} N_s \\ N_\vartheta \end{bmatrix} + \begin{bmatrix} p_s \\ q \end{bmatrix} = \begin{bmatrix} 0 \\ 0 \end{bmatrix} \tag{3.54b}$$

It is remarkable that the thin domes appear to be statically determinate (square matrix operators) as occurs in the case of beams.

From the second of Equations 3.54b, we obtain the fundamental algebraic relation that links the forces N_s and N_ϑ:

$$\frac{N_s}{R_1} + \frac{N_\vartheta}{R_2} = q \tag{3.55a}$$

while, from the first, we obtain the following differential equation:

$$\frac{dN_s}{ds} + \frac{\sin\alpha}{r}N_s - \frac{\sin\alpha}{r}N_\vartheta + p_s = 0 \tag{3.55b}$$

On the other hand, by means of Equation 3.55a, we can express N_ϑ as a function of N_s:

$$N_\vartheta = R_2\left(q - \frac{N_s}{R_1}\right) \tag{3.56}$$

and this expression, inserted into Equation 3.55b, gives

$$\frac{dN_s}{ds} + \left(\frac{1}{R_1} + \frac{1}{R_2}\right)\tan\alpha\, N_s = q\tan\alpha - p_s \tag{3.57}$$

which is a differential equation with ordinary derivatives in the unknown function $N_s(s)$.

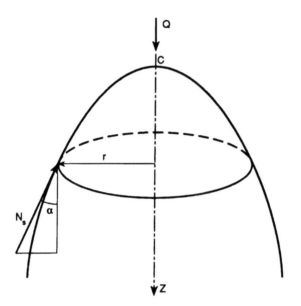

Figure 3.10

Instead of resolving the foregoing differential equation, alternatively we can consider equilibrium to translation in the Z direction of the portion of a thin dome (or membrane) that remains above a generic parallel (Figure 3.10):

$$Q = N_s \cos\alpha \left(2\pi r\right) \tag{3.58}$$

where Q is the integral of the vertical loads acting on that portion. From Equation 3.58, we obtain immediately

$$N_s = \frac{Q}{2\pi r \cos\alpha} \tag{3.59}$$

Via Equation 3.56, we then obtain the corresponding force along the parallel.

If σ_s and σ_ϑ denote the internal forces transmitted per unit area of the cross section (N_s and N_ϑ are forces per unit length), Equation 3.55a is transformed as follows:

$$\frac{\sigma_s}{R_1} + \frac{\sigma_\vartheta}{R_2} = \frac{p}{h} \tag{3.60}$$

where p denotes now the pressure acting normally to the middle surface (i.e., $p_s = 0$ and $q = p$), and h denotes the thickness of the thin dome (or membrane).

In the case of an indefinitely long **cylindrical membrane** subjected to the internal pressure p (Figure 3.11), we have $R_1 \to \infty$, $R_2 = r$, and thus the circumferential stress is

$$\sigma_\vartheta = \frac{pr}{h} \tag{3.61}$$

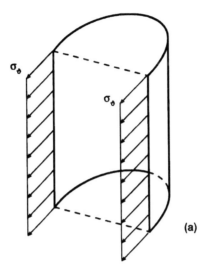
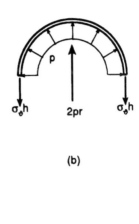

(b)

(a)

Figure 3.11

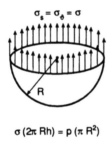

$\sigma_s = \sigma_\vartheta = \sigma$

$\sigma (2\pi\, Rh) = p\, (\pi\, R^2)$

Figure 3.12

This internal reaction increases naturally with the increase in the pressure p and the radius r, and with the decrease in the thickness h.

In the case of a **spherical membrane** subjected to the internal pressure p (Figure 3.12), we have $R_1 = R_2 = R$, and thus the state of stress is isotropic:

$$\sigma_s = \sigma_\vartheta = \frac{pR}{2h} \tag{3.62}$$

Also in this case, the internal reaction increases with pressure and radius, and decreases with the increase in thickness.

In the case of an indefinite **conical membrane** subjected to the internal pressure p (Figure 3.13), we have $R_1 \to \infty$, $R_2 = r/\cos\alpha$, and hence the circumferential stress is

$$\sigma_\vartheta = \frac{pr}{h\cos\alpha} \tag{3.63}$$

The conditions of internal loading in the case of finite cylinders and cones are equal to those mentioned, obtained only at sufficiently large distances from the externally constrained zones. This amount is, however, to be evaluated in relation to the thickness h of the shell.

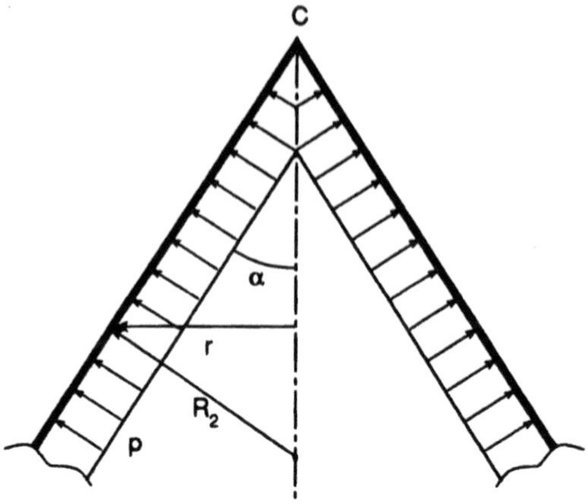

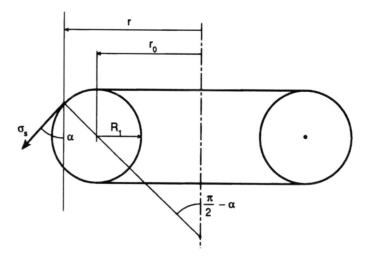

Figure 3.13

In the case of a **toroidal membrane** under pressure (Figure 3.14), Equations 3.60 and 3.59 become

$$\frac{\sigma_s}{R_1} + \frac{\sigma_\vartheta}{r}\cos\alpha = \frac{p}{h} \tag{3.64a}$$

$$\sigma_s h = \frac{\pi(r^2 - r_0^2)p}{2\pi r \cos\alpha} \tag{3.64b}$$

Since $r = r_0 + R_1 \cos\alpha$ (Figure 3.14), Equation 3.64b offers

$$\sigma_s = \frac{pR_1}{2h}\frac{2r_0 + R_1\cos\alpha}{r_0 + R_1\cos\alpha} \tag{3.65}$$

On the crown of the toroidal surface, we have $\alpha = \pi/2$, and hence

Figure 3.14

$$\sigma_s = \frac{pR_1}{h} \tag{3.66}$$

This stress corresponds to the circumferential stress of a pressurized cylinder of radius R_1. The minimum stress σ_s occurs on the maximum parallel of the torus for $\alpha = 0$:

$$\sigma_s = \frac{pR_1}{2h} \frac{2r_0 + R_1}{r_0 + R_1} \tag{3.67}$$

while the maximum occurs on the minimum parallel of the torus for $\alpha = \pi$:

$$\sigma_s = \frac{pR_1}{2h} \frac{2r_0 - R_1}{r_0 - R_1} \tag{3.68}$$

The stress along the parallels is obtained from Equation 3.64a, and in each point of the torus amounts to

$$\sigma_\vartheta = \frac{pR_1}{2h} \tag{3.69}$$

It is equal to the longitudinal stress of a pressurized cylinder of radius R_1 having, for example, hemispherical caps (Figure 3.12).

3.8 CIRCULAR PLATES

The case of shells of revolution loaded symmetrically reduces to the particular case of **circular plates** for $R_1 \to \infty$, $\alpha = \pi/2$. The curvilinear coordinate along the meridian, s, coincides with the radial coordinate r, so that the kinematic Equation 3.51 transforms as follows:

$$
\begin{bmatrix} \varepsilon_r \\ \varepsilon_\vartheta \\ \gamma_r \\ \chi_r \\ \chi_\vartheta \end{bmatrix}
=
\begin{bmatrix}
\dfrac{d}{dr} & 0 & 0 \\[2mm]
\dfrac{1}{r} & 0 & 0 \\[2mm]
0 & \dfrac{d}{dr} & +1 \\[2mm]
0 & 0 & \dfrac{d}{dr} \\[2mm]
0 & 0 & \dfrac{1}{r}
\end{bmatrix}
\begin{bmatrix} u \\ w \\ \varphi_r \end{bmatrix}
\tag{3.70}
$$

The static Equation 3.52, on the other hand, becomes

$$
\begin{bmatrix}
\left(\dfrac{d}{dr}+\dfrac{1}{r}\right) & -\dfrac{1}{r} & 0 & 0 & 0 \\[2mm]
0 & 0 & \left(\dfrac{d}{dr}+\dfrac{1}{r}\right) & 0 & 0 \\[2mm]
0 & 0 & -1 & \left(\dfrac{d}{dr}+\dfrac{1}{r}\right) & -\dfrac{1}{r}
\end{bmatrix}
\begin{bmatrix} N_r \\ N_\vartheta \\ T_r \\ M_r \\ M_\vartheta \end{bmatrix}
+
\begin{bmatrix} p_r \\ q \\ m_r \end{bmatrix}
=
\begin{bmatrix} 0 \\ 0 \\ 0 \end{bmatrix}
\tag{3.71}
$$

Restricting the analysis to the flexural regime only, we obtain the following equations, which are kinematic and static, respectively:

$$
\begin{bmatrix} \gamma_r \\ \chi_r \\ \chi_\vartheta \end{bmatrix} = \begin{bmatrix} \dfrac{d}{dr} & +1 \\ 0 & \dfrac{d}{dr} \\ 0 & \dfrac{1}{r} \end{bmatrix} \begin{bmatrix} w \\ \varphi_r \end{bmatrix}
$$
(3.72)

$$
\begin{bmatrix} \left(\dfrac{d}{dr}+\dfrac{1}{r}\right) & 0 & 0 \\ -1 & \left(\dfrac{d}{dr}+\dfrac{1}{r}\right) & -\dfrac{1}{r} \end{bmatrix} \begin{bmatrix} T_r \\ M_r \\ M_\vartheta \end{bmatrix} + \begin{bmatrix} -q \\ 0 \end{bmatrix} = \begin{bmatrix} 0 \\ 0 \end{bmatrix}
$$
(3.73)

where the distributed moment has been set equal to zero and the distributed load has been taken positive if downward. The indefinite equation of equilibrium (Equation 3.73) represents a system of two differential equations in the three unknowns T_r, M_r, M_ϑ. The polar symmetry thus reduces the degree of static indeterminacy of the deflected plates from two to one.

The first of Equations 3.73 represents the condition of equilibrium with regard to the transverse translation of a plate element identified by two radii forming the angle $d\vartheta$, and by two circumferences of radius r and $r+dr$ (Figure 3.15a):

$$
dT_r r\, d\vartheta + T_r\, dr\, d\vartheta - qr\, dr\, d\vartheta = 0
$$
(3.74a)

The second term of the foregoing equation is due to the greater length presented by the outermost arc of circumference. Dividing by the elementary area $r dr d\vartheta$, we once more obtain Equation 3.73.

The second of Equations 3.73 represents the condition of equilibrium with regard to rotation of the same plate element about the circumference of radius r (Figure 3.15b):

$$
-T_r r\, d\vartheta\, dr + dM_r r\, d\vartheta + M_r\, dr\, d\vartheta - M_\vartheta dr\, d\vartheta = 0
$$
(3.74b)

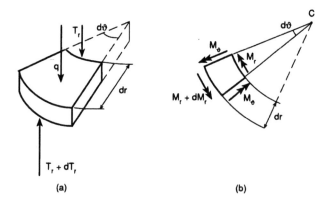

(a)   (b)

Figure 3.15

Also in this case, the contribution of the third term is due to the greater length of the outermost arc of circumference.

Using the elastic constitutive equations that link bending moments and curvatures:

$$M_r = D(\chi_r + \nu\chi_\vartheta) \tag{3.75a}$$

$$M_\vartheta = D(\chi_\vartheta + \nu\chi_r) \tag{3.75b}$$

and assuming zero shearing strain γ_r:

$$\gamma_r = \frac{dw}{dr} + \varphi_r = 0 \tag{3.76a}$$

whereby

$$\chi_r = \frac{d\varphi_r}{dr} = -\frac{d^2w}{dr^2} \tag{3.76b}$$

$$\chi_\vartheta = \frac{\varphi_r}{r} = -\frac{1}{r}\frac{dw}{dr} \tag{3.76c}$$

the second of the indefinite equations of equilibrium (Equation 3.73) transforms into a third-order differential equation in the unknown function w:

$$-T_r - D\frac{d}{dr}\left(\frac{d^2w}{dr^2} + \frac{\nu}{r}\frac{dw}{dr}\right) - \frac{D}{r}\left(\frac{d^2w}{dr^2} + \frac{\nu}{r}\frac{dw}{dr}\right) + \frac{D}{r}\left(\frac{1}{r}\frac{dw}{dr} + \nu\frac{d^2w}{dr^2}\right) = 0 \tag{3.77}$$

Reordering the terms, we obtain

$$\frac{d^3w}{dr^3} + \frac{1}{r}\frac{d^2w}{dr^2} - \frac{1}{r^2}\frac{dw}{dr} = -\frac{T_r}{D} \tag{3.78}$$

The foregoing equation is equivalent to the following one, in which the unknown function is the radial rotation φ_r:

$$\frac{d^2\varphi_r}{dr^2} + \frac{1}{r}\frac{d\varphi_r}{dr} - \frac{1}{r^2}\varphi_r = \frac{T_r}{D} \tag{3.79}$$

If $Q(r)$ denotes the integral of the normal loads acting on the plate within the circumference of radius r, from equilibrium we have

$$2\pi r T_r = Q(r) \tag{3.80}$$

whereby Equation 3.79 can be cast in the following form:

$$\frac{d}{dr}\left[\frac{1}{r}\frac{d}{dr}(r\varphi_r)\right] = \frac{Q(r)}{2\pi Dr} \tag{3.81}$$

A first integration yields

$$\frac{1}{r}\frac{d}{dr}(r\varphi_r) = \int_0^r \frac{Q(r)}{2\pi Dr}dr + C_1 \tag{3.82}$$

Multiplying by r and integrating again, we obtain

$$r\varphi_r = \int_0^r \left[r\int_0^r \frac{Q(r)}{2\pi Dr}dr \right]dr + C_1\frac{r^2}{2} + C_2 \tag{3.83}$$

from which, on further integration, we find the equation of the elastic deformed configuration $w(r)$.

Consider, for instance, a circular plate of radius R clamped at the boundary and uniformly loaded with a pressure q. In this case, we have $Q(r) = q\pi\, r^2$, so that Equation 3.83 becomes

$$r\varphi_r = \int_0^r \frac{qr^3}{4D}dr + C_1\frac{r^2}{2} + C_2 \tag{3.84}$$

and hence

$$\varphi_r = \frac{qr^3}{16D} + C_1\frac{r}{2} + \frac{C_2}{r} \tag{3.85}$$

For reasons of symmetry, we must have $\varphi_r(0) = 0$, and hence the constant C_2 is zero. The condition of a built-in constraint at the edge, on the other hand, furnishes the relation:

$$\varphi_r(R) = \frac{qR^3}{16D} + C_1\frac{R}{2} = 0 \tag{3.86}$$

from which we obtain the constant C_1:

$$C_1 = -\frac{qR^2}{8D} \tag{3.87}$$

The equation of the elastic deformed configuration is thus drawn from the integration of the following equation:

$$-\frac{dw}{dr} = \frac{qr^3}{16D} - \frac{qR^2}{16D}r \tag{3.88}$$

The displacement w is thus defined, but for a constant C_3:

$$w = -\frac{qr^4}{64D} + \frac{qR^2}{32D}r^2 + C_3 \tag{3.89}$$

which may be determined by imposing the annihilation of the displacement at the built-in constraint:

$$C_3 = -\frac{qR^4}{64D} \tag{3.90}$$

Finally, therefore, the displacement normal to the middle plane and the radial rotation are expressible as follows:

$$w = -\frac{q}{64D}(R^2 - r^2)^2 \tag{3.91a}$$

$$\varphi_r = -\frac{qr}{16D}(R^2 - r^2) \tag{3.91b}$$

The transverse displacement at the center of the plate is therefore equal to

$$f = |w(0)| = \frac{qR^4}{64D} \tag{3.92}$$

The bending moments in Equations 3.75 are obtained taking into account Equations 3.76b and c and Equations 3.91:

$$M_r = -\frac{q}{16}\left[(1+v)R^2 - (3+v)r^2\right] \tag{3.93a}$$

$$M_\vartheta = -\frac{q}{16}\left[(1+v)R^2 - (1+3v)r^2\right] \tag{3.93b}$$

In the center, the two moments, the radial one and the circumferential one, are equal to one another:

$$M_r(0) = M_\vartheta(0) = -(1+v)\frac{qR^2}{16} \tag{3.94}$$

On the clamped edge, we have

$$M_r(R) = \frac{qR^2}{8}, \quad M_\vartheta(R) = v\frac{qR^2}{8} \tag{3.95}$$

The maximum moment is the radial one at the built-in constraint. Figure 3.16 shows the elastic deformed configuration and the internal reactions M_r and M_ϑ.

If this plate is loaded by the concentrated force Q, Equation 3.83 becomes

$$r\varphi_r = \frac{Q}{2\pi D}\int_0^r r\log r\,dr + C_1\frac{r^2}{2} + C_2 \tag{3.96}$$

and hence

$$\varphi_r = \frac{Qr}{8\pi D}(2\log r - 1) + C_1\frac{r}{2} + \frac{C_2}{r} \tag{3.97}$$

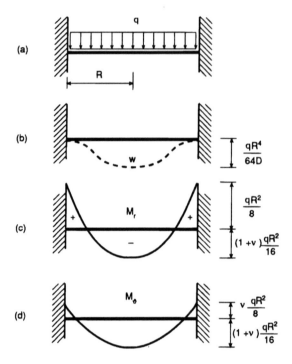

Figure 3.16

The symmetry condition and the boundary condition, respectively,

$$\varphi_r(0) = 0 \tag{3.98a}$$

$$\varphi_r(R) = 0 \tag{3.98b}$$

furnish the corresponding values of the two constants:

$$C_2 = 0 \tag{3.99a}$$

$$C_1 = -\frac{Q}{4\pi D}(2\log R - 1) \tag{3.99b}$$

We thus find

$$\varphi_r = -\frac{Q}{4\pi D}r\log\frac{R}{r} \tag{3.100a}$$

$$w = -\frac{Q}{16\pi D}\left(R^2 - r^2 - 2r^2\log\frac{R}{r}\right) \tag{3.100b}$$

The normal displacement at the center of the plate is, therefore,

$$f = |w(0)| = \frac{QR^2}{16\pi D} \tag{3.101}$$

In the case where $Q = q\pi R^2$, this latter transverse displacement is four times that of Equation 3.92. The bending moments are given by

$$M_r = -\frac{Q}{4\pi}\left[(1+v)\log\frac{R}{r} - 1\right] \tag{3.102a}$$

$$M_{\vartheta} = -\frac{Q}{4\pi}\left[(1+v)\log\frac{R}{r} - v\right] \tag{3.102b}$$

In the center, they are theoretically infinite, while at the clamped edge they are equal to

$$M_r(R) = \frac{Q}{4\pi}, \quad M_{\vartheta}(R) = v\frac{Q}{4\pi} \tag{3.103}$$

Figure 3.17 depicts the elastic deformed configuration and presents the M_r and M_{ϑ} bending moment diagrams.

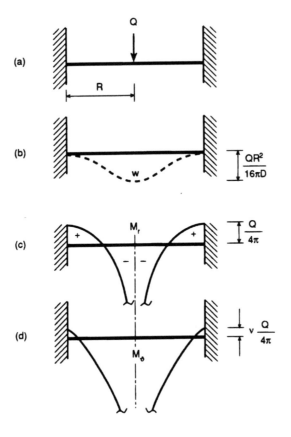

Figure 3.17

3.9 CYLINDRICAL SHELLS

In **cylindrical shells,** the principal radius of curvature $R_1 \to \infty$, while the angle α between the meridian and the axis of symmetry vanishes. Furthermore, the curvilinear coordinate along the meridian s coincides with the longitudinal coordinate x, and the second principal radius of curvature $R_2 = r$ coincides with the radius R of the circular directrix.

The kinematic Equation 3.51 is thus transformed as follows (Figure 3.18):

$$
\begin{bmatrix} \varepsilon_x \\ \varepsilon_\vartheta \\ \gamma_x \\ \chi_x \\ \chi_\vartheta \end{bmatrix} =
\begin{bmatrix}
\dfrac{d}{dx} & 0 & 0 \\[2mm]
0 & \dfrac{1}{R} & 0 \\[2mm]
0 & \dfrac{d}{dx} & +1 \\[2mm]
0 & 0 & \dfrac{d}{dx} \\[2mm]
0 & 0 & 0
\end{bmatrix}
\begin{bmatrix} u \\ w \\ \varphi_x \end{bmatrix}
\tag{3.104}
$$

while the static Equation 3.52 becomes

$$
\begin{bmatrix}
\dfrac{d}{dx} & 0 & 0 & 0 & 0 \\[2mm]
0 & -\dfrac{1}{R} & \dfrac{d}{dx} & 0 & 0 \\[2mm]
0 & 0 & -1 & \dfrac{d}{dx} & 0
\end{bmatrix}
\begin{bmatrix} N_x \\ N_\vartheta \\ T_x \\ M_x \\ M_\vartheta \end{bmatrix}
+ \begin{bmatrix} 0 \\ q \\ 0 \end{bmatrix}
= \begin{bmatrix} 0 \\ 0 \\ 0 \end{bmatrix}
\tag{3.105}
$$

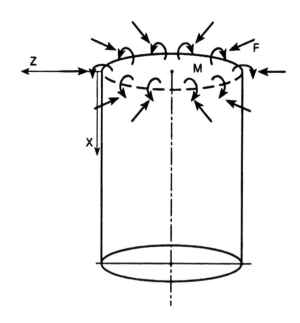

Figure 3.18

in the case where the only external force acting on the shell is a load $q(x)$, normal to the middle surface. Notice that the variation in curvature χ_ϑ vanishes, just as the moment M_ϑ is not involved in any of the three equations of equilibrium.

The first of Equation 3.105 is the equation of equilibrium with regard to longitudinal translation:

$$\frac{dN_x}{dx} = 0 \tag{3.106a}$$

which gives N_x = constant. The second equation is the equation of equilibrium with regard to normal translation:

$$-\frac{N_\vartheta}{R} + \frac{dT_x}{dx} = -q \tag{3.106b}$$

while the third is the equation of equilibrium with regard to rotation about the parallel:

$$T_x = \frac{dM_x}{dx} \tag{3.106c}$$

Substituting Equation 3.106c into Equation 3.106b, we obtain

$$-\frac{N_\vartheta}{R} + \frac{d^2M_x}{dx^2} = -q \tag{3.107}$$

In the case where the longitudinal dilation ε_x is zero, we have

$$N_\vartheta = \frac{Eh}{(1-v^2)}\varepsilon_\vartheta = \frac{Eh}{R(1-v^2)}w \tag{3.108}$$

The moment M_x is, on the other hand, proportional to the variation in curvature χ_x as $\chi_\vartheta = 0$:

$$M_x = D\chi_x = D\frac{d\varphi_x}{dx} \tag{3.109}$$

If we disregard the shearing strain:

$$\gamma_x = \frac{dw}{dx} + \varphi_x = 0 \tag{3.110}$$

we obtain

$$M_x = -D\frac{d^2w}{dx^2} \tag{3.111}$$

Substituting Equations 3.108 and 3.111 into Equation 3.107, we obtain the following differential equation in the unknown function w:

$$D\frac{d^4w}{dx^4} + \frac{Eh}{R^2(1-v^2)}w = q \tag{3.112}$$

Equation 3.112 is formally identical to the differential equation of the beam on an elastic foundation. In the case where $q = 0$, Equation 3.112 can be cast in the form:

$$\frac{\mathrm{d}^4 w}{\mathrm{d}x^4} + 4\beta^4 w = 0 \tag{3.113}$$

where β denotes the parameter:

$$\beta = \sqrt[4]{\frac{Eh}{4DR^2(1 - \nu^2)}} \tag{3.114}$$

From Equation 3.10, we obtain

$$\beta = \sqrt[4]{\frac{3}{h^2 R^2}} \tag{3.115}$$

3.10 CYLINDRICAL PRESSURIZED VESSELS WITH BOTTOMS

The displacement and rotation at the edge of a cylindrical shell, produced by forces and moments distributed along the edge itself (Figure 3.18), can be obtained on the basis of the analogy between cylindrical shells and beams on an elastic foundation. The equations related to radial displacement and rotation at the end of the beam can be expressed in the following form:

$$w(0) = -\lambda_{FF}F + \lambda_{FM}M \tag{3.116a}$$

$$\varphi_x(0) = -\lambda_{MF}F + \lambda_{MM}M \tag{3.116b}$$

where the elastic coefficients λ_{ij}, which for the beam on an elastic foundation are found to be equal to

$$\lambda_{FF} = \frac{2\beta}{k} = \frac{1}{2\beta^3 EI} \tag{3.117a}$$

$$\lambda_{FM} = \lambda_{MF} = \frac{2\beta^2}{k} = \frac{1}{2\beta^2 EI} \tag{3.117b}$$

$$\lambda_{MM} = \frac{4\beta^3}{k} = \frac{1}{\beta EI} \tag{3.117c}$$

for the semi-infinite cylinder are likewise equal to

$$\lambda_{FF} = \frac{1}{2\beta^3 D} \tag{3.118a}$$

$$\lambda_{FM} = \lambda_{MF} = \frac{1}{2\beta^2 D} \tag{3.118b}$$

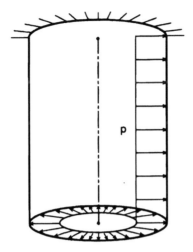

Figure 3.19

$$\lambda_{MM} = \frac{1}{\beta D} \tag{3.118c}$$

If the edges of a pressurized cylinder are free, we should have, exclusively, the circumferential stress, given by Equation 3.61, and the radial displacement:

$$w = \frac{pR^2}{Eh} \tag{3.119}$$

If, instead, the edges of the cylinder are clamped (Figure 3.19), and the cylinder is assumed as being sufficiently long, a localized flexural regime will be produced, which will be superposed in the surroundings of the edge on the aforementioned membrane regime. Since the constraint prevents both the radial displacement and the rotation of the edge, we have the following two equations of congruence (Figures 3.18 and 3.19), arising from the double static indeterminacy of the shells of revolution:

$$w(0) = -\lambda_{FF}F + \lambda_{FM}M + \frac{pR^2}{Eh} = 0 \tag{3.120a}$$

$$\varphi_x(0) = -\lambda_{MF}F + \lambda_{MM}M = 0 \tag{3.120b}$$

where F and M are the statically indeterminate reactions that act along the edge. Using Equations 3.118 and 3.115, we obtain

$$M = \frac{p}{2\beta^2} \tag{3.121a}$$

$$F = \frac{p}{\beta} \tag{3.121b}$$

For a strength assessment, it will be possible to consider, at the edge, the circumferential stress:

$$\sigma_\vartheta = \frac{pR}{h} \pm v\frac{6}{h^2}M \qquad\qquad (3.122a)$$

where the second term is due to the Poisson effect along the parallels.

The longitudinal stress is

$$\sigma_x = \pm\frac{6}{h^2}M \qquad\qquad (3.122b)$$

as well as the shearing stress:

$$\tau = \frac{F}{h} \qquad\qquad (3.122c)$$

In the case of a pressurized flat-faced cylinder, the edges of the cylinder and of the circular plate exchange a distributed force and a distributed moment (Figures 3.18 and 3.20), the shell and plate system having two degrees of static indeterminacy, as in the previous case.

The equation of angular congruence takes the form

$$-\lambda_{MF}F + \lambda_{MM}M = -\frac{MR}{(1+v)D} + \frac{pR^3}{8(1+v)D} \qquad\qquad (3.123)$$

where the left-hand side of the equation represents the rotation of the edge of the cylinder, while the two terms on the right represent the rotations of the edge of the bottom plate, due, respectively, to the redundant moment and to the internal pressure. The first rotation is deduced in the case of uniform bending. If $M_x = M_y = M$, the curvatures also equal one another, $\chi_x = \chi_y = \chi$, so that the relation in Equation 3.9a gives

$$\chi = \frac{M}{(1+v)D} \qquad\qquad (3.124)$$

and thus the angle of rotation at the edge (Figure 3.21) is

$$\varphi_x = R\chi \qquad\qquad (3.125)$$

which appears in Equation 3.123. The second rotation can then be obtained from the foregoing one by substituting, in place of M, the radial moment at the built-in constraint in Equation 3.95.

The second equation of congruence is the one corresponding to the radial displacement:

$$-\lambda_{FF}F + \lambda_{FM}M + \frac{pR^2}{2Eh}(2-v) = \frac{FR}{Eh}(1-v) \qquad\qquad (3.126)$$

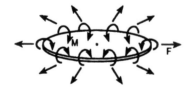

Figure 3.20

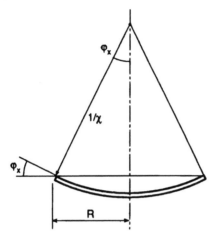

Figure 3.21

The third term on the left-hand side represents the radial displacement of the cylinder, which, when pressurized, is subject to a biaxial stress condition:

$$\sigma_\vartheta = \frac{pR}{h}, \quad \sigma_x = \frac{pR}{2h} \tag{3.127}$$

whereby the circumferential dilation

$$\varepsilon_\vartheta = \frac{1}{E}(\sigma_\vartheta - \nu\sigma_x) \tag{3.128}$$

produces the radial displacement

$$w = \varepsilon_\vartheta R = \frac{pR^2}{2Eh}(2 - \nu) \tag{3.129}$$

The right-hand side of Equation 3.126 represents, on the other hand, the radial displacement of the edge of the circular plate, since this is in a condition of uniform stress $\sigma = F/h$.

In the case of a pressurized cylinder with hemispherical caps (Figures 3.18 and 3.22), the equation of angular congruence takes the following form:

$$-\lambda_{MF}F + \lambda_{MM}M = -\lambda_{MF}F - \lambda_{MM}M \tag{3.130}$$

if we assume, as is approximately the case, that the elastic coefficients of the cylinder and the hemisphere are equal. From Equation 3.130, it follows that the redundant moment M vanishes.

On the other hand, the equation of congruence for the radial displacement is

$$-\lambda_{FF}F + \frac{pR^2}{2Eh}(2 - \nu) = \lambda_{FF}F + \frac{pR^2}{2Eh}(1 - \nu) \tag{3.131}$$

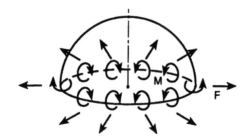

Figure 3.22

where the second terms on both sides of the equation take into account the biaxial stress condition of the shells, which are cylindrical and spherical, respectively. From Equation 3.131, we obtain

$$2\lambda_{FF}F = \frac{pR^2}{2Eh} \tag{3.132}$$

and thus, applying Equations 3.118a and 3.114, we have

$$F = \frac{pR^2\beta^3 D}{2Eh} = \frac{p}{8\beta} \tag{3.133}$$

3.11 THREE-DIMENSIONAL BODIES OF REVOLUTION

In the case of a three-dimensional body of revolution, not loaded symmetrically, the kinematic and static equations appear as follows (Figure 3.23):

$$
\begin{bmatrix} \varepsilon_r \\ \varepsilon_\vartheta \\ \varepsilon_z \\ \gamma_{r\vartheta} \\ \gamma_{rz} \\ \gamma_{\vartheta z} \end{bmatrix}
=
\begin{bmatrix}
\dfrac{\partial}{\partial r} & 0 & 0 \\[2mm]
\dfrac{1}{r} & \dfrac{1}{r}\dfrac{\partial}{\partial \vartheta} & 0 \\[2mm]
0 & 0 & \dfrac{\partial}{\partial z} \\[2mm]
\dfrac{1}{r}\dfrac{\partial}{\partial \vartheta} & \left(\dfrac{\partial}{\partial r} - \dfrac{1}{r}\right) & 0 \\[2mm]
\dfrac{\partial}{\partial z} & 0 & \dfrac{\partial}{\partial r} \\[2mm]
0 & \dfrac{\partial}{\partial z} & \dfrac{1}{r}\dfrac{\partial}{\partial \vartheta}
\end{bmatrix}
\begin{bmatrix} u \\ v \\ w \end{bmatrix}
\tag{3.134a}
$$

$$
\begin{bmatrix}
\left(\dfrac{\partial}{\partial r}+\dfrac{1}{r}\right) & -\dfrac{1}{r} & 0 & \dfrac{1}{r}\dfrac{\partial}{\partial \vartheta} & \dfrac{\partial}{\partial z} & 0 \\[2mm]
0 & \dfrac{1}{r}\dfrac{\partial}{\partial \vartheta} & 0 & \left(\dfrac{\partial}{\partial r}+\dfrac{2}{r}\right) & 0 & \dfrac{\partial}{\partial z} \\[2mm]
0 & 0 & \dfrac{\partial}{\partial z} & 0 & \left(\dfrac{\partial}{\partial r}+\dfrac{1}{r}\right) & \dfrac{1}{r}\dfrac{\partial}{\partial \vartheta}
\end{bmatrix}
\begin{bmatrix} \sigma_r \\ \sigma_\vartheta \\ \sigma_z \\ \tau_{r\vartheta} \\ \tau_{rz} \\ \tau_{\vartheta z} \end{bmatrix}
+
\begin{bmatrix} \mathcal{F}_r \\ \mathcal{F}_\vartheta \\ \mathcal{F}_z \end{bmatrix}
=
\begin{bmatrix} 0 \\ 0 \\ 0 \end{bmatrix}
\tag{3.134b}
$$

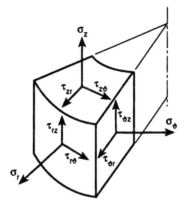

Figure 3.23

The static Equation 3.134b represents the three indefinite equations of equilibrium with regard to translation, in the radial, circumferential, and axial directions, respectively (Figure 3.23). Notice that, once again, the terms $1/r$ in the static matrix are due to the difference between the areas of the two parallel curved faces of the element of Figure 3.23, as well as to the different action lines of the stresses σ_ϑ and $\tau_{\vartheta r}$ acting on the two opposite faces.

In the case where the body of revolution is also loaded symmetrically with respect to its axis, the degrees of internal redundancy become two instead of three:

$$
\begin{bmatrix} \varepsilon_r \\ \varepsilon_\vartheta \\ \varepsilon_z \\ \gamma_{rz} \end{bmatrix} =
\begin{bmatrix} \dfrac{\partial}{\partial r} & 0 \\[2mm] \dfrac{1}{r} & 0 \\[2mm] 0 & \dfrac{\partial}{\partial z} \\[2mm] \dfrac{\partial}{\partial z} & \dfrac{\partial}{\partial r} \end{bmatrix}
\begin{bmatrix} u \\ w \end{bmatrix}
\tag{3.135a}
$$

$$
\begin{bmatrix} \left(\dfrac{\partial}{\partial r}+\dfrac{1}{r}\right) & -\dfrac{1}{r} & 0 & \dfrac{\partial}{\partial z} \\[3mm] 0 & 0 & \dfrac{\partial}{\partial z} & \left(\dfrac{\partial}{\partial r}+\dfrac{1}{r}\right) \end{bmatrix}
\begin{bmatrix} \sigma_r \\ \sigma_\vartheta \\ \sigma_z \\ \tau_{rz} \end{bmatrix} +
\begin{bmatrix} \mathcal{F}_r \\ \mathcal{F}_z \end{bmatrix} =
\begin{bmatrix} 0 \\ 0 \end{bmatrix}
\tag{3.135b}
$$

When the problem presents a plane stress condition, we have $\sigma_z=\tau_{rz}=0$, and hence only the first of Equations 3.135b remains significant:

$$
\frac{d\sigma_r}{dr}+\frac{\sigma_r-\sigma_\vartheta}{r}+\mathcal{F}_r=0
\tag{3.136}
$$

Chapter 4

Finite element method

4.1 INTRODUCTION

The finite element method is illustrated here as a method of discretization and interpolation for the approximate solution of elastic problems. This method is introduced in an altogether general manner, without specifying the structural element to which it is applied, whether it is of one, two, or three dimensions, and in the first two cases, whether it does or does not have an intrinsic curvature. On the other hand, the two dimensions that characterize the element are brought into the forefront: that of the generalized displacement vector and that common to the two vectors of static and deformation characteristics.

Applying the principle of minimum total potential energy and the Ritz–Galerkin numerical approximation, we arrive at the analytical and variational definition of the finite element method. Furthermore, applying the principle of virtual work, the alternative definition of the method is also given—the one more widely known in the engineering field, namely, the mechanical and matrix one. Via the definition of shape functions, we arrive at the notion of the local stiffness matrix of the individual element. This matrix is thus expanded and assembled, that is, added to all the other similar matrices, to provide, finally, the global stiffness matrix. In this context, an explanation is given of the change of algebraic sign shown by the unity and algebraic terms of the static and kinematic matrix operators, which, in the case of beams, plates, and shells, are each the adjoint of the other.

The application of the finite element method to physical problems different from those examined in this chapter is illustrated in Appendix II. Two further complementary topics are then dealt with: the problem of initial strains and residual stresses (Appendix III), and the problem of plane elasticity with couple stresses (Appendix IV).

4.2 SINGLE-DEGREE-OF-FREEDOM SYSTEM

Consider a material point subjected to an external force F and to the elastic restoring force of a linear spring having stiffness k (Figure 4.1). Since the restoring force is proportional to the elongation x of the spring and is acting in the opposite direction to that of the external force, the condition of static equilibrium will be expressed by the following equation:

$$F - kx = 0 \tag{4.1}$$

from which the abscissa of the **position of equilibrium** is deduced:

$$x = F/k \tag{4.2}$$

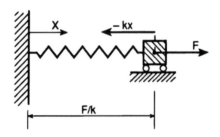

Figure 4.1

This simple result may be obtained in principle by also considering the **total potential energy** of the system, which is equal to the sum of the potential energy of the spring and the potential energy of the nonpositional force *F*:

$$W(x) = \frac{1}{2}kx^2 - Fx \qquad (4.3)$$

As is known from rational mechanics, the derivative of the total potential energy, with change of sign, yields the total force acting on the system:

$$\text{Total force} = -\frac{dW}{dx} = -kx + F \qquad (4.4)$$

from which we deduce that the system is in equilibrium when Equation 4.2 is satisfied. The condition:

$$-\frac{dW}{dx} = 0 \qquad (4.5)$$

also defines a point of stationarity, namely, the **minimum total potential energy**, which can be represented by a parabolic curve as a function of the elongation *x* (Figure 4.2). This parabola is also known as the **potential well**.

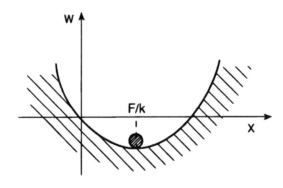

Figure 4.2

The position of equilibrium in Equation 4.2 may also be obtained by applying the principle of virtual work: if we impose a virtual displacement Δx on the system in a condition of equilibrium, the work thus produced must be zero:

$$F\Delta x - kx\Delta x = 0 \tag{4.6}$$

from which, by canceling Δx, the already known result of relation in Equation 4.2 follows.

The finite element method, even though it concerns multiple-degree-of-freedom systems, can be introduced by following the two paths indicated for the single-degree-of-freedom system:

1. Principle of minimum total potential energy
2. Principle of virtual work

The finite element method is basically a **discretization method**, in the sense that, instead of the continuous function of the displacements $\{\eta\}$, it considers as unknowns only the displacements $\{\delta\}$ of a discrete number n of points called **nodes**. It is, at the same time, an **interpolation method**, in the sense that, once the displacements $\{\delta\}$ are determined, it connects them with sufficiently regular functions. The problem is thus reduced to the determination of the **equilibrium configuration** $\{\delta\}$, from which we then obtain by interpolation the displacement field $\{\eta\}$; and by derivation, the strain field $\{\varepsilon\}$ or the deformation characteristics field $\{q\}$; and, via the constitutive equations, the stress field $\{\sigma\}$ or the static characteristics field $\{Q\}$.

For each of the structural geometries considered hitherto, there are two numbers that characterize the system: (1) the **degrees of freedom** g, which are represented by the dimension of the displacement vector $\{\eta\}$; (2) the dimension d of the kinematic characteristics vector, $\{\varepsilon\}$ or $\{q\}$, and the static characteristics vector, $\{\sigma\}$ or $\{Q\}$. Table 4.1 gives the characteristic numbers g and d for all the one-, two-, and three-dimensional elastic solids.

Note that only in the cases of beams and thin domes do the degrees of freedom g coincide with the dimension d of the characteristics vectors. In all other cases, we have $g < d$.

Table 4.1

Structural element	g	d
Beam in the plane	3	3
Beam in space	6	6
Plate in flexure	3	5
Shell with double curvature	5	8
Shell of revolution	3	5
Thin dome	2	2
Circular plate	2	3
Cylindrical shell	3	4
Two-dimensional solid	2	3
Three-dimensional solid	3	6
Solid of revolution	2	4

4.3 PRINCIPLE OF MINIMUM TOTAL POTENTIAL ENERGY

Every elastic problem, in one, two, or three dimensions, can be referred to **Lamé's equation**, of the sort

$$\underset{(g \times g)}{[\mathcal{L}]} \underset{(g \times 1)}{\{\eta\}} = -\underset{(g \times 1)}{\{\mathcal{F}\}} \tag{4.7}$$

where:

$[\mathcal{L}]$ is the differential and matrix operator corresponding to the geometry in question
$\{\eta\}$ is the vector of the generalized displacements
$\{\mathcal{F}\}$ is the vector of the external forces acting in the elastic domain

In the case where the structural element does not present an intrinsic curvature (arches and shells), the **boundary conditions** of **equivalence** assume the following form:

$$(\underset{(g \times d)}{[\mathcal{N}]^{\mathrm{T}}} \underset{(d \times d)}{[H]} \underset{(d \times g)}{[\partial]}) \underset{(g \times 1)}{\{\eta\}} = \underset{(g \times 1)}{\{p\}} \tag{4.8}$$

where:

$[\mathcal{N}]^{\mathrm{T}}$ is the matrix that transforms the static characteristics vector into the vector of the external forces acting on the boundary
$[H]$ is the Hessian matrix of the elastic potential Φ
$[\partial]$ is the kinematic operator
$\{p\}$ is the vector of the external forces acting on the boundary S of the elastic domain

In the case where the external forces $\{\mathcal{F}\}$ and $\{p\}$ are not self-balanced, boundary conditions of a kinematic type are necessary:

$$\underset{(g \times 1)}{\{\eta\}} = \underset{(g \times 1)}{\{\eta_0\}} \tag{4.9}$$

valid over a part of the boundary, or else over the entire boundary.

In the following, it will be shown how the **operator formulation** of the elastic problem, that is, Equations 4.7 and 4.8, imply the principle of minimum total potential energy, and *vice versa*.

The total potential energy, in the case of an elastic structural element, is defined as follows:

$$W(\eta) = \int_V \Phi(q) dV - \int_V \{\eta\}^{\mathrm{T}} \{\mathcal{F}\} dV - \int_S \{\eta\}^{\mathrm{T}} \{p\} dS \tag{4.10}$$

Applying Clapeyron's theorem, we have

$$W(\eta) = \frac{1}{2} \left(\int_V \{\eta\}^{\mathrm{T}} \{\mathcal{F}\} dV + \int_S \{\eta\}^{\mathrm{T}} \{p\} dS \right)$$

$$- \left(\int_V \{\eta\}^{\mathrm{T}} \{\mathcal{F}\} dV + \int_S \{\eta\}^{\mathrm{T}} \{p\} dS \right) \tag{4.11}$$

where the terms in brackets are identical and both represent twice the strain energy. Denoting by $[\mathcal{L}_0]$ the operator for the boundary conditions of equivalence in Equation 4.8:

$$\underset{(g\times g)}{[\mathcal{L}_0]}\underset{(g\times 1)}{\{\eta\}} = \underset{(g\times 1)}{\{p\}} \tag{4.12}$$

and substituting Equation 4.7 and Equation 4.12 only in the first term of Equation 4.11, we have

$$W(\eta) = \frac{1}{2}\left(-\int_V \{\eta\}^T[\mathcal{L}]\{\eta\}dV + \int_S \{\eta\}^T[\mathcal{L}_0]\{\eta\}dS\right)$$
$$-\left(\int_V \{\eta\}^T\{\mathcal{F}\}dV + \int_S \{\eta\}^T\{p\}dS\right) \tag{4.13}$$

The total potential energy, for the external forces $\{\mathcal{F}\}$, $\{p\}$ and for an incremented displacement vector $\{\eta + \Delta\eta\}$, will be written as follows:

$$W(\eta + \Delta\eta) = \frac{1}{2}\left(-\int_V \{\eta + \Delta\eta\}^T[\mathcal{L}]\{\eta + \Delta\eta\}dV + \int_S \{\eta + \Delta\eta\}^T[\mathcal{L}_0]\{\eta + \Delta\eta\}dS\right)$$
$$-\left(\int_V \{\eta + \Delta\eta\}^T\{\mathcal{F}\}dV + \int_S \{\eta + \Delta\eta\}^T\{p\}dS\right) \tag{4.14}$$

The associative property allows us to dismember each integral into a sum of different integrals:

$$W(\eta + \Delta\eta) = \frac{1}{2}\left(-\int_V \{\eta\}^T[\mathcal{L}]\{\eta\}dV - \int_V \{\Delta\eta\}^T[\mathcal{L}]\{\eta\}dV - \int_V \{\eta\}^T[\mathcal{L}]\{\Delta\eta\}dV\right.$$
$$-\int_V \{\Delta\eta\}^T[\mathcal{L}]\{\Delta\eta\}dV + \int_S \{\eta\}^T[\mathcal{L}_0]\{\eta\}dS + \int_S \{\Delta\eta\}^T[\mathcal{L}_0]\{\eta\}dS$$
$$\left.+\int_S \{\eta\}^T[\mathcal{L}_0]\{\Delta\eta\}dS + \int_S \{\Delta\eta\}^T[\mathcal{L}_0]\{\Delta\eta\}dS\right)$$
$$-\left(\int_V \{\eta\}^T\{\mathcal{F}\}dV + \int_V \{\Delta\eta\}^T\{\mathcal{F}\}dV + \int_S \{\eta\}^T\{p\}dS + \int_S \{\Delta\eta\}^T\{p\}dS\right) \tag{4.15}$$

The application of Betti's reciprocal theorem yields

$$W(\eta + \Delta\eta) = W(\eta) - \int_V \{\Delta\eta\}^T[\mathcal{L}]\{\eta\}dV + \int_S \{\Delta\eta\}^T[\mathcal{L}_0]\{\eta\}dS$$
$$+\frac{1}{2}\left(-\int_V \{\Delta\eta\}^T[\mathcal{L}]\{\Delta\eta\}dV + \int_S \{\Delta\eta\}^T[\mathcal{L}_0]\{\Delta\eta\}dS\right)$$
$$-\int_V \{\Delta\eta\}^T\{\mathcal{F}\}dV - \int_S \{\Delta\eta\}^T\{p\}dS \tag{4.16}$$

On the basis of the field Equation 4.7 and the boundary Equation 4.12, four of the six integrals of Equation 4.16 cancel each other out, thus yielding

$$W(\eta + \Delta\eta) = W(\eta) + \frac{1}{2}\left(-\int_V \{\Delta\eta\}^T [\mathcal{L}]\{\Delta\eta\}dV + \int_S \{\Delta\eta\}^T [\mathcal{L}_0]\{\Delta\eta\}dS\right) \tag{4.17}$$

The integrals in parentheses represent the work that the body forces $\{\Delta\mathcal{F}\}$ and surface forces $\{\Delta p\}$ perform by the displacements caused by them, where

$$[\mathcal{L}]\{\Delta\eta\} = -\{\Delta\mathcal{F}\} \tag{4.18a}$$

$$[\mathcal{L}_0]\{\Delta\eta\} = \{\Delta p\} \tag{4.18b}$$

Hence, by virtue of Clapeyron's theorem, we obtain

$$W(\eta + \Delta\eta) = W(\eta) + \int_V \Phi(\Delta q)dV \tag{4.19}$$

The total potential energy for the displacement field $\{\eta\}$ that resolves the elastic problem is thus the minimum, with respect to any other arbitrarily chosen field $\{\eta + \Delta\eta\}$. The elastic potential Φ is in fact a positive definite quadratic form of the deformation characteristics $\{\Delta q\}$. We have thus shown how the operator formulation expressed by Equations 4.7 and 4.8 implies the so-called **variational formulation**:

$$W(\eta) = minimum \tag{4.20}$$

On the other hand, by virtue of the arbitrariness of the incremental vector $\{\Delta\eta\}$, the reverse also holds. In fact, taking the foregoing formulation in the inverse direction, we arrive at the implications of orthogonality (cf. Equation 4.16):

$$([\mathcal{L}]\{\eta\} + \{\mathcal{F}\}) \perp \{\Delta\eta\} \tag{4.21a}$$

$$([\mathcal{L}_0]\{\eta\} - \{p\}) \perp \{\Delta\eta\} \tag{4.21b}$$

which hold for any incremental vector $\{\Delta\eta\}$. From this, it follows that the vectors on the left-hand sides of Equations 4.21 vanish and hence the operator formulation holds good.

4.4 RITZ–GALERKIN METHOD

When the Ritz–Galerkin numeric approximation method is used, the **functional** $W(\eta)$ is assumed to be stationary, expressing the unknown function $\{\eta\}$ as the sum of known and linearly independent functions $\{\eta_i\}$, with $i = 1, 2, \ldots, (g \times n)$:

$$\{\eta\} = \sum_{i=1}^{g\times n} \alpha_i \{\eta_i\} \tag{4.22}$$

To express it using the customary language of functional analysis, the functional $W(\eta)$ is rendered stationary on a subspace of finite dimension, subtended by a set of known linearly independent functions. The problem thus emerges as discretized, since, instead of the vector function $\{\eta\}$, the new unknowns are now the $(g\times n)$ coefficients α_i, where n is the number of the nodes and g the degrees of freedom of each node.

Inserting the linear combination of Equation 4.22 in the expression of total potential energy in Equation 4.13, and applying the associative property, we obtain

$$W\{\eta\} = \frac{1}{2}\left(-\sum_{i=1}^{g\times n}\sum_{j=1}^{g\times n} \alpha_i \alpha_j \int_V \{\eta_i\}^{\mathrm{T}}[\mathcal{L}]\{\eta_j\}\mathrm{d}V \right.$$

$$+ \sum_{i=1}^{g\times n}\sum_{j=1}^{g\times n} \alpha_i \alpha_j \int_S \{\eta_i\}^{\mathrm{T}}[\mathcal{L}_0]\{\eta_j\}\mathrm{d}S \right)$$

$$\left. -\left(\sum_{i=1}^{g\times n} \alpha_i \int_V \{\eta_i\}^{\mathrm{T}}\{\mathcal{F}\}\mathrm{d}V + \sum_{i=1}^{g\times n} \alpha_i \int_S \{\eta_i\}^{\mathrm{T}}\{p\}\mathrm{d}S \right) \right. \tag{4.23}$$

In a more synthetic form, we have

$$W(\alpha) = \frac{1}{2}\{\alpha\}^{\mathrm{T}}[L]\{\alpha\} - \{\alpha\}^{\mathrm{T}}\{F\} \tag{4.24}$$

where $\{\alpha\}$ is the vector of the unknown coefficients of the linear combination. The square matrix $[L]$ has the dimension $(g\times n)$, and as elements, the following integrals:

$$L_{ij} = -\int_V \{\eta_i\}^{\mathrm{T}}[\mathcal{L}]\{\eta_j\}\mathrm{d}V + \int_S \{\eta_i\}^{\mathrm{T}}[\mathcal{L}_0]\{\eta_j\}\mathrm{d}S \tag{4.25}$$

while the vector $\{F\}$ has the dimension $(g\times n)$, and as elements:

$$F_i = \int_V \{\eta_i\}^{\mathrm{T}}\{\mathcal{F}\}\mathrm{d}V + \int_S \{\eta_i\}^{\mathrm{T}}\{p\}\mathrm{d}S \tag{4.26}$$

The matrix $[L]$ is symmetrical by virtue of Betti's reciprocal theorem, and is called the **Ritz–Galerkin matrix**.

The minimum of the total potential energy is obtained by deriving the expression of Equation 4.23 with respect to each coefficient α_i, and equating the result to zero:

$$\sum_{j=1}^{g\times n} L_{ij}\alpha_j - F_i = 0, \quad \text{for } i=1, 2, \ldots, (g\times n) \tag{4.27}$$

We have therefore arrived at a system of $(g \times n)$ linear algebraic equations in the $(g \times n)$ unknowns α_j, which in synthetic form may be written thus:

$$[L]\{\alpha\} = \{F\} \qquad (4.28)$$

That the condition of stationarity in Equation 4.28 is also a condition of minimum is guaranteed by the fact that the quadratic form present in Equation 4.23 is positive definite, representing as it does the strain energy of the solid in a discretized form.

In the case where the functions $\{\eta_i\}$ are defined over the entire domain V, the matrix $[L]$ is ill-conditioned, and thus the resolving numerical algorithm presents problems of instability. With the **isoparametric finite element method**, the so-called **splines** are used as $\{\eta_i\}$ functions. These are functions defined only on subsets of the domain V (whence the term *finite elements*) that present a value of unity in one node and zero values in all the other nodes that belong to their own domain of definition. The splines can be linear or of a higher order. A number of examples are shown in Figure 4.3. The simplest are, of course, the linear splines. To each node k there corresponds a spline η_k and hence, if the degrees of freedom are g, then there correspond g vectors of dimension g:

$$\begin{matrix} 1- \\ 2- \\ \vdots \\ \vdots \\ g- \end{matrix} \begin{bmatrix} \eta_k \\ 0 \\ 0 \\ \vdots \\ 0 \end{bmatrix} \begin{bmatrix} 0 \\ \eta_k \\ 0 \\ \vdots \\ 0 \end{bmatrix} , \cdots, \begin{bmatrix} 0 \\ 0 \\ 0 \\ \vdots \\ \eta_k \end{bmatrix} , \text{ for } k = 1, 2, \ldots, n \qquad (4.29)$$

$$\begin{matrix} 1- & 2- & & g- \end{matrix}$$

Therefore, it is evident that, with splines, the coefficients α_i of the linear combination in Equation 4.22 coincide with the nodal values of the generalized displacements. Ordering

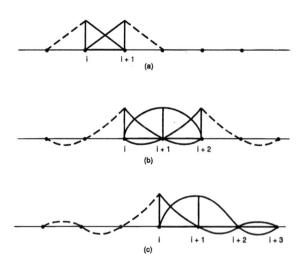

Figure 4.3

these values in the vector $\{\delta\}$ of dimension $(g \times n)$, the resolving Equation 4.28 becomes more expressive, no longer presenting simple coefficients as unknowns, but rather the nodal displacements themselves:

$$[L]\{\delta\} = \{F\} \tag{4.30}$$

4.5 APPLICATION OF THE PRINCIPLE OF VIRTUAL WORK

In this section, we shall define again the finite element method on the basis of the principle of virtual work, and we shall show how this is equivalent to the definition proposed in the previous section and based on the principle of minimum total potential energy.

Let the elastic domain V be divided into subdomains V_e, called **finite elements** of the domain V, and let each element contain m nodal points (Figure 4.4). Usually, in the two-dimensional cases (plane stress or strain conditions, plates or shells, axisymmetrical solids, etc.), the elements are triangular or quadrangular, with the nodes at the vertices, on the sides and, in some cases, inside. In three-dimensional cases, the elements are usually tetrahedrons or prisms with quadrangular sides. A number of specific examples are presented in Appendix V.

To each of the nodal points of the element V_e, let there correspond a spline, defined on the sole element V_e if the node is internal, also on the adjacent element if the node is on one side, and also on all the other elements to which the node belongs if this coincides with a vertex (Figure 4.5). To each node k of the element V_e, let there then correspond a diagonal matrix made up of the g vectors of Equation 4.29:

$$\underset{(g \times g)}{[\eta_k]} = \begin{bmatrix} \eta_k & & & \\ & \eta_k & & \\ & & \ddots & \\ & & & \eta_k \end{bmatrix}, \text{ for } k = 1, 2, \ldots, m \tag{4.31}$$

These matrices are referred to as **shape functions**, and have the following properties:

$$[\eta_k]_k = [1] \tag{4.32a}$$

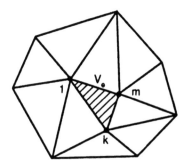

Figure 4.4

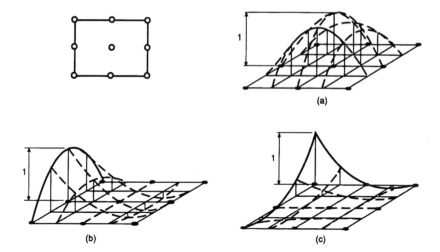

Figure 4.5

$$[\eta_k]_j = [0], \text{ for } k \neq j \tag{4.32b}$$

Using the Kronecker symbol δ_{kj}, we can write more synthetically:

$$[\eta_k]_j = [\delta_{kj}] \tag{4.33}$$

The displacement vector can be expressed by interpolation, via the shape functions and on the basis of the nodal displacements:

$$\underset{(g\times 1)}{\{\eta\}} = \overset{(g\times g)}{[\eta_1]}\cdots\overset{(g\times g)}{[\eta_k]}\cdots\overset{(g\times g)}{[\eta_m]}\begin{bmatrix} \delta_1 \\ \vdots \\ \delta_k \\ \vdots \\ \delta_m \end{bmatrix}\begin{matrix} {\scriptstyle(g\times 1)} \\ \\ {\scriptstyle(g\times 1)} \\ \\ {\scriptstyle(g\times 1)} \end{matrix} \tag{4.34}$$

In compact form, the **displacement vector field** defined on the element V_e may be represented as

$$\underset{(g\times 1)}{\{\eta_e\}} = \underset{g\times(g\times m)}{[\eta_e]}\ \underset{(g\times m)\times 1}{[\delta_e]} \tag{4.35}$$

The **deformation characteristics vector** is obtained by derivation:

$$\underset{(d\times 1)}{\{q_e\}} = \underset{(d\times g)}{[\partial]}\ \underset{(g\times 1)}{\{\eta_e\}} \tag{4.36}$$

whence, applying the Equation 4.35, we obtain

$$\{q_e\} = [\partial] \ [\eta_e] \ \{\delta_e\}$$
$$\underset{(d\times1)}{} \quad \underset{(d\times g)}{} \underset{g\times(g\times m)}{} \underset{(g\times m)\times1}{}$$
(4.37)

or, in synthetic form

$$\{q_e\} = [B_e] \ \{\delta_e\}$$
$$\underset{(d\times1)}{} \quad \underset{d\times(g\times m)}{} \underset{(g\times m)\times1}{}$$
(4.38)

where the matrix

$$[B_e] = [\partial] \ [\eta_e]$$
$$\underset{d\times(g\times m)}{} \quad \underset{(d\times g)}{} \underset{g\times(g\times m)}{}$$
(4.39)

is calculated by derivation of the splines.

The **static characteristics vector** is obtained by premultiplying the deformation characteristics vector by the Hessian matrix of strain energy:

$$\{Q_e\} = [H] \ [B_e] \ \{\delta_e\}$$
$$\underset{(d\times1)}{} \quad \underset{(d\times d)}{} \underset{d\times(g\times m)}{} \underset{(g\times m)\times1}{}$$
(4.40)

Let the principle of virtual work now be applied to the element V_e. This fundamental principle can be demonstrated in the cases considered in Table 4.1. In the cases where there is the presence of an intrinsic curvature, it is sufficient to substitute the operators $[\partial]$, $[\partial]^*$, $[\mathcal{N}]$, and $[\mathcal{N}]^T$, respectively, with

$$[\partial][N], \ [N]^T[\partial]^*, \ [\mathcal{N}][N], \ [N]^T[\mathcal{N}]^T$$
(4.41)

Let a field of virtual displacements $\{\Delta\eta\}$ be imposed on the element V_e. The principle of virtual work implies the following equality:

$$\int_{V_e} \{\Delta q\}^T \{Q_e\} dV = \int_{V_e} \{\Delta\eta\}^T \{\mathcal{F}\} dV + \int_{S_e} \{\Delta\eta\}^T \{p\} dS$$
(4.42)

On the basis of Equations 4.35, 4.38, and 4.40, we deduce

$$\int_{V_e} \{\Delta\delta\}^T [B_e]^T [H][B_e]\{\delta_e\} dV$$
$$= \int_{V_e} \{\Delta\delta\}^T [\eta_e]^T \{\mathcal{F}\} dV + \int_{S_e} \{\Delta\delta\}^T [\eta_e]^T \{p\} dS$$
(4.43)

Canceling on both sides the virtual nodal displacements $\{\Delta\delta\}^T$, we obtain

$$\int_{V_e \ (g\times m)\times d \ (d\times d) \ d\times(g\times m)} [B_e]^T \ [H] \ [B_e] \ dV \cdot \{\delta_e\}_{(g\times m)\times 1}$$

$$= \int_{V_e \ (g\times m)\times g \ (g\times 1)} [\eta_e]^T \ \{\mathcal{F}\} dV + \int_{S_e \ (g\times m)\times g \ (g\times 1)} [\eta_e]^T \ \{p\} \ dS \tag{4.44}$$

The **vector of the nodal displacements of the element** V_e, $\{\delta_e\}$, has been carried out from the integral since it is constant. The integral on the left-hand side is called the **local stiffness matrix:**

$$[K_e]_{(g\times m)(g\times m)} = \int_{V_e \ (g\times m)\times d \ (d\times d) \ d\times(g\times m)} [B_e]^T \ [H] \ [B_e] \ dV \tag{4.45}$$

Equation 4.44 therefore takes on the following form:

$$[K_e]_{(g\times m)(g\times m)} \{\delta_e\}_{(g\times m)\times 1} = \{F_e\}_{(g\times m)\times 1} + \{p_e\}_{(g\times m)\times 1} \tag{4.46}$$

The two vectors on the right-hand side are the **vectors of the equivalent nodal forces,** and represent the integrated effect of the forces distributed in the domain and on the boundary of the element V_e. Once the local stiffness matrix is calculated, it would be possible to determine the vector of the nodal displacements $\{\delta_e\}$ on the basis of the Equation 4.46 only if the forces $\{p\}$ acting on the boundary of the element were known beforehand, hence the vector of the equivalent forces $\{p_e\}$ was obtained by integration. Whereas the body forces $\{\mathcal{F}\}$ are a datum of the problem, the forces $\{p\}$, which exchange between them the elements at the reciprocal boundaries, are *a priori* unknown.

To get round this obstacle and, at the same time, to resolve the general problem of the determination of the vector of all the nodal displacements $\{\delta\}$ of the solid, one must add Equation 4.46, valid for the element V_e, to all the similar relations valid for the other elements of the mesh. In this way, the surface contributions $\{p_e\}$ all cancel out, except for those that do not belong to interfaces between elements, but which belong to the outer boundary. This operation is called **assemblage,** and involves a prior **expansion** of the vectors $\{\delta_e\}$, $\{F_e\}$, $\{p_e\}$ from the local dimension $(g\times m)$ to the global dimension $(g\times n)$, where n is the global number of nodal points of the mesh. The procedure will therefore be to order all the nodes of the mesh of finite elements, so as to be able to insert the nodes of the generic element V_e in the positions that they should have. This may be achieved by premultiplying the vector of the local nodal displacements $\{\delta_e\}$ by a suitable assemblage matrix $[A_e]^T$, of dimensions $(g\times n) \times (g\times m)$, where all the elements are zero, except for $(g\times m)$ elements having the value of unity set in the $(g\times m)$ different rows to be filled, and corresponding to the $(g\times m)$ columns:

$$\{\delta^e\} = [A_e]^T\{\delta_e\} \tag{4.47a}$$

$$\{F^e\} = [A_e]^T\{F_e\} \tag{4.47b}$$

$$\{p^e\}_{(g\times n)\times 1} = [A_e]^T_{(g\times n)(g\times m)} \{p_e\}_{(g\times m)\times 1} \tag{4.47c}$$

Substituting the inverse relations in Equation 4.46, we obtain

$$[K_e][A_e]\{\delta^e\} = [A_e]\{F^e\} + [A_e]\{p^e\} \tag{4.48}$$

which, premultiplied by $[A_e]^T$, yields

$$([A_e]^T[K_e][A_e])\underset{(g\times n)(g\times n)}{} \{\delta^e\}\underset{(g\times n)\times 1}{} = \{F^e\}\underset{(g\times n)\times 1}{} + \{p^e\}\underset{(g\times n)\times 1}{} \tag{4.49}$$

Equation 4.49 remains valid even if the expanded vector of local displacements $\{\delta^e\}$ is substituted with the global vector of nodal displacements $\{\delta\}$:

$$[K^e]\{\delta\} = \{F^e\} + \{p^e\} \tag{4.50}$$

where $[K^e]$ is the local stiffness matrix in expanded form:

$$[K^e]\underset{(g\times n)(g\times n)}{} = [A_e]^T\underset{(g\times n)(g\times m)}{} [K_e]\underset{(g\times m)(g\times m)}{} [A_e]\underset{(g\times m)(g\times n)}{} \tag{4.51}$$

The local relation, but in the expanded form of Equation 4.50, may be added to the similar relations for the other finite elements:

$$\left(\sum_e [K^e]\right)\underset{(g\times n)(g\times n)}{} \{\delta\}\underset{(g\times n)\times 1}{} = \{F\}\underset{(g\times n)\times 1}{} \tag{4.52}$$

having gathered to a common factor the vector of the nodal displacements $\{\delta\}$, and where

$$\{F\} = \sum_e (\{F^e\} + \{p^e\}) \tag{4.53}$$

From Equations 4.44 and 4.47, we deduce

$$\{F\} = \int_V [A_e]^T[\eta_e]^T\{\mathcal{F}\}dV + \int_S [A_e]^T[\eta_e]^T\{p\}dS \tag{4.54}$$

where the integrals extended to the boundaries of the elements cancel out two by two, since the forces that the interfaces of the elements exchange are equal and opposite. It is easy to verify that the vector in Equation 4.54 has Equation 4.26 as its components.

Finally, we thus derive the equation:

$$[K]\{\delta\} = \{F\} \tag{4.55}$$

which coincides with Equation 4.30, once the equality of the global stiffness matrix $[K]$ with the Ritz–Galerkin matrix $[L]$ has been demonstrated. However, this is possible on the basis of Equations 4.45 and 4.39:

$$[K_e] = \int_{V_e} [B_e]^T [H][B_e] dV = \int_{V_e} ([\partial][\eta_e])^T [H][\partial][\eta_e] dV \qquad (4.56)$$

Applying the rule of integration by parts on a three-dimensional domain, we have

$$[K_e] = -\int_{V_e} [\eta_e]^T [\partial]^* [H][\partial][\eta_e] dV$$

$$+ \int_{S_e} [\eta_e]^T [\mathcal{N}]^T [H][\partial][\eta_e] dS \qquad (4.57)$$

The minus sign in front of the first integral, which derives from the rule of integration by parts and is necessary for the terms of the matrix $[\partial]$ that are differential operators, constitutes the reason why the algebraic terms of the matrix $[\partial]$ change their sign in the adjoint matrix $[\partial]^*$.

In Equation 4.57, the presence of the Lamé operators $[\mathcal{L}]$ and $[\mathcal{L}_0]$ can be recognized:

$$[K_e] = -\int_{V_e} [\eta_e]^T [\mathcal{L}][\eta_e] dV$$

$$+ \int_{S_e} [\eta_e]^T [\mathcal{L}_0][\eta_e] dS \qquad (4.58)$$

The global stiffness matrix is therefore obtained by summing up all the contributions of Equation 4.58, after premultiplying them by the matrices $[A_e]^T$ and postmultiplying them by the matrices $[A_e]$:

$$[K] = \sum_e [K^e] = -\int_V [A_e]^T [\eta_e]^T [\mathcal{L}][\eta_e][A_e] dV$$

$$+ \int_S [A_e]^T [\eta_e]^T [\mathcal{L}_0][\eta_e][A_e] dS \qquad (4.59)$$

The contributions corresponding to the interface between elements cancel each other out. It is easy to verify that the matrix in Equation 4.59 has Equation 4.25 as its elements, and hence the identity between the global stiffness matrix and the Ritz–Galerkin matrix holds:

$$[K]=[L] \qquad (4.60)$$

On the basis of Equations 4.24 and 4.60, the total potential energy can thus be expressed as follows:

$$W(\delta) = \frac{1}{2} \{\delta\}^T [K]\{\delta\} - \{\delta\}^T \{F\} \qquad (4.61)$$

where the first term represents the strain energy of the discretized elastic solid, and the second term represents the potential energy of the external (body and surface) forces.

4.6 KINEMATIC BOUNDARY CONDITIONS

So far, we have not considered the boundary conditions of a kinematic type, as given by Equation 4.9. However, the principle of minimum total potential energy can be reproposed in the case where the external forces do not constitute a self-balanced system, so that we arrive at the same resolving Equation 4.55. Some of the elements of the vector $\{\delta\}$ are in this case known, rather than unknown, terms, just as the constraint reactions now play a role of unknowns, and no longer of known terms, as hitherto assumed.

Partitioning the vectors and the stiffness matrix in such a way as to separate the free displacements from the constrained ones, we obtain

$$\begin{bmatrix} K_{LL} & K_{LV} \\ K_{VL} & K_{VV} \end{bmatrix} \begin{bmatrix} \delta_L \\ \delta_V \end{bmatrix} = \begin{bmatrix} F_L \\ F_V \end{bmatrix} \tag{4.62}$$

While the constrained displacements $\{\delta_V\}$ are zero, or are predetermined in the case of imposed displacements, the free displacements $\{\delta_L\}$ represent the unknowns of the problem:

$$[K_{LL}]\{\delta_L\} = \{F_L\} - [K_{LV}]\{\delta_V\} \tag{4.63}$$

from which we obtain

$$\{\delta_L\} = [K_{LL}]^{-1}(\{F_L\} - [K_{LV}]\{\delta_V\}) \tag{4.64}$$

The external constraint reactions are thus expressible as follows:

$$\{Q_V\} = \{F_V\} - \{F_V^0\} \tag{4.65}$$

where

$$\{F_V\} = [K_{VL}]\{\delta_L\} + [K_{VV}]\{\delta_V\} \tag{4.66}$$

represents the vector of the equivalent nodal forces, acting in the constrained nodes, while

$$\{F_V^0\} = \sum_e \{F_V^e\} \tag{4.67}$$

represents the vector of the nodal forces, equivalent only to the body forces.

Chapter 5

Dynamics of discrete systems

5.1 INTRODUCTION

The main goal of the **dynamics of structures** is to present methods for analyzing stresses and deflections in any given structural system, when it is subjected to arbitrary dynamic loadings. This objective can be considered as an extension of standard methods of structural analysis, which are usually concerned with static loadings. These thus become functions of time, as well as the structural response.

The dynamic loadings acting on a structure can be periodic or nonperiodic. The simplest periodic loading has the sinusoidal variation of Figure 5.1a, which is also termed as **harmonic**. Loadings of this type are characteristic of unbalanced-mass effects in rotating machinery. Other forms of periodic loadings are, for instance, those caused by hydrodynamic pressures generated by a propeller at the stern of a ship (Figure 5.1b). On the other hand, nonperiodic loadings may be either **short-duration** or **impulsive** loadings, as those generated by explosions (Figure 5.1c), or **long-duration** loadings, as might result from an earthquake (Figure 5.1d).

If a structure is subjected to a static load, the internal moments and shearing forces, as well as the deflection shape, depend only on this load, by established principles of internal force equilibrium.

In contrast, if the load is applied dynamically, the structural response depends also on the inertial forces, which oppose the accelerations producing them. If the motion is so slow to neglect both damping and inertial forces, the analysis can be considered to be static instant-by-instant, although loading and structural response are both time dependent.

5.2 FREE VIBRATIONS

The equation of motion of a single mass, subjected to an elastic force and a damping force (Figure 5.2), can be expressed as

$$m\ddot{x}(t) + c\dot{x}(t) + kx(t) = 0 \tag{5.1}$$

where:
- x is the linear elastic spring elongation, which depends on time t (the dot over the function represents the time derivative)
- m is the mass
- c is the damping constant
- k is the spring stiffness

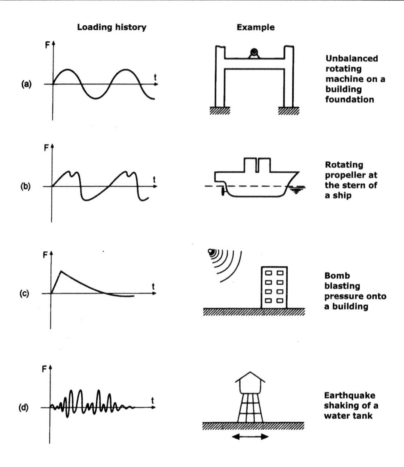

Figure 5.1

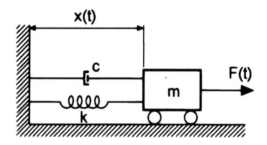

Figure 5.2

Equation 5.1 represents the well-known dynamic equation: *force = mass × acceleration*. In fact, both the active forces, $-kx$ and $-c\dot{x}$, turn out to be negative in cases of positive elongations and velocities, respectively. Another interpretation that can be given to Equation 5.1 is by means of D'Alembert's principle, according to which each mass is in equilibrium in its frame of reference, once subjected to all the active and **inertial forces**. The latter oppose the acceleration and are equal to the product of the acceleration itself times the mass.

When the forces applied to the mass are not external, but only internal (elastic and damping forces) and inertial, the motion of the system is called **free vibration**. The solution of Equation 5.1 takes the following form:

$$x(t) = C e^{st} \tag{5.2}$$

Substituting Equation 5.2 into Equation 5.1 yields

$$\left(ms^2 + cs + k\right) C e^{st} = 0 \tag{5.3}$$

Dividing by mCe^{st} and introducing the notation:

$$\omega^2 = \frac{k}{m} \tag{5.4}$$

Equation 5.3 becomes

$$s^2 + \frac{c}{m}s + \omega^2 = 0 \tag{5.5}$$

5.2.1 Undamped free vibrations (c = 0)

In this case, the two solutions of Equation 5.5 are

$$s = \pm i\omega \tag{5.6}$$

where i is the imaginary unit. The system response is thus given by

$$x(t) = C_1 e^{i\omega t} + C_2 e^{-i\omega t} \tag{5.7}$$

Since, according to Euler's formula, we have

$$e^{\pm i\omega t} = \cos \omega t \pm i \sin \omega t \tag{5.8}$$

Equation 5.7 can be rewritten as

$$x(t) = A \sin \omega t + B \cos \omega t \tag{5.9}$$

where the two constants A and B can be expressed by means of the initial conditions. In fact, since

$$x(0) = B \tag{5.10a}$$

$$\dot{x}(0) = A\omega \tag{5.10b}$$

Equation 5.9 becomes (Figure 5.3)

$$x(t) = \frac{\dot{x}(0)}{\omega} \sin \omega t + x(0) \cos \omega t \tag{5.11}$$

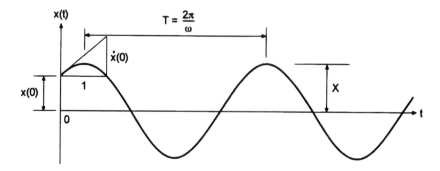

Figure 5.3

This solution represents a simple harmonic motion and it is homogeneous from a dimensional point of view, since the **angular frequency** (or **angular velocity**) ω has dimensions $[T]^{-1}$ and is measured in radians per second. The **ordinary frequency** is measured in hertz (cycles per second):

$$f = \frac{\omega}{2\pi} \tag{5.12}$$

whereas its reciprocal represents the **period** T:

$$T = \frac{1}{f} = \frac{2\pi}{\omega} \tag{5.13}$$

In addition to Equation 5.11, the motion can be described by the following expression:

$$x(t) = X\cos(\omega t - \varphi) \tag{5.14}$$

where the **amplitude** is given by

$$X = \sqrt{\left[x(0)\right]^2 + \left[\frac{\dot{x}(0)}{\omega}\right]^2} \tag{5.15}$$

and the **phase angle** by

$$\varphi = \arctan\frac{\dot{x}(0)}{\omega x(0)} \tag{5.16}$$

5.2.2 Damped free vibrations (c > 0)

In this case, the two solutions of Equation 5.5 are

$$s_{1,2} = -\frac{c}{2m} \pm \sqrt{\left(\frac{c}{2m}\right)^2 - \omega^2} \tag{5.17}$$

Three types of motion are represented by this expression, according to the quantity under the square root, whether it is positive, negative, or equal to zero. It is convenient to discuss first the case when the radical term vanishes.

First case: c=2mω (critically damped system)
The critical value of the damping coefficient is

$$c_c = 2m\omega = 2m\sqrt{\frac{k}{m}} = 2\sqrt{km} \qquad (5.18)$$

Then, it follows from Equation 5.17 that

$$s_1 = s_2 = -\frac{c}{2m} = -\omega \qquad (5.19)$$

according to which, Equation 5.2 takes the form:

$$x(t) = (C_1 + C_2 t)e^{-\omega t} \qquad (5.20)$$

where the second term must contain t, since the two roots s are identical.

Using the initial conditions, the final form of the dynamic response can be written as

$$x(t) = \left[x(0)(1+\omega t) + \dot{x}(0)t\right]e^{-\omega t} \qquad (5.21)$$

which is represented graphically in Figure 5.4 for positive values of $x(0)$ and $\dot{x}(0)$. Note that this response does not include oscillation about the equilibrium position, but only an exponential decay toward this position. It can be stated that the critically damped condition represents the smallest amount of damping for which no oscillation occurs in the free-motion response.

Second case: c<2mω (undercritically damped system)
Since the quantity under the radical sign in Equation 5.17 is negative, it is convenient to describe damping in terms of a **damping ratio** ξ, which is the ratio of the given damping coefficient c to the critical value c_c:

$$\xi = \frac{c}{c_c} = \frac{c}{2m\omega} \qquad (5.22)$$

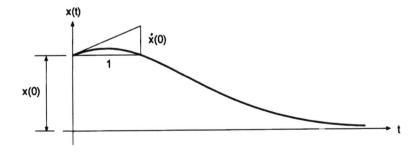

Figure 5.4

Introducing Equation 5.22 into Equation 5.17 leads to

$$s_{1,2} = -\xi\omega \pm \sqrt{(\xi\omega)^2 - \omega^2} \tag{5.23}$$

with $0 < \xi < 1$.

Equation 5.23 can be expressed as

$$s_{1,2} = -\xi\omega \pm i\,\omega_D \tag{5.24}$$

where

$$\omega_D = \omega\sqrt{1-\xi^2} \tag{5.25}$$

is the **damped frequency**. Its value is close to that related to the undamped frequency ω in practical cases, where, generally, $\xi < 1/4$.

The dynamic response of an undercritically damped system is obtained by substituting Equation 5.24 into Equation 5.2:

$$x(t) = C_1 e^{-\xi\omega t + i\omega_D t} + C_2 e^{-\xi\omega t - i\omega_D t} = e^{-\xi\omega t}\left(C_1 e^{i\omega_D t} + C_2 e^{-i\omega_D t}\right) \tag{5.26}$$

The term within parentheses represents a simple harmonic motion. Equation 5.26 can thus be rewritten as

$$x(t) = e^{-\xi\omega t}\left(A\sin\omega_D t + B\cos\omega_D t\right) \tag{5.27}$$

Evaluating the two arbitrary constants A and B, by means of initial conditions, yields

$$x(t) = e^{-\xi\omega t}\left[\frac{\dot{x}(0) + x(0)\xi\omega}{\omega_D}\sin\omega_D t + x(0)\cos\omega_D t\right] \tag{5.28}$$

Alternatively, this response can be written in the following form:

$$x(t) = X e^{-\xi\omega t}\cos\left(\omega_D t - \varphi\right) \tag{5.29}$$

with

$$X = \sqrt{\left[\frac{\dot{x}(0) + x(0)\xi\omega}{\omega_D}\right]^2 + \left[x(0)\right]^2} \tag{5.30}$$

$$\varphi = \frac{\dot{x}(0) + x(0)\xi\omega}{x(0)\omega_D} \tag{5.31}$$

Observe the analogy between Equations 5.30 and 5.31 and between Equations 5.15 and 5.16.

A plot of the response of an undercritically damped system subjected to an initial displacement $x(0)$ and starting with zero velocity is shown in Figure 5.5. The mass oscillates about the neutral position with an exponentially decreasing amplitude. Note that the

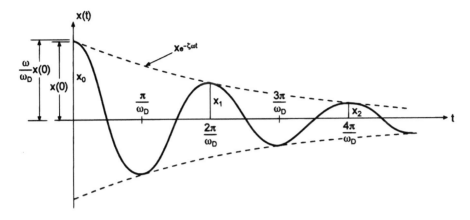

Figure 5.5

maximum and minimum elongations observed do not coincide exactly with the instants when $\cos(\omega_D t - \varphi) = \pm 1$, but with the instants when the velocity $\dot{x}(t)$ vanishes to change sign (analytical stationary points).

To evaluate experimentally the damping ratio ξ, consider any two consecutive positive peaks, x_n and x_{n+1} (Figure 5.5). From Equation 5.29, the ratio of these successive values is given by

$$\frac{x_n}{x_{n+1}} \simeq e^{2\pi\xi(\omega/\omega_D)} \tag{5.32}$$

Taking the natural logarithm of both sides of this equation, one obtains the so-called **logarithmic damping decrement** δ:

$$\delta = \ln\frac{x_n}{x_{n+1}} \simeq 2\pi\xi\frac{\omega}{\omega_D} \tag{5.33}$$

which can be rewritten as

$$\delta \simeq \frac{2\pi\xi}{\sqrt{1-\xi^2}} \tag{5.34}$$

For low values of the damping constant, Equation 5.34 can be approximated by

$$\delta \simeq 2\pi\xi \tag{5.35}$$

and thus

$$\xi \simeq \frac{1}{2\pi}\ln\frac{x_n}{x_{n+1}} \tag{5.36}$$

A simple method for estimating the damping ratio is to count the number of cycles required to give a 50% reduction in amplitude. The relationship to be used in this case is presented graphically in Figure 5.6. As a quick rule of thumb, it is convenient to emphasize that, for

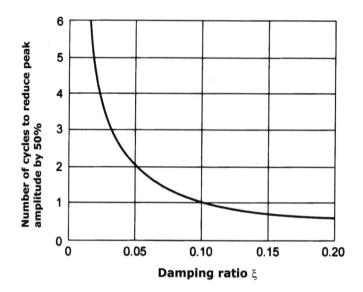

Figure 5.6

$\xi = 0.1$ (typical value for practical applications), the amplitude is reduced by 50% in approximately one cycle.

Let us consider, for instance, a shear-type frame and let us suppose that we will displace the rigid cross member (or girder) 0.5 cm by means of a force equal to 50 t. After releasing the initial displacement, the maximum displacement on the first return swing is 0.4 cm and the period is $T = 2$ s. From these data, the following dynamic properties are determined:

1. *Effective weight of the rigid cross member*
   The period is equal to

$$T = \frac{2\pi}{\omega} = 2\pi \sqrt{\frac{k}{m}} = 2\,\mathrm{s}$$

   from which

$$m = k\left(\frac{T}{2\pi}\right)^2 = \left(\frac{10}{0.5}\right)\left(\frac{2}{6.28}\right)^2 \times 980 \approx 2000\,\mathrm{t}$$

2. *Damping ratio*
   The logarithmic decrement is equal to (Equation 5.33)

$$\delta = \ln\frac{x_0}{x_1} = \ln\frac{0.5}{0.4} \approx 0.223$$

   Eventually, Equation 5.35 provides

$$\xi = \frac{\delta}{2\pi} \approx 3.55\%$$

Third case: c>2mω (overcritically damped system)
In this case, $\xi = c/c_c > 1$, and the response is similar to the motion of a critically damped system. However, the asymptotic return to the neutral position is slower, depending on the amount of damping. Note that it is very unusual, under normal conditions, to have overcritically damped structural systems.

5.3 HARMONIC LOADING AND RESONANCE

5.3.1 Undamped systems

In the case where a harmonic oscillator is subjected to a harmonically varying load $F\sin\omega_F t$, the dynamic equation becomes nonhomogeneous:

$$m\ddot{x}(t) + kx(t) = F\sin\omega_F t \tag{5.37}$$

The particular solution $x_p(t)$ of Equation 5.37 represents the specific response generated by the external loading, whereas the complementary solution describes the free vibration of the system:

$$x_p(t) = C\sin\omega_F t \tag{5.38a}$$

$$x_c(t) = A\sin\omega t + B\cos\omega t \tag{5.38b}$$

The amplitude C of the particular solution is obtained by substituting Equation 5.38a into Equation 5.37 and dividing both sides by $\sin\omega_F t$:

$$-m\omega_F^2 C + kC = F \tag{5.39}$$

from which there follows:

$$C\left(1 - \frac{\omega_F^2}{\omega^2}\right) = \frac{F}{k} \tag{5.40}$$

After some rearrangement, one obtains

$$C = \frac{F}{k}\frac{1}{1-\beta^2} \tag{5.41}$$

where β is the **frequency ratio**, defined as the ratio of the applied loading frequency to the natural free-vibration angular frequency:

$$\beta = \frac{\omega_F}{\omega} \tag{5.42}$$

The general solution of Equation 5.37 is now obtained by combining the complementary and particular solutions and making use of Equation 5.41:

$$x(t) = x_c(t) + x_p(t) = A \sin \omega t + B \cos \omega t + \frac{F}{k} \frac{1}{1-\beta^2} \sin \omega_F t \tag{5.43}$$

The values of A and B depend on the initial conditions. For a system starting from rest, $x(0) = \dot{x}(0) = 0$, it is easily shown that

$$A = -\frac{F\beta}{k} \frac{1}{1-\beta^2}, \quad B = 0 \tag{5.44}$$

In this case, the response (Equation 5.43) becomes

$$x(t) = \frac{F}{k} \frac{1}{1-\beta^2} (\sin \omega_F t - \beta \sin \omega t) \tag{5.45}$$

Since, in a practical case, damping causes the last term to vanish, it is termed the **transient response**. On the other hand, the first term holds, being related to the applied loading. It is called the **steady-state response** and it is amplified by the **resonance factor** $1/(1-\beta^2)$.

5.3.2 Systems with viscous damping

In the presence of viscous forces, the differential equation of motion (Equation 5.37) becomes

$$m\ddot{x}(t) + c\dot{x}(t) + kx(t) = F \sin \omega_F t \tag{5.46}$$

Dividing by m and noting that $c/m = 2\xi\omega$ yields

$$\ddot{x}(t) + 2\xi\omega\dot{x}(t) + \omega^2 x(t) = \frac{F}{m} \sin \omega_F t \tag{5.47}$$

The complementary solution of this equation is the damped free-vibration response given by Equation 5.27, while the particular solution is of the form:

$$x_p(t) = C_1 \sin \omega_F t + C_2 \cos \omega_F t \tag{5.48}$$

where the cosine term is required because the response of a damped system is generally not in phase with the loading.

Substituting Equation 5.48 into Equation 5.47 and separating the multiples of $\sin \omega_F t$ from the multiples of $\cos \omega_F t$ leads to

$$\left[-C_1\omega_F^2 - C_2\omega_F(2\xi\omega) + C_1\omega^2 \right] \sin \omega_F t = \frac{F}{m} \sin \omega_F t \tag{5.49a}$$

$$\left[-C_2\omega_F^2 + C_1\omega_F(2\xi\omega) + C_2\omega^2 \right] \cos \omega_F t = 0 \tag{5.49b}$$

Dividing both equations by ω^2, one obtains

$$C_1\left(1-\beta^2\right)-C_2\left(2\xi\beta\right)=\frac{F}{k} \tag{5.50a}$$

$$C_2\left(1-\beta^2\right)+C_1\left(2\xi\beta\right)=0 \tag{5.50b}$$

from which

$$C_1 = \frac{F}{k}\frac{1-\beta^2}{\left(1-\beta^2\right)^2+\left(2\xi\beta\right)^2} \tag{5.51a}$$

$$C_2 = -\frac{F}{k}\frac{2\xi\beta}{\left(1-\beta^2\right)^2+\left(2\xi\beta\right)^2} \tag{5.51b}$$

Introducing these expressions into Equation 5.48 and combining the result with the complementary solution (Equation 5.27), the total response is obtained in the form

$$x(t) = e^{-\xi\omega t}\left(A\sin\omega_D t + B\cos\omega_D t\right)$$

$$+\frac{F}{k}\frac{1}{\left(1-\beta^2\right)^2+\left(2\xi\beta\right)^2}\left[\left(1-\beta^2\right)\sin\omega_F t - 2\xi\beta\cos\omega_F t\right] \tag{5.52}$$

The first term on the right-hand side represents the **transient response** (A and B depend on the initial conditions), which damps out in accordance with $\exp(-\xi\omega t)$, whereas the second term represents the **steady-state response**, which has the same frequency of loading, but not the same phase.

The steady-state response can be expressed as follows:

$$x_p(t) = X\sin\left(\omega_F t - \varphi\right) \tag{5.53}$$

with

$$X = \frac{F}{k}\frac{1}{\sqrt{\left(1-\beta^2\right)^2+\left(2\xi\beta\right)^2}} \tag{5.54}$$

$$\varphi = \arctan\frac{2\xi\beta}{1-\beta^2} \tag{5.55}$$

The ratio of the steady-state response amplitude to the static displacement that would be produced by the force F is called the **dynamic magnification factor** D:

$$D = \frac{X}{F/k} = \frac{1}{\sqrt{\left(1-\beta^2\right)^2+\left(2\xi\beta\right)^2}} \tag{5.56}$$

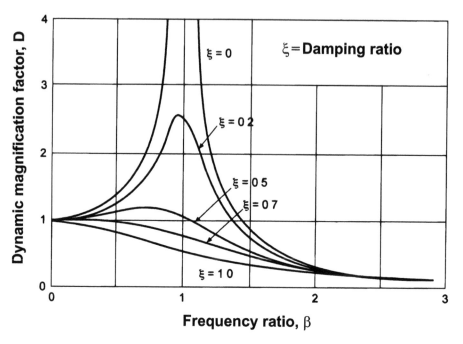

Figure 5.7

This factor tends to infinity (resonance condition) when $\xi \to 0$ (undamped case) and $\beta \to 1$ ($\omega_F \to \omega$). In general, this turns out to be a function of the damping ratio ξ and the frequency ratio β (Figure 5.7). For sufficiently low values of ξ ($\xi \le 1/\sqrt{2}$), the frequency ratio that provides the peak amplitude of the steady-state response is

$$\beta_{peak} = \sqrt{1 - 2\xi^2} \tag{5.57}$$

while the peak is equal to

$$D_{max} = \frac{1}{2\xi\sqrt{1-\xi^2}} \tag{5.58}$$

At the resonant exciting frequency ($\beta = 1$), Equation 5.52 becomes

$$x(t) = e^{-\xi\omega t}\left[A\sin\omega_D t + B\cos\omega_D t\right] - \frac{F}{k}\frac{\cos\omega t}{2\xi} \tag{5.59}$$

Assuming that the system starts from rest ($x(0) = \dot{x}(0) = 0$), the constants are

$$A = \frac{F}{k}\frac{\omega}{2\omega_D} = \frac{F}{k}\frac{1}{2\sqrt{1-\xi^2}}, \quad B = \frac{F}{k}\frac{1}{2\xi} \tag{5.60}$$

and Equation 5.59 becomes

$$x(t) = \frac{1}{2\xi}\frac{F}{k}\left[e^{-\xi\omega t}\left(\frac{\xi}{\sqrt{1-\xi^2}}\sin\omega_D t + \cos\omega_D t\right) - \cos\omega t\right] \tag{5.61}$$

For a low amount of damping, it follows that $\omega \simeq \omega_D$ and the denominator term is nearly equal to unity. In this case, the equation can be written in the approximate form:

$$x(t) = \frac{1}{2\xi}\frac{F}{k}\left(e^{-\xi\omega t} - 1\right)\cos\omega t \tag{5.62}$$

For zero damping, this equation is indeterminate, but when L'Hôpital's rule is applied, one obtains

$$x(t) = -\frac{1}{2}\frac{F}{k}\omega t \cos\omega t \tag{5.63}$$

Equation 5.63 is plotted in Figure 5.8, describing the linear increase in the oscillation amplitude.

The notions investigated in the present section, and particularly the graphic of Figure 5.7, are particularly useful in the (stiffness and damping) design of machine suspensions, of anti-seismic building foundations, and, more generally, in all those cases where it is necessary to dynamically isolate a structural element from an adjacent vibrating element. The bodywork of a car must be isolated from wheel motion, as well as a building structure from ground motion. In contrast, the floor must be isolated from machine motion in the case of rotating machines.

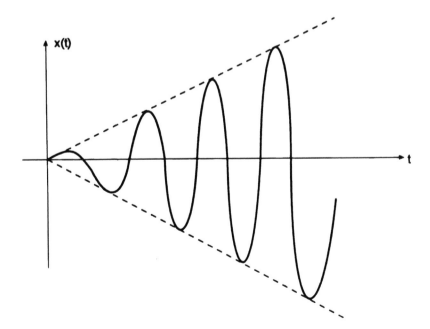

Figure 5.8

5.4 PERIODIC LOADING

Let us consider a harmonic oscillator subjected to a periodic loading $F(t)$ (Figure 5.9). It is convenient to express this function in a Fourier series: the response to each term of the series will be similar to that obtained for the harmonic loading. The global solution will be the sum of each partial response, by the use of the superposition principle.

The periodic function $F(t)$ can be expressed as

$$F(t) = a_0 + \sum_{n=1}^{\infty} a_n \cos \frac{2\pi n}{T_F} t + \sum_{n=1}^{\infty} b_n \sin \frac{2\pi n}{T_F} t \tag{5.64}$$

where T_F is the loading period and the harmonic amplitude coefficients can be evaluated using the expressions:

$$a_0 = \frac{1}{T_F} \int_0^{T_F} F(t) dt \tag{5.65a}$$

$$a_n = \frac{2}{T_F} \int_0^{T_F} F(t) \cos \frac{2\pi n}{T_F} t \, dt \tag{5.65b}$$

$$b_n = \frac{2}{T_F} \int_0^{T_F} F(t) \sin \frac{2\pi n}{T_F} t \, dt \tag{5.65c}$$

By denoting the frequency of the loading $F(t)$ with $\omega_F = 2\pi/T_F$, the angular frequency related to the n-order harmonic is $\omega_n = n\omega_F$. The steady-state response generated in an undamped harmonic system due to each harmonic load of the series is provided by Equation 5.45, with the exception of the transient term:

$$x_n(t) = \frac{a_n}{k} \frac{1}{1-\beta_n^2} \cos \omega_n t \tag{5.66a}$$

or

$$x_n(t) = \frac{b_n}{k} \frac{1}{1-\beta_n^2} \sin \omega_n t \tag{5.66b}$$

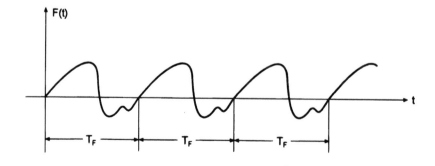

Figure 5.9

where $\beta_n = \omega_n/\omega = n(\omega_F/\omega)$.

The steady-state response to the constant load a_0 is simply the static displacement:

$$x_0 = \frac{a_0}{k} \tag{5.66c}$$

The total periodic response is given by the sum of the individual responses (Equations 5.66):

$$x(t) = \frac{1}{k}\left[a_0 + \sum_{n=1}^{\infty} \frac{1}{1-\beta_n^2}\left(a_n \cos\omega_n t + b_n \sin\omega_n t\right)\right] \tag{5.67}$$

where the coefficients a_n and b_n are provided by Equations 5.65.

To take viscous damping into account in evaluating the steady-state response of systems to periodic loading, it is necessary to consider Equation 5.62 for each harmonic, so that the total response is given by

$$\begin{aligned}x(t) = \frac{1}{k}\Bigg(a_0 + \sum_{n=1}^{\infty} &\frac{1}{\left(1-\beta_n^2\right)^2 + \left(2\xi\beta_n\right)^2} \\ &\times \left\{\left[a_n 2\xi\beta_n + b_n\left(1-\beta_n^2\right)\right]\sin\omega_n t\right. \\ &\left.+ \left[a_n\left(1-\beta_n^2\right) - b_n 2\xi\beta_n\right]\cos\omega_n t\right\}\Bigg)\end{aligned} \tag{5.68}$$

5.5 IMPULSIVE LOADING

The impulsive loading is of relatively short duration, such that damping has much less importance in controlling the maximum response. This will be reached in a very short time, before the damping forces can absorb much energy from the structure. For this reason, only the undamped response to impulsive loads will be considered in this section.

Consider the single half-sine-wave impulse shown in Figure 5.10. During the first phase, the structure is subjected to the single half-sine-wave loading, starting from rest. The free vibration that occurs during the second phase depends on the displacement $x(t_F)$ and velocity $\dot{x}(t_F)$ at the end of the first phase. This motion can be described by means of Equation 5.11:

$$x(t - t_F) = \frac{\dot{x}(t_F)}{\omega}\sin\omega(t - t_F) + x(t_F)\cos\omega(t - t_F) \tag{5.69}$$

The entity of the dynamic response depends on the ratio of the impulsive loading duration, t_F, and to the fundamental period of the structure, T. The ratio $x(t)/(F/k)$ thus depends on t_F/T. The case $t_F/T = 3/4$ is reported in Figure 5.11: the maximum dynamic response is equal to 1.77 times the static response. By differentiating Equation 5.45 with respect to time and setting the resulting expression to zero, the instant providing the peak can be estimated:

$$\dot{x}(t) = \frac{F}{k}\frac{1}{1-\beta^2}\left(\omega_F \cos\omega_F t - \omega_F \cos\omega t\right) = 0 \tag{5.70}$$

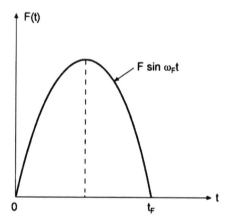

Figure 5.10

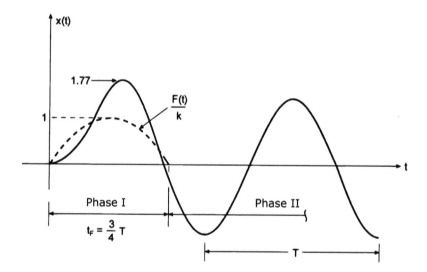

Figure 5.11

From Equation 5.70, it follows that

$$\cos \omega_F t = \cos \omega t \qquad (5.71)$$

and thus

$$\omega_F t = 2\pi n \pm \omega t, \qquad n = 0, \pm 1, 2, 3, \ldots \qquad (5.72)$$

For $\beta < 1$ ($\omega_F < \omega$, rigid oscillator), one obtains

$$\omega_F t = \frac{2\pi}{1 + \dfrac{\omega}{\omega_F}} \qquad (5.73)$$

For $\beta > 1$ ($\omega_F > \omega$, prevailing mass) the maximum response occurs during the second phase, which is related to free vibration. The initial displacement and velocity for this phase are obtained by setting $\omega_F t_F = \pi$ in Equation 5.45:

$$x(t_F) = \frac{F}{k}\frac{1}{1-\beta^2}\left(0 - \beta\sin\frac{\pi}{\beta}\right) \qquad (5.74a)$$

$$\dot{x}(t_F) = \frac{F}{k}\frac{\omega_F}{1-\beta^2}\left(-1 - \cos\frac{\pi}{\beta}\right) \qquad (5.74b)$$

The amplitude of the free vibration is furnished by Equation 5.15:

$$X = \frac{F}{k}\frac{\beta}{\beta^2 - 1}\sqrt{2\left(1 + \cos\frac{\pi}{\beta}\right)} = \frac{F}{k}\frac{2\beta}{\beta^2 - 1}\cos\frac{\pi}{2\beta} \qquad (5.75)$$

Hence, the **dynamic magnification factor** $D = X/(F/k)$ depends only on the ratio β.

The **response spectrum**, related to impulsive loading, shows the magnification factor D as a function of the ratio t_F/T. In Figure 5.12, three different spectra related to three different impulsive loading shapes (half-sine, rectangular, and triangular, respectively) are reported. It can be observed that, for short-duration impulsive loadings, the magnification factor is low because inertial forces oppose most of the applied loading. Consequently, stresses lower than those generated by longer loadings arise in the structure.

For long-duration loadings ($t_F/T > 1$), the dynamic magnification factor depends principally on the rate of increase in the load up to its maximum value. A step (rectangular) loading of sufficient duration produces a factor of 2, whereas a more gradual increase, also called **quasi-static** loading, produces a factor of 1.

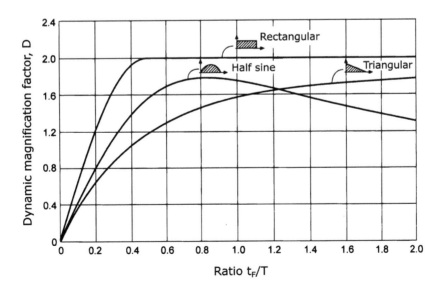

Figure 5.12

For short-duration loads ($t_F/T < 0.25$), the maximum displacement amplitude X depends principally on the magnitude of the applied impulse $\mathcal{I} = \int_0^{t_F} F(t)\,dt$ and is not particularly influenced by the shape of the loading impulse. The dynamic response after the impulse is represented by the free vibration described by Equation 5.69, which can be approximated by neglecting the displacement $x(t_F)$ and by setting $\dot{x}(t_F) = \Delta\dot{x}$:

$$x(t - t_F) \simeq \frac{\Delta\dot{x}}{\omega}\sin\omega(t - t_F) \tag{5.76}$$

with $\Delta\dot{x} = \mathcal{I}/m$.

The maximum response is reached when $\sin\omega(t - t_F) = 1$:

$$x_{max} \simeq \frac{\mathcal{I}}{m\omega} \tag{5.77}$$

The maximum elastic force, generated by the spring with stiffness k, is equal to kx_{max}, and is of fundamental importance for structural engineering applications.

5.6 GENERAL DYNAMIC LOADING

The procedure described in the previous section for approximating the response of a structure subjected to short-duration impulse loads can be used as the basis for evaluating the response to a general dynamic loading.

Consider an arbitrary generic loading $F(t)$ as illustrated in Figure 5.13 and, for the moment, concentrate on the intensity of the loading $F(\tau)$ acting at time $t = \tau$. This loading acting during the interval of time $d\tau$ represents a very short-duration impulse $F(\tau)d\tau$ on the structure, so that Equation 5.76 can be used to evaluate the resulting response. Although Equation 5.76 is approximate for finite-duration impulsive loadings, it becomes exact for very short-duration loadings:

$$dx(t - \tau) = \frac{F(\tau)d\tau}{m\omega}\sin\omega(t - \tau), \qquad \text{for } t > \tau \tag{5.78}$$

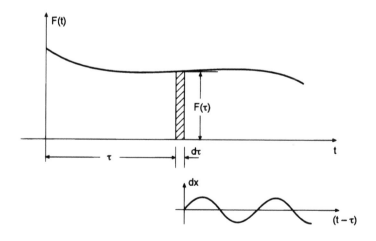

Figure 5.13

The entire loading history can be considered to consist of a succession of short impulses, each producing its own differential response in the form of Equation 5.78. For a linear elastic system, the total response can then be obtained by summing all these contributions:

$$x(t) = \frac{1}{m\omega} \int_0^t F(\tau) \sin \omega(t - \tau) d\tau \qquad (5.79)$$

Equation 5.79 is generally known as **Duhamel integral equation** for undamped systems.
  By exploiting the trigonometric identity

$$\sin(\omega t - \omega \tau) = \sin \omega t \cos \omega \tau - \cos \omega t \sin \omega \tau \qquad (5.80)$$

it is possible to express the convolution integral (Equation 5.79) in the form

$$x(t) = \frac{\sin \omega t}{m\omega} \int_0^t F(\tau) \cos \omega \tau \, d\tau - \frac{\cos \omega t}{m\omega} \int_0^t F(\tau) \sin \omega \tau \, d\tau \qquad (5.81)$$

or, equivalently

$$x(t) = A \sin \omega t - B \cos \omega t \qquad (5.82)$$

where the integrals

$$A(t) = \frac{1}{m\omega} \int_0^t F(\tau) \cos \omega \tau \, d\tau \qquad (5.83a)$$

$$B(t) = \frac{1}{m\omega} \int_0^t F(\tau) \sin \omega \tau \, d\tau \qquad (5.83b)$$

need to be evaluated analytically or numerically.
  In the damped structural case, by means of Equation 5.28, Equation 5.79 can be generalized as

$$x(t) = \frac{1}{m\omega_D} \int_0^t F(\tau) e^{-\xi\omega(t-\tau)} \sin \omega_D(t - \tau) d\tau \qquad (5.84)$$

Although the time-domain analysis developed earlier may be used to determine the response to any arbitrary loading, it is sometimes more convenient to perform the analysis in the frequency domain. This approach is conceptually similar to that presented in Section 5.4, related to a periodic loading. Both the procedures involve expressing the applied loading in terms of harmonic components, estimating the response of the structure to each component and then superposing the harmonic responses to obtain the total structural response.
  The Fourier series representation of the loading depicted in Figure 5.14 can be expressed in exponential form

$$F(t) = \frac{\omega_F}{2\pi} \sum_{n=-\infty}^{+\infty} c(\omega_n) \exp(i\omega_n t) \qquad (5.85)$$

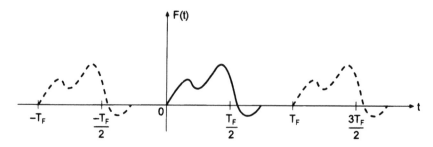

Figure 5.14

with

$$c(\omega_n) = \int_{-T_F/2}^{+T_F/2} F(t)\exp(-i\omega_n t)\,dt \tag{5.86}$$

If the loading period is extended to infinity ($T_F \to \infty$), the loading tends to become non-periodic, so that only the wave described by the continuous line in Figure 5.14 proves to be effective. For $T_F \to \infty$, one obtains $\omega_F \to 0$, and thus $\omega_n = n\omega_F \to 0$. The Fourier series Equation 5.85 takes the integral form

$$F(t) = \frac{1}{2\pi} \int_{-\infty}^{+\infty} c(\omega)\exp(i\omega t)\,d\omega \tag{5.87}$$

$$c(\omega) = \int_{-\infty}^{+\infty} F(t)\exp(-i\omega t)\,dt \tag{5.88}$$

The two integral Equations 5.87 and 5.88 are known as the inverse and direct Fourier transforms, respectively. As can be noted, the time function $F(t)$ can be obtained from the frequency function $c(\omega)$ and *vice versa*.

By substituting the complex form of the unit load into the equation of motion, one obtains

$$m\ddot{x}(t) + kx(t) = \exp(i\omega_F t) \tag{5.89}$$

which presents a solution in the form

$$x(t) = H(\omega_F)\exp(i\omega_F t) \tag{5.90}$$

Introducing Equation 5.90 into Equation 5.89 yields

$$H(\omega_F) = \frac{1}{-\omega_F^2 m + k} \tag{5.91}$$

or

$$H(\omega_F) = \frac{1}{k(1-\beta^2)} \tag{5.92}$$

The total response in the frequency domain can thus be obtained from Equations 5.87 and 5.90:

$$x(t) = \frac{1}{2\pi} \int_{-\infty}^{+\infty} H(\omega)c(\omega)\exp(i\omega t)\,d\omega \tag{5.93}$$

where the functions H and c are produced by Equations 5.91 and 5.88, respectively.

5.7 NONLINEAR ELASTIC SYSTEMS

In the investigation of linear elastic structures subjected to arbitrary dynamic loadings, the Duhamel integral or frequency domain analysis, described in the previous section, turn out to be the most convenient solving techniques.

These procedures are based on the superposition principle and can be applied to all systems in which the properties remain constant during structural response. On the other hand, there exist important structural dynamic problems that cannot be considered linear. In these systems, the stiffness and/or the mass vary during vibration motion. To face non-linear problems, the most suitable technique turns out to be **step-by-step integration**. The condition for dynamic equilibrium is established at the beginning and the end of each step, while the global response is obtained using the displacement and the velocity evaluated at the end of each interval, as initial conditions for the subsequent interval.

The forces acting on the mass of the system are indicated in Figure 5.15. At each time instant t, the dynamic equilibrium of forces requires

$$F_I(t) + F_D(t) + F_S(t) = F(t) \tag{5.94}$$

where $F_I(t)$, $F_D(t)$, $F_S(t)$, and $F(t)$ are the inertial, damping, elastic, and external forces, respectively. Since, after a short time Δt, Equation 5.94 becomes

$$F_I(t + \Delta t) + F_D(t + \Delta t) + F_S(t + \Delta t) = F(t + \Delta t) \tag{5.95}$$

subtracting Equation 5.94 from Equation 5.95 yields the incremental equation of motion:

$$\Delta F_I(t) + \Delta F_D(t) + \Delta F_S(t) = \Delta F(t) \tag{5.96}$$

The incremental forces may be expressed as

$$\Delta F_I(t) = F_I(t + \Delta t) - F_I(t) = m(t)\Delta\ddot{x}(t) \tag{5.97a}$$

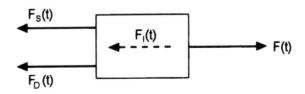

Figure 5.15

$$\Delta F_D(t) = F_D(t + \Delta t) - F_D(t) = c(t)\Delta \dot{x}(t) \tag{5.97b}$$

$$\Delta F_S(t) = F_S(t + \Delta t) - F_S(t) = k(t)\Delta x(t) \tag{5.97c}$$

$$\Delta F(t) = F(t + \Delta t) - F(t) \tag{5.97d}$$

where $k(t)$ and $c(t)$ represent the tangential characteristics:

$$k(t) = \left(\frac{dF_S}{dx}\right)_t, \quad c(t) = \left(\frac{dF_D}{d\dot{x}}\right)_t \tag{5.98}$$

By substituting Equations 5.97 into Equation 5.96, one obtains the incremental equilibrium equation:

$$m(t)\Delta \ddot{x}(t) + c(t)\Delta \dot{x}(t) + k(t)\Delta x(t) = \Delta F(t) \tag{5.99}$$

The characteristics $k(t)$ and $c(t)$ can describe every form of nonlinearity. It is not necessary, for instance, that the elastic force $F_S(t)$ depends only on the displacement, as happens for nonlinear elastic springs. A nonlinear hysteretic spring, in which the elastic force depends on past strain history, in addition to the present displacement value, may also be considered.

The fundamental hypothesis of the evaluation procedure is that the acceleration changes linearly over each time increment, while the structural characteristics remain constant during the same interval. A quadratic velocity variation and a cubic displacement variation correspond to a linear acceleration variation. Expanding the former two functions in the Taylor series around the instant t yields

$$\Delta \dot{x}(t) = \ddot{x}(t)\Delta t + \frac{\Delta \ddot{x}(t)}{\Delta t}\frac{\Delta t^2}{2} \tag{5.100a}$$

$$\Delta x(t) = \dot{x}(t)\Delta t + \ddot{x}(t)\frac{\Delta t^2}{2} + \frac{\Delta \ddot{x}(t)}{\Delta t}\frac{\Delta t^3}{6} \tag{5.100b}$$

From Equation 5.100b, one obtains

$$\Delta \ddot{x}(t) = \frac{6}{\Delta t^2}\Delta x(t) - \frac{6}{\Delta t}\dot{x}(t) - 3\ddot{x}(t) \tag{5.101}$$

which, inserted into Equation 5.100a produces

$$\Delta \dot{x}(t) = \frac{3}{\Delta t}\Delta x(t) - 3\dot{x}(t) - \frac{\Delta t}{2}\ddot{x}(t) \tag{5.102}$$

By substituting Equations 5.101 and 5.102 into Equation 5.99, we have

$$m(t)\left[\frac{6}{\Delta t^2}\Delta x(t) - \frac{6}{\Delta t}\dot{x}(t) - 3\ddot{x}(t)\right]$$

$$+ c(t)\left[\frac{3}{\Delta t}\Delta x(t) - 3\dot{x}(t) - \frac{\Delta t}{2}\ddot{x}(t)\right] + k(t)\Delta x(t) = \Delta F(t) \tag{5.103}$$

By rearranging all the terms so that the known initial conditions are on the right-hand side, Equation 5.103 takes the following form:

$$k^*(t)\Delta x(t) = \Delta F^*(t) \tag{5.104}$$

where

$$k^*(t) = k(t) + \frac{6}{\Delta t^2}m(t) + \frac{3}{\Delta t}c(t) \tag{5.105a}$$

$$\Delta F^*(t) = \Delta F(t) + m(t)\left[\frac{6}{\Delta t}\dot{x}(t) + 3\ddot{x}(t)\right] + c(t)\left[3\dot{x}(t) + \frac{\Delta t}{2}\ddot{x}(t)\right] \tag{5.105b}$$

Inertial and damping effects are thus taken into account both in the stiffness term $k^*(t)$ and in the effective loading $\Delta F^*(t)$.

Summarizing, the following operations must be carried out:

1. The initial displacement $x(t)$ and velocity $\dot{x}(t)$ values are known from the preceding time increment or from the initial conditions.
2. The stiffness $k(t)$ and the damping constant $c(t)$ can be evaluated from the $x(t)$ and $\dot{x}(t)$ values.
3. The initial acceleration value is given by the dynamic equilibrium Equation 5.94:

$$\ddot{x}(t) = \frac{1}{m(t)}\left[F(t) - F_D(t) - F_S(t)\right] \tag{5.106}$$

where $F_S(t)$ and $F_D(t)$ depend on $x(t)$ and $\dot{x}(t)$, respectively.
4. The effective values $k^*(t)$ and $\Delta F^*(t)$ are evaluated through Equations 5.105.
5. The displacement increment $\Delta x(t)$ is given by Equation 5.104.
6. The velocity increment $\Delta \dot{x}(t)$ is given by Equation 5.102.
7. The acceleration increment $\Delta \ddot{x}(t)$ is given by Equation 5.101.
8. The displacement, velocity, and acceleration at the end of the time increment are obtained as:

$$x(t + \Delta t) = x(t) + \Delta x(t) \tag{5.107a}$$

$$\dot{x}(t + \Delta t) = \dot{x}(t) + \Delta \dot{x}(t) \tag{5.107b}$$

$$\ddot{x}(t + \Delta t) = \ddot{x}(t) + \Delta \ddot{x}(t) \tag{5.107c}$$

Whereas the first two expressions represent the initial conditions for the next step, Equation 5.107c is replaced by Equation 5.106 when evaluated at time $t + \Delta t$. By doing so, numerical error accumulations are avoided.

Linear systems can also be treated by this procedure, which becomes simplified due to the physical properties remaining constant over the entire time-history response. In general, using an increment-period ratio of $\Delta t/T \leq 10$ will give reliable results.

5.8 ELASTIC–PERFECTLY PLASTIC SPRING

In the case of an elastic–perfectly plastic spring (Figure 5.16), it is possible to obtain an analytical solution in addition to the numerical one described in the previous section. Equation 5.1 still holds during elastic phases, provided that the elastic force $kx(t)$ is replaced by $k[x(t) - x^*]$, where x^* is the permanent displacement related to the preceding plastic phase (Figure 5.16). On the other hand, during the perfectly plastic phases, the elastic force is equal to $\pm kx_P$, keeping this value until $\dot{x}(t)$ changes sign.

Let us consider an undamped system, subjected to an initial impulsive loading $m\dot{x}_0 > 0$.

During the elastic phases, the equation of motion is

$$m\ddot{x}(t) + kx(t) = kx^* \tag{5.108}$$

Since $x^* = 0$ at the initial elastic phase, the solution is given by Equation 5.11, with $x(0) = 0$, $\dot{x}(0) = \dot{x}_0$:

$$x(t) = \frac{\dot{x}_0}{\omega}\sin \omega t \tag{5.109}$$

If $\dot{x}(0) > \omega x_P$, the elastic response reaches and exceeds the plastic displacement value x_P. The beginning of the plastic phase is reached at time $t = t_1$, so that

$$x_P = \frac{\dot{x}_0}{\omega}\sin \omega t_1 \tag{5.110}$$

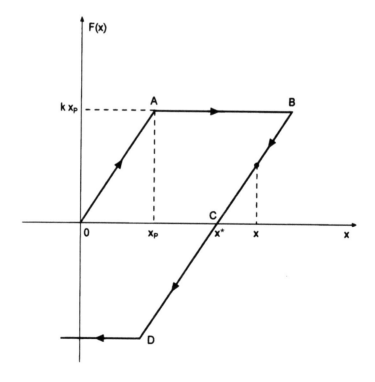

Figure 5.16

from which it follows that

$$t_1 = \frac{1}{\omega} \arcsin \frac{\omega x_P}{\dot{x}_0} \tag{5.111}$$

For $t > t_1$, the equation of motion becomes

$$m\ddot{x}(t) = -kx_P \tag{5.112}$$

the solution of which is parabolic:

$$x(t) = -\omega^2 x_P \frac{t^2}{2} + At + B \tag{5.113}$$

The arbitrary constants A and B are determined by imposing the initial conditions at the instant t_1:

$$x(t_1) = -\omega^2 x_P \frac{t_1^2}{2} + At_1 + B = x_P \tag{5.114a}$$

$$\dot{x}(t_1) = -\omega^2 x_P t_1 + A = \dot{x}_0 \cos \omega t_1 \tag{5.114b}$$

which produce

$$A = \omega^2 x_P t_1 + \dot{x}_0 \cos \omega t_1 \tag{5.115a}$$

$$B = x_P \left(1 - \omega^2 \frac{t_1^2}{2} \right) - \dot{x}_0 t_1 \cos \omega t_1 \tag{5.115b}$$

During the plastic phase, the solution thus produces

$$x(t) = x_P \left[1 - \frac{\omega^2}{2} (t - t_1)^2 \right] + \dot{x}_0 (t - t_1) \cos \omega t_1 \tag{5.116a}$$

$$\dot{x}(t) = -\omega^2 x_P (t - t_1) + \dot{x}_0 \cos \omega t_1 \tag{5.116b}$$

The return to the elastic phase occurs when $\dot{x}(t)$ changes sign; that is, at the instant $t = t_2$, so that

$$-\omega^2 x_P (t_2 - t_1) + \dot{x}_0 \cos \omega t_1 = 0 \tag{5.117}$$

from which it follows that

$$t_2 = t_1 + \frac{\dot{x}_0}{\omega^2 x_P} \cos \omega t_1 \tag{5.118}$$

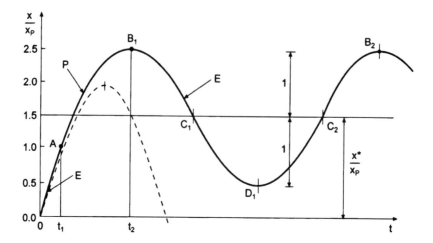

Figure 5.17

The displacement at this instant is equal to

$$x_2 = x_P \left(1 + \frac{\dot{x}_0^2}{2\omega^2 x_P^2} \cos^2 \omega t_1 \right) \qquad (5.119)$$

The permanent displacement thus takes the following form:

$$x^* = x_2 - x_P = \frac{\dot{x}_0^2}{2\omega^2 x_P} \cos^2 \omega t_1 \qquad (5.120)$$

and Equation 5.108 becomes

$$m\ddot{x}(t) + kx(t) = \frac{m\dot{x}_0^2}{2x_P} \cos^2 \omega t_1 \qquad (5.121)$$

For $t > t_2$, there will be some elastic free vibrations about the permanent position x^*, with amplitudes equal to the plasticization elongation x_P. The time history is represented in Figure 5.17, by setting $\dot{x}_0 = 2\omega x_P$; that is, twice the minimum value of the initial velocity providing plasticization. The dotted line represents the response of an indefinitely elastic structure with the same stiffness.

5.9 LINEAR ELASTIC SYSTEMS WITH TWO OR MORE DEGREES OF FREEDOM

In general, structures must be described by discretized models with several degrees of freedom, and not by a single degree of freedom model. Indeed, the structures are continuous systems and would present an infinite number of degrees of freedom.

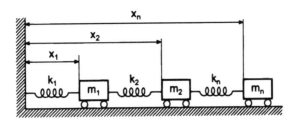

Figure 5.18

Consider a vibrating system formed by n masses m_i and by n springs in series, with stiffness k_i, $i = 1, 2, ..., n$ (Figure 5.18). There are n equations of motion, one for each vibrating mass:

$$m_1\ddot{x}_1 + k_1 x_1 - k_2 (x_2 - x_1) = 0$$

$$m_2\ddot{x}_2 + k_2 (x_2 - x_1) - k_3 (x_3 - x_2) = 0$$

$$m_3\ddot{x}_3 + k_3 (x_3 - x_2) - k_4 (x_4 - x_3) = 0 \qquad (5.122)$$

$$\vdots$$

$$m_n\ddot{x}_n + k_n (x_n - x_{n-1}) = 0$$

In Equations 5.122, viscous damping forces are neglected, as well as the presence of forcing actions. Equations 5.122 can be put in compact form:

$$[M]\{\ddot{x}\} + [K]\{x\} = \{0\} \qquad (5.123)$$

where $[M]$ and $[K]$ are the mass and stiffness matrices, respectively:

$$[M] = \begin{bmatrix} m_1 & 0 & 0 & \cdots & 0 \\ 0 & m_2 & 0 & \cdots & 0 \\ 0 & 0 & m_3 & \cdots & 0 \\ \vdots & \vdots & \vdots & \ddots & \vdots \\ 0 & 0 & 0 & \cdots & m_n \end{bmatrix} \qquad (5.124a)$$

$$[K] = \begin{bmatrix} k_1 + k_2 & -k_2 & 0 & 0 & \cdots & 0 \\ -k_2 & k_2 + k_3 & -k_3 & 0 & \cdots & 0 \\ 0 & -k_3 & k_3 + k_4 & -k_4 & \cdots & 0 \\ \vdots & \vdots & \vdots & \vdots & \ddots & \vdots \\ 0 & 0 & 0 & & -k_n & k_n \end{bmatrix} \qquad (5.124b)$$

Note that the mass matrix is diagonal, while the stiffness matrix turns out to be tridiagonal. Let us search for a solution of Equation 5.123 in the form

$$\{x\} = \{X\}\sin(\omega t - \varphi) \qquad (5.125)$$

By introducing Equation 5.125 into Equation 5.123, one obtains

$$-\omega^2[M]\{X\}\sin(\omega t - \varphi) + [K]\{X\}\sin(\omega t - \varphi) = \{0\} \tag{5.126}$$

that is

$$\left([K] - \omega^2[M]\right)\{X\} = \{0\} \tag{5.127}$$

Equation 5.127 represents an **eigenvalue problem**, since the linear algebraic equation system is homogeneous and the trivial solution lacks a physical meaning. The determinant of the term in parentheses must then be equal to zero:

$$\mathrm{Det}\left([K] - \omega^2[M]\right) = 0 \tag{5.128}$$

The n-order polynomial equation in the unknown ω^2 that arises from Equation 5.128 represents the **characteristic equation** of the elastic system. To each **eigenvalue** ω_i^2, $i = 1, 2, ..., n$, there corresponds an **eigenvector** $\{X_i\}$, but for a multiplicative constant.

It can thus be deduced that a vibrating system with n degrees of freedom has n **eigenfrequencies**, as well as n **mode shapes**. Each vibrating mode shape corresponds to a different eigenfrequency. The lowest frequency is named as **fundamental frequency**. The related mode shape is called the **fundamental mode**.

This topic will be discussed more in detail in Chapter 6, where discrete systems, with a finite number of degrees of freedom, and continuous systems, with infinite degrees of freedom, will be investigated. In both cases, the concepts of frequencies and vibrating mode shapes will be recalled, forming the basis for so-called **modal analysis**.

Equation 5.128 will remain the fundamental condition for the analysis, although matrices $[M]$ and $[K]$ will be no longer diagonal or tridiagonal, but will be more complex. This complexity reflects the real and effective connection between vibrating masses, which are generally not in series as in Figure 5.18. The only significant case that can be described by masses in series is that related to frames with rigid cross members and flexible and inextensible columns (**shear-type** frames), which will be studied in Chapter 6.

5.10 RAYLEIGH RATIO

Premultiplying Equation 5.127 by $\{X\}^\mathrm{T}$, one obtains a scalar equation, which after some simple analytical manipulations produces

$$\omega^2 = \frac{\{X\}^\mathrm{T}[K]\{X\}}{\{X\}^\mathrm{T}[M]\{X\}} \tag{5.129}$$

The right-hand side is the **Rayleigh ratio**, which generalizes the ratio k/m related to single-degree-of-freedom systems. Equation 5.129 is verified only if $\{X\}$ coincides with one eigenvector. Otherwise, the ratio will not coincide with one eigenvalue. It can be proved that the minimum Rayleigh ratio with respect to any possible choice of $\{X\}$ coincides with the minimum eigenvalue. This case is verified when $\{X\}$ coincides with the first eigenvector. Concerning all other cases, the angular frequency evaluated through Equation 5.129 is higher, since it is as if the structure were stiffened by additional constraints. A method to estimate the fundamental frequency, by assuming the shape of the first eigenvector *a priori*, comes from this property.

An alternative way to obtain Equation 5.129 is by considering the maximum potential energy related to an eigenvector:

$$W_{max} = \frac{1}{2}\{X\}^T [K]\{X\} \tag{5.130a}$$

as well as the maximum kinetic energy:

$$T_{max} = \frac{1}{2}\omega^2 \{X\}^T [M]\{X\} \tag{5.130b}$$

and equaling these quantities.

As an application of Equation 5.129, consider the shear-type frame depicted in Figure 5.19. The mass of each cross member is denoted by m, while k is the shearing stiffness of each couple of columns. Assume $\{X\}^T = (1,2)$ as a test vector, supposing thus a lateral displacement of the upper girder equal to twice that of the lower girder. Equation 5.129 produces

$$\omega^2 = \frac{\begin{bmatrix} 1 & 2 \end{bmatrix}\begin{bmatrix} 2k & -k \\ -k & k \end{bmatrix}\begin{bmatrix} 1 \\ 2 \end{bmatrix}}{\begin{bmatrix} 1 & 2 \end{bmatrix}\begin{bmatrix} m & 0 \\ 0 & m \end{bmatrix}\begin{bmatrix} 1 \\ 2 \end{bmatrix}} \tag{5.131}$$

After some manipulations, one obtains

$$\omega^2 = \frac{2k}{5m} = 0.4\frac{k}{m} \tag{5.132}$$

which results in an acceptable (over-)estimation with respect to the exact value:

$$\omega^2 = 0.382\frac{k}{m} \tag{5.133}$$

The relative percentage difference between Equation 5.132 and Equation 5.133 is 4.7%, while the difference between the two frequencies is 2.3% (nearly half of the previous value).

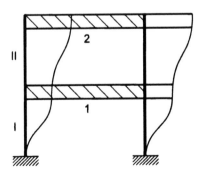

Figure 5.19

5.11 STODOLA–VIANELLO METHOD

In the previous section, an approximate method to evaluate the fundamental frequency of a discrete elastic system was described. In this section, an approximating procedure to obtain the first eigenvector will be presented. This method, known in numerical analysis as the **power method**, is an iterative procedure and was applied by Stodola and Vianello in this specific case.

By premultiplying Equation 5.127 by the inverse of the stiffness matrix $[K]^{-1}$ (also called the compliance matrix), one obtains

$$\left([1] - \omega^2 [K]^{-1}[M]\right)\{X\} = \{0\} \tag{5.134}$$

from which it follows that

$$\{X\} = \omega^2 [\mathcal{D}]\{X\} \tag{5.135}$$

where

$$[\mathcal{D}] = [K]^{-1}[M] \tag{5.136}$$

is called the **dynamic matrix**.

If $\{X_0\}$ denotes a first test-eigenvector, a new estimation can be obtained from Equation 5.135:

$$\{X_1\} = \omega^2 [\mathcal{D}]\{X_0\} \tag{5.137}$$

By applying Equation 5.135 again to $\{X_1\}$, an additional value for the eigenvector can be estimated:

$$\{X_2\} = \omega^2 [\mathcal{D}]\{X_1\} = \omega^4 [\mathcal{D}]^2 \{X_0\} \tag{5.138}$$

and so on through an iterative procedure:

$$\{X_h\} = \omega^{2h} [\mathcal{D}]^h \{X_0\} \tag{5.139}$$

It can be proved that, as the number of iterations h increases, $\{X_h\}$ converges to the first eigenvector. Expressing the test vector $\{X_0\}$ as a linear combination of the eigenvectors $\{\psi_i\}$, which are supposed to be known:

$$\{X_h\} = \omega_1^{2h} [\mathcal{D}]^h \sum_{i=1}^{n} x_i \{\psi_i\} = \omega_1^{2h} \sum_{i=1}^{n} x_i [\mathcal{D}]^h \{\psi_i\} = \sum_{i=1}^{n} x_i \left(\frac{\omega_1}{\omega_i}\right)^{2h} \{\psi_i\} \tag{5.140}$$

Assuming $\omega_1 < \omega_2 < \ldots < \omega_i < \ldots < \omega_n$ and representing the fundamental frequency ω_1, the following limit holds

$$\lim_{h \to \infty} \{X_h\} = x_1 \{\psi_1\} \tag{5.141}$$

Indeed, since the fundamental angular frequency ω_1 is unknown and it is a scalar quantity, it does not modify the direction of $\{X_b\}$. It can thus be assumed that

$$\{X_b\} = [\mathcal{D}]^b \{X_0\} \tag{5.142}$$

instead of Equation 5.139.

As an application example of Equations 5.141 and 5.142, consider again the shear-type frame of Figure 5.19. In this case, one obtains

$$[\mathcal{D}] = [K]^{-1}[M] = \begin{bmatrix} \dfrac{1}{k} & \dfrac{1}{k} \\ \dfrac{1}{k} & \dfrac{2}{k} \end{bmatrix} \begin{bmatrix} m & 0 \\ 0 & m \end{bmatrix} = \begin{bmatrix} \dfrac{m}{k} & \dfrac{m}{k} \\ \dfrac{m}{k} & \dfrac{2m}{k} \end{bmatrix} \tag{5.143}$$

By assuming $\{X_0\}^T = (1,2)$, the first four iterations are

$$\{X_1\} = \begin{bmatrix} 1 & 1 \\ 1 & 2 \end{bmatrix} \begin{bmatrix} 1 \\ 2 \end{bmatrix} = \begin{bmatrix} 3 \\ 5 \end{bmatrix} \tag{5.144a}$$

$$\{X_2\} = \begin{bmatrix} 1 & 1 \\ 1 & 2 \end{bmatrix} \begin{bmatrix} 3 \\ 5 \end{bmatrix} = \begin{bmatrix} 8 \\ 13 \end{bmatrix} \tag{5.144b}$$

$$\{X_3\} = \begin{bmatrix} 1 & 1 \\ 1 & 2 \end{bmatrix} \begin{bmatrix} 8 \\ 13 \end{bmatrix} = \begin{bmatrix} 21 \\ 34 \end{bmatrix} \tag{5.144c}$$

$$\{X_4\} = \begin{bmatrix} 1 & 1 \\ 1 & 2 \end{bmatrix} \begin{bmatrix} 21 \\ 34 \end{bmatrix} = \begin{bmatrix} 55 \\ 89 \end{bmatrix} \tag{5.144d}$$

Since the eigenvectors are determined unless a multiplicative constant, this can be written as

$$\{X_4\} = \begin{bmatrix} 0.6179 \\ 1 \end{bmatrix} \tag{5.145}$$

which coincides with the exact eigenvector, as will be shown in Section 6.8.

On the other hand, supposing that $\{X_0\}^T = (1,1)$ yields

$$\{X_1\} = \begin{bmatrix} 1 & 1 \\ 1 & 2 \end{bmatrix} \begin{bmatrix} 1 \\ 1 \end{bmatrix} = \begin{bmatrix} 2 \\ 3 \end{bmatrix} \tag{5.146a}$$

$$\{X_2\} = \begin{bmatrix} 1 & 1 \\ 1 & 2 \end{bmatrix} \begin{bmatrix} 2 \\ 3 \end{bmatrix} = \begin{bmatrix} 5 \\ 8 \end{bmatrix} \tag{5.146b}$$

$$\{X_3\} = \begin{bmatrix} 1 & 1 \\ 1 & 2 \end{bmatrix} \begin{bmatrix} 5 \\ 8 \end{bmatrix} = \begin{bmatrix} 13 \\ 21 \end{bmatrix} \tag{5.146c}$$

$$\{X_4\} = \begin{bmatrix} 1 & 1 \\ 1 & 2 \end{bmatrix} \begin{bmatrix} 13 \\ 21 \end{bmatrix} = \begin{bmatrix} 34 \\ 55 \end{bmatrix} \tag{5.146d}$$

and thus

$$\{X_4\} = \begin{bmatrix} 0.6182 \\ 1 \end{bmatrix} \tag{5.147}$$

which, once again, is extremely close to the exact eigenvector.

It is possible to evaluate the fundamental frequency approximately as

$$\omega_1^2 = \lim_{h \to \infty} \frac{|X_{h-1}|}{|X_h|} \tag{5.148}$$

Using Equations 5.146c and d, for instance, yields

$$\omega^2 \simeq \frac{|X_3|}{|X_4|} = \frac{\sqrt{13^2 + 21^2}}{\sqrt{34^2 + 55^2}} = 0.382$$

which coincides with the exact frequency but for the factor k/m.

Equation 5.148 can be justified by applying Equation 5.142 and expressing $\{X_0\}$ as a linear combination of the eigenvectors $\{\psi_i\}$, which are supposed to be known:

$$\{X_h\} = [\mathcal{D}]^h \{X_0\} = [\mathcal{D}]^h \sum_{i=1}^n x_i \{\psi_i\} = \sum_{i=1}^n x_i \left(\frac{1}{\omega_i}\right)^{2h} \{\psi_i\} \tag{5.149}$$

Equation 5.149 produces

$$\frac{|X_{h-1}|}{|X_h|} = \frac{\omega_1^{2h}}{\omega_1^{2(h-1)}} \frac{\left| \sum_{i=1}^n x_i \left(\frac{\omega_1}{\omega_i}\right)^{2(h-1)} \{\psi_i\} \right|}{\left| \sum_{i=1}^n x_i \left(\frac{\omega_1}{\omega_i}\right)^{2h} \{\psi_i\} \right|} \tag{5.150}$$

from which, since $\omega_1 < \omega_2 < \dots < \omega_i < \dots < \omega_n$, Equation 5.148 can be derived.

Once $\{\psi_1\}$ and ω_1 are evaluated, it is possible to estimate $\{\psi_2\}$ and ω_2 by means of a similar procedure. It will be sufficient to choose a test vector $\{X_1\}$ orthogonal to $\{\psi_1\}$ and to express it as a linear combination of the remaining eigenvectors $\{\psi_i\}$, $i=2, 3,\dots, n$. For instance, it can be imposed as

$$\{X_0\} = \{Z\} - \frac{\{\psi_1\}^T \{Z\}}{\{\psi_1\}^T \{\psi_1\}} \{\psi_1\} \tag{5.151}$$

where $\{Z\}$ is a generic vector. In fact, one obtains

$$\{\psi_1\}^T \{X_0\} = 0 \tag{5.152}$$

Chapter 6

Dynamics of continuous elastic systems

6.1 INTRODUCTION

Whereas Chapter 5 was devoted to the dynamics of discrete systems, that is, to those structures described by one or more degrees of freedom, the present chapter will consider the dynamics of continuous elastic systems, possessing infinite degrees of freedom. By considering a modal analysis similar to that introduced for discrete systems, it will be possible to emphasize the countable infinity of the natural modes, according to which the structural elements such as beams, membranes, or plates can vibrate.

Each natural mode is described by the corresponding deformed shape of the element, which is defined but for a multiplicative (either positive or negative) constant, and by the natural frequency. The countable infinity of the natural deformed shapes establishes the eigenfunction set related to the considered dynamic problem. As the eigenfunction order increases, so does the number of nodal points (beams) or lines (membranes and plates) where the transverse displacement of the deformed shape vanishes and remains null during the natural vibration. It is as if additional constraints are added as the eigenfunction order increases. As the eigenvalue order increases, so does the corresponding frequency, since it is related to a more constrained (and thus stiffer) system.

The eigenfunction set is orthogonal and complete. This means that any oscillation can be described by a finite summation or a series of eigenfunctions and that, if the system is perturbed according to a configuration proportional to one of the eigenfunctions with zero initial velocity, then the element will oscillate proportionally to the initial deformed shape once left free to oscillate.

In this latter case, the continuous system behaves like a harmonic system with a single degree of freedom. During the motion, the deformed shape always remains similar to itself and needs just a single parameter to be described in time. If applied to the first natural mode, this property allows the definition of an approximating method to determine the first natural frequency of the system (Rayleigh–Ritz method).

The first natural frequency represents the lowest and most dangerous one, since it refers to the least constrained system, with which the most relevant displacements and solicitations are associated. It is for this reason that approximating methods, such as Rayleigh–Ritz or Stodola–Vianello, focus their attention in particular on the first-order eigensolutions. It should be noted, however, that for very compliant structures, the determination of additional eigensolutions (e.g., the first 3 or 10) can be revealed as important.

Ultimately, concerning the resonance problem, the oscillator with n (or infinite) degrees of freedom must avoid interactions with forces possessing frequencies close to its natural ones, whereas for the simple oscillator, the dynamic loading must differ from only a single frequency. In other words, the frequency spectrum of the dynamic loading must present no peaks corresponding to the natural frequencies of the system, and in particular to the lowest frequencies, which represent the most dangerous ones, as already stated.

6.2 MODAL ANALYSIS OF DEFLECTED BEAMS

To analyze the free flexural oscillations of beams, let us consider the differential equation of the elastic line:

$$\frac{d^4 v}{dz^4} = \frac{q(z)}{EI} \tag{6.1}$$

and let us replace the transverse distributed load $q(z)$ with the force of inertia:

$$q_\mu(z) = -\mu \frac{d^2 v}{dt^2} \tag{6.2}$$

where μ denotes the linear density of the beam (mass per unit length).

By inserting Equation 6.2 into Equation 6.1 and considering that the transverse displacement v in the dynamic regime is a function of both variables z (axial coordinate) and t (time), it follows that

$$\frac{\partial^4 v}{\partial z^4} = -\frac{\mu}{EI} \frac{\partial^2 v}{\partial t^2} \tag{6.3}$$

Equation 6.3 represents an equation with separable variables, the solution being representable as the product of two different functions, each one with a single variable:

$$v(z,t) = \eta(z) f(t) \tag{6.4}$$

By substituting Equation 6.4 into Equation 6.3, it follows that

$$\frac{d^4 \eta}{dz^4} f + \frac{\mu}{EI} \eta \frac{d^2 f}{dt^2} = 0 \tag{6.5}$$

Dividing Equation 6.5 by the product ηf yields

$$-\frac{\left(\dfrac{d^2 f}{dt^2} \right)}{f} = \frac{EI}{\mu} \frac{\left(\dfrac{d^4 \eta}{dz^4} \right)}{\eta} = \omega^2 \tag{6.6}$$

where ω^2 represents a positive constant, the first two terms of Equation 6.6 being, at most, functions of the time t and of the coordinate z, respectively.

From Equation 6.6, two ordinary differential equations follow:

$$\frac{d^2 f}{dt^2} + \omega^2 f = 0 \tag{6.7a}$$

$$\frac{d^4\eta}{dz^4} - \alpha^4\eta = 0 \tag{6.7b}$$

with

$$\alpha = \sqrt[4]{\frac{\mu\omega^2}{EI}} \tag{6.8}$$

Whereas Equation 6.7a is the equation of the harmonic oscillator, with the well-known complete integral:

$$f(t) = A\cos \omega t + B\sin \omega t \tag{6.9a}$$

Equation 6.7b presents a solution in the form:

$$\eta(z) = C\cos \alpha z + D\sin \alpha z + E\cosh \alpha z + F\sinh \alpha z \tag{6.9b}$$

As will be shown later, the constants A, B may be determined on the basis of the initial conditions, whereas the constants C, D, E, F may be determined on the basis of the boundary conditions. However, the parameter ω remains, at present, undetermined and, therefore, so does the parameter α, according to Equation 6.8. This represents the eigenvalue of the problem from the mathematical standpoint, which is to say, the angular frequency of the system from a mechanical point of view. It will be demonstrated later how the angular frequency ω may also be obtained on the basis of the boundary conditions. An infinite number of eigenvalues ω_i will be obtained, and thus α_i, as well as an infinite number of eigenfunctions f_i, and thus η_i. On the basis of the principle of superposition, the complete integral of differential Equation 6.3 may therefore be expressed as

$$v(z,t) = \sum_{i=1}^{\infty} \eta_i(z)f_i(t) \tag{6.10}$$

with

$$f_i(t) = A_i\cos\omega_i t + B_i\sin \omega_i t \tag{6.11a}$$

$$\eta_i(z) = C_i \cos\alpha_i z + D_i\sin\alpha_i z + E_i\cosh\alpha_i z + F_i\sinh\alpha_i z \tag{6.11b}$$

The eigenfunctions η_i are **orthonormal** functions. Equation 6.7b can, in fact, be written for two different eigensolutions:

$$\eta_j^{IV} = \alpha_j^4\eta_j \tag{6.12a}$$

$$\eta_k^{IV} = \alpha_k^4\eta_k \tag{6.12b}$$

By multiplying Equation 6.12a by η_k, Equation 6.12b by η_j, and integrating over the length of the beam, it follows that

$$\int_0^l \eta_k \eta_j^{IV} dz = \alpha_j^4 \int_0^l \eta_k \eta_j \, dz \tag{6.13a}$$

$$\int_0^l \eta_j \eta_k^{IV} dz = \alpha_k^4 \int_0^l \eta_j \eta_k \, dz \tag{6.13b}$$

Integrating by parts the left-hand sides, the previous equations become

$$[\eta_k \eta_j''']_0^l - [\eta_k' \eta_j'']_0^l + \int_0^l \eta_k'' \eta_j'' dz = \alpha_j^4 \int_0^l \eta_k \eta_j \, dz \tag{6.14a}$$

$$[\eta_j \eta_k''']_0^l - [\eta_j' \eta_k'']_0^l + \int_0^l \eta_j'' \eta_k'' dz = \alpha_k^4 \int_0^l \eta_j \eta_k \, dz \tag{6.14b}$$

When both the beam ends are constrained by a built-in support ($\eta = \eta' = 0$), by a hinge ($\eta = \eta'' = 0$), or by a double rod ($\eta' = \eta''' = 0$), or when the beam is unconstrained ($\eta'' = \eta''' = 0$), the quantities within square brackets vanish. Subtracting member by member, we thus have

$$(\alpha_j^4 - \alpha_k^4) \int_0^l \eta_j \eta_k \, dz = 0 \tag{6.15}$$

from which the condition of orthonormality follows:

$$\int_0^l \eta_j \eta_k \, dz = \delta_{jk} \tag{6.16}$$

δ_{jk} being the **Kronecker symbol**. When the eigenvalues are distinct, the integral of the product of the corresponding eigenfunctions vanishes. On the other hand, when the indices j and k coincide, the condition of normality reminds us that the eigenfunctions are defined but for a factor of proportionality, as follows from the homogeneity of Equation 6.7b.

As already mentioned, the constants A_i, B_i of Equation 6.9a are determined by the initial conditions:

$$v(z,0) = v_0(z) \tag{6.17a}$$

$$\frac{\partial v}{\partial t}(z,0) = \dot{v}_0(z) \tag{6.17b}$$

which, on the basis of Equations 6.10 and 6.11a, become

$$\sum_{i=1}^{\infty} A_i \eta_i(z) = v_0(z) \tag{6.18a}$$

$$\sum_{i=1}^{\infty} \omega_i B_i \eta_i(z) = \dot{v}_0(z) \tag{6.18b}$$

Multiplying by any desired eigenfunction η_j and integrating over the complete length of the beam, it follows that

$$\sum_{i=1}^{\infty} A_i \int_0^l \eta_i \eta_j dz = \int_0^l \eta_j \upsilon_0 dz \tag{6.19a}$$

$$\sum_{i=1}^{\infty} \omega_i B_i \int_0^l \eta_i \eta_j dz = \int_0^l \eta_j \dot{\upsilon}_0 dz \tag{6.19b}$$

Taking the condition of orthonormality (Equation 6.16) into account, one obtains

$$A_j = \int_0^l \eta_j \upsilon_0 dz \tag{6.20a}$$

$$B_j = \frac{1}{\omega_j} \int_0^l \eta_j \dot{\upsilon}_0 dz \tag{6.20b}$$

When the system is initially perturbed, by assigning to the beam a deformation that is proportional to one of the eigenfunctions, with initial zero velocity, the beam, once left free to oscillate, continues to do so proportionally to the initial deformed configuration. In this case, we have

$$\upsilon_0(z) = a\eta_i(z) \tag{6.21a}$$

$$\dot{\upsilon}_0(z) = 0 \tag{6.21b}$$

where a is an arbitrary constant of proportionality. Equations 6.20 then give

$$A_j = a\delta_{ij} \tag{6.22a}$$

$$B_j = 0 \tag{6.22b}$$

and hence the complete integral (Equation 6.10) takes the following form:

$$\upsilon(z,t) = a\eta_i(z)\cos\omega_i t = \upsilon_0(z)\cos\omega_i t \tag{6.23}$$

The beam therefore oscillates proportionally to the initial deformation and with an angular frequency that corresponds to the same eigenfunction. These oscillations are called the **natural modes of vibration of the system.**

6.3 DIFFERENT BOUNDARY CONDITIONS FOR THE SINGLE BEAM

6.3.1 Simply supported beam

Regarding, in particular, a beam of length l supported at both ends (Figure 6.1a), the boundary conditions imposed on Equation 6.11b, corresponding to the end A, are written as

$$\eta(0) = C + E = 0 \tag{6.24a}$$

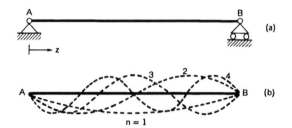

Figure 6.1

$$\eta''(0) = -\alpha^2(C - E) = 0 \tag{6.24b}$$

whence it is obtained:

$$C = E = 0 \tag{6.25}$$

On the other hand, the boundary conditions corresponding to the end B yield

$$\eta(l) = D\sin\alpha l + F\sinh\alpha l = 0 \tag{6.26a}$$

$$\eta''(l) = -\alpha^2(D\sin\alpha l - F\sinh\alpha l) = 0 \tag{6.26b}$$

from which it follows that

$$D\sin\alpha l = 0 \tag{6.27a}$$

$$F\sinh\alpha l = 0 \tag{6.27b}$$

From Equation 6.27b, one obtains

$$F = 0 \tag{6.28}$$

whereas from Equation 6.27a, once the trivial solution $D = 0$ has been ruled out, it follows that

$$\alpha = n\frac{\pi}{l} \tag{6.29}$$

where n is a natural number.
   Equation 6.8 provides

$$\alpha_n^4 = n^4\frac{\pi^4}{l^4} = \frac{\mu\omega_n^2}{EI} \tag{6.30}$$

whereby the **natural angular frequencies** of the system follow:

$$\omega_n = n^2\frac{\pi^2}{l^2}\sqrt{\frac{EI}{\mu}} \tag{6.31}$$

and hence its natural periods:

$$T_n = \frac{2\pi}{\omega_n} = \frac{2l^2}{\pi n^2}\sqrt{\frac{\mu}{EI}} \tag{6.32}$$

The normalized eigenfunctions are thus represented by the following sinusoidal succession (Figure 6.1b):

$$\eta_n(z) = \sqrt{\frac{2}{l}}\,\sin n\pi\frac{z}{l} \tag{6.33}$$

6.3.2 Cantilever beam

In the case of the cantilever beam of Figure 6.2, the boundary conditions imposed on Equation 6.11b are

$$\eta(0) = C + E = 0 \tag{6.34a}$$

$$\eta'(0) = \alpha(D + F) = 0 \tag{6.34b}$$

$$\eta''(l) = -\alpha^2\left(C\cos\alpha l + D\sin\alpha l - E\cosh\alpha l - F\sinh\alpha l\right) = 0 \tag{6.34c}$$

$$\eta'''(l) = \alpha^3\left(C\sin\alpha l - D\cos\alpha l + E\sinh\alpha l + F\cosh\alpha l\right) = 0 \tag{6.34d}$$

Whereas from the first two equations, we have

$$E = -C, \quad F = -D \tag{6.35}$$

from the last two, it follows that

$$C(\cos\alpha l + \cosh\alpha l) + D(\sin\alpha l + \sinh\alpha l) = 0 \tag{6.36a}$$

$$C(\sin\alpha l - \sinh\alpha l) - D(\cos\alpha l + \cosh\alpha l) = 0 \tag{6.36b}$$

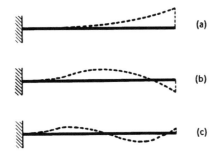

(a)

(b)

(c)

Figure 6.2

The system of algebraic Equations 6.36 give, on the other hand, a solution that is different from the trivial one, if and only if the determinant of the coefficients is null:

$$(\cos\alpha l + \cosh\alpha l)^2 + (\sin^2\alpha l - \sinh^2\alpha l) = 0 \tag{6.37}$$

Solving Equation 6.37, the transcendental equation that provides the set or *spectrum* of eigenvalues is obtained:

$$\cos\alpha_n l \cosh\alpha_n l = -1 \tag{6.38}$$

The first three roots of Equation 6.38 are

$$\alpha_1 l = 1.875, \quad \alpha_2 l = 4.694, \quad \alpha_3 l = 7.885$$

The natural angular frequencies and periods of the cantilever beam are given by

$$\omega_n = \alpha_n^2 \sqrt{\frac{EI}{\mu}} \tag{6.39a}$$

$$T_n = \frac{2\pi}{\alpha_n^2} \sqrt{\frac{\mu}{EI}} \tag{6.39b}$$

The fundamental period thus results as

$$T_1 = 1.79 \, l^2 \sqrt{\frac{\mu}{EI}} \tag{6.40}$$

and is approximately three times higher than that of the supported beam provided by Equation 6.32:

$$T_1 = 0.64 \, l^2 \sqrt{\frac{\mu}{EI}} \tag{6.41}$$

The first three eigenfunctions are depicted in Figure 6.2.

6.3.3 Rope in tension

In the case of a rope in tension (Figure 6.1), the flexural stiffness EI tends to zero. The bending moment, in the case of large displacements, is therefore given by the product of the axial force by the transverse displacement:

$$M = -Nv \tag{6.42}$$

By applying the indefinite equation of equilibrium:

$$\frac{d^2 M}{dz^2} = -q(z) \tag{6.43}$$

it is possible to obtain the equivalent transverse load, so that the equation of dynamic equilibrium to vertical translation is as follows:

$$N\frac{\partial^2 v}{\partial z^2} = \mu\frac{\partial^2 v}{\partial t^2} \tag{6.44}$$

Equation 6.44 transforms into the wave equation:

$$\frac{\partial^2 v}{\partial t^2} = c^2\frac{\partial^2 v}{\partial z^2} \tag{6.45}$$

where

$$c^2 = \frac{N}{\mu} \tag{6.46}$$

is the square of the velocity of the transverse wave in the rope in tension.

Equation 6.45 is formally identical to the equation of longitudinal waves in elastic bars. If, in the indefinite equation of equilibrium:

$$EA\frac{d^2 u}{dx^2} = -\mathcal{F}_x(x) \tag{6.47}$$

the distributed longitudinal force $\mathcal{F}_x(x)$ is replaced by the force of inertia $-\mu(\partial^2 u/\partial t^2)$, it follows that

$$EA\frac{\partial^2 u}{\partial x^2} = \mu\frac{\partial^2 u}{\partial t^2} \tag{6.48}$$

and thus

$$\frac{\partial^2 u}{\partial t^2} = c^2\frac{\partial^2 u}{\partial x^2} \tag{6.49}$$

where

$$c^2 = \frac{EA}{\mu} \tag{6.50}$$

is the square of the velocity of the longitudinal wave in the elastic bar.

6.3.4 Unconstrained beam

In the case of an unconstrained beam, the following boundary conditions are imposed:

$$\eta''(0) = 0 \tag{6.51a}$$

$$\eta'''(0) = 0 \tag{6.51b}$$

$$\eta''(l) = 0 \tag{6.51c}$$

$$\eta'''(l) = 0 \tag{6.51d}$$

Setting the general solution (Equation 6.9b) in the equivalent form:

$$\eta(z) = C_1(\cos\alpha z + \cosh\alpha z) + C_2(\cos\alpha z - \cosh\alpha z)$$
$$+ C_3(\sin\alpha z + \sinh\alpha z) + C_4(\sin\alpha z - \sinh\alpha z) \tag{6.52}$$

Equations 6.51a and b give $C_2=C_4=0$, whereas Equations 6.51c and d provide the linear and homogeneous algebraic equations:

$$C_1(-\cos\alpha l + \cosh\alpha l) + C_3(-\sin\alpha l + \sinh\alpha l) = 0 \tag{6.53a}$$

$$C_1(\sin\alpha l + \sinh\alpha l) + C_3(-\cos\alpha l + \cosh\alpha l) = 0 \tag{6.53b}$$

By imposing the coefficient determinant equal to zero, one obtains

$$(-\cos\alpha l + \cosh\alpha l)^2 - (\sinh^2\alpha l - \sin^2\alpha l) = 0 \tag{6.54}$$

Recalling that

$$\cosh^2\alpha l - \sinh^2\alpha l = 1 \tag{6.55a}$$

$$\cos^2\alpha l + \sin^2\alpha l = 1 \tag{6.55b}$$

the **eigenfrequency equation** can be written as

$$\cos\alpha_n l \ \cosh\alpha_n l = 1 \tag{6.56}$$

The first three roots of Equation 6.56 are

$$\alpha_1 l = 4.730, \quad \alpha_2 l = 7.853, \quad \alpha_3 l = 10.996 \tag{6.57}$$

Equation 6.8 gives the relevant angular frequencies ω_1, ω_2, ω_3.

On substitution of Equation 6.57 into Equations 6.53, the ratio C_1/C_3 is obtained for the relevant vibration modes; in the same way, Equation 6.52, with $C_2=C_4=0$, provides the relevant natural deformed configurations. Figure 6.3 represents the first three natural vibration modes of a free beam.

6.3.5 Double clamped beam

In the case of a beam clamped at both ends, the following boundary conditions must be assigned:

$$\eta(0) = 0 \tag{6.58a}$$

$$\eta'(0) = 0 \tag{6.58b}$$

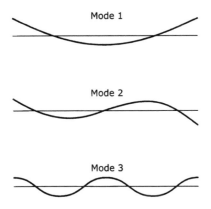

Figure 6.3

$$\eta(l) = 0 \tag{6.58c}$$

$$\eta'(l) = 0 \tag{6.58d}$$

The first two conditions are satisfied if $C_1 = C_3 = 0$ in the general Equation 6.52. From the two remaining conditions, one obtains

$$C_2\left(\cos\alpha l - \cosh\alpha l\right) + C_4\left(\sin\alpha l - \sinh\alpha l\right) = 0 \tag{6.59a}$$

$$C_2\left(\sin\alpha l + \sinh\alpha l\right) + C_4\left(\cos\alpha l + \cosh\alpha l\right) = 0 \tag{6.59b}$$

Setting the coefficient determinant to zero, the same **eigenfrequency** Equation 6.56 applying to the unconstrained beam is obtained. Figure 6.4 represents the first three natural vibration modes of a double clamped beam.

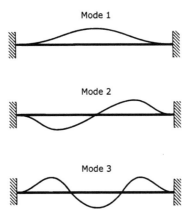

Figure 6.4

6.3.6 Clamped–hinged beam

In the case of a beam clamped at one end and hinged at the other one, the following boundary conditions are written as

$$\eta(0) = 0 \tag{6.60a}$$

$$\eta'(0) = 0 \tag{6.60b}$$

$$\eta(l) = 0 \tag{6.60c}$$

$$\eta''(l) = 0 \tag{6.60d}$$

The first two conditions are satisfied if $C_1 = C_3 = 0$ in the general Equation 6.52. From the two remaining conditions, the following **eigenfrequency equation** is obtained:

$$\tan \alpha l = \tanh \alpha l \tag{6.61}$$

The first three roots of this equation are

$$\alpha_1 l = 3.927, \quad \alpha_2 l = 7.069, \quad \alpha_3 l = 10.210 \tag{6.62}$$

Compared with the previous case of a double clamped beam, this structure appears coherently less stiff and, hence, with lower natural frequencies than Equation 6.57. Figure 6.5 represents the first three natural vibration modes of a clamped–hinged beam.

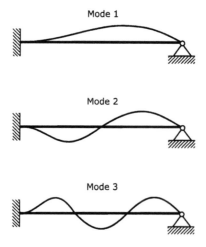

Figure 6.5

6.4 CONTINUOUS BEAM ON THREE OR MORE SUPPORTS

Let us consider an n-span continuous beam with uniform stiffness EI, constrained by $(n-1)$ intermediate supports (Figure 6.6). Let l_1, l_2,..., l_n be the lengths of the single consecutive spans. Supposing the origin of each local system of reference to be coincident with the left end of the respective span, let us implement Equation 6.9b for each single deformed beam. Since $\eta_i(0)=0$, for the generic ith span one has

$$\eta_i(z) = C_i(\cos\alpha z - \cosh\alpha z) + D_i\sin\alpha z + F_i\sinh\alpha z \qquad (6.63)$$

where C_i, D_i, and F_i are arbitrary constants. The derivatives of Equation 6.63 are

$$\eta_i'(z) = -C_i\alpha(\sin\alpha z + \sinh\alpha z) + D_i\alpha\cos\alpha z + F_i\alpha\cosh\alpha z \qquad (6.64a)$$

$$\eta_i''(z) = -C_i\alpha^2(\cos\alpha z + \cosh\alpha z) - D_i\alpha^2\sin\alpha z + F_i\alpha^2\sinh\alpha z \qquad (6.64b)$$

For $z=0$, Equations 6.64 become

$$\eta_i'(0) = \alpha(D_i + F_i) \qquad (6.65a)$$

$$\eta_i''(0) = -2\alpha^2 C_i \qquad (6.65b)$$

Considering the boundary conditions at the right end of the ith span, we have

$$\eta_i(l_i) = 0 \qquad (6.66a)$$

$$\eta_i'(l_i) = \eta_{i+1}'(0) \qquad (6.66b)$$

$$\eta_i''(l_i) = \eta_{i+1}''(0) \qquad (6.66c)$$

because the first derivatives of transverse displacement represent the rotation angles of the section (and also of the axis of the beam), whereas the second derivatives represent the bending moments, but for one proportional factor. Equations 6.63 and 6.64 provide the following conditions:

$$C_i(\cos\alpha l_i - \cosh\alpha l_i) + D_i\sin\alpha l_i + F_i\sinh\alpha l_i = 0 \qquad (6.67a)$$

$$-C_i(\sin\alpha l_i + \sinh\alpha l_i) + D_i\cos\alpha l_i + F_i\cosh\alpha l_i = D_{i+1} + F_{i+1} \qquad (6.67b)$$

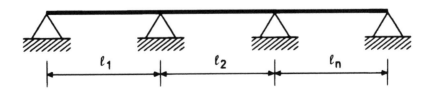

Figure 6.6

$$C_i\left(\cos\alpha l_i + \cosh\alpha l_i\right) + D_i \sin\alpha l_i - F_i \sinh\alpha l_i = 2C_{i+1} \tag{6.67c}$$

By adding and subtracting Equations 6.67a and c, it follows that

$$C_i \cos\alpha l_i + D_i \sin\alpha l_i = C_i \cosh\alpha l_i - F_i \sinh\alpha l_i = C_{i+1}$$

which gives

$$D_i = \frac{C_{i+1} - C_i \cos\alpha l_i}{\sin\alpha l_i} \tag{6.68a}$$

$$F_i = \frac{-C_{i+1} + C_i \cos\alpha l_i}{\sin\alpha l_i} \tag{6.68b}$$

and hence

$$D_i + F_i = C_i\left(\coth\alpha l_i - \cot\alpha l_i\right) - C_{i+1}\left(\operatorname{cosech}\alpha l_i - \operatorname{cosec}\alpha l_i\right) \tag{6.69}$$

Using the notation

$$\coth\alpha l_i - \cot\alpha l_i = \varphi_i \tag{6.70a}$$

$$\operatorname{cosech}\alpha l_i - \operatorname{cosec}\alpha l_i = \psi_i \tag{6.70b}$$

Equation 6.69 can be rewritten in the following form:

$$D_i + F_i = C_i\varphi_i - C_{i+1}\psi_i \tag{6.71}$$

In the same way, for the $(i+1)$th span, one obtains

$$D_{i+1} + F_{i+1} = C_{i+1}\varphi_{i+1} - C_{i+2}\psi_{i+1} \tag{6.72}$$

Substituting Equations 6.68 and 6.72 into Equation 6.67b yields

$$C_i\psi_i - C_{i+1}\left(\varphi_i + \varphi_{i+1}\right) + C_{i+2}\psi_{i+1} = 0 \tag{6.73}$$

Writing similar equations for each intermediate support provides the following system of $(n-1)$ equations:

$$-C_2\left(\varphi_1 + \varphi_2\right) + C_3\psi_2 = 0$$

$$C_2\psi_2 - C_3\left(\varphi_2 + \varphi_3\right) + C_4\psi_3 = 0 \tag{6.74}$$

$$\vdots$$

$$C_{n-1}\psi_{n-1} - C_n\left(\varphi_{n-1} + \varphi_n\right) = 0$$

By setting the determinant of coefficients to zero, the equation of natural frequencies for the continuous beam on multiple supports is obtained.

In the case of a beam on three supports, for instance, only one of Equations 6.74 remains valid, and the equation of frequencies becomes the following:

$$\varphi_1 + \varphi_2 = 0 \tag{6.75}$$

The natural frequencies will be given by the following condition:

$$\coth \alpha l_1 - \cot \alpha l_1 = -\coth \alpha l_2 + \cot \alpha l_2 \tag{6.76}$$

This expression represents a transcendental equation in the unknown α, and in the particular case of $l_1 = (4/3)l_2$, it provides the following eigenvalues:

$$\alpha_1 l_1 = 3.416, \ \alpha_2 l_1 = 4.787, \ \alpha_3 l_1 = 6.690 \tag{6.77}$$

Therefore, the first three frequencies have a ratio equal to 1:1.96:3.82.

6.5 METHOD OF APPROXIMATION OF RAYLEIGH–RITZ

A vibrating system with one degree of freedom, with no damping, has a natural angular frequency equal to $\omega = \sqrt{k/m}$, where k represents the stiffness and m is the mass. This expression can be applied to any structure vibrating proportionally to a known elastic deformed shape. This is the principle of Rayleigh–Ritz approximation method, which will be discussed in this section.

The total energy of a free-vibrating system remains constant in time if damping forces are absent. In case of a spring–mass elementary system one has

$$x(t) = x_0 \sin \omega t \tag{6.78a}$$

$$\dot{x}(t) = x_0 \omega \cos \omega t \tag{6.78b}$$

The elastic potential energy of the spring is

$$W(t) = \frac{1}{2} k x^2 = \frac{1}{2} k x_0^2 \sin^2 \omega t \tag{6.79}$$

whereas the mass kinetic energy can be expressed as

$$T(t) = \frac{1}{2} m \dot{x}^2 = \frac{1}{2} m x_0^2 \omega^2 \cos^2 \omega t \tag{6.80}$$

Since the maximum values for W and T must coincide:

$$\frac{1}{2} k x_0^2 = \frac{1}{2} m \omega^2 x_0^2 \tag{6.81}$$

the well-known expression for the angular frequency ω is derived from the previous equation.

Let us consider, for instance, a simply supported beam with nonuniform stiffness (Figure 6.7). Such a beam has infinite degrees of freedom. To apply the Rayleigh–Ritz

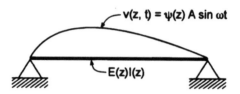

Figure 6.7

method, the elastic deformed shape that the beam will assume during its fundamental mode of vibration must be defined:

$$v(z,t) = \psi(z) A \sin \omega t \tag{6.82}$$

The elastic potential energy of the system is

$$W = \frac{1}{2}\int_0^l EI(z)\left(\frac{\partial^2 v}{\partial z^2}\right)^2 dz \tag{6.83}$$

and its kinetic energy is

$$T = \frac{1}{2}\int_0^l \mu(z)\dot{v}^2 dz \tag{6.84}$$

The maximum values that Equations 6.83 and 6.84 can assume, according to Equation 6.82, are

$$W_{max} = \frac{1}{2}A^2 \int_0^l EI(z)(\psi'')^2\, dz \tag{6.85a}$$

$$T_{max} = \frac{1}{2}A^2\omega^2 \int_0^l \mu(z)\, \psi^2 dz \tag{6.85b}$$

From Equations 6.85, it follows that the approximating value for the fundamental angular frequency immediately becomes

$$\omega^2 = \frac{\displaystyle\int_0^l EI(z)(\psi'')^2\, dz}{\displaystyle\int_0^l \mu(z)\psi^2 dz} \tag{6.86}$$

The numerator of Equation 6.86 represents a generalized stiffness, whereas the denominator represents a generalized mass. The stiffness and the linear density values are integrated after being weighted by the square values of the second-order derivative ψ'' and of the function ψ itself, respectively.

The Rayleigh–Ritz method gives the fundamental frequency of a continuous vibrating system with an accuracy that depends entirely on the $\psi(z)$-shape function. In principle, any shape can be chosen, provided that boundary conditions are satisfied. Moreover, any shape that is different from the exact one would require additional constraints: these would increase the stiffness of the system and hence the evaluated frequency. Therefore, it can be said that the real shape function will provide the lowest obtainable frequency by the Rayleigh–Ritz method. When having to choose between different approximating results, the lowest frequency always represents the best estimate.

Let us consider, for example, a simply supported beam with uniform stiffness and linear density variation. As a first approximation, let us assume a parabolic elastic deformed shape:

$$\psi(z) = \left(\frac{z}{l}\right)\left(\frac{z}{l} - 1\right) \tag{6.87}$$

from which

$$\psi''(z) = \frac{2}{l^2} \tag{6.88}$$

Applying Equation 6.86 yields

$$\omega^2 = \frac{EI\int_0^l \frac{4}{l^4}\,dz}{\mu\int_0^l \frac{z^2}{l^2}\left(\frac{z}{l} - 1\right)^2 dz} = 120\frac{EI}{\mu l^4} \tag{6.89}$$

By assuming the elastic deformed shape as a sinusoidal function:

$$\psi(z) = \sin\pi\frac{z}{l} \tag{6.90}$$

one has

$$\omega^2 = \pi^4\frac{EI}{\mu l^4} \tag{6.91}$$

which actually matches the square of the real fundamental frequency, being Equation 6.90 the correct shape function. Note that $\pi^4 < 120$.

For beams that show only flexural vibrations, and hence only transverse displacements, the inertial forces (which are equal to the product of masses by accelerations) appear to be proportional to the transverse displacements. A very good approximation would, therefore, be that of assuming, as shape function $\bar{\psi}(z)$, the elastic deformed shape generated by the transverse distributed load:

$$q(z) = \mu(z)\psi(z)$$

where $\psi(z)$ represents a reasonable approximation of the right shape. Since this calculation is often revealed to be complex, it is preferred simply to consider $q(z) = \mu(z)$. The frequency is therefore evaluated on the basis of the elastic deformed shape $\psi_\mu(z)$, caused exclusively by the weight of the beam.

According to Clapeyron's theorem, it follows that

$$W_{max} = \frac{1}{2}\int_0^l \mu(z)\,\psi_\mu(z)\,dz \tag{6.92a}$$

whereas, from Equation 6.85b, one has

$$T_{max} = \frac{1}{2}\omega^2 \int_0^l \mu(z)\ \psi_\mu^2(z)dz \qquad (6.92b)$$

Equaling the two last expressions yields the square of the angular frequency:

$$\omega^2 = \frac{\int_0^l \mu(z)\psi_\mu(z)dz}{\int_0^l \mu(z)\psi_\mu^2(z)dz} \qquad (6.93)$$

In addition to horizontal beams, Equation 6.93 can be accurately applied to rotating-node frames, where the components of rigid motion are absent. On the other hand, in the case of the portal frame of Figure 6.8, which is a classic example of a translating-node frame, the load distribution to be assumed to approximate the fundamental frequency is the skew-symmetric one (Figure 6.8a), whereas the symmetrical load distribution (Figure 6.8b), which provides a stiffer scheme, will approximate a natural frequency of higher order. It should be noted that, in the first scheme, the kinetic energy contribution due to the horizontal rigid motion of the girder is significant and cannot be neglected.

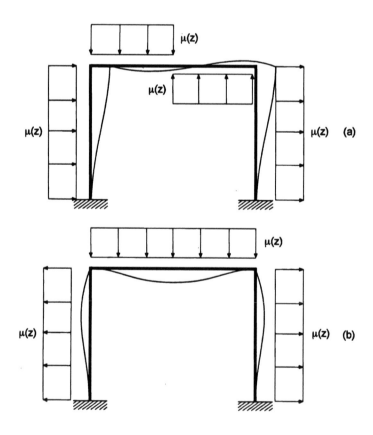

Figure 6.8

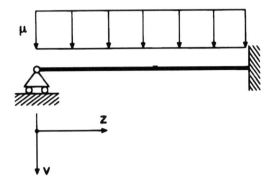

Figure 6.9

As an application example of Equation 6.93, let us consider the supported and clamped beam in Figure 6.9. The bending moment due to the weight μ is

$$M(z) = -\frac{1}{2}\mu z\left(z - \frac{3}{4}l\right) \tag{6.94}$$

The equation of the elastic line,

$$\frac{d^2v}{dz^2} = \frac{\mu z}{2EI}\left(z - \frac{3}{4}l\right) \tag{6.95}$$

once integrated, gives

$$\frac{dv}{dz} = \frac{\mu}{2EI}\frac{z^3}{3} - \frac{3\mu l}{8EI}\frac{z^2}{2} + C_1 \tag{6.96a}$$

$$v = \frac{\mu}{6EI}\frac{z^4}{4} - \frac{3\mu l}{16EI}\frac{z^3}{3} + C_1 z + C_2 \tag{6.96b}$$

where the arbitrary constants C_1 and C_2 can be obtained by the boundary conditions:

$$v(0) = C_2 = 0 \tag{6.97a}$$

$$v(l) = \frac{\mu l^4}{24EI} - \frac{\mu l^4}{16EI} + C_1 l = 0 \tag{6.97b}$$

The result is

$$C_1 = \frac{\mu l^3}{48EI} \tag{6.98a}$$

$$C_2 = 0 \tag{6.98b}$$

Furthermore, it is possible to verify that

$$v'(l) = \frac{\mu l^3}{6EI} - \frac{3\mu l^3}{16EI} + \frac{\mu l^3}{48EI} = 0 \qquad (6.99)$$

Concerning the application of Equation 6.93, it follows that

$$\psi_\mu(z) = v(z) = \frac{\mu}{24EI} z^4 - \frac{\mu l}{16EI} z^3 + \frac{\mu l^3}{48EI} z \qquad (6.100a)$$

$$\psi_\mu^2(z) = v^2(z) = \frac{\left(2\mu z^4 - 3\mu l z^3 + \mu l^3 z\right)^2}{\left(48EI\right)^2} \qquad (6.100b)$$

After some manipulations, one obtains

$$\int_0^l \psi_\mu(z)\,dz = \frac{1}{48EI}\left(\frac{2}{5}\mu l^5 - \frac{3}{4}\mu l^5 + \frac{1}{2}\mu l^5\right) = \frac{1}{320}\frac{\mu l^5}{EI} \qquad (6.101a)$$

$$\int_0^l \psi_\mu^2(z)\,dz = \frac{1}{\left(48EI\right)^2}\int_0^l \left(4\mu^2 z^8 + 9\mu^2 l^2 z^6 + \mu^2 l^6 z^2 - 12\mu^2 l z^7 + 4\mu^2 l^3 z^5 - 6\mu^2 l^4 z^4\right)dz$$

$$= \frac{\mu^2 l^9}{\left(48EI\right)^2}\left(\frac{4}{9} + \frac{9}{7} + \frac{1}{3} - \frac{12}{8} + \frac{4}{6} - \frac{6}{5}\right) = \frac{\mu^2 l^9}{\left(48EI\right)^2}\frac{19}{630} \qquad (6.101b)$$

Equation 6.93, on the basis of Equations 6.101 and assuming a uniform linear density, gives

$$\omega^2 = \frac{\int_0^l \psi_\mu(z)\,dz}{\int_0^l \psi_\mu^2(z)\,dz} = \frac{\dfrac{1}{320}\dfrac{\mu l^5}{EI}}{\dfrac{19}{630}\dfrac{\mu^2 l^9}{\left(48EI\right)^2}} \cong 238.74\frac{EI}{\mu l^4} \qquad (6.102)$$

On the other hand, with reference to the first expression of Equation 6.62, the result would be

$$\omega^2 = (3.927)^4 \frac{EI}{\mu l^4} = 237.82\frac{EI}{\mu l^4} \qquad (6.103)$$

It can be seen that the Rayleigh–Ritz approximation method, applied according to Equation 6.93, gives a very good excess approximation, with a deviation of just 0.387%. The percentage deviation on simple frequency values is equal to about one half that of the previous one (i.e., lower than 2‰).

6.6 DYNAMICS OF BEAM SYSTEMS

If the distributed masses of a plane frame are assumed to be concentrated in the nodes, the equilibrium Equation 2.41 can be transformed into the **equation of free motion**:

$$[K]\{\delta\} + [M]\{\ddot{\delta}\} = \{0\} \qquad (6.104)$$

where the **mass matrix** $[M]$ is a diagonal matrix that shows an equivalent mass corresponding to each nodal translation and a null mass corresponding to each nodal rotation. The equivalent mass can, for instance, be calculated by adding the weights, divided by two, of the beams that converge at the node. However, it should be noted that, if one wishes to take into account the actual mass distribution along beams and columns, it is always possible to apply the **finite element method** already introduced in Chapter 4, subdividing each beam or column into one or more finite elements. According to this method, of course, the mass matrix would no longer be diagonal.

On the other hand, in practice, the procedure often adopted is to carry out a further simplification and approximation with respect to the two methods just outlined, which is to say the finite element method and the **method of concentrated masses in the nodes**. Since the moment of inertia of the horizontal beams is usually much greater than that of columns, the horizontal cross members are considered to be infinitely rigid and the masses concentrated just in the cross elements. This scheme, previously introduced (Figure 2.16), called the **shear-type frame**, and the relevant procedure that will be described, is known as the **method of rigid cross members**.

Let us consider, as an example, the two-story frame of Figure 5.19. Denoting the masses of the two cross members as m_1, m_2, and designating the shear stiffnesses of the two uprights as k_1, k_2:

$$k_i = \frac{24EI_i}{h_i^3} \tag{6.105}$$

the two motion equations of the cross elements are

$$m_1\ddot{\delta}_1 = -k_1\delta_1 + k_2(\delta_2 - \delta_1) \tag{6.106a}$$

$$m_2\ddot{\delta}_2 = -k_2(\delta_2 - \delta_1) \tag{6.106b}$$

where δ_1 and δ_2 are the horizontal translations of the cross elements. Equations 6.106 may be written in matrix form:

$$\begin{bmatrix} (k_1 + k_2) & -k_2 \\ -k_2 & k_2 \end{bmatrix} \begin{bmatrix} \delta_1 \\ \delta_2 \end{bmatrix} + \begin{bmatrix} m_1 & 0 \\ 0 & m_2 \end{bmatrix} \begin{bmatrix} \ddot{\delta}_1 \\ \ddot{\delta}_2 \end{bmatrix} = \begin{bmatrix} 0 \\ 0 \end{bmatrix} \tag{6.107}$$

To study the **free motion** of the system with two degrees of freedom, let us suppose that the coordinates δ_1, δ_2 of the system vary harmonically in time with equal angular frequency and without phase shift:

$$\delta_1(t) = \delta_1 \sin \omega t \tag{6.108a}$$

$$\delta_2(t) = \delta_2 \sin \omega t \tag{6.108b}$$

where the angular frequency ω and the maximum amplitudes δ_1 and δ_2 are to be determined via an eigenvalue problem. By substituting Equations 6.108 into Equation 6.107, the following homogeneous algebraic equation is obtained:

$$\begin{bmatrix} (k_1 + k_2 - m_1\omega^2) & -k_2 \\ -k_2 & (k_2 - m_2\omega^2) \end{bmatrix} \begin{bmatrix} \delta_1 \\ \delta_2 \end{bmatrix} = \begin{bmatrix} 0 \\ 0 \end{bmatrix} \tag{6.109}$$

This equation has a solution different from the trivial one if and only if the determinant of the coefficient matrix is equal to zero:

$$\omega^4 - \left(\frac{k_1+k_2}{m_1} + \frac{k_2}{m_2}\right)\omega^2 + \frac{k_1 k_2}{m_1 m_2} = 0 \qquad (6.110)$$

In the case where the columns have the same moment of inertia and the same height, it follows that $k_1=k_2=k$. If it is further assumed that the masses of the cross elements also present the same value, $m_1=m_2=m$, the characteristic Equation 6.110 simplifies as

$$\omega^4 - \frac{3k}{m}\omega^2 + \frac{k^2}{m^2} = 0 \qquad (6.111)$$

and yields the following two eigenvalues:

$$\omega^2 = \frac{3\pm\sqrt{5}}{2}\frac{k}{m} \qquad (6.112)$$

whence it follows that

$$\omega_1 = 0.618\sqrt{\frac{k}{m}} \qquad (6.113a)$$

$$\omega_2 = 1.618\sqrt{\frac{k}{m}} \qquad (6.113b)$$

The eigenvectors are obtained by resolving the homogeneous system (Equation 6.109), after substituting the corresponding eigenvalues (Equations 6.113):

$$\delta_{11} = 0.618\delta_{12} \qquad (6.114a)$$

$$\delta_{22} = -0.618\delta_{21} \qquad (6.114b)$$

The eigenvectors (Equations 6.114) are determined but for a factor of proportionality, and are represented in Figure 6.10, normalizing the maximum absolute value of the coordinates. The second modal deformed configuration presents translations of opposite sign (Figure 6.10b).

6.7 FORCED OSCILLATIONS OF SHEAR-TYPE MULTISTORY FRAMES

Generally, in multistory frames, the oscillation modes of rank higher than the first one (the **fundamental mode**) show alternate signs in the cross beam displacements. The equation of motion can be written as Equation 6.104 for these systems as well, with [M] being the diagonal matrix of the cross beam masses, and [K] the global stiffness matrix, already defined in Section 5.9.

The **forced oscillation** problem, due to external constraint shaking, will now be analyzed. Let us suppose that the column base of a multistory frame with rigid cross beams undergoes

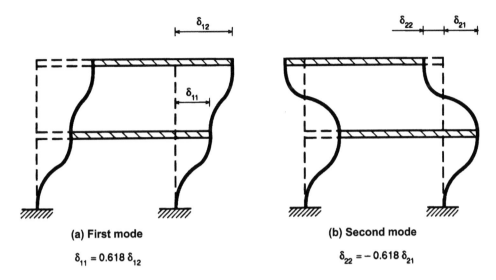

(a) First mode

$\delta_{11} = 0.618\,\delta_{12}$

(b) Second mode

$\delta_{22} = -0.618\,\delta_{21}$

Figure 6.10

a seismic motion with displacement $x(t)$, velocity $\dot{x}(t)$, and acceleration $\ddot{x}(t)$. Equation 6.104 will, therefore, become

$$[K]\{\delta\}+[M]\{\ddot{\delta}+\ddot{x}\}=\{0\} \tag{6.115}$$

since the elastic forces depend on the displacements evaluated with respect to the reference system moving with the ground, whereas the inertia forces are proportional to the absolute system accelerations. From Equation 6.115, it follows that

$$[K]\{\delta\}+[M]\{\ddot{\delta}\}=-[M]\{\ddot{x}\} \tag{6.116}$$

This is a dynamic nonhomogeneous equation, and it shows an acting force on the right-hand side that comes from the ground acceleration $\ddot{x}$. By introducing the system **normal coordinates** f_i (see Section 6.10) and the eigenvectors $\{\delta_i\}$, and multiplying by $\{\delta_i\}^T$, one obtains

$$f_i\{\delta_i\}^T[K]\{\delta_i\}+\ddot{f_i}\{\delta_i\}^T[M]\{\delta_i\}=-\{\delta_i\}^T[M]\{\ddot{x}\} \tag{6.117}$$

Dividing both members by the coefficient of $\ddot{f_i}$ yields

$$\ddot{f_i}+f_i\frac{\{\delta_i\}^T[K]\{\delta_i\}}{\{\delta_i\}^T[M]\{\delta_i\}}=-\frac{\{\delta_i\}^T[M]\{\ddot{x}\}}{\{\delta_i\}^T[M]\{\delta_i\}} \tag{6.118}$$

Finally, recalling Equation 5.128 and naming the coefficient of f_i as the **Rayleigh ratio** (see Section 5.10), the following decoupled equation for each oscillating mode, and hence for each normal coordinate f_i, is obtained:

$$\ddot{f_i}+\omega_i^2 f_i=-g_i\ddot{x}, \quad \text{for } i=1,2,\ldots,n \tag{6.119}$$

Since, for multistory frames with rigid cross beams, the mass matrix is diagonal, the factor g_i takes the following form:

$$g_i = \frac{\sum_{j=1}^{n} m_j \delta_{ij}}{\sum_{j=1}^{n} m_j \delta_{ij}^2} \tag{6.120}$$

and is usually called the **participation coefficient** of the system to the ith oscillation mode.

The analysis of the forced oscillations in the framed structures has therefore been reduced to the analysis of n elementary oscillators, each of them subjected to a fraction g_i of the excitation at the base.

Once the differential Equation 6.119 has been solved, it is possible to work out the cross-member displacements as functions of time:

$$\{\delta(t)\} = \sum_{i=1}^{n} f_i \{\delta_i\} \tag{6.121}$$

via the eigenvectors $\{\delta_i\}$, and then, still by linear transformation, it is possible to find out the internal characteristics acting on the frame columns.

The maximum acceleration that a single-degree-of-freedom system can undergo due to a seismic event is established by standard rules through the formula:

$$a = \bar{C} R(\omega) g \tag{6.122}$$

where:

$\bar{C}$ is the **seismic intensity coefficient** related to the seismicity of the area where the structure is located

$R(\omega)$ is the **response coefficient** related to the angular frequency of the system

g is the gravity acceleration

If earthquake effects are considered on the ith oscillation mode of a frame with multiple degrees of freedom, the maximum acceleration that can be associated to such elementary scheme is

$$\left| \ddot{f_i} \right| = \bar{C} R(\omega_i) g g_i \tag{6.123}$$

The maximum acceleration suffered by the jth story due to the ith oscillation mode can be deduced from Equation 6.123:

$$a_{ij} = \bar{C} R(\omega_i) g g_i \delta_{ij} \tag{6.124}$$

and, consequently, the maximum force is

$$F_{ij} = a_{ij} m_j \tag{6.125}$$

Equation 6.125 can be written as

$$F_{ij} = \bar{C}R(\omega_i)W_j\gamma_{ij} \tag{6.126}$$

W_j being the jth story weight and

$$\gamma_{ij} = g_i\delta_{ij} \tag{6.127}$$

the **distribution coefficient** of the ith oscillation mode on the jth story.

The maximum story forces (Equation 6.126) act simultaneously within the same ith mode, but are phase shifted when different oscillation modes are considered. Once the maximum of a characteristic has been calculated—for example, the bending moment M_i—as the ith oscillation mode varies, the problem of combining these values together arises. The following alternative empirical formulas are applicable when obtaining the maximum equivalent moment:

$$M_{max} = \sum_{i=1}^{n}|M_i| \tag{6.128a}$$

$$M_{max} = |M_1| \tag{6.128b}$$

$$M_{max} = \sqrt{\sum_{i=1}^{n}M_i^2} \tag{6.128c}$$

Equation 6.128a, which corresponds to the hypothesis that the maximum values occur simultaneously, is excessively conservative. Equation 6.128b, which considers only the fundamental mode, is not applicable to slender structures, where modes higher than the first one play a fundamental role. Equation 6.128c thus turns out to be the most suitable one, and is known as the **composition in quadrature** relationship.

It should be noted that the participation coefficients g_i (Equation 6.120) and, hence, the distribution coefficients γ_{ij} (Equation 6.127) clearly tend to decrease when the oscillation mode index i increases, due to the sign variations in δ_{ij} displacements. The reason for the usual limitation of seismic analysis to the lower index modes is therefore understandable, avoiding in such a way the effort of a whole modal analysis.

Let us consider a shear-type 10-story frame (Figure 6.11). Assume the mass and stiffness matrices of the equivalent oscillator (Figure 6.11c) with 10 degrees of freedom as

$$[M] = 10^3\text{kg}\begin{bmatrix} 430 & 0 & 0 & 0 & 0 & 0 & 0 & 0 & 0 & 0 \\ 0 & 406 & 0 & 0 & 0 & 0 & 0 & 0 & 0 & 0 \\ 0 & 0 & 382 & 0 & 0 & 0 & 0 & 0 & 0 & 0 \\ 0 & 0 & 0 & 358 & 0 & 0 & 0 & 0 & 0 & 0 \\ 0 & 0 & 0 & 0 & 334 & 0 & 0 & 0 & 0 & 0 \\ 0 & 0 & 0 & 0 & 0 & 310 & 0 & 0 & 0 & 0 \\ 0 & 0 & 0 & 0 & 0 & 0 & 286 & 0 & 0 & 0 \\ 0 & 0 & 0 & 0 & 0 & 0 & 0 & 262 & 0 & 0 \\ 0 & 0 & 0 & 0 & 0 & 0 & 0 & 0 & 238 & 0 \\ 0 & 0 & 0 & 0 & 0 & 0 & 0 & 0 & 0 & 215 \end{bmatrix} \tag{6.129}$$

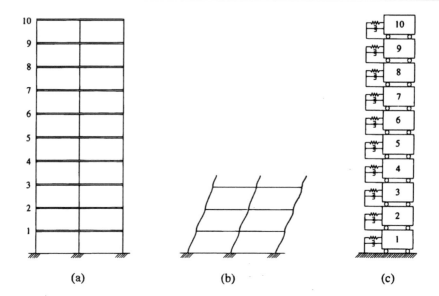

Figure 6.11

and

$$[K] = 10^6 \frac{\text{N}}{\text{m}} \begin{bmatrix} 661 & -321 & 0 & 0 & 0 & 0 & 0 & 0 & 0 & 0 \\ -321 & 623 & -302 & 0 & 0 & 0 & 0 & 0 & 0 & 0 \\ 0 & -302 & 585 & -283 & 0 & 0 & 0 & 0 & 0 & 0 \\ 0 & 0 & -283 & 548 & -264 & 0 & 0 & 0 & 0 & 0 \\ 0 & 0 & 0 & -264 & 510 & -246 & 0 & 0 & 0 & 0 \\ 0 & 0 & 0 & 0 & -246 & 472 & -227 & 0 & 0 & 0 \\ 0 & 0 & 0 & 0 & 0 & -227 & 434 & -208 & 0 & 0 \\ 0 & 0 & 0 & 0 & 0 & 0 & -208 & 397 & -189 & 0 \\ 0 & 0 & 0 & 0 & 0 & 0 & 0 & -189 & 359 & -170 \\ 0 & 0 & 0 & 0 & 0 & 0 & 0 & 0 & -170 & 170 \end{bmatrix}$$

(6.130)

The system of linear equations that governs the free oscillations provides 10 eigenvalues. Let us limit the analysis to the three lowest ones, which turn out to be

$$\omega_1 = 4.738 \text{ rad/s}, \quad \omega_2 = 12.578 \text{ rad/s}, \quad \omega_3 = 20.294 \text{ rad/s}$$

(6.131a)

and thus

$$T_1 = 1.326 \text{ s}, \quad T_2 = 0.499 \text{ s}, \quad T_3 = 0.309 \text{ s}$$

(6.131b)

The three corresponding eigenvectors, normalized with respect to the mass, are depicted in Figure 6.12.

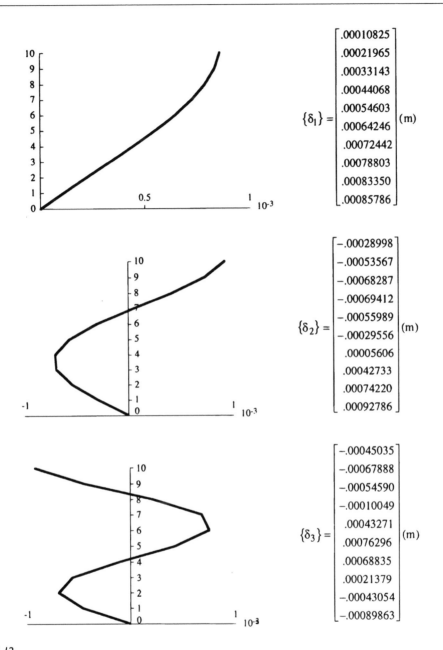

Figure 6.12

Since the mass matrix is diagonal, Equation 6.120 simplifies as follows:

$$g_i = \sum_{j=1}^{10} m_j \delta_{ij}, \quad i = 1, 2, 3 \tag{6.132}$$

The first three participation factors can be then evaluated as

$$g_1 = 1598.09 \text{ m}^{-1}, \quad g_2 = -626.03 \text{ m}^{-1}, \quad g_3 = -375.53 \text{ m}^{-1} \tag{6.133}$$

Note once again that the absolute value of g_i decreases as i increases, due to the sign changes in δ_{ij}.

The displacement can then be evaluated by means of

$$\{u_i\} = \{\delta_i\} g_i S_{di}, \qquad i = 1, 2, 3 \tag{6.134}$$

where S_{di} is the spectrum of response in terms of displacement. It depends on the **seismic intensity coefficient,** the **response coefficient,** and other parameters defined by standard seismic rules. By assuming

$$S_{d1} = 0.0405 \text{ m}, \quad S_{d2} = 0.0080 \text{ m}, \quad S_{d3} = 0.0031 \text{ m} \tag{6.135}$$

the story displacements $\{u_i\}$, for $i = 1, 2, 3$, are represented in Figures 6.13a through c.

The composition in quadrature of Equation 6.134 is computed by the following relationship:

$$\{u_{max}\} = \sqrt{\sum_{i=1}^{3} \{u_i^2\}} \tag{6.136}$$

which provides the maximum displacements.

Based on the displacements provided by Equation 6.134, the bending moment $\{M_i\}$, the shearing force $\{T_i\}$, and the axial force $\{N_i\}$ can be evaluated in the extreme sections of

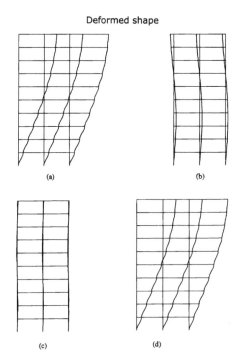

Deformed shape

(a)　　　　　　(b)

(c)　　　　　　(d)

Figure 6.13

Bending moment

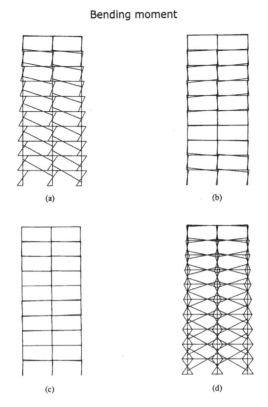

(a)

(b)

(c)

(d)

Figure 6.14

beams and columns. The diagrams of $\{M_i\}$, for $i = 1, 2, 3$, and that related to the composition in quadrature expressed by

$$\{M_{\max}\} = \sqrt{\sum_{i=1}^{3}\{M_i^2\}} \qquad (6.137)$$

are reported in Figure 6.14. The analogous diagrams for $\{T_i\}$ and $\{N_i\}$ are plotted in Figures 6.15 and 6.16, respectively.

6.8 VIBRATING MEMBRANES

Membranes are two-dimensional structural elements with no flexural rigidity, subjected to a uniform tensile stress s. Let us consider a rectangular membrane of width a and length b (Figure 6.17) lying on the XY plane, and a transverse displacement w of its points along the Z axis.

The increase in the membrane elastic potential energy during its deflection can be expressed as

$$W = \frac{s}{2}\int_{A}\left[\left(\frac{\partial w}{\partial x}\right)^2 + \left(\frac{\partial w}{\partial y}\right)^2\right]dxdy \qquad (6.138)$$

Shearing force

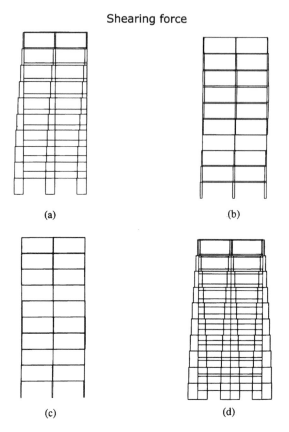

(a)   (b)

(c)   (d)

Figure 6.15

whereas the membrane kinetic energy during its own vibration is

$$T = \frac{1}{2}\rho \int_A \dot{w}^2 dxdy$$ (6.139)

where ρ is the membrane surface density.

Let us expand the unknown function w in a double series of trigonometrical functions:

$$w = \sum_{m=1}^{\infty} \sum_{n=1}^{\infty} A_{mn} \sin m\pi \frac{x}{a} \sin n\pi \frac{y}{b}$$ (6.140)

so that the coefficients A_{mn} become the generalized coordinates of the problem. It can be easily verified that each term of Equation 6.140 satisfies the boundary conditions:

$$w = 0, \text{ for } x = 0, \ x = a$$ (6.141a)

$$w = 0, \text{ for } y = 0, \ y = b$$ (6.141b)

Axial force

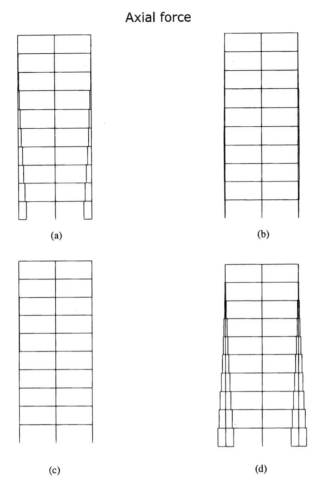

Figure 6.16

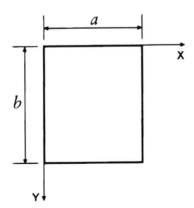

Figure 6.17

On substitution of Equation 6.140 into Equation 6.138, the potential energy in discrete form is obtained:

$$
W = \frac{\pi^2}{2} s \int_0^a \int_0^b \left[\left(\sum_{m=1}^{\infty} \sum_{n=1}^{\infty} A_{mn} \frac{m}{a} \cos m\pi \frac{x}{a} \sin n\pi \frac{y}{b} \right)^2 \right.
$$

$$
\left. + \left(\sum_{m=1}^{\infty} \sum_{n=1}^{\infty} A_{mn} \frac{n}{b} \sin m\pi \frac{x}{a} \cos n\pi \frac{y}{b} \right)^2 \right] dx dy \tag{6.142}
$$

By integrating Equation 6.142 on the membrane area, it follows that

$$
W = \frac{\pi^2}{8} abs \sum_{m=1}^{\infty} \sum_{n=1}^{\infty} \left(\frac{m^2}{a^2} + \frac{n^2}{b^2} \right) A_{mn}^2 \tag{6.143}
$$

Similarly, if Equation 6.140 is substituted into Equation 6.139, the membrane kinetic energy in discrete form is written as

$$
T = \frac{1}{2} \rho \frac{ab}{4} \sum_{m=1}^{\infty} \sum_{n=1}^{\infty} \dot{A}_{mn}^2 \tag{6.144}
$$

Since Equations 6.143 and 6.144 do not include the products of the generalized coordinates A_{mn} nor those of the corresponding velocities $\dot{A}_{mn}$, but only the values of the respective squared values, it can be said that the chosen **coordinates** are the **principal** ones, and that the relevant modes of vibration are the **normal** ones for the membrane.

The differential motion equation for a normal vibration will therefore be

$$
\rho \frac{ab}{4} \ddot{A}_{mn} + \frac{\pi^2}{4} sab \left(\frac{m^2}{a^2} + \frac{n^2}{b^2} \right) A_{mn} = 0 \tag{6.145}
$$

which provides the natural frequencies:

$$
f_{mn} = \frac{1}{2} \sqrt{\frac{s}{\rho} \left(\frac{m^2}{a^2} + \frac{n^2}{b^2} \right)} \tag{6.146}
$$

The first vibration mode is obtained for $m=n=1$:

$$
f_{11} = \frac{1}{2} \sqrt{\frac{s}{\rho} \left(\frac{1}{a^2} + \frac{1}{b^2} \right)} \tag{6.147a}
$$

$$
w_{11} = C \sin \pi \frac{x}{a} \sin \pi \frac{y}{b} \tag{6.147b}
$$

In this case, the elastic deflection does not have null points, except on the boundary.

For a square-shaped membrane ($a=b$), one obtains

$$
f_{11} = \frac{1}{a\sqrt{2}} \sqrt{\frac{s}{\rho}} \tag{6.148}
$$

Table 6.1

Membrane shape	Coefficient α
Circle	$2.404\sqrt{\pi} = 4.261$
Square	$\sqrt{2}\pi = 4.443$
Rectangle 3×2	$\sqrt{\dfrac{13}{6}}\pi = 4.624$
Equilateral triangle	$2\pi\sqrt{\tan\dfrac{\pi}{6}} = 4.774$
Rectangle 2×1	$\pi\sqrt{\dfrac{5}{2}} = 4.967$
Rectangle 3×1	$\pi\sqrt{\dfrac{10}{3}} = 5.736$

and the fundamental frequency is directly proportional to the square root of the stress, as well as inversely proportional to the square root of the density and to the length of the membrane side.

In the case of a circular membrane, it can be proved that the fundamental frequency is

$$f = \frac{2.404}{r}\sqrt{\frac{s}{\rho}} \tag{6.149}$$

where r is the radius of the membrane. More generally, for generically shaped membranes, the fundamental frequency can be expressed as

$$f = \alpha\sqrt{\frac{s}{\rho A}} \tag{6.150}$$

where A is the membrane area and the coefficient α, listed in Table 6.1, represents the effect of lower or higher deviation with respect to the circular shape.

In the case of a rope in tension, already discussed in Section 6.3, formulas similar to the ones presented here are applicable, taking into account that the stress s has to be replaced by the axial force N, the surface density ρ by the linear one μ, and the membrane characteristic dimension by the rope length.

6.9 VIBRATING PLATES

Plates are two-dimensional structural elements with a relevant flexural rigidity. The potential energy due to the plate deflection can be expressed as

$$
\begin{aligned}
W &= \frac{1}{2}\left(\int_A M_x\chi_x + M_y\chi_y + M_{xy}\chi_{xy}\right)dxdy \\
&= \frac{1}{2}D\int_A\left\{\left(\frac{\partial^2 w}{\partial x^2} + \frac{\partial^2 w}{\partial y^2}\right)^2 - \left[\left(\frac{\partial^2 w}{\partial x^2}\right)\left(\frac{\partial^2 w}{\partial y^2}\right) - \left(\frac{\partial^2 w}{\partial x\partial y}\right)^2\right]\right\}dxdy
\end{aligned} \tag{6.151}
$$

where D represents the flexural rigidity (see Equation 3.10). The kinetic energy of the vibrating plate is

$$T = \frac{1}{2}\rho h \int_A \dot{w}^2 dx dy \tag{6.152}$$

where ρ represents the material bulk density.

Analyzing the case of a rectangular plate simply supported on its edges, one can proceed in the same way as for a rectangular membrane and therefore consider Equation 6.140 for the transverse displacement w. It is easy to verify how each term of the expression satisfies the boundary conditions, which require the functions w, $\partial^2 w/\partial x^2$, $\partial^2 w/\partial y^2$ to be null on the boundary.

By inserting Equation 6.140 into Equation 6.151, the potential energy in discrete form is obtained:

$$W = \frac{\pi^4}{8} abD \sum_{m=1}^{\infty} \sum_{n=1}^{\infty} \left(\frac{m^2}{a^2} + \frac{n^2}{b^2} \right)^2 A_{mn}^2 \tag{6.153}$$

Similarly, substituting Equation 6.140 into Equation 6.152 yields the plate kinetic energy in discrete form:

$$T = \frac{1}{2} \rho h \frac{ab}{4} \sum_{m=1}^{\infty} \sum_{n=1}^{\infty} \dot{A}_{mn}^2 \tag{6.154}$$

Coefficients A_{mn} represent the normal coordinates; therefore, the differential equation of motion for a normal vibration will be

$$\rho h \ddot{A}_{mn} + \pi^4 D \left(\frac{m^2}{a^2} + \frac{n^2}{b^2} \right)^2 A_{mn} = 0 \tag{6.155}$$

which provides the natural angular frequencies:

$$\omega_{mn} = \pi^2 \sqrt{\frac{D}{\rho h}} \left(\frac{m^2}{a^2} + \frac{n^2}{b^2} \right) \tag{6.156}$$

In the case of a square plate $(a=b)$, the fundamental frequency related to the first vibration mode is

$$f_{11} = \frac{\omega_{11}}{2\pi} = \frac{\pi}{a^2} \sqrt{\frac{D}{\rho h}} \tag{6.157}$$

The fundamental frequency is thus directly proportional to the square root of the flexural rigidity D, as well as inversely proportional to the square root of the surface density (i.e., equal to bulk density multiplied by thickness) and to the length of membrane side.

For a circular plate clamped on its edge, it can be proved that the fundamental angular frequency is

$$\omega = \frac{10.21}{r^2} \sqrt{\frac{D}{\rho h}} \tag{6.158}$$

By comparing the fundamental frequencies of a simply supported square plate and a circular plate with the same area but clamped along its edges, it can be seen that

$$f_\square = \frac{\pi}{A}\sqrt{\frac{D}{\rho h}}$$

(6.159)

$$f_\bigcirc = \frac{5.1}{A}\sqrt{\frac{D}{\rho h}}$$

(6.160)

and thus

$$f_\bigcirc = 1.625 f_\square$$

(6.161)

Hence, the results for the clamped plate show that it is stiffer than the simply supported one.

6.10 DYNAMICS OF SHELLS AND THREE-DIMENSIONAL ELASTIC SOLIDS

If the body force $\{\mathcal{F}\}$ in Equation 4.7 is replaced by the force of inertia:

$$-[\rho]\frac{\partial^2}{\partial t^2}\{\eta\}$$

(6.162)

the equation of free oscillations for the elastic solid under examination is obtained:

$$\left([\mathcal{L}]-[\rho]\frac{\partial^2}{\partial t^2}\right)\{\eta\} = \{0\}$$

(6.163)

where $[\rho]$ denotes the **density matrix,** which is a diagonal matrix of dimension $(g \times g)$ where the density ρ of the material corresponds to translations, and the moment of inertia $(1/12)$ ρh^2 to rotations, h being the thickness of the beam or plate.

In the absence of static body and surface forces of static type and in the presence of inertial forces, Equation 4.58 becomes

$$[K_{ed}] = -\int_{V_e}[\eta_e]^T[\mathcal{L}][\eta_e]dV + \int_{V_e}[\eta_e]^T[\rho][\eta_e]dV \cdot \frac{\partial^2}{\partial t^2}$$
$$+ \int_{S_e}[\eta_e]^T[\mathcal{L}_0][\eta_e]dS$$

(6.164)

from which the **dynamic stiffness matrix** is obtained:

$$[K_{ed}] = [K_e] + [M_e]\frac{\partial^2}{\partial t^2}$$

(6.165)

with

$$[M_e] = \int_{V_e} [\eta_e]^{\mathrm{T}} [\rho][\eta_e] \mathrm{d}V \tag{6.166}$$

representing the **local matrix of masses**.

Similarly, replacing the body forces $\{\mathcal{F}\}$ in Equation 4.44 with the inertial forces (Equation 6.162) yields

$$[K_e]\{\delta_e\} = -\int_{V_e} [\eta_e]^{\mathrm{T}} [\rho][\eta_e] \mathrm{d}V \cdot \frac{\partial^2}{\partial t^2}\{\delta_e\} + \int_{S_e} [\eta_e]^{\mathrm{T}} \{p\} \mathrm{d}S \tag{6.167}$$

Expanding and assembling the local dynamic stiffness matrices, it follows that

$$\sum_e \left[K^{ed}\right] = \sum_e [A_e]^{\mathrm{T}} [K_{ed}][A_e] \tag{6.168}$$

which is a relation that is a generalization of Equations 4.51 and 4.52, and gives the **global matrix of masses**:

$$[M] = \sum_e [A_e]^{\mathrm{T}} [M_e][A_e] \tag{6.169}$$

Eventually, the following equation is obtained:

$$[K]\{\delta\} + [M]\{\ddot{\delta}\} = \{0\} \tag{6.170}$$

which is formally identical to the equation of a harmonic oscillator with one degree of freedom, in the absence of viscous forces and forcing loads. It should be noted, once again, that $[M]$ is not, in general, a diagonal matrix.

It would have been possible to obtain the same equation by considering the inertial forces as body and surface forces of static type and applying Equations 4.53, 4.54, 4.35, 4.47a, and 6.122, sequentially.

In relation to Equation 6.170, let us choose a solution in the form:

$$\{\delta(t)\} = \{\delta\} f(t) \tag{6.171}$$

separating the temporal variable t and considering the same oscillatory law for all the generalized coordinates of the system.

Substituting Equation 6.171 into Equation 6.170 yields

$$-\frac{\ddot{f}}{f} = \frac{\{\delta\}^{\mathrm{T}} [K]\{\delta\}}{\{\delta\}^{\mathrm{T}} [M]\{\delta\}} = \lambda \tag{6.172}$$

and hence the separation of the temporal problem from the spatial one:

$$\ddot{f} + \lambda f = 0 \tag{6.173a}$$

$$\{\delta\}^{T}\left([K]-\lambda[M]\right)\{\delta\}=0 \qquad\qquad (6.173b)$$

From Equation 6.173b, one obtains

$$\mathrm{Det}\left([K]-\lambda[M]\right)=0 \qquad\qquad (6.174)$$

which is an algebraic equation with the unknown λ, of a degree equal to the number of degrees of freedom of the system. This equation is called **characteristic**, and its solutions are the eigenvalues of the problem. The eigenvectors are obtained, but for a factor of proportionality, from the equation:

$$\left([K]-\lambda[M]\right)\{\delta\}=0 \qquad\qquad (6.175)$$

The eigenvectors possess the property of orthogonality with respect to the stiffness and mass matrices. Let Equation 6.175 be written for two different eigenvectors:

$$[K]\{\delta_{j}\}=\lambda_{j}[M]\{\delta_{j}\} \qquad\qquad (6.176a)$$

$$[K]\{\delta_{k}\}=\lambda_{k}[M]\{\delta_{k}\} \qquad\qquad (6.176b)$$

Multiplying the former by $\{\delta_{k}\}^{T}$ and the latter by $\{\delta_{j}\}^{T}$, taking the symmetry of $[K]$ and $[M]$ into account, and subtracting member by member, it follows that

$$0=\left(\lambda_{j}-\lambda_{k}\right)\{\delta_{k}\}^{T}[M]\{\delta_{j}\} \qquad\qquad (6.177)$$

and hence, in normal form with respect to the mass:

$$\{\delta_{k}\}^{T}[M]\{\delta_{j}\}=\delta_{jk} \qquad\qquad (6.178a)$$

where δ_{jk} is the Kronecker symbol. Equations 6.176a and 6.178a also imply

$$\{\delta_{k}\}^{T}[K]\{\delta_{j}\}=\lambda_{j}\delta_{jk} \qquad\qquad (6.178b)$$

Since the matrices $[M]$ and $[K]$ are symmetrical and positive definite, it is possible to prove that the eigenvalues λ_{j}, for $j=1,2,\ldots, (g\times n)$ are all real and positive. Equation 6.173a can thus be written as

$$\ddot{f_{i}}+\omega_{i}^{2}f_{i}=0, \qquad i=1,2,\ldots,(g\times n) \qquad\qquad (6.179)$$

and it has the following integral:

$$f_{i}(t)=A_{i}\cos\omega_{i}t+B_{i}\sin\omega_{i}t \qquad\qquad (6.180)$$

The complete integral (Equation 6.170) can therefore be put in the form:

$$\{\delta(t)\}=\sum_{i=1}^{g\times n}\{\delta_{i}\}\left(A_{i}\cos\omega_{i}t+B_{i}\sin\omega_{i}t\right) \qquad\qquad (6.181)$$

The $2 \times (g \times n)$ constants A_i and B_i are determined by imposing the initial conditions, in a manner similar to that adopted in Section 6.2 in the case of deflected beams:

$$\{\delta(0)\} = \{\delta_0\} \tag{6.182a}$$

$$\{\dot{\delta}(0)\} = \{\dot{\delta}_0\} \tag{6.182b}$$

From Equations 6.181 and 6.182, it is deduced that

$$\sum_{i=1}^{g \times n} A_i \{\delta_i\} = \{\delta_0\} \tag{6.183a}$$

$$\sum_{i=1}^{g \times n} B_i \omega_i \{\delta_i\} = \{\dot{\delta}_0\} \tag{6.183b}$$

Two distinct systems of equations are thus obtained with the unknowns A_i and B_i, respectively, which can be solved by transposition of the single members:

$$\sum_{i=1}^{g \times n} A_i \{\delta_i\}^T = \{\delta_0\}^T \tag{6.184a}$$

$$\sum_{i=1}^{g \times n} B_i \omega_i \{\delta_i\}^T = \{\dot{\delta}_0\}^T \tag{6.184b}$$

Postmultiplying by $[M]\{\delta_j\}$, exploiting the property of orthogonality, and normalizing with respect to the mass, it follows that

$$A_j = \{\delta_0\}^T [M]\{\delta_j\} \tag{6.185a}$$

$$B_j = \frac{1}{\omega_j} \{\dot{\delta}_0\}^T [M]\{\delta_j\} \tag{6.185b}$$

for $j = 1, 2, \ldots, (g \times n)$.

As in the case of deflected beams, in the more general framework of the finite element method, a system initially perturbed according to an eigenvector, with zero initial velocity, keeps oscillating indefinitely in proportion to that deformed configuration. Let us assume that

$$\{\delta_0\} = a\{\delta_i\} \tag{6.186a}$$

$$\{\dot{\delta}_0\} = \{0\} \tag{6.186b}$$

From Equations 6.185, it follows that

$$A_j = a\delta_{ij} \tag{6.187a}$$

$$B_j = 0 \tag{6.187b}$$

and hence the complete integral is written as

$$\{\delta(t)\} = a\{\delta_i\}\cos\omega_i t = \{\delta_0\}\cos\omega_i t \tag{6.188}$$

The eigenvectors being known, it is possible to consider as generalized coordinates of the system the temporal functions f_i. Ordering these functions, called **normal coordinates**, in the vector $\{f\}$, let us perform the following coordinate transformation:

$$\{\delta(t)\} = [\Delta]\{f\} \tag{6.189}$$

which is an alternative way of writing Equation 6.181, $[\Delta]$ being the **modal matrix**, which has the eigenvectors as its columns:

$$[\Delta] = \left[\delta_1|\delta_2|...|\delta_{g\times n}\right] \tag{6.190}$$

Substituting Equation 6.189 into Equation 6.170, and premultiplying by $[\Delta]^T$, one obtains

$$\left([\Delta]^T[M][\Delta]\right)\{\ddot{f}\} + \left([\Delta]^T[K][\Delta]\right)\{f\} = \{0\} \tag{6.191}$$

Taking Equations 6.178 into account, we have, on the other hand:

$$[\Delta]^T[M][\Delta] = [1] \tag{6.192a}$$

$$[\Delta]^T[K][\Delta] = [\Lambda] \tag{6.192b}$$

where $[\Lambda]$ is the diagonal matrix of the eigenvalues. Equation 6.191 reduces, therefore, to the vector form:

$$\{\ddot{f}\} + [\Lambda]\{f\} = \{0\} \tag{6.193}$$

and thus to the scalar form (Equation 6.179). The equations of motion are thus decoupled, each containing a single unknown, which is one of the normal coordinates. The transformation of coordinates (Equation 6.189) allows the expressions of elastic potential and kinetic energy to be reduced to the so-called **canonical form**:

$$W = \frac{1}{2}\{\delta\}^T[K]\{\delta\} = \frac{1}{2}\{f\}^T[\Lambda]\{f\} \tag{6.194a}$$

$$T = \frac{1}{2}\{\dot{\delta}\}^T[M]\{\dot{\delta}\} = \frac{1}{2}\{\dot{f}\}^T[1]\{\dot{f}\} \tag{6.194b}$$

6.11 DYNAMICS OF ELASTIC SOLIDS WITH LINEAR VISCOUS DAMPING

When the dynamic behavior of elastic solids with linear viscous damping is considered, in addition to the inertia forces (Equation 6.162) obtained by the well-known D'Alembert principle, the friction forces that oppose the vibrating motion must be taken into account as well. They can be expressed as

$$-[\mu]\frac{\partial}{\partial t}\{\eta\} \tag{6.195}$$

where $[\mu]$ is the matrix of damping coefficients.

As previously seen, the equivalent static problem can be discretized at any instant by replacing the static Lamé operator with its corresponding dynamic and damping operator:

$$[\mathcal{L}]-[\mu]\frac{\partial}{\partial t}-[\rho]\frac{\partial^2}{\partial t^2} \tag{6.196}$$

When both inertia forces and viscous forces are present, Equation 4.58 becomes

$$[K_{edd}]=-\int_{V_e}[\eta_e]^{\mathrm{T}}[\mathcal{L}][\eta_e]\mathrm{d}V+\int_{V_e}[\eta_e]^{\mathrm{T}}[\mu][\eta_e]\mathrm{d}V\cdot\frac{\partial}{\partial t}$$

$$+\int_{V_e}[\eta_e]^{\mathrm{T}}[\rho][\eta_e]\mathrm{d}V\cdot\frac{\partial^2}{\partial t^2}+\int_{S_e}[\eta_e]^{\mathrm{T}}[\mathcal{L}_0][\eta_e]\mathrm{d}S \tag{6.197}$$

which can be rewritten in compact form as

$$[K_{edd}]=[K_e]+[C_e]\frac{\partial}{\partial t}+[M_e]\frac{\partial^2}{\partial t^2} \tag{6.198}$$

with

$$[C_e]=\int_{V_e}[\eta_e]^{\mathrm{T}}[\mu][\eta_e]\mathrm{d}V \tag{6.199}$$

which represents the **local damping matrix**, and

$$[M_e]=\int_{V_e}[\eta_e]^{\mathrm{T}}[\rho][\eta_e]\mathrm{d}V \tag{6.200}$$

which represents the **local mass matrix**.

Expanding and assembling the local matrices, one obtains

$$\sum_e[K^{edd}]=\sum_e[A_e]^{\mathrm{T}}[K_{edd}][A_e] \tag{6.201}$$

which is similar to Equations 4.51 and 4.52, and provides the **global matrices**:

$$[C]=\sum_e[A_e]^{\mathrm{T}}[C_e][A_e] \tag{6.202a}$$

$$[M] = \sum_e [A_e]^{\mathrm{T}} [M_e][A_e] \qquad (6.202b)$$

Eventually, the following equation is obtained:

$$[K]\{\delta\} + [C]\{\dot{\delta}\} + [M]\{\ddot{\delta}\} = \{F\} \qquad (6.203)$$

which is formally analogous to the equation of a simple oscillator, subjected to both damping and forcing loads. It should be noted that $[C]$ is not, generally, a diagonal matrix.

Chapter 7

Buckling instability in slender, thin, and shallow structures

7.1 INTRODUCTION

The hypothesis of small displacements so far advanced considers the cardinal equations of statics in relation to undeformed structural configurations. In other words, the elastic displacements have been hypothesized as being so small as to make it possible for the deformed configuration to be confused with the undeformed one when the static characteristics are to be evaluated. In this chapter, this hypothesis will be removed, and it will be shown how the solution of an elastic problem can represent, in actual fact, a condition of stable, neutral, or unstable equilibrium, according to the magnitude of the load applied. Moreover, there exist, around the condition of neutral equilibrium, an infinite number of other similar conditions, characterized by different static parameters (applied loads) and kinematic parameters (configuration of the system).

The instability of elastic equilibrium occurs in general for slender or thin structural elements subjected to compressive loads, such as columns of buildings, machine shafts, struts of trusses, thin arches and shells, and cylindrical and spherical shells subjected to external pressure. But, also, other cases that are more complex as regards both their geometry and the loading conditions can equally be considered. It will suffice to think of the lateral torsional buckling of beams of thin rectangular cross section, where the disparity between the orders of magnitude of the two central moments of inertia can cause, in a deflected beam, a sudden torsional deformation. The instability of elastic equilibrium is, moreover, a critical phenomenon that may affect an entire beam system before it involves a particular element of the system. This occurs in the case of metal trussed or framed structures, which are frequently made up of extremely slender rods or beams.

The loss of stability of elastic equilibrium is commonly referred to as **buckling**. This is one of the three fundamental phenomena of structural collapse, the other two being **general yielding** and **brittle fracturing**, which we shall discuss in Chapters 10 and 12, respectively. These phenomena do not, in general, occur separately, but interact during the phases of collapse. In this chapter, we shall see how yielding can interact with buckling in the context of a transition from one to the other as the slenderness of the structure increases. In Chapter 12, we shall consider, instead, the interaction between plastic collapse and brittle fracture, as well as the ductile–brittle transition as the size scale increases, the geometrical shape of the structure remaining the same.

In the final part of the present chapter, the interaction between the instability due to flexural deformation (buckling) and the instability due to axial or membrane deformation (snap-through) will be considered for the cases of shallow arches and domes. In Chapter 8, we will consider another important interaction between buckling and resonance (elastic instability in the dynamic regime).

7.2 DISCRETE MECHANICAL SYSTEMS WITH ONE DEGREE OF FREEDOM

Let us consider the mechanical system of Figure 7.1a, consisting of two rigid rods connected by an elastic hinge of rotational rigidity k, and constrained at one end by a hinge and at the other by a roller support. When the system is loaded with an axial force N and the absolute rotation φ of the two arms is assumed as the generalized coordinate, the total potential energy of the whole system is

$$W(\varphi) = \frac{1}{2}k(2\varphi)^2 - 2Nl(1 - \cos\varphi) \tag{7.1}$$

The conditions of equilibrium are identified by imposing the stationarity of Equation 7.1:

$$W'(\varphi) = 4k\varphi - 2Nl\sin\varphi = 0 \tag{7.2}$$

from which we obtain the relation

$$N = \frac{2k\varphi}{l\sin\varphi} \tag{7.3}$$

which links loading condition and deformed configuration along the branch of equilibrium presented in Figure 7.1b. The plane $N-\varphi$ is thus divided into two sectors by the curve of Equation 7.3: the points of the upper sector represent conditions of instability, while those of the lower sector represent conditions of stability. Starting from the initial condition $\varphi = N = 0$, it will thus be possible to traverse, in a stable manner, the vertical segment of the axis N up to the point C ($\varphi = 0$, $N = N_c = 2k/l$), then to deviate onto one of the two branches of equilibrium of Figure 7.1b. Alternatively, it would be possible to proceed along the vertical axis beyond the branching point C, although in this case the equilibrium is of an unstable type.

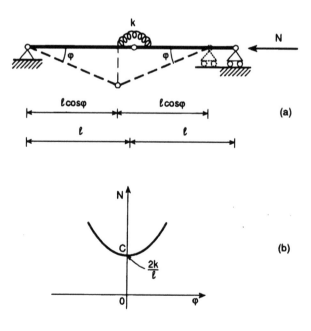

Figure 7.1

Note how the global behavior of the system is then of a hardening type, with the increase in deformation requiring a further increase in the external load. A nonlinear and postcritical behavior of this sort is said to be stable. To verify the stability of the postcritical branch, it is possible to consider the concavity of the potential as expressed by Equation 7.1 and, hence, its derivatives of a higher order, calculated for $\varphi = 0$:

$$W'(0) = W''(0) = W'''(0) = 0 \tag{7.4a}$$

$$W^{IV}(0) = 4k > 0 \tag{7.4b}$$

The determination of the critical load N_c can also be made via the simple **method of direct equilibrium**, that is, by equating the destabilizing moment:

$$M_i = Nl\sin\varphi \simeq Nl\varphi \tag{7.5a}$$

and the stabilizing moment:

$$M_s = 2k\varphi \tag{7.5b}$$

Note that, in Equation 7.5a, recourse has been made to the hypothesis of linearized kinematics. This hypothesis simplifies the calculations, even though it prevents the definition of the postcritical behavior.

Let us now consider the mechanical system of Figure 7.2a, consisting of two rigid rods connected by a hinge, and constrained externally by a hinge and two roller supports,

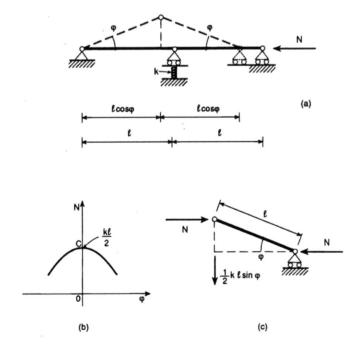

Figure 7.2

the intermediate one resting on an elastic foundation of rigidity k. The total potential energy is

$$W(\varphi) = \frac{1}{2}k(l\sin\varphi)^2 - 2Nl(1 - \cos\varphi) \qquad (7.6)$$

where the first term represents the potential energy of the spring, while the second represents the potential energy of the axial force N. The conditions of equilibrium are obtained by imposing the stationarity of Equation 7.6:

$$W'(\varphi) = l\sin\varphi(kl\cos\varphi - 2N) = 0 \qquad (7.7)$$

from which we obtain (Figure 7.2b)

$$N = \frac{kl}{2}\cos\varphi \qquad (7.8)$$

In this second example, the global behavior of the system is of a softening type, a decrease in the external load corresponding to an increase in deformation. A postcritical behavior of this sort is said to be unstable. To verify the instability of the postcritical branch, it is possible to consider the convexity of the potential expressed by Equation 7.6 and, hence, its derivatives of a higher order, calculated for $\varphi = 0$:

$$W'(0) = W''(0) = W'''(0) = 0 \qquad (7.9a)$$

$$W^{IV}(0) = -3kl^2 < 0 \qquad (7.9b)$$

Also in this case, the determination of the critical load N_c may be made by equating the destabilizing moment acting on each rod, evaluated around the end supports (Figure 7.2c):

$$M_i = Nl\sin\varphi \simeq Nl\varphi \qquad (7.10a)$$

and the stabilizing moment due to the reaction of the central support (Figure 7.2c):

$$M_s = \frac{1}{2}kl^2\sin\varphi\cos\varphi \simeq \frac{1}{2}kl^2\varphi \qquad (7.10b)$$

7.3 DISCRETE MECHANICAL SYSTEMS WITH TWO OR MORE DEGREES OF FREEDOM

Let us consider the mechanical system with two degrees of freedom in Figure 7.3a, consisting of three rigid rods connected by two elastic hinges of rotational rigidity k, and constrained at one end by a hinge and at the other by a roller support. When the system is loaded with an axial force N and the transverse displacements x_1 and x_2 of the elastic hinges

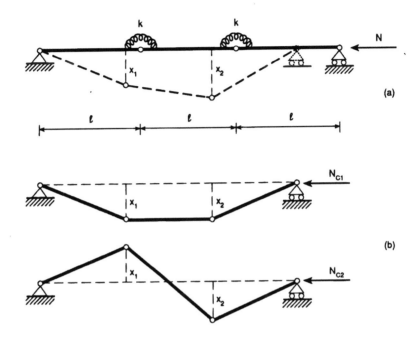

k   k

N

x_1 x_2

(a)

ℓ ℓ ℓ

N_{C1}

x_1 x_2

(b)

N_{C2}

x_1

x_2

Figure 7.3

are assumed to be generalized coordinates, the total potential energy of the entire system is expressible as follows:

$$W(x_1, x_2) = \frac{1}{2}k\left[\left(\arcsin\frac{x_1}{l} - \arcsin\frac{x_2 - x_1}{l}\right)^2\right.$$

$$+\left.\left(\arcsin\frac{x_2}{l} + \arcsin\frac{x_2 - x_1}{l}\right)^2\right]$$

$$-Nl\left[3 - \cos\left(\arcsin\frac{x_1}{l}\right) - \cos\left(\arcsin\frac{x_2}{l}\right)\right.$$

$$\left.-\cos\left(\arcsin\frac{x_2 - x_1}{l}\right)\right] \tag{7.11}$$

Performing a Taylor series expansion of Equation 7.11 about the origin, we obtain

$$W(x_1, x_2) \simeq \frac{k}{2l^2}\left(5x_1^2 + 5x_2^2 - 8x_1x_2\right)$$

$$-\frac{N}{l}\left(x_1^2 + x_2^2 - x_1x_2\right) \tag{7.12}$$

The conditions of equilibrium are identified by imposing the stationarity of Equation 7.12:

$$\frac{\partial W}{\partial x_1} = x_1\left(\frac{5k}{l^2} - \frac{2N}{l}\right) - x_2\left(\frac{4k}{l^2} - \frac{N}{l}\right) = 0 \tag{7.13a}$$

$$\frac{\partial W}{\partial x_2} = -x_1\left(\frac{4k}{l^2} - \frac{N}{l}\right) + x_2\left(\frac{5k}{l^2} - \frac{2N}{l}\right) = 0 \tag{7.13b}$$

Equations 7.13 constitute a homogeneous system of linear algebraic equations, and possess a solution different from the trivial one when the determinant of the coefficients is equal to zero:

$$\begin{vmatrix} \left(\dfrac{5k}{l^2} - \dfrac{2N}{l}\right) & -\left(\dfrac{4k}{l^2} - \dfrac{N}{l}\right) \\[3mm] -\left(\dfrac{4k}{l^2} - \dfrac{N}{l}\right) & \left(\dfrac{5k}{l^2} - \dfrac{2N}{l}\right) \end{vmatrix} = 0 \tag{7.14}$$

Evaluating this determinant, we obtain a second-degree algebraic equation in N:

$$\left(\frac{5k}{l^2} - \frac{2N}{l}\right)^2 - \left(\frac{4k}{l^2} - \frac{N}{l}\right)^2 = 0 \tag{7.15}$$

and hence

$$N^2 - \frac{4k}{l}N + 3\frac{k^2}{l^2} = 0 \tag{7.16}$$

which yields the two eigenvalues:

$$N_{c1} = \frac{k}{l} \tag{7.17a}$$

$$N_{c2} = 3\frac{k}{l} \tag{7.17b}$$

From the system of Equations 7.13, we then obtain the corresponding eigenvectors:

$$x_1 = x_2 \tag{7.18a}$$

$$x_1 = -x_2 \tag{7.18b}$$

The eigenvectors (Equations 7.18) represent the two modes of deformation corresponding to the two critical conditions and are shown, but for a factor of proportionality, in Figure 7.3b.

To analyze the postcritical branch corresponding to the first eigenvector, the following changes of variable are useful:

$$x_1 = \varepsilon + y_1$$

$$x_2 = \varepsilon + y_2$$

where y_1 and y_2 are infinitesimals of a higher order with respect to the displacement ε that may even be finite. The condition of equilibrium is identified by imposing the stationarity of the function $W(\varepsilon + y_1, \varepsilon + y_2)$ and reconsidering the original Equation 7.11:

$$N = \frac{k}{\varepsilon} \arcsin \frac{\varepsilon}{l} \tag{7.19}$$

Equation 7.19, for $\varepsilon \rightarrow 0$, tends to the first eigenvalue k/l in accordance with Equation 7.17a. Substituting Equation 7.19 into Equation 7.11 and computing the elements of the Hessian:

$$[H] = \begin{bmatrix} \dfrac{\partial^2 W}{\partial y_1^2} & \dfrac{\partial^2 W}{\partial y_1 \partial y_2} \\[3ex] \dfrac{\partial^2 W}{\partial y_2 \partial y_1} & \dfrac{\partial^2 W}{\partial y_2^2} \end{bmatrix} \tag{7.20}$$

we obtain

$$\frac{\partial^2 W}{\partial y_1^2} = \frac{k}{l^2}\left[3 + \frac{7}{6}\left(\frac{\varepsilon}{l}\right)^2 + \frac{137}{120}\left(\frac{\varepsilon}{l}\right)^4 + \frac{629}{560}\left(\frac{\varepsilon}{l}\right)^6 + \cdots \right] \tag{7.21a}$$

$$\det[H] = \frac{k^2}{l^4}\left[2\left(\frac{\varepsilon}{l}\right)^2 + \frac{52}{15}\left(\frac{\varepsilon}{l}\right)^4 + \frac{6043}{1260}\left(\frac{\varepsilon}{l}\right)^6 + \cdots \right] \tag{7.21b}$$

These two series expansions continue with the even powers only of ε/l and with all the coefficients positive. It may thus be concluded that, since the Hessian is positive definite, the postcritical branch is stable, analogously to what has already been seen in the example of Figure 7.1.

The eigenvalue problem just illustrated can be solved rapidly, considering directly the equations of equilibrium with regard to rotation about the elastic hinges, in the framework of linearized kinematics (Figure 7.3a):

$$N x_1 = k\left(\frac{x_1}{l} - \frac{x_2 - x_1}{l} \right) \tag{7.22a}$$

$$N x_2 = k\left(\frac{x_2}{l} + \frac{x_2 - x_1}{l} \right) \tag{7.22b}$$

From Equations 7.22, we find

$$x_1\left(N - \frac{2k}{l}\right) + \frac{k}{l}x_2 = 0 \tag{7.23a}$$

$$\frac{k}{l}x_1 + x_2\left(N - \frac{2k}{l}\right) = 0 \tag{7.23b}$$

Also in this case, the homogeneous system of Equations 7.23 possesses other solutions besides the trivial one, when the determinant of the coefficients becomes zero:

$$\begin{vmatrix} \left(N - \dfrac{2k}{l}\right) & \dfrac{k}{l} \\ \dfrac{k}{l} & \left(N - \dfrac{2k}{l}\right) \end{vmatrix} = 0 \tag{7.24}$$

Consequently, we again obtain the characteristic Equation 7.16.

Note how, for $N = 0$, the matrix (Equation 7.14) coincides with the stiffness matrix of the discrete system being considered (Figure 7.3a). The columns of the stiffness matrix are obtained, in fact, by setting one of the two generalized coordinates equal to unity and equating the other to zero (Figure 7.4). The elements of each individual column are furnished by

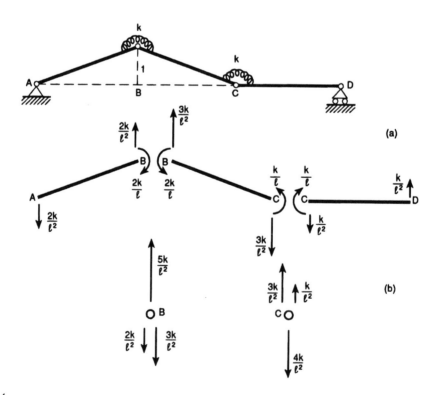

Figure 7.4

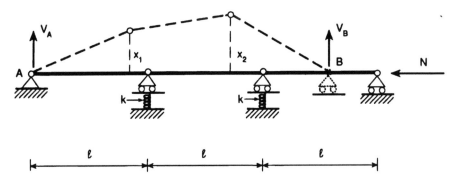

Figure 7.5

the vertical forces that produce this situation. The vertical forces are, on the other hand, the loads dual with respect to the generalized coordinates chosen.

The physical meaning of the matrix in Equation 7.24 is not, however, equally evident. For $N = 0$, it represents, in fact, the stiffness matrix of the system, when the reactive moments of the elastic hinges are assumed to be the static characteristics. These moments are not, however, the loads dual to the coordinates x_1 and x_2.

As a second example of a system with two degrees of freedom, let us examine that of Figure 7.5, which consists of three rigid rods on four supports, of which the central ones are assumed to be elastically compliant with rigidity k. The total potential energy may be expressed as follows:

$$W(x_1, x_2) = \frac{1}{2}k\left(x_1^2 + x_2^2\right) - Nl\left[3 - \cos\left(\arcsin\frac{x_1}{l}\right)\right.$$

$$\left. - \cos\left(\arcsin\frac{x_2}{l}\right) - \cos\left(\arcsin\frac{x_2 - x_1}{l}\right)\right] \tag{7.25}$$

Expanding Equation 7.25 into a Taylor series about the origin, we obtain

$$W(x_1, x_2) \simeq \frac{1}{2}k\left(x_1^2 + x_2^2\right) - \frac{N}{l}\left(x_1^2 + x_2^2 - x_1 x_2\right) \tag{7.26}$$

The stationarity of the potential W requires that the two first partial derivatives be equal to zero:

$$\frac{\partial W}{\partial x_1} = x_1\left(k - \frac{2N}{l}\right) + \frac{N}{l}x_2 = 0 \tag{7.27a}$$

$$\frac{\partial W}{\partial x_2} = \frac{N}{l}x_1 + x_2\left(k - \frac{2N}{l}\right) = 0 \tag{7.27b}$$

Making the determinant of the coefficient matrix zero:

$$\begin{vmatrix} \left(k - \dfrac{2N}{l}\right) & \dfrac{N}{l} \\[3mm] \dfrac{N}{l} & \left(k - \dfrac{2N}{l}\right) \end{vmatrix} = 0 \tag{7.28}$$

yields the characteristic equation:

$$\frac{3}{l^2}N^2 - \frac{4k}{l}N + k^2 = 0 \tag{7.29}$$

and, hence, the eigenvalues:

$$N_{c1} = \frac{1}{3}kl \tag{7.30a}$$

$$N_{c2} = kl \tag{7.30b}$$

From the system of Equations 7.27, the two corresponding eigenvectors are found:

$$x_1 = -x_2 \tag{7.31a}$$

$$x_1 = x_2 \tag{7.31b}$$

which are the same, in reverse order, as in the case previously considered (Figure 7.3b).

To analyze the postcritical branch corresponding to the first eigenvector, the following changes of variable are useful:

$$x_1 = \varepsilon + y_1$$
$$x_2 = -\varepsilon + y_2$$

where y_1 and y_2 are infinitesimals of a higher order with respect to the displacement ε that may even be finite. The condition of equilibrium is identified by imposing the stationarity of the function $W(\varepsilon + y_1, -\varepsilon + y_2)$ and reconsidering the original Equations 7.27:

$$N = kl \, \frac{\left[1 - \left(\dfrac{\varepsilon}{l}\right)^2\right]^{1/2} \left[1 - 4\left(\dfrac{\varepsilon}{l}\right)^2\right]^{1/2}}{2\left[1 - \left(\dfrac{\varepsilon}{l}\right)^2\right]^{1/2} + \left[1 - 4\left(\dfrac{\varepsilon}{l}\right)^2\right]^{1/2}} \tag{7.32}$$

Equation 7.32, for $\varepsilon \rightarrow 0$, tends to the first eigenvalue $kl/3$ in accordance with Equation 7.30a. Substituting Equation 7.32 into Equations 7.27 and computing the elements of the Hessian (Equation 7.20), we obtain

$$\frac{\partial^2 W}{\partial y_1^2} = k\left[\frac{1}{3} - \frac{3}{2}\left(\frac{\varepsilon}{l}\right)^2 - \frac{45}{8}\left(\frac{\varepsilon}{l}\right)^4 - \frac{381}{16}\left(\frac{\varepsilon}{l}\right)^6 - \cdots\right]$$ (7.33a)

$$\det[H] = k^2\left[-2\left(\frac{\varepsilon}{l}\right)^2 - 8\left(\frac{\varepsilon}{l}\right)^4 - \frac{143}{4}\left(\frac{\varepsilon}{l}\right)^6 - \cdots\right]$$ (7.33b)

These two series expansions continue with the even powers only of ε/l and with all the coefficients negative. It may thus be concluded that, since the Hessian is not positive definite, the postcritical branch is unstable, analogously to what has already been seen in the example of Figure 7.2.

A more rapid solution to the problem may be obtained by imposing equilibrium with regard to rotation of the two end rods, with respect to the intermediate hinges:

$$N x_1 = V_A l$$ (7.34a)

$$N x_2 = V_B l$$ (7.34b)

where V_A and V_B denote the vertical reactions of the end supports (Figure 7.5). The equations of equilibrium with regard to vertical translation and to rotation about the point A of the entire structure:

$$V_A + V_B = k(x_1 + x_2)$$ (7.35a)

$$k x_1 l + 2 k x_2 l = 3 V_B l$$ (7.35b)

make it possible to obtain these reactions as functions of the displacements:

$$V_B = \frac{k}{3}(x_1 + 2x_2)$$ (7.36a)

$$V_A = \frac{k}{3}(2x_1 + x_2)$$ (7.36b)

When Equations 7.36 are substituted into Equations 7.34, we find

$$\begin{bmatrix} \left(\frac{2}{3}kl - N\right) & \frac{kl}{3} \\ \frac{kl}{3} & \left(\frac{2}{3}kl - N\right) \end{bmatrix}\begin{bmatrix} x_1 \\ x_2 \end{bmatrix} = \begin{bmatrix} 0 \\ 0 \end{bmatrix}$$ (7.37)

Making the determinant of the coefficient matrix zero, we obtain once more the characteristic Equation 7.29.

Also in this case, for $N = 0$, the matrix in Equation 7.28 coincides with the stiffness matrix of the system.

In general, when a discrete system with n degrees of freedom is considered, as occurs in the case of the finite element method, the problem of the stability of elastic equilibrium can always be cast in the form:

$$\left([K] - \lambda [K_g]\right)\{\delta\} = \{0\} \tag{7.38}$$

where:

$[K]$ designates the **elastic stiffness matrix**, already defined in Chapter 4
$[K_g]$ designates the **geometric stiffness matrix**
$\{\delta\}$ denotes the nodal displacement vector
λ indicates a multiplier of the loads, which are assumed to increase proportionally

The eigenvalues of the problem are obtained via the condition

$$\det\left([K] - \lambda [K_g]\right) = 0 \tag{7.39}$$

The minimum eigenvalue λ is said to be the **critical multiplier of the loads,** and represents the load of incipient collapse.

As an example, it is emphasized that, in the case of both the systems just considered, the geometric stiffness matrix is the same and takes the following form:

$$[K_g] = \begin{bmatrix} \dfrac{2}{l} & -\dfrac{1}{l} \\ -\dfrac{1}{l} & \dfrac{2}{l} \end{bmatrix} \tag{7.40}$$

7.4 RECTILINEAR ELASTIC BEAMS WITH DIFFERENT CONSTRAINT CONDITIONS

Let us consider a slender beam of constant cross section, inextensible, and not deformable in shear, though deformable in bending, constrained at one end by a hinge and at the other by a roller support, loaded by an axial force N and by an orthogonal distributed load $q(z)$ (Figure 7.6a).

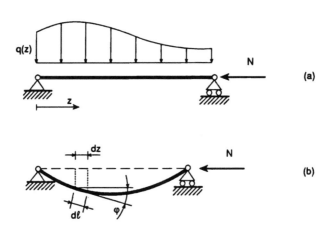

Figure 7.6

The total potential energy in a deformed configuration $\upsilon(z)$ is

$$W = \frac{1}{2} \int_0^l \frac{M^2}{EI} dz - Nw - \int_0^l q(z)\upsilon(z) dz \tag{7.41}$$

Using the relationship bending moment versus curvature, and noting that the axial displacement w of the point of application of the force N is (Figure 7.6b)

$$w = \int_0^l (dl - dz) = \int_0^l (1 - \cos\varphi) dl \tag{7.42}$$

and, hence, with the expansion of the cosine into a Taylor series:

$$w \approx \frac{1}{2} \int_0^l \varphi^2 dz \approx \frac{1}{2} \int_0^l \upsilon'^2 dz \tag{7.43}$$

the total potential energy can be expressed as follows:

$$W = \int_0^l \left[\frac{1}{2}\left(EI\upsilon''^2 - N\upsilon'^2\right) - q\upsilon \right] dz \tag{7.44}$$

Enforcing stationarity of the functional $W(\upsilon)$, we obtain

$$\delta W = \int_0^l \left(EI\upsilon''\delta\upsilon'' - N\upsilon'\delta\upsilon' - q\delta\upsilon\right) dz = 0 \tag{7.45}$$

where $\delta\upsilon$ is referred to as a **variation** of function υ and indicates a function with infinitesimal values (perturbation) contained in the class of the solutions υ. Integration by parts gives

$$-\left[\left(EI\upsilon''' + N\upsilon'\right)\delta\upsilon\right]_0^l + \left[EI\upsilon''\delta\upsilon'\right]_0^l$$

$$+ \int_0^l \left(EI\upsilon^{IV} + N\upsilon'' - q\right)\delta\upsilon\, dz = 0 \tag{7.46}$$

Since Equation 7.46 must hold for any $\delta\upsilon$, the following equations are identically satisfied:

$$EI\upsilon^{IV} + N\upsilon'' - q = 0 \tag{7.47a}$$

$$\left(EI\upsilon''' + N\upsilon'\right)\delta\upsilon = 0, \quad \text{for } z = 0, l \tag{7.47b}$$

$$\left(EI\upsilon''\right)\delta\upsilon' = 0, \quad \text{for } z = 0, l \tag{7.47c}$$

Equation 7.47a is called the **equation of the elastic line with second-order effects** and, if we neglect the term $N\upsilon''$, coincides with the usual form. For the simply supported beam (Figure 7.6), we have the boundary conditions $\upsilon(0) = \upsilon(l) = 0$, which imply $\delta\upsilon = 0$ at the ends and, hence, that Equation 7.47b is satisfied. On the other hand, υ'' is zero at the ends, because the bending moment is zero at the hinges and, hence, Equation 7.47c is also satisfied for the specific case considered.

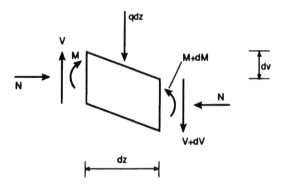

Figure 7.7

An alternative, and more immediate, way of obtaining Equation 7.47a is that of considering the equilibrium of a rotated beam element delimited by two vertical sections (Figure 7.7). Equilibrium with regard to vertical translation furnishes

$$\frac{dV}{dz} = -q \tag{7.48}$$

where V represents the vertical component of the internal reaction, which is not to be confused in this case with the transverse or shearing component.

On the other hand, equilibrium with regard to rotation furnishes

$$V = \frac{dM}{dz} - N\frac{dv}{dz} \tag{7.49}$$

from which, via Equation 7.48, we find

$$\frac{d^2M}{dz^2} - N\frac{d^2v}{dz^2} + q = 0 \tag{7.50}$$

Finally, using the relationship bending moment versus curvature, we arrive back at Equation 7.47a.

Consider the case of a uniformly distributed load $q(z) = q$ (Figure 7.8a). The integral of the Equation 7.47a assumes the form:

$$v(z) = A\cos\alpha z + B\sin\alpha z + Cz + D + \frac{qz^2}{2N} \tag{7.51}$$

where we have set:

$$\alpha^2 = \frac{N}{EI} \tag{7.52}$$

The four constants, A, B, C, D are determined by the boundary conditions:

$$v(0) = v(l) = EIv''(0) = EIv''(l) = 0 \tag{7.53}$$

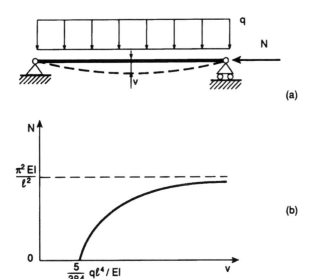

Figure 7.8

which yield

$$A = -D = \frac{q}{\alpha^2 N} \tag{7.54a}$$

$$B = \frac{q}{\alpha^2 N} \frac{1 - \cos \alpha l}{\sin \alpha l} \tag{7.54b}$$

$$C = -\frac{ql}{2N} \tag{7.54c}$$

Equation 7.51 therefore becomes

$$\upsilon(z) = \frac{q}{N} \left\{ \frac{1}{\alpha^2} \left[(1 - \cos \alpha l) \frac{\sin \alpha z}{\sin \alpha l} - (1 - \cos \alpha z) \right] - \frac{z(l - z)}{2} \right\} \tag{7.55}$$

It is important to note that, for $\alpha l \rightarrow \pi$, that is, for

$$N \rightarrow N_c = \pi^2 \frac{EI}{l^2} \tag{7.56}$$

we have $\sin \alpha l \rightarrow 0$ and, hence, a deformed configuration that tends to infinity (Figure 7.8b). This means that the flexural stiffness of a compressed beam is less than that of the same beam not loaded in compression, if we take into account the geometrical nonlinearities. This stiffness even becomes zero when the compressive force equals its critical value N_c. The same occurs, for instance, in the case where the supported beam is loaded, in addition to the compressive force N, also by an end moment m (Figure 7.9).

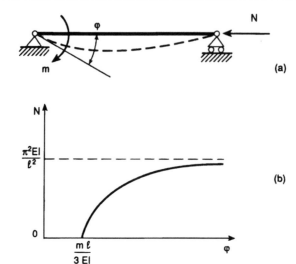

Figure 7.9

In the case where the distributed load is absent, that is, $q = 0$, the equation of the elastic line with geometrical nonlinearities (Equation 7.47a) simplifies as follows:

$$EIv^{IV} + Nv'' = 0 \tag{7.57}$$

The integral of Equation 7.57 is

$$v(z) = A\cos\alpha z + B\sin\alpha z + Cz + D \tag{7.58}$$

Enforcing the boundary conditions (Equation 7.53), we have

$$
\begin{bmatrix}
1 & 0 & 0 & 1 \\
\cos\alpha l & \sin\alpha l & l & 1 \\
-\alpha^2 & 0 & 0 & 0 \\
-\alpha^2\cos\alpha l & -\alpha^2\sin\alpha l & 0 & 0
\end{bmatrix}
\begin{bmatrix}
A \\ B \\ C \\ D
\end{bmatrix}
=
\begin{bmatrix}
0 \\ 0 \\ 0 \\ 0
\end{bmatrix}
\tag{7.59}
$$

The system possesses a solution different from the trivial one if and only if the determinant of the coefficient matrix is zero, and hence when $\sin\alpha l = 0$. This condition coincides with the one that makes the flexural stiffness of the beam zero (Equation 7.55).

It is possible to arrive at the same solution by imposing the condition that, in each section of the beam, the destabilizing moment:

$$M_i = Nv \tag{7.60a}$$

should be equal to the stabilizing moment:

$$M_s = -EI \frac{d^2 v}{dz^2} \qquad (7.60b)$$

Therefore, letting $M_i = M_s$, we obtain the differential equation:

$$v'' + \alpha^2 v = 0 \qquad (7.61)$$

the second derivative of which coincides with Equation 7.57. The complete integral of Equation 7.61 is

$$v(z) = A \cos \alpha z + B \sin \alpha z \qquad (7.62)$$

and, since we must have $v(0) = v(l) = 0$, it follows that

$$A = 0, \qquad \sin \alpha l = 0 \qquad (7.63)$$

and the coefficient B can assume any value.

From the second of Equations 7.63, we obtain the succession of the eigenvalues of the problem

$$\alpha_n = \frac{n\pi}{l}, \quad n = \text{natural number} \qquad (7.64)$$

and, hence, from Equation 7.52:

$$N_{cn} = n^2 \pi^2 \frac{EI}{l^2} \qquad (7.65)$$

To each eigenvalue N_{cn}, there corresponds an eigenfunction:

$$v_n(z) = B \sin \alpha_n z \qquad (7.66)$$

which represents the critical mode of deformation for that force. This deformed configuration consists of a number n of sinusoidal half-waves (Figure 7.10). Of course, if there are no further constraints on the beam apart from the two end supports, the critical load is that corresponding to $n = 1$:

$$N_{c1} = \pi^2 \frac{EI}{l^2} \qquad (7.67)$$

This force, called **Euler's critical load**, is the force that determines the buckling of the beam. For $N < N_{c1}$, the equilibrium is stable; for $N = N_{c1}$, the equilibrium is neutral; while for $N > N_{c1}$, the equilibrium is unstable. It should be noted that Euler's critical load increases in proportion to the rigidity EI of the beam, and decreases in inverse proportion to the square of the length of the beam.

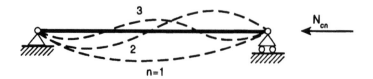

Figure 7.10

Euler's formula shows, on the other hand, limits of validity in the case of insufficiently slender beams, for which the inelastic behavior of the material can come to interact with the mechanism of buckling.

Let us indicate **Euler's critical pressure** by

$$\sigma_c = \frac{N_{c1}}{A} \tag{7.68}$$

which, on the basis of Equation 7.67, can be cast in the form:

$$\sigma_c = \pi^2 \frac{EI}{l^2 A} = \pi^2 E \frac{\rho^2}{l^2} \tag{7.69}$$

where ρ denotes the radius of gyration of the cross section in the direction of the bending axis. If λ designates the slenderness l/ρ, it is possible to express Equation 7.69 in the following form:

$$\sigma_c = \frac{\pi^2 E}{\lambda^2} \tag{7.70}$$

Drawing the diagram of Equation 7.70 on the plane $\sigma_c - \lambda^2$, we obtain the so-called **Euler's hyperbola** (Figure 7.11). This hyperbola envisages critical loads tending to zero as the slenderness tends to infinity and, conversely, critical loads tending to infinity as the slenderness tends to zero. The latter tendency is unlikely, since for stubby beams, the failure due to yielding:

$$\sigma_c = \sigma_P \tag{7.71}$$

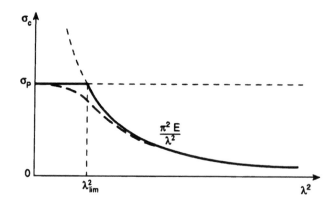

Figure 7.11

can precede, even markedly, that due to buckling (Equation 7.70). If there were no interaction between the two critical phenomena, there would be a point of discontinuity in the passage from one to the other, corresponding to the **limit slenderness**:

$$\lambda_{\lim} = \pi \left(\frac{E}{\sigma_P} \right)^{1/2} \tag{7.72}$$

which proves to be a function of the elastic modulus E and the yielding stress σ_P of the material. For steel, $E/\sigma_P \sim 10^3$ and, hence, $\lambda_{\lim} \sim 10^2$.

In actual fact, the two critical phenomena interact and, hence, there is a gradual transition from one to the other as the slenderness of the beam varies. The critical pressure is thus furnished by the dashed curve of Figure 7.11, which connects the two critical curves corresponding to Equations 7.70 and 7.71, rounding off the cusp that these form at their point of intersection. This curve joining the two is normally given in tabulated form, putting

$$\sigma < \sigma_P / \omega \tag{7.73}$$

where ω is a safety factor greater than unity that depends on the material and on the slenderness of the beam.

So far, we have only examined the case of a beam that is constrained by a hinge and a roller support. Equation 7.57 represents, on the other hand, the equilibrium equation of a beam, whatever the means of constraint. The boundary conditions vary, instead, according to the constraints at the ends. Since there are four degrees of freedom—free or restrained—at the two ends (two deflections and two rotations), there are, likewise, four boundary conditions. These are partly **kinematic** (or **essential**) **conditions** and partly **static** (or **natural**) **conditions**. Table 7.1 illustrates the different possible cases: a beam supported at either end; a cantilever beam; a beam built in at one end and supported at the other; a beam with one end built in and the other constrained with a transverse double rod; a beam with one end built in and the other constrained by an axial double rod; and a beam supported at one end and constrained at the other by an axial double rod. For each of these cases, the kinematic and static boundary conditions are given, with the reminder that the second derivative of the deflection υ'' is proportional to the bending moment, while the third derivative υ''' is proportional to the shearing force. In the case of the cantilever, the static condition:

$$EI\upsilon'''(l) + N\upsilon'(l) = 0 \tag{7.74}$$

or

$$T(l) = -N\varphi(l) \tag{7.75}$$

yields the shear at the end as the transverse component of the horizontal force N (Figure 7.12).

For each case, Table 7.1 then gives the critical load, which is always expressible in the form:

$$N_{c1} = \pi^2 \frac{EI}{l_0^2} \tag{7.76}$$

Table 7.1

Kinematic conditions	Static conditions	Critical load (N_{cl})	Free length of deflection (l_0)
l = Beam length			
$v(0) = 0$ $v(l) = 0$	$v''(0) = 0$ $v''(l) = 0$	$\pi^2 \dfrac{EI}{l^2}$	l
$v(0) = 0$ $v'(0) = 0$	$v'(l) = 0$ $EIv'''(l) + Nv'(l) = 0$	$\pi^2 \dfrac{EI}{4\,l^2}$	$2l$
$v(0) = 0$ $v(l) = 0$ $v'(0) = 0$	$v''(l) = 0$	$\sim 2\pi^2 \dfrac{EI}{l^2}$	$\dfrac{\sim l}{\sqrt{2}}$
$v(0) = 0$ $v(l) = 0$ $v'(0) = 0$ $v'(l) = 0$	None	$4\pi^2 \dfrac{EI}{l^2}$	$\dfrac{l}{2}$
$v(0) = 0$ $v'(l) = 0$ $v'(0) = 0$	$v'''(l) = 0$	$\pi^2 \dfrac{EI}{l^2}$	l
$v(0) = 0$ $v'(l) = 0$	$v''(0) = 0$ $v'''(l) = 0$	$\pi^2 \dfrac{EI}{4l^2}$	$2l$

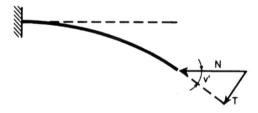

Figure 7.12

The dimension l_0 is the so-called **free length of deflection**, which represents the distance between two successive points of inflection in the critical deformed configuration.

Finally, notice how the static (or natural) conditions may also be deduced from the boundary conditions (Equations 7.47b and c), once the kinematic (or geometrical) conditions are applied to the perturbation $\delta\upsilon$ and to its derivative $\delta\upsilon'$.

7.5 FRAMED BEAM SYSTEMS

In some cases, beam systems, owing to their simplicity, can be accommodated within the elementary schemes of Table 7.1. In particular, portal frames with rigid cross members can be referred directly to the last four cases, according to whether or not wind bracing is present, and to whether the feet of the columns are hinged or built in (Figure 7.13).

In other cases, the axial redundant reactions, obtainable by the usual equations of congruence, can cause instability of equilibrium. A classic case is that of bars hinged or built in at the ends (Figure 7.14), subjected to an increase in temperature and hence preventing dilation. If the bar is only hinged at the ends, the critical temperature increase is (Figure 7.14a)

$$\Delta T_c = \pi^2/\alpha\lambda^2 \tag{7.77}$$

while it is quadrupled if the bar is built in (Figure 7.14b).

When the beam system cannot be reduced to the schemes already seen, it is possible to apply the finite element method, considering the elastic and geometrical stiffness matrices, already introduced in Section 7.3.

For the ith beam, we can assume

$$\upsilon_i(z) = \{\eta_i\}^{\mathrm{T}}_{(1\times4)} \{\delta_i\}_{(4\times1)} \tag{7.78}$$

where:

    υ_i represents the transverse displacement
    $\{\eta_i\}$ denotes the shape function vector
    $\{\delta_i\}$ indicates the nodal displacement vector (two transverse displacements and two rotations)

The shape functions $\{\eta_i\}$ must be chosen in such a way that

$$\upsilon_i'(0) = -\delta_{i1}, \ \upsilon_i(0) = \delta_{i2}, \ \upsilon_i'(l_i) = -\delta_{i3}, \ \upsilon_i(l_i) = \delta_{i4}$$

For beams of constant cross section, the shape functions $\{\eta_i\}$ are cubic, and are obtained by imposing, in turn, one of the nodal displacements $\delta_{ij} = 1$, and leaving the others equal to zero (Figure 7.15).

The total potential energy of the ith beam in a generic deformed configuration $\upsilon_i(z)$ is equal to

$$W(\upsilon_i) = \int_0^{l_i} \left[\frac{1}{2} \left(EI_i \upsilon_i''^2 - N_i \upsilon_i'^2 \right) - q_i \upsilon_i \right] dz \tag{7.79}$$

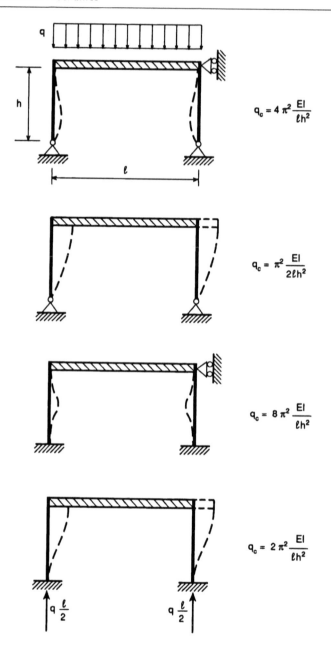

Figure 7.13

Figure 7.14

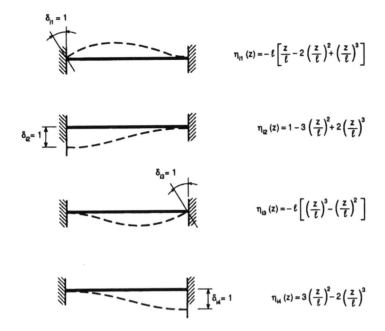

$$\eta_{i1}(z) = -\ell\left[\frac{z}{\ell} - 2\left(\frac{z}{\ell}\right)^2 + \left(\frac{z}{\ell}\right)^3\right]$$

$$\eta_{i2}(z) = 1 - 3\left(\frac{z}{\ell}\right)^2 + 2\left(\frac{z}{\ell}\right)^3$$

$$\eta_{i3}(z) = -\ell\left[\left(\frac{z}{\ell}\right)^3 - \left(\frac{z}{\ell}\right)^2\right]$$

$$\eta_{i4}(z) = 3\left(\frac{z}{\ell}\right)^2 - 2\left(\frac{z}{\ell}\right)^3$$

Figure 7.15

From Equation 7.78 we have

$$v_i'(z) = \{\eta_i'\}^{\mathrm{T}}\{\delta_i\} \tag{7.80a}$$

$$v_i''(z) = \{\eta_i''\}^{\mathrm{T}}\{\delta_i\} \tag{7.80b}$$

and hence

$$v_i'^2 = \{\delta_i\}^{\mathrm{T}}\{\eta_i'\}\{\eta_i'\}^{\mathrm{T}}\{\delta_i\} \tag{7.81a}$$

$$v_i''^2 = \{\delta_i\}^{\mathrm{T}}\{\eta_i''\}\{\eta_i''\}^{\mathrm{T}}\{\delta_i\} \tag{7.81b}$$

Substituting Equations 7.81 into Equation 7.79, we obtain

$$W(v_i) = \frac{1}{2}\{\delta_i\}^{\mathrm{T}}\left[\int_0^{l_i} EI_i\{\eta_i''\}\{\eta_i''\}^{\mathrm{T}}\,dz\right]\{\delta_i\}$$

$$-\frac{1}{2}N_i\{\delta_i\}^{\mathrm{T}}\left[\int_0^{l_i}\{\eta_i'\}\{\eta_i'\}^{\mathrm{T}}\,dz\right]\{\delta_i\}$$

$$-\left[\int_0^{l_i} q_i\{\eta_i\}^{\mathrm{T}}\,dz\right]\{\delta_i\} \tag{7.82}$$

Equation 7.82 may be cast in the form

$$W(v_i) = \frac{1}{2}\{\delta_i\}^T \left([K_i] - N_i [K_{gi}]\right)\{\delta_i\} - \{F_i\}^T \{\delta_i\} \tag{7.83}$$

which, compared with Equation 4.61, highlights the **elastic stiffness matrix**:

$$[K_i] = \int_0^{l_i} EI_i \{\eta_i''\}\{\eta_i''\}^T dz \tag{7.84a}$$

and the **geometrical stiffness matrix** of the ith beam:

$$[K_{gi}] = \int_0^{l_i} \{\eta_i'\}\{\eta_i'\}^T dz \tag{7.84b}$$

as well as the **equivalent nodal force vector**:

$$\{F_i\} = \int_0^{l_i} q_i \{\eta_i\}^T dz \tag{7.84c}$$

Computing with the shape functions given in Figure 7.15, we obtain

$$[K_i] = EI_i \begin{bmatrix} \dfrac{4}{l_i} & -\dfrac{6}{l_i^2} & \dfrac{2}{l_i} & \dfrac{6}{l_i^2} \\[2mm] -\dfrac{6}{l_i^2} & \dfrac{12}{l_i^3} & -\dfrac{6}{l_i^2} & -\dfrac{12}{l_i^3} \\[2mm] \dfrac{2}{l_i} & -\dfrac{6}{l_i^2} & \dfrac{4}{l_i} & \dfrac{6}{l_i^2} \\[2mm] \dfrac{6}{l_i^2} & -\dfrac{12}{l_i^3} & \dfrac{6}{l_i^2} & \dfrac{12}{l_i^3} \end{bmatrix} \tag{7.85}$$

which corresponds to Equation 2.24, and in addition:

$$[K_{gi}] = \frac{1}{l_i} \begin{bmatrix} \dfrac{2}{15}l_i^2 & -\dfrac{1}{10}l_i & -\dfrac{1}{30}l_i^2 & \dfrac{1}{10}l_i \\[2mm] -\dfrac{1}{10}l_i & \dfrac{6}{5} & -\dfrac{1}{10}l_i & -\dfrac{6}{5} \\[2mm] -\dfrac{1}{30}l_i^2 & -\dfrac{1}{10}l_i & \dfrac{2}{15}l_i^2 & \dfrac{1}{10}l_i \\[2mm] \dfrac{1}{10}l_i & -\dfrac{6}{5} & \dfrac{1}{10}l_i & \dfrac{6}{5} \end{bmatrix} \tag{7.86}$$

Basically, then, the presence of the axial force N_i decreases the stiffness of the ith element.

As regards the subsequent operations of the finite element method, the procedure is exactly as outlined in Chapter 4, with the **rotation** and the **expansion** of the local stiffness matrices. Finally, the assemblage operation provides the global stiffness matrices, so that the eigenvalue problem for seeking the critical loads is formulated as follows:

$$\det\left(\left[K\right] - \lambda\left[K_g\right]\right) = 0 \tag{7.87}$$

where λ represents the multiplier of the external loads.

In actual fact, the axial forces in the single beams do not increase in proportion to the loads. It may be assumed, on the other hand, in first approximation, that the axial forces are maintained proportionally to the values that are obtained from a geometrically linear analysis:

$$N_i\left(\lambda\right) = \lambda N_i\left(\lambda = 1\right) \tag{7.88}$$

7.6 RINGS AND CYLINDRICAL SHELLS SUBJECTED TO EXTERNAL PRESSURE

Let us consider a beam with curvilinear axis that is inextensible and not deformable in shear. The kinematic equations, on the hypothesis that $\gamma = \varepsilon = 0$, yield

$$\varphi = -\frac{dv}{ds} + \frac{w}{r} \tag{7.89a}$$

$$\frac{dw}{ds} = -\frac{v}{r} \tag{7.89b}$$

$$\chi = \frac{d\varphi}{ds} \tag{7.89c}$$

Substituting Equation 7.89a into Equation 7.89c, we obtain

$$\chi = -\frac{d^2v}{ds^2} + \frac{d}{ds}\left(\frac{w}{r}\right) \tag{7.90}$$

and, hence, neglecting the variation in the intrinsic curvature and applying Equation 7.89b:

$$\chi = -\frac{d^2v}{ds^2} - \frac{v}{r^2} \tag{7.91}$$

Finally, recalling the relation that links the variation of curvature χ and the bending moment M, we derive the **equation of the elastic line for curvilinear beams** (Boussinesq gives an analogous treatment):

$$\frac{d^2v}{ds^2} + \frac{v}{r^2} = -\frac{M}{EI} \tag{7.92}$$

Consider a cylindrical shell of radius R, subjected to an external pressure q, in a deformed configuration, symmetrical with respect to two orthogonal diameters (Figure 7.16). If the

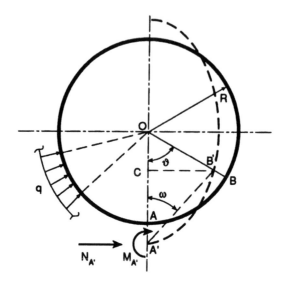

Figure 7.16

static characteristics in the point A' are $N_{A'} = -q\overline{A'O}$ and $M_{A'}$, the bending moment in the generic point B' is equal to

$$M_{B'} = M_{A'} + N_{A'}\overline{A'C} + \frac{1}{2}q\overline{A'B'}^2 \tag{7.93}$$

and hence

$$M_{B'} = M_{A'} - q\left(\overline{A'O}\ \overline{A'C} - \frac{1}{2}\overline{A'B'}^2\right) \tag{7.94}$$

Between the sides of the triangle $OA'B'$, there holds the relation

$$\overline{B'O}^2 = \overline{A'O}^2 + \overline{A'B'}^2 - 2\overline{A'O}\ \overline{A'B'}\cos\omega \tag{7.95}$$

whence

$$\overline{A'O}\ \overline{A'C} = \frac{1}{2}\left(\overline{A'O}^2 + \overline{A'B'}^2 - \overline{B'O}^2\right) \tag{7.96}$$

Having substituted Equation 7.96 into Equation 7.94, we obtain

$$M_{B'} = M_{A'} - \frac{1}{2}q\left(\overline{A'O}^2 - \overline{B'O}^2\right) \tag{7.97}$$

Since we have

$$\overline{A'O} = R + \upsilon_0 \tag{7.98a}$$

$$\overline{B'O} = R + \upsilon \tag{7.98b}$$

Equation 7.97, once infinitesimals of a higher order have been neglected, becomes

$$M_{B'} = M_{A'} + qR(\upsilon - \upsilon_0) \tag{7.99}$$

Taking into account Equation 7.99 and that $ds = Rd\vartheta$, Equation 7.92 is transformed as follows:

$$\frac{d^2\upsilon}{d\vartheta^2} + \upsilon = -\frac{R^2}{EI}\left[M_{A'} + qR(\upsilon - \upsilon_0)\right] \tag{7.100}$$

or

$$\frac{d^2\upsilon}{d\vartheta^2} + \upsilon\left(\frac{qR^3}{EI} + 1\right) = \frac{R^2}{EI}(qR\upsilon_0 - M_{A'}) \tag{7.101}$$

Setting

$$\alpha^2 = \frac{qR^3}{EI} + 1 \tag{7.102}$$

the eigenvalue equation becomes

$$\frac{d^2\upsilon}{d\vartheta^2} + \alpha^2\upsilon = \frac{R^2}{EI}(qR\upsilon_0 - M_{A'}) \tag{7.103}$$

The integral of Equation 7.103 is

$$\upsilon(\vartheta) = A\sin\alpha\vartheta + B\cos\alpha\vartheta + \frac{qR^3\upsilon_0 - M_{A'}R^2}{qR^3 + EI} \tag{7.104}$$

Enforcing the two conditions of symmetry:

$$\frac{d\upsilon}{d\vartheta} = \alpha A\cos\alpha\vartheta - \alpha B\sin\alpha\vartheta = 0, \quad \text{for } \vartheta = 0, \frac{\pi}{2} \tag{7.105}$$

we obtain the two equations:

$$\alpha A = 0 \tag{7.106a}$$

$$\alpha B\sin\frac{\alpha\pi}{2} = 0 \tag{7.106b}$$

which give

$$A = 0 \tag{7.107a}$$

$$\alpha\frac{\pi}{2} = n\pi, \ n = \text{natural number} \tag{7.107b}$$

Equation 7.107b yields the succession of eigenvalues:

$$\alpha_n = 2n, \quad n = \text{natural number} \tag{7.108}$$

For $n = 1$, from Equation 7.102 we obtain the first critical load:

$$q_c = \frac{3EI}{R^3} \tag{7.109}$$

and the deformed configuration:

$$\upsilon(\vartheta) = \frac{M_{A'}R^2 + \upsilon_0 EI}{qR^3 + EI} \cos 2\vartheta + \frac{qR^3\upsilon_0 - M_{A'}R^2}{qR^3 + EI} \tag{7.110}$$

On the other hand, from the condition of inextensibility (Equation 7.89b), we obtain

$$w(\vartheta) = -\frac{M_{A'}R^2 + \upsilon_0 EI}{qR^3 + EI} \frac{1}{2} \sin 2\vartheta - \frac{qR^3\upsilon_0 - M_{A'}R^2}{qR^3 + EI} \vartheta \tag{7.111}$$

For $\vartheta = 0, \pi/2$, the axial displacement w must vanish by symmetry, and hence

$$M_{A'} = qR\upsilon_0 \tag{7.112}$$

Substituting Equation 7.112 into Equation 7.110, we finally obtain the following deformed configuration:

$$\upsilon(\vartheta) = \upsilon_0 \cos 2\vartheta \tag{7.113}$$

which represents an **ovalization** of the tube (Figure 7.17a). It is possible, then, to demonstrate that the second eigenshape of even order consists of the four-lobed configuration of Figure 7.17b. In general, the nth eigenshape of even order will present $2n$ lobes.

Enforcing the more general conditions of symmetry:

$$\frac{d\upsilon}{d\vartheta} = \alpha A \cos \alpha\vartheta - \alpha B \sin \alpha\vartheta = 0, \quad \text{for } \vartheta = 0, \frac{\pi}{m+1}, \quad m = \text{natural number} \tag{7.114}$$

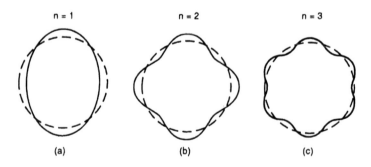

n = 1   n = 2   n = 3

(a)   (b)   (c)

Figure 7.17

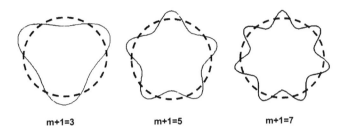

Figure 7.18

we obtain

$$\alpha A = 0 \tag{7.115}$$

$$\alpha B \sin \frac{\alpha \pi}{m+1} = 0 \tag{7.116}$$

from which there follows

$$A = 0 \tag{7.117}$$

$$\frac{\alpha}{m+1} = n, \quad n = \text{natural number} \tag{7.118a}$$

and, hence, the eigenshapes of odd order are also now included (Figure 7.18):

$$\alpha_c = m+1, \quad m+1 = \text{total number of lobes} \tag{7.118b}$$

7.7 LATERAL TORSIONAL BUCKLING

Consider a beam of thin rectangular cross section, constrained at the ends so that rotation about the longitudinal axis Z is prevented. Let this beam be subjected to uniform bending by means of the application at the ends of two moments m contained in the plane YZ of greater flexural rigidity (Figure 7.19a).

Consider a deformed configuration of the beam with deflection thereof in the XZ plane of smaller flexural rigidity and simultaneous torsion about the axis Z (Figure 7.19b). The deflection $u(z)$ and the torsional rotation $\varphi_z(z)$ generate components of the external moment m in the axial direction Z (Figure 7.19c) and in the transverse direction Y (Figure 7.19d), respectively:

$$M_{zi} = m \frac{du}{dz} \tag{7.119a}$$

$$M_{yi} = -m\varphi_z \tag{7.119b}$$

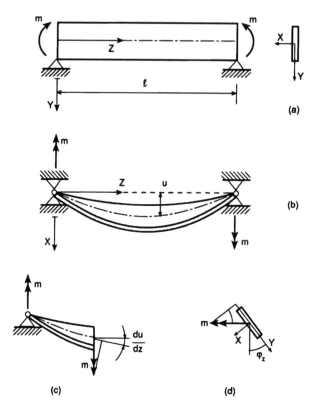

Figure 7.19

Both the loads M_{zi} and M_{yi} are destabilizing, because they tend to increase the torsional rotation φ_z and the flexural deflection u, respectively. On the other hand, as for Euler's beam, the corresponding stabilizing loads are present:

$$M_{zs} = GI_t \frac{\mathrm{d}\varphi_z}{\mathrm{d}z} \tag{7.120a}$$

$$M_{ys} = EI_y \frac{\mathrm{d}^2 u}{\mathrm{d}z^2} \tag{7.120b}$$

The equilibrium is neutral when Equations 7.119 are, respectively, equal to Equations 7.120:

$$GI_t \frac{\mathrm{d}\varphi_z}{\mathrm{d}z} = m \frac{\mathrm{d}u}{\mathrm{d}z} \tag{7.121a}$$

$$EI_y \frac{\mathrm{d}^2 u}{\mathrm{d}z^2} = -m\varphi_z \tag{7.121b}$$

Differentiating Equation 7.121a with respect to z and substituting the result into Equation 7.121b, we obtain

$$\frac{d^2\varphi_z}{dz^2} + \frac{m^2}{EGI_yI_t}\varphi_z = 0 \tag{7.122}$$

If we set

$$\alpha^2 = \frac{m^2}{EGI_yI_t} \tag{7.123}$$

the equation:

$$\varphi_z'' + \alpha^2\varphi_z = 0 \tag{7.124}$$

assumes the same form as Equation 7.61, corresponding to the problem of the axially compressed slender beam. As usual, the complete integral of Equation 7.124:

$$\varphi_z(z) = A\cos\alpha z + B\sin\alpha z \tag{7.125}$$

satisfies the boundary conditions:

$$\varphi_z(0) = \varphi_z(l) = 0 \tag{7.126}$$

for $A = 0$ and $\sin\alpha l = 0$. The eigenvalues of the problem are thus

$$\alpha_n = n\frac{\pi}{l}, \ n = \text{natural number} \tag{7.127a}$$

and the first, $\alpha_1 = \pi/l$, yields the critical load:

$$m_c = \frac{\pi}{l}\sqrt{EGI_yI_t} \tag{7.127b}$$

Equation 7.127b is commonly known as **Prandtl's formula.**

As an alternative, deriving Equation 7.121b with respect to z and substituting the result into Equation 7.121a, we obtain a differential equation in the variable u:

$$\frac{d^3u}{dz^3}EI_y = -\frac{m^2}{GI_t}\frac{du}{dz} \tag{7.128}$$

In this case, the same parameter α defined in Equation 7.123 can be recognized. The obtained differential equation has the same solutions as the characteristic equation of the one in the variable φ_z. As a consequence, the analogy between the previous and the actual boundary conditions, $u(0) = u(l) = 0$, leads to Equation 7.127b.

The phenomenon of lateral torsional buckling is especially relevant to deep beams, while it is virtually present for beams of compact cross section, for which the critical moment, expressed by Equation 7.127b, is so high as to exceed the plastic moment of the cross section (see Chapter 10). It is possible to note, on the other hand, how beams of compact cross section can also undergo lateral torsional buckling, in the case where they are particularly slender ($l \to \infty$).

7.8 PLATES SUBJECTED TO COMPRESSION

On the basis of Equations 3.2 and 3.3, the strain energy per unit surface of a deflected plate is

$$\Phi = -\frac{1}{2}\left(M_x \frac{\partial^2 w}{\partial x^2} + M_y \frac{\partial^2 w}{\partial y^2} + 2M_{xy} \frac{\partial^2 w}{\partial x \partial y} \right) \tag{7.129}$$

Using Equations 3.25, we obtain

$$\Phi = \frac{1}{2}D\left[\left(\frac{\partial^2 w}{\partial x^2}\right)^2 + \left(\frac{\partial^2 w}{\partial y^2}\right)^2 + 2\nu\left(\frac{\partial^2 w}{\partial x^2}\right)\left(\frac{\partial^2 w}{\partial y^2}\right) \right] + D(1-\nu)\left(\frac{\partial^2 w}{\partial x \partial y}\right)^2 \tag{7.130}$$

If, in addition to being considered undeformable in shear, the plate is also considered as inextensible and subjected to a membrane regime, N_x, N_y, N_{xy}, the potential energy of these loads in a deflected configuration is

$$\Phi_N = \frac{1}{2}\left[N_x\left(\frac{\partial w}{\partial x}\right)^2 + N_y\left(\frac{\partial w}{\partial y}\right)^2 + 2N_{xy}\left(\frac{\partial w}{\partial x}\right)\left(\frac{\partial w}{\partial y}\right) \right] \tag{7.131}$$

As regards the first two terms of Equation 7.131, these contributions are analogous to those calculated for the rectilinear beam (see Equation 7.44); while the third term represents the work of the shearing stresses acting through the shearing strains due to the deflection w, and can be justified as follows. Let us consider two infinitesimal segments OA and OB in the directions of the two coordinate axes X and Y (Figure 7.20). Because of the deflection

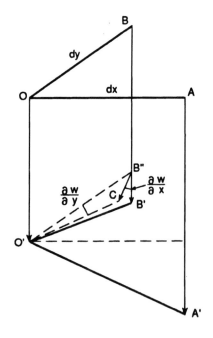

Figure 7.20

w, these segments are transformed into $O'A'$ and $O'B'$. The difference between the angle $A'O'B'$ and $\pi/2$ represents the shearing strain sought. For the purpose of determining this difference, let us consider the right angle $B''O'A'$. Rotating this angle about the side $O'A'$ by the amount $\partial w/\partial y$, the plane $B''O'A'$ comes to coincide with the plane $B'O'A'$, with the point B'' assuming the position C. The displacement $B''C$ is equal to $(\partial w/\partial y)dy$ and is inclined with respect to the vertical $B''B'$ by the angle $\partial w/\partial x$. Consequently, the segment CB' is equal to $(\partial w/\partial x)(\partial w/\partial y)dy$ and the angle $CO'B'$, which represents the shearing strain due to the deflection w, is equal to $(\partial w/\partial x)(\partial w/\partial y)$.

The total potential energy of the deflected plate is, therefore, equal to the sum of the integrals of the strain energy expressed by Equation 7.130 and of the potential energy of the membrane forces expressed by Equation 7.131:

$$W = \frac{1}{2}D\int_A \left\{ \left(\frac{\partial^2 w}{\partial x^2} + \frac{\partial^2 w}{\partial y^2} \right)^2 \right.$$

$$\left. -2(1-v)\left[\left(\frac{\partial^2 w}{\partial x^2} \right)\left(\frac{\partial^2 w}{\partial y^2} \right) - \left(\frac{\partial^2 w}{\partial x \partial y} \right)^2 \right] \right\} dx\, dy$$

$$+ \frac{1}{2}\int_A \left[N_x \left(\frac{\partial w}{\partial x} \right)^2 + N_y \left(\frac{\partial w}{\partial y} \right)^2 + 2N_{xy}\left(\frac{\partial w}{\partial x} \right)\left(\frac{\partial w}{\partial y} \right) \right] dx\, dy \tag{7.132}$$

The **variational equation of equilibrium** can be obtained by imposing the stationarity of W in a way similar to that already seen for the rectilinear beam:

$$D\left(\frac{\partial^4 w}{\partial x^4} + 2\frac{\partial^4 w}{\partial x^2 \partial y^2} + \frac{\partial^4 w}{\partial y^4} \right)$$

$$+ N_x \frac{\partial^2 w}{\partial x^2} + N_y \frac{\partial^2 w}{\partial y^2} + 2N_{xy}\frac{\partial^2 w}{\partial x \partial y} = 0 \tag{7.133}$$

Note that Equation 7.133 is formally analogous to Equation 7.57.

In the case of a rectangular plate of sides a, b, supported on the four sides and compressed by a force N per unit length of the edge, acting orthogonally to the side of length b, Equation 7.133 assumes the following form:

$$\left(\frac{\partial^4 w}{\partial x^4} + 2\frac{\partial^4 w}{\partial x^2 \partial y^2} + \frac{\partial^4 w}{\partial y^4} \right) + \frac{N}{D}\left(\frac{\partial^2 w}{\partial x^2} \right) = 0 \tag{7.134}$$

The constraints impose $w = 0$ on the four sides and the disappearance of the bending moment on the edge:

$$w = 0, \quad \left(\frac{\partial^2 w}{\partial y^2} \right) + v\left(\frac{\partial^2 w}{\partial x^2} \right) = 0, \;\; \text{for } y = 0, b \tag{7.135a}$$

$$w = 0, \quad \left(\frac{\partial^2 w}{\partial x^2} \right) + v\left(\frac{\partial^2 w}{\partial y^2} \right) = 0, \;\; \text{for } x = 0, a \tag{7.135b}$$

Each function:

$$w(x,y) = A_{nm} \sin n\pi \frac{x}{a} \sin m\pi \frac{y}{b} \qquad (7.136)$$

satisfies the preceding boundary conditions for n,m = natural numbers. Substituting Equation 7.136 into Equation 7.134 and dividing by the common factor $A_{nm} \sin(n\pi x/a)$ $\sin(m\pi y/b)$, we obtain

$$\left(\frac{n\pi}{a}\right)^4 + 2\left(\frac{n\pi}{a}\right)^2\left(\frac{m\pi}{b}\right)^2 + \left(\frac{m\pi}{b}\right)^4 = \frac{N}{D}\left(\frac{n\pi}{a}\right)^2 \qquad (7.137)$$

and hence

$$N_c^{nm} = \pi^2 D \frac{a^2}{n^2}\left(\frac{n^2}{a^2} + \frac{m^2}{b^2}\right)^2 \qquad (7.138)$$

The smallest value of N_c^{nm} is to be considered the critical load for instability of the elastic equilibrium of the plate. This value is obtained for $m = 1$, since m appears only in the numerator in Equation 7.138:

$$N_c^{n1} = \pi^2 \frac{D}{b^2}\left(n\frac{b}{a} + \frac{1}{n}\frac{a}{b}\right)^2 \qquad (7.139)$$

and corresponds to a deformed configuration with only one half-wave along the side of length b and n half-waves along the side of length a.

Figure 7.21 presents the diagram of the nondimensional critical load as a function of the ratio a/b between the sides of the rectangle. In actual fact, a succession of curves is obtained as n varies, but, for each value a/b, we have a certain value of n for which N_c^{n1} is a minimum. For $a/b < \sqrt{2}$, the minimum is obtained for $n = 1$. The critical deformed configuration thus

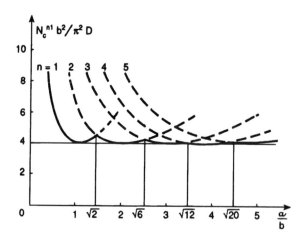

Figure 7.21

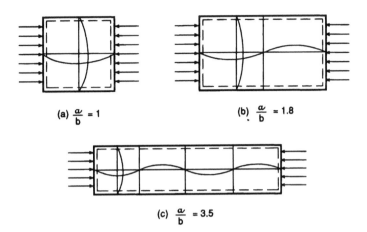

(a) $\frac{a}{b} = 1$ (b) $\frac{a}{b} = 1.8$

(c) $\frac{a}{b} = 3.5$

Figure 7.22

presents a half-wave in each direction (Figure 7.22a). For $\sqrt{2} < a/b < \sqrt{6}$, we have $n = 2$, and the critical deformed configuration presents two half-waves in the X direction and one half-wave in the Y direction (Figure 7.22b). For $a/b > \sqrt{6}$, we have $N_c^{n1} \simeq 4\pi^2 D / b^2$, and n is such as to give rise to half-waves of comparable amplitude along X and along Y (Figure 7.22c).

The behavior of the plate previously analyzed is analogous to that of a beam on an elastic foundation. In fact, it may be assimilated to that of a system of longitudinal beams constrained laterally through a system of transverse beams. This prevents the value of N_c^{n1} from dropping below the value $4\pi^2 D/b^2$, whatever the value of a may be.

The total potential energy of a beam on an elastic foundation is (Figure 7.23)

$$W = \frac{1}{2}\int_0^l \left(EIv''^2 - Nv'^2 + Kv^2\right)dz \tag{7.140}$$

where K is the elastic modulus of the foundation.

In the framework of the Ritz–Galerkin method, let us assume for the deflection v the following series expansion:

$$v(z) = \sum_n A_n \sin n\pi\frac{z}{l} \tag{7.141}$$

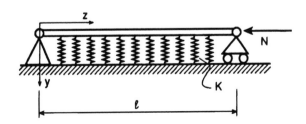

Figure 7.23

Substituting Equation 7.140 into Equation 7.139 and recalling the orthonormality of the trigonometric functions:

$$\int_0^l \sin n\pi \frac{z}{l} \sin m\pi \frac{z}{l} \, dz$$

$$= \int_0^l \cos n\pi \frac{z}{l} \cos m\pi \frac{z}{l} \, dz = \frac{l}{2} \delta_{nm} \tag{7.142}$$

where δ_{nm} is the Kronecker symbol, we obtain

$$W = \frac{l}{4} \sum_n A_n^2 \left(EI \frac{n^4 \pi^4}{l^4} - N \frac{n^2 \pi^2}{l^2} + K \right) \tag{7.143}$$

Equation 7.143 is a diagonal quadratic form in the coefficients A_n that ceases to be positive definite as soon as N is such as to cause one of the terms within parentheses to vanish:

$$N_{cn} = EI \frac{n^2 \pi^2}{l^2} + K \frac{l^2}{n^2 \pi^2} \tag{7.144}$$

Figure 7.24 presents the diagram of the critical load as a function of the length l of the beam. As in the case of the plate, we have a succession of curves according to the variation in n, but, for each value of l, we have a certain value of n for which N_{cn} is a minimum. These curves present local minima for values of l equal to

$$l = n\pi \left(\frac{EI}{K} \right)^{1/4} \tag{7.145}$$

and these minima are all equal to

$$N_{cn} = 2 \left(KEI \right)^{1/2} \tag{7.146}$$

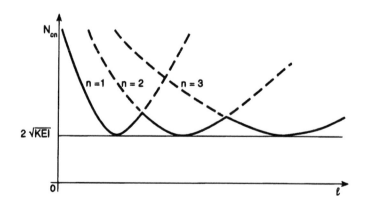

Figure 7.24

7.9 SHALLOW ARCHES AND SHELLS SUBJECTED TO VERTICAL LOADING: INTERACTION BETWEEN BUCKLING AND SNAP-THROUGH

Let us consider the shallow arch of Figure 7.25, consisting of two axially deformable rods of stiffness K, both hinged in the crown as well as in the foundation. Let the distance between the two springers of the arch be $2l$, and the angle that the two rods AC and BC form initially with the horizontal be α. Under the action of the force F, let this angle diminish by the infinitesimal quantity φ.

If only symmetrical deformations are considered, the system will have only one degree of freedom, and the strain energy of the arch will then be expressible as follows:

$$\Phi(\varphi) = K\left[\frac{l}{\cos\alpha} - \frac{l}{\cos(\alpha - \varphi)}\right]^2 \tag{7.147}$$

In the hypothesis of a flat arch, we can set

$$\cos\alpha \simeq 1 - \frac{\alpha^2}{2} \tag{7.148a}$$

$$\cos(\alpha - \varphi) \simeq 1 - \frac{1}{2}(\alpha - \varphi)^2 \tag{7.148b}$$

and, hence, with a further application of the Taylor series expansion:

$$\frac{1}{\cos\alpha} \simeq 1 + \frac{\alpha^2}{2} \tag{7.149a}$$

$$\frac{1}{\cos(\alpha - \varphi)} \simeq 1 + \frac{1}{2}(\alpha - \varphi)^2 \tag{7.149b}$$

Substituting Equations 7.149 into Equation 7.147, we obtain

$$\Phi(\varphi) = \frac{1}{4}Kl^2\varphi^2(2\alpha - \varphi)^2 \tag{7.150}$$

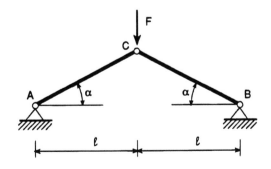

Figure 7.25

The deflection caused by the load F is equal, on the other hand, to

$$\eta(\varphi) = l\tan\alpha - l\tan(\alpha - \varphi) \qquad (7.151)$$

whereby, to a first approximation:

$$\eta(\varphi) \cong l\varphi \qquad (7.152)$$

Finally, the total potential energy of the system is found from Equations 7.150 and 7.152:

$$W(\varphi) = \Phi(\varphi) - F\eta(\varphi) \qquad (7.153)$$

whereby it is equal to

$$W(\varphi) = \frac{1}{4}Kl^2\varphi^2(2\alpha - \varphi)^2 - Fl\varphi \qquad (7.154)$$

The conditions of equilibrium are all those and only those for which Equation 7.154 is stationary:

$$W'(\varphi) = Kl^2\varphi(2\alpha^2 + \varphi^2 - 3\alpha\varphi) - Fl = 0 \qquad (7.155)$$

from which we obtain

$$F = Kl\varphi(\varphi - \alpha)(\varphi - 2\alpha) \qquad (7.156)$$

Equation 7.156 is displayed in Figure 7.26. There thus exist three positions of equilibrium with $F = 0$, when $\varphi = 0$, α, 2α. While the first and the last represent conditions of stable equilibrium with the connecting rods unloaded, the intermediate one is a condition of unstable equilibrium represented by the configuration with the connecting rods aligned and compressed. A rigorous study of the stability is conducted by examining the second derivative of the total potential energy:

$$W''(\varphi) = Kl^2\left(3\varphi^2 - 6\alpha\varphi + 2\alpha^2\right) \qquad (7.157)$$

which is greater than zero for

$$\varphi < \alpha\left(1 - \frac{\sqrt{3}}{3}\right) \quad \text{or} \quad \varphi > \alpha\left(1 + \frac{\sqrt{3}}{3}\right) \qquad (7.158)$$

Equation 7.155 is hence stationary for $\varphi = \alpha\left(1 - \frac{1}{3}\sqrt{3}\right)$, where it presents a maximum, and for $\varphi = \alpha\left(1 + \frac{1}{3}\sqrt{3}\right)$, where it presents a minimum (Figure 7.26). Since the third derivative of the total potential energy:

$$W'''(\varphi) = 6Kl^2(\varphi - \alpha) \qquad (7.159)$$

is different from zero for $\varphi \neq \alpha$, it may be concluded that the maximum of the curve $F(\varphi)$ represents a state of unstable equilibrium.

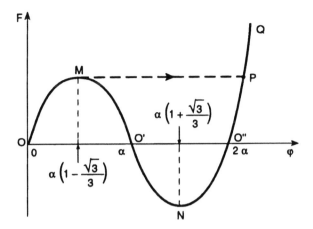

Figure 7.26

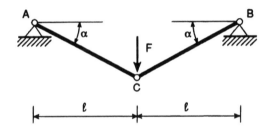

Figure 7.27

Therefore, if the flat arch ACB of Figure 7.25 is loaded, the portion OM of the curve $F(\varphi)$ of Figure 7.26 is traversed in a stable manner until the stationary point M is reached. If the load F continues to be increased, there is an abrupt jump on the stable branch PQ that, with the force F being equal, presents an angle φ which is much greater, and a configuration of the system that is inverted with respect to the initial one (Figure 7.27).

If, instead, we wish to go along the virtual branch MNP, it is necessary to control the phenomenon by imposing an angle φ that presents a continuous and monotonic growth. In this case, the force F can be interpreted as a constraint reaction that decreases between M and N, becoming even negative beyond the point O'. This means that, beyond the aligned connecting rod configuration, a force is necessary in the upward direction so as to proceed along the curve $F(\varphi)$ in a controlled manner.

The energy released by the system in the jump MP (Figure 7.26) is equal to the area $MO'NO''P$ multiplied by the length l. This energy will thus be transformed into the vibrational kinetic energy of the system about the condition represented by the point P. The instability phenomenon just described, and in particular the jump MP at constant load, is termed **snap-through**, and is analogous to the phenomenon of **snap-back** that will be looked at in Chapter 12.

Also in the more complex cases of flat arches or shells, which are both flexurally and axially or membranally compliant, the phenomenon of snap-through can develop, so giving rise to a sudden change of configuration. Figure 7.28 shows the load versus deflection curves corresponding to spherical thin domes built in at the edge, loaded by a uniform pressure q. The dashed curves correspond to a linear elastic buckling analysis, while the continuous line curves correspond to a geometrically nonlinear step-by-step analysis, considering both

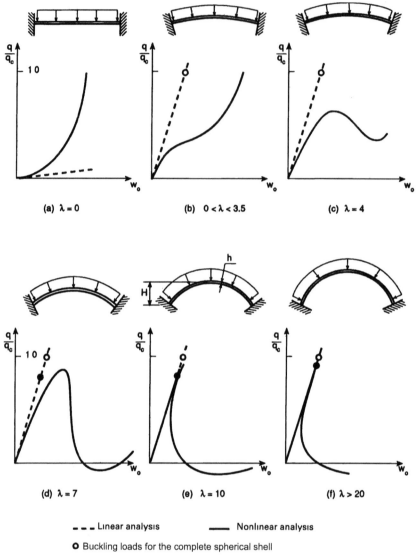

(a) $\lambda = 0$ (b) $0 < \lambda < 3.5$ (c) $\lambda = 4$

(d) $\lambda = 7$ (e) $\lambda = 10$ (f) $\lambda > 20$

- - - Linear analysis   —— Nonlinear analysis

○ Buckling loads for the complete spherical shell
● Buckling loads for the built-in spherical dome

Figure 7.28

flexural and membrane deformabilities. On the linear response curves, the buckling instability loads are also marked, corresponding, respectively, to a complete spherical shell of equal radius and thickness, or to the same built-in spherical dome. The parameter λ represents the slenderness to shallowness (or flatness) ratio of the dome:

$$\lambda = 2\left[3\left(1 - \nu^2\right)\right]^{1/4} \left(H / h\right)^{1/2} \tag{7.160}$$

where H is the rise of the dome with respect to the edge plane and h is the thickness.

For $\lambda \leq 3.5$, the behavior of the shell does not present the phenomenon of snap-through. In particular, for $\lambda = 0$, the load versus deflection curve presents a tension stiffening due to the intervention of the tensile forces as the plate is deflected. For $3.5 \leq \lambda \leq 7$, the phenomenon of snap-through emerges in the curves of Figure 7.28. Finally, for $\lambda \geq 7$, both the phenomena of snap-through and snap-back present themselves. Note that, in this latter interval, the behavior of the dome prior to instability tends to be increasingly linear as λ increases. In conclusion, a compression-buckling to tension-stiffening transition occurs, passing through snap-through instabilities, as the slenderness-to-flatness ratio decreases, that is, as the slenderness R/h decreases and the flatness R/H increases.

Very interestingly, a cusp catastrophe (or snap-back) interpretation of buckling instability emerges, analogously to the cusp catastrophe interpretations of fracture instability (Carpinteri 1989) and friction stick-slip instability (Carpinteri et al. 2009). Therefore, the second of René Thom's seven categories (the first one is folding and represents softening behavior) represents the most significant mechanical instabilities: fracture, friction, and buckling.

Snap-through and snap-back phenomena occur during the different phases of cracking in high-strength concrete reinforced beams (Figure 7.29). Whereas snap-back is due basically to brittle fracturing of the concrete, snap-through is due to pulling out, yielding, and hardening of the steel reinforcing bars.

Finally, it is worthwhile recalling how the phenomena of snap-through and snap-back are theoretically predicted, both for complete spherical shells subjected to external pressure (Figure 7.30a) and for cylindrical shells subjected to axial compression (Figure 7.30b). On the other hand, it is difficult to bring out such phenomena experimentally, in view of the considerable sensitivity to initial imperfections displayed by the abovementioned geometries. With the increase in the initial inherent imperfections, the structural response tends to become less unstable and the phenomena of snap-through and snap-back both disappear. Figure 7.31 represents the load versus axial contraction response in the case of an axially loaded cylindrical shell, as the eccentricity of the cross section varies. For particularly high eccentricities, the phenomenon of snap-through also disappears.

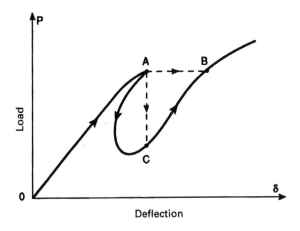

Figure 7.29

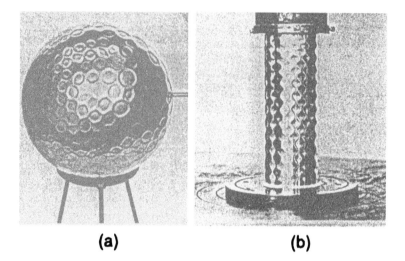

(a)   **(b)**

Figure 7.30

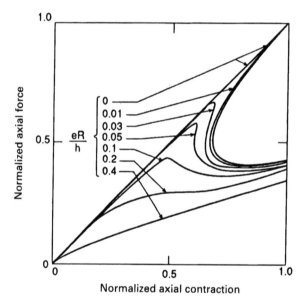

Figure 7.31

7.10 TRUSSED VAULTS AND DOMES: THE CASE OF PROGRESSIVE SNAP-THROUGH

Nowadays, different light and high-strength materials lead to the realization of long-span and slender roofing structures. The most usual typology is the trussed space structure, which offers technological advantages (prefabrication and assemblage) together with convenient architectural aspects, such as cost efficiency and sustainability.

If all the elements that constitute the roof are contained in a plane, the trussed roof is commonly called a **single-layer grid**. The pattern that defines the disposal of the elements is generally chosen to satisfy the required rigidity as well as the architectural aspects; some of these patterns are shown in Figure 7.32a. In some cases, it is necessary to vertically connect two planar grids to improve the global rigidity of the structure, which becomes a double-layer grid (Figure 7.32b). Otherwise, a trussed roof can be geometrically defined by the introduction of one or two curvatures into the generating surface. In these cases, the resultant structure is, respectively, a **vault** (Figure 7.32c) or a **dome** (Figure 7.32d). The use of a curved structure leads to better efficiency in terms of structural strength and stiffness.

When the total span is large compared with the rise, the probability of local and/or global instabilities increases. One of the most impressive collapses of space trussed structures was that of the Bucharest Meeting Hall dome (Romania), built in 1961 and failed two years later (Figure 7.33a). In this case, the structure was subjected to a severe snap-through, hanging, almost undamaged, from the robust concrete ring at its edge (Figure 7.33a). The dome collapsed due to the action of a modest snow load.

The snap-through and snap-back phenomena can be observed at the meso- and nanoscales. Regarding the mesoscale, we can refer to an electromechanical switch that creates a contact by a sudden change in its geometrical configuration (Figure 7.33b). The nanoscale can be, on the other hand, related to cellular biology, where a protein can be seen as a complex structure that modifies its geometry by a series of drastic snaps (folding) to explicate the metabolic functions (Figures 7.33c and d).

Let us consider the possibility of sustaining the three-hinged flat arch already examined with a continuous straight horizontal beam that is able to support the crown, as illustrated in Figure 7.34a. This configuration can be assumed as the modular scheme of a trussed roofing vault (Figure 7.34b). The determination of the structural response can be investigated through a nonlinear step-by-step analysis. It is possible to observe the influence of the geometrical properties of the supporting beam on the structural stability. Different supporting beam sections are considered with an increasing moment of inertia from I-1 to I-4.

Curves force versus deflection are reported in Figure 7.34c. It is evident from the results that the snap-through disappears when the supporting beam is present (dashed lines). The first part of the curves presents a hardening response, whereas in the second part, tension stiffening prevails.

The structure illustrated in Figure 7.35a is a shallow trussed system of six bars converging to the crown, and connected at the ground to an equilateral hexagonal bracing ring. This simple structure could represent the modular scheme of a diamatic trussed dome. As in the previous case, the stability of the structure is studied by a nonlinear step-by-step analysis. In Figure 7.35c, curves F versus δ are reported when the connection between all the bars is realized by spherical hinges. Six different shallownesses ($\lambda = l/z$) are considered, and it can be easily seen how they substantially influence the structural response.

The same analysis is conducted considering fixed-joints instead of hinges as internal and external constraints (Figure 7.35d). It is evident that the mechanical behavior drastically changes. If the shallowness ratio is greater than 15, snap-through instability can occur even with fixed-joints. In conclusion, the constraints and the flatness play a crucial role in the behavior of a shallow roofing structure.

Finally, the case of a diamatic dome is analyzed. The structure is similar to the ones illustrated in Figure 7.32d and the shallowness ratio is equal to 5. The internal and external constraints are assumed to be spherical hinges. A nonlinear step-by-step analysis,

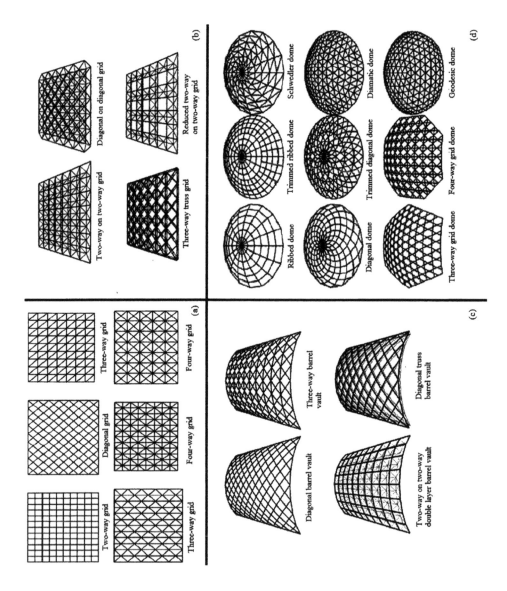

Figure 7.32

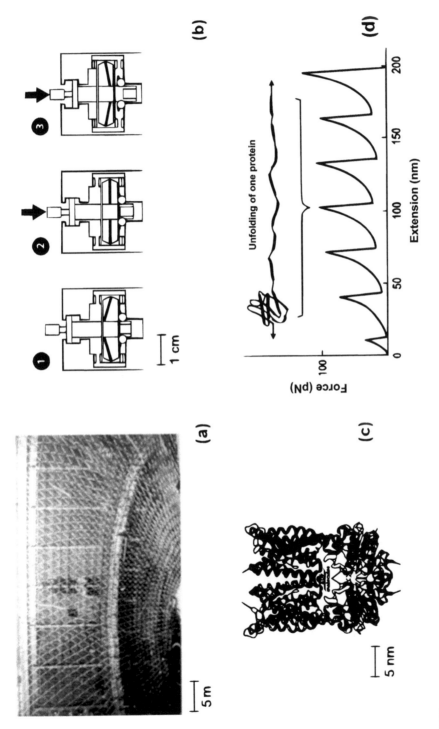

Figure 7.33

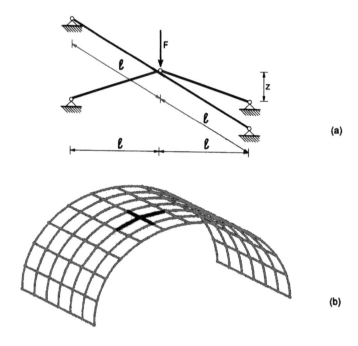

(a)

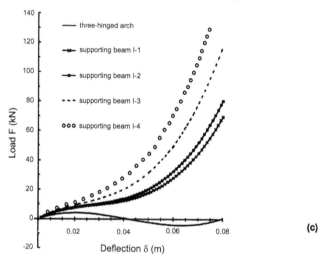

(b)

Continuous beam sustained arch (vault modulus)

(c)

Figure 7.34

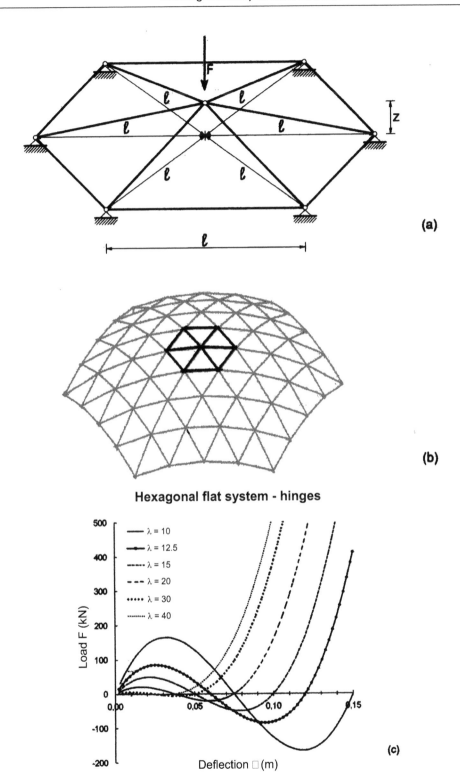

Hexagonal flat system - higes

Figure 7.35

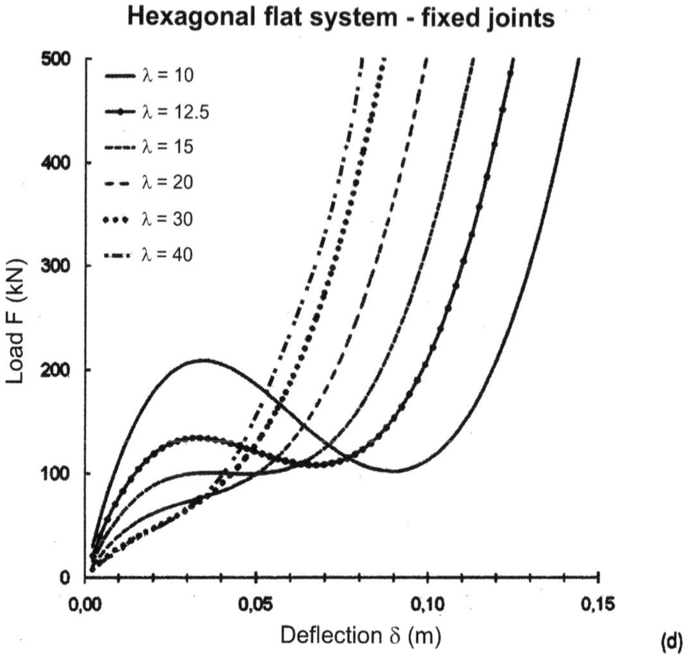

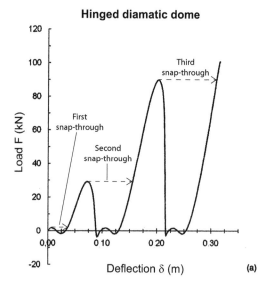

Hexagonal flat system - fixed joints

Figure 7.35 (Continued)

Hinged diamatic dome

Figure 7.36

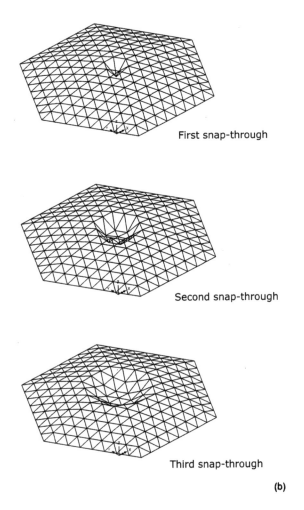

First snap-through

Second snap-through

Third snap-through

(b)

Figure 7.36 (Continued)

controlling the vertical deflection of the crown, gives the equilibrium path toward a vertical concentrated reactive force (Figure 7.36a). After a dent or corrugation of the inner part, the structure is still globally stable, although it has already collapsed locally. If the controlling parameter of the analysis were the force, the structure would change its configuration via a snap-through and a dent up to the second ring of the mesh, and so on with the following rings (Figures 7.36a and b).

Chapter 8

Long-span structures

Dynamics and buckling

8.1 INTRODUCTION

In Chapters 5 and 6, where we dealt with the dynamics of discrete and continuous systems, respectively, we showed how to determine the natural frequencies of a linear elastic structure according to modal analysis. In addition, we remarked on the importance of keeping the vibration frequencies of an external periodic excitation as far as possible from such natural frequencies to avoid resonance phenomena.

On the other hand, in Chapter 7, we showed how to obtain the buckling loads of a structure according to a linearized analysis of elastic stability. The smallest among these loads (or load multipliers) defines the critical load (or critical load multiplier) that separates the region of stability (load smaller than the critical one) from that of instability (load greater than the critical one).

From a mathematical point of view, both modal analysis and buckling analysis appear as eigenvalue problems: they both lead to two formally identical equations written in symbolic form, each one having, of course, a different physical meaning. In this chapter, we will show how the dynamic problem and the buckling problem are connected and how the related collapses may interact. We will also discuss the dependence of the natural frequencies on the applied loads. Beginning with the analysis of structures subjected to conservative loads (such as gravity loads), we will derive a **generalized eigenvalue problem**, where both the buckling loads and the natural frequencies of the system are unknown and represent the eigenvalues. In particular, we will consider numerous examples: discrete mechanical systems with one or two degrees of freedom; continuous mechanical systems, such as oscillating beams subjected to a compressive axial load; as well as oscillating beams subjected to a load that may cause lateral buckling. Furthermore, a general finite element formulation will be outlined, with the possibility to be applied to two- or three-dimensional beam systems, as well as to plates and shells.

Afterward, we will introduce a particular case of mechanical instability that may be encountered when a structure is subjected to nonconservative loads, such as, for example, aerodynamic and hydrodynamic forces. We will emphasize the fact that the **static method** (i.e., the direct equilibrium method or Euler method) and the **energy method** are not universal, and they cannot be applied to the analysis of the elastic stability of a system subjected to nonconservative loads. We will, therefore, introduce a more general method of analysis, the so-called **dynamic method**, as well as the concept of dynamic instability (**flutter**) of a mechanical system, in addition to the already studied static instability (**divergence**).

Lastly, in the final part of the present chapter, we will give a brief description of the main forms of **aeroelastic instability** occurring in long-span suspension or cable-stayed bridges subjected to wind loads: torsional divergence, galloping, and flutter.

8.2 INFLUENCE OF DEAD LOADS ON NATURAL FREQUENCIES

In engineering applications, there are a lot of situations where structures that undergo flexural vibrations are also loaded by a static axial load. Very common examples are structural members as columns, struts, and towers. In other cases, instead, beams carrying transverse loads may undergo lateral–torsional vibrations.

As we pointed out in Chapter 7, the stiffness of a slender structure subjected to static loads is influenced by geometric nonlinearities. This variation in stiffness produces, in turn, a variation in the natural vibration frequencies that will, therefore, differ from those of the unloaded structure. Consequently, an external harmonic excitation should have a frequency that matches one of the natural frequencies of the loaded structure to produce resonance, instead of those of the unloaded one.

One example is given by the transverse vibrations of a cable in tension: if the tension is increased, the geometric contribution to the elastic stiffness makes the cable more rigid, and so its natural frequencies increase. Similarly, a tension on a beam will increase its natural frequencies of bending, while a compression will decrease them. As we will see later on, there are more complicated cases, in which an external load increases certain frequencies, whereas others are reduced. Even though the natural frequencies of a structure depend on the applied load, this does not have to be necessarily regarded as negative: it could even become a powerful tool. For instance, one way to avoid the excitation of a particular natural frequency (and mode) of bending vibration of a beam could be to apply a suitable tension or compression in the axial direction.

In the first part of this chapter, we will focus our analysis on structures subjected to static loads that may cause instability. Thus, taking into account the effect of geometric nonlinearity in the equations of motion through the geometric stiffness matrix, the problem will be reduced to a generalized eigenproblem, where the natural frequencies can be obtained as functions of the load multiplier. According to this approach, all the **interaction curves** between the buckling loads and the natural frequencies, furnished, respectively, by the usual buckling and free dynamic analyses, can be obtained.

8.3 DISCRETE SYSTEMS WITH ONE OR TWO DEGREES OF FREEDOM

Let us consider the mechanical system shown in Figure 7.1a, consisting of two rigid rods connected by an elastic hinge of rotational stiffness k and constrained at one end by a hinge and at the other by a roller support. A mass m is now considered in correspondence to the intermediate elastic hinge and the system is loaded by an axial force N. Assuming the absolute rotation φ of the two arms as the generalized coordinate, the total potential energy W of the whole system is given by Equation 7.1, whereas its kinetic energy $\mathcal{T}$ is (dotted symbols indicate derivatives with respect to time t)

$$\mathcal{T}(\dot{\varphi}) = \frac{1}{2}m\left[\frac{d}{dt}(l\sin\varphi)\right]^2 + \frac{1}{2}m\left[\frac{d}{dt}(l - l\cos\varphi)\right]^2 \simeq \frac{1}{2}ml^2\dot{\varphi}^2 \tag{8.1}$$

The equation of motion can be determined by writing **Lagrange's equation:**

$$\frac{d}{dt}\left(\frac{\partial \mathcal{T}}{\partial \dot{\varphi}}\right) - \frac{\partial \mathcal{T}}{\partial \varphi} = -\frac{\partial W}{\partial \varphi} \tag{8.2}$$

In the present case, this yields

$$ml^2\ddot{\varphi} = -4k\varphi + 2Nl\sin\varphi \qquad (8.3)$$

which can be suitably linearized in correspondence to $\varphi = 0$:

$$ml^2\ddot{\varphi} = -4k\varphi + 2Nl\varphi \qquad (8.4)$$

Looking for the solution to Equation 8.4 in the general form $\varphi = \varphi_0 e^{i\omega t}$, where ω denotes the **natural angular frequency** of the system, we obtain the following equation that provides the conditions of dynamic equilibrium of the system:

$$\left(4k - 2Nl - \omega^2 ml^2\right)\varphi_0 = 0 \qquad (8.5)$$

A nontrivial solution to Equation 8.5 exists if and only if the term within parentheses is equal to zero. This condition establishes a one-to-one relationship between the applied axial force N and the angular frequency ω:

$$N = \frac{2k}{l} - \frac{ml}{2}\omega^2 \qquad (8.6)$$

Equation 8.6 admits two important limit cases for $N = 0$ and $m = 0$, respectively. In the former case, Equation 8.6 gives the natural angular frequency of the system according to classic modal analysis, that is, $\omega_1 = \sqrt{4k/ml^2}$. In the latter, the critical load for buckling instability is obtained, that is, $N_c = 2k/l$ (see Equation 7.3 and Figure 7.1b). Notice that the last limit case can also be obtained, with a nonzero value of m, for $\omega = 0$. In fact, Equation 8.6 can be rewritten as $\omega = \sqrt{((4k/l^2) - (2N/l))/m}$, showing that the natural frequency vanishes for $N = N_c$, the condition at which the system has no stiffness with respect to transverse displacements. In fact, the vanishing of the natural frequency in correspondence to the buckling load denotes the loss of static elastic stability, as will be discussed further in Section 8.7.

Dividing Equation 8.6 by N_c, we obtain the following relationship between N and ω in a nondimensional form:

$$\frac{N}{N_c} + \left(\frac{\omega}{\omega_1}\right)^2 = 1 \qquad (8.7)$$

A graphical representation of Equation 8.7 in Figure 8.1 shows that the resonance frequency is a decreasing function of the compression axial load. This demonstrates, for the analyzed mechanical system with a single degree of freedom, that resonance can take place for $\omega < \omega_1$, provided that the system is loaded by an axial compression force N given by Equation 8.7.

As a second example, let us consider the mechanical system shown in Figure 7.2a, consisting of two rigid rods on three supports, of which the intermediate one is assumed to be elastically compliant with stiffness k. As in the previous case, a mass m is now considered in correspondence to the intermediate hinge and the system is loaded by an axial force N. Assuming the absolute rotation φ of the two arms as the generalized coordinate, the total

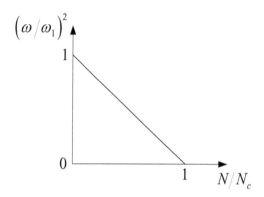

Figure 8.1

potential energy W is given by Equation 7.6, whereas the kinetic energy T is still given by Equation 8.1.

Following the procedure discussed, we determine the equation of motion by employing Lagrange's Equation 8.2:

$$ml^2\ddot{\varphi} = -l\sin\varphi(kl\cos\varphi - 2N) \tag{8.8}$$

which can be suitably linearized in correspondence to $\varphi = 0$:

$$ml^2\ddot{\varphi} = -l\varphi(kl - 2N) \tag{8.9}$$

Looking for the solution to Equation 8.9 in the general form $\varphi = \varphi_0 e^{i\omega t}$, where ω denotes the natural angular frequency of the system, we obtain the following condition of dynamic equilibrium of the system:

$$\left(kl^2 - 2Nl - \omega^2 ml^2\right)\varphi_0 = 0 \tag{8.10}$$

As in the previous example, by setting the term within parentheses equal to zero, we obtain a one-to-one relationship between the applied axial force N and the angular frequency ω:

$$N = \frac{kl}{2} - \frac{ml}{2}\omega^2 \tag{8.11}$$

Analogously to the previous case, this equation admits two important limit cases for $N = 0$ and $m = 0$ (or, alternatively, $\omega = 0$), respectively. In the former case, Equation 8.11 gives the natural angular frequency of the system according to classic modal analysis, that is, $\omega_1 = \sqrt{k/m}$. In the latter, the critical load for buckling instability is obtained, that is, $N_c = kl/2$ (see Equation 7.8 and Figure 7.2b). Dividing Equation 8.11 by N_c, we obtain the same relationship between the nondimensional terms N/N_c and $(\omega/\omega_1)^2$ as in the previous example (see Equation 8.7).

As a third example, let us now consider the mechanical system with two degrees of freedom shown in Figure 7.3a, consisting of three rigid rods connected by two elastic hinges of rotational stiffness k, and constrained at one end by a hinge and at the other by a roller support. A mass m is now considered in correspondence to both the elastic hinges and the system is loaded by an axial force N. Assuming the transverse displacements x_1 and x_2 of the elastic hinges as the generalized coordinates, the total potential energy W of the whole system is given by Equation 7.11, whereas its kinetic energy T is given by

$$T\left(\dot{x}_1, \dot{x}_2\right) = \frac{1}{2}m\left[\frac{d}{dt}\left(l\sin\left(\arcsin\left(\frac{x_1}{l}\right)\right)\right)\right]^2 + \frac{1}{2}m\left[\frac{d}{dt}\left(l - l\cos\left(\arcsin\left(\frac{x_1}{l}\right)\right)\right)\right]^2$$

$$+ \frac{1}{2}m\left[\frac{d}{dt}\left(l\sin\left(\arcsin\left(\frac{x_1}{l}\right)\right) + l\sin\left(\arcsin\left(\frac{x_2 - x_1}{l}\right)\right)\right)\right]^2$$

$$+ \frac{1}{2}m\left[\frac{d}{dt}\left(2l - l\cos\left(\arcsin\left(\frac{x_1}{l}\right)\right) - l\cos\left(\arcsin\left(\frac{x_2 - x_1}{l}\right)\right)\right)\right]^2$$

$$\cong \frac{1}{2}m\dot{x}_1^2 + \frac{1}{2}m\dot{x}_1^2\frac{x_1^2}{l^2} + \frac{1}{2}m\dot{x}_2^2 + \frac{1}{2}m\left(\frac{2x_1\dot{x}_1}{l} + \frac{x_2\dot{x}_2}{l} - \frac{x_2\dot{x}_1}{l} - \frac{x_1\dot{x}_2}{l}\right)^2 \quad (8.12)$$

where the following approximations are adopted:

$$\frac{d}{dt}\left[\arcsin\left(\frac{x_1}{l}\right)\right] = \frac{1}{\sqrt{1 - \left(\frac{x_1}{l}\right)^2}}\frac{\dot{x}_1}{l} \cong \frac{\dot{x}_1}{l} \quad (8.13a)$$

$$\frac{d}{dt}\left[\arcsin\left(\frac{x_2 - x_1}{l}\right)\right] = \frac{1}{\sqrt{1 - \left(\frac{x_2 - x_1}{l}\right)^2}}\frac{\dot{x}_2 - \dot{x}_1}{l} \cong \frac{\dot{x}_2 - \dot{x}_1}{l} \quad (8.13b)$$

Performing a Taylor series expansion of Equations 7.11 and 8.12 about the origin, and assuming $x_1/l < 1/10$ and $x_2/l < 1/10$, we obtain Equation 7.12 and the following equation, respectively:

$$T\left(\dot{x}_1, \dot{x}_2\right) \cong \frac{1}{2}m\dot{x}_1^2 + \frac{1}{2}m\dot{x}_2^2 \quad (8.14)$$

The equations of motion can be identified by considering **Lagrange's equations:**

$$\frac{d}{dt}\left(\frac{\partial T}{\partial \dot{x}_i}\right) - \frac{\partial T}{\partial x_i} = -\frac{\partial W}{\partial x_i}, \quad i = 1, 2 \quad (8.15)$$

In matrix form, they appear as follows:

$$\begin{bmatrix} m & 0 \\ 0 & m \end{bmatrix}\begin{bmatrix} \ddot{x}_1 \\ \ddot{x}_2 \end{bmatrix} + \begin{bmatrix} \dfrac{5k}{l^2} & -\dfrac{4k}{l^2} \\ -\dfrac{4k}{l^2} & \dfrac{5k}{l^2} \end{bmatrix}\begin{bmatrix} x_1 \\ x_2 \end{bmatrix} - N\begin{bmatrix} \dfrac{2}{l} & -\dfrac{1}{l} \\ -\dfrac{1}{l} & \dfrac{2}{l} \end{bmatrix}\begin{bmatrix} x_1 \\ x_2 \end{bmatrix} = \begin{bmatrix} 0 \\ 0 \end{bmatrix} \quad (8.16)$$

Looking for the solution to Equation 8.16 in the general form $\{q\} = \{q_0\}e^{i\omega t}$, where ω denotes the natural angular frequency of the system, we obtain the following equation, written in compact form:

$$\left(-\omega^2 [M] + [K] - N[K_g]\right)\{q_0\} = \{0\} \tag{8.17}$$

where $[M]$, $[K]$, and $[K_g]$ denote, respectively, the **mass matrix**, the **elastic stiffness matrix**, and the **geometric stiffness matrix** of the mechanical system.

Their expressions can be simply obtained by comparing Equation 8.17 with Equation 8.16.

A nontrivial solution to Equation 8.17 exists if and only if the determinant of the resultant coefficient matrix of the vector $\{q_0\}$ vanishes. This yields the following **generalized eigenvalue problem**:

$$\det\left([K] - N[K_g] - \omega^2 [M]\right) = 0 \tag{8.18}$$

where N and ω^2 represent the **eigenvalues**. For this example, Equation 8.18 provides the following relationships between the eigenvalues ω^2 and N:

$$\omega^2 = \frac{k}{ml^2} - \frac{N}{ml} \tag{8.19a}$$

$$\omega^2 = \frac{9k}{ml^2} - \frac{3N}{ml} \tag{8.19b}$$

As limit cases, if $m = 0$ (or, alternatively, $\omega = 0$), then we obtain the critical buckling loads N_{c1} and N_{c2}, given, respectively, by Equations 7.17a and b; whereas, if $N = 0$, we obtain the natural frequencies of the unloaded system:

$$\omega_1 = \sqrt{\frac{k}{ml^2}} \tag{8.20a}$$

$$\omega_2 = \sqrt{\frac{9k}{ml^2}} \tag{8.20b}$$

As far as the eigenvectors are concerned, Equation 8.17 yields the **eigenvectors** corresponding, respectively, to the eigenfrequencies (Equations 8.19a and b) as functions of N:

$$x_1 = \frac{4k/l - N}{6k/l - 3N} x_2 \tag{8.21a}$$

$$x_1 = \frac{4k/l - N}{14k/l - 5N} x_2 \tag{8.21b}$$

Dividing Equations 8.19a and b, respectively, by ω_1^2 and ω_2^2, we derive the following non-dimensional relationships between the eigenvalues:

$$\frac{N}{N_{c1}} + \left(\frac{\omega}{\omega_1}\right)^2 = 1 \tag{8.22a}$$

$$\frac{N}{N_{c2}} + \left(\frac{\omega}{\omega_2}\right)^2 = 1 \tag{8.22b}$$

In analogy with the results obtained for the single-degree-of-freedom systems, a graphical representation of Equations 8.22a and b is provided in Figure 8.2 (the values of ω^2 and N are nondimensionalized with respect to ω_1^2 and N_{c1} for both curves). We notice that both the eigenfrequencies are decreasing functions of the compression axial load. Entering the diagram with a value of the nondimensional compression axial force in the range $0 < N/N_{c1} < 1$, the coordinates of the points of the two curves provide the two modified resonance frequencies of the mechanical system. Axial forces larger than N_{c1} in the range $1 < N/N_{c1} < N_{c2}/N_{c1}$ can only be experienced if an additional constraint is introduced into the system to prevent transverse displacement of the midpoint, while allowing, at the same time, rotation and axial displacement.

As the last example, let us consider the system with two degrees of freedom reported in Figure 7.5 that consists of three rigid rods on four supports, of which the central ones are assumed to be elastically compliant with stiffness k. A mass m is now considered in correspondence to the intermediate hinges and the system is loaded by an axial force N. Assuming the vertical displacements x_1 and x_2 of the central hinges as the generalized coordinates, the total potential energy W and the kinetic energy T, after a Taylor series expansion about the origin ($x_1/l < 1/10$ and $x_2/l < 1/10$), are given by Equations 7.26 and 8.14, respectively.

In this case, Lagrange's Equation 8.15 yields the following matrix form:

$$\begin{bmatrix} m & 0 \\ 0 & m \end{bmatrix} \begin{bmatrix} \ddot{x}_1 \\ \ddot{x}_2 \end{bmatrix} + \begin{bmatrix} k & 0 \\ 0 & k \end{bmatrix} \begin{bmatrix} x_1 \\ x_2 \end{bmatrix} - N \begin{bmatrix} \dfrac{2}{l} & -\dfrac{1}{l} \\ -\dfrac{1}{l} & \dfrac{2}{l} \end{bmatrix} \begin{bmatrix} x_1 \\ x_2 \end{bmatrix} = \begin{bmatrix} 0 \\ 0 \end{bmatrix} \tag{8.23}$$

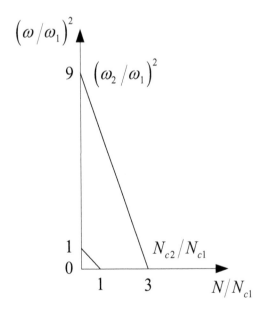

Figure 8.2

Looking for the solution to Equation 8.23 in the general form $\{q\} = \{q_0\}e^{i\omega t}$, we obtain again the compact form (Equation 8.17) and, therefore, the generalized eigenvalue problem (Equation 8.18). As can be readily seen by comparing Equations 8.23 and 8.16, the mass matrix and the geometric stiffness matrix for this problem are the same as those of the previous example, the only difference being represented by the elastic stiffness matrix.

For the present example, Equation 8.18 provides the following relationships between the eigenvalues:

$$\omega^2 = \frac{k}{m} - 3\frac{N}{ml} \tag{8.24a}$$

$$\omega^2 = \frac{k}{m} - \frac{N}{ml} \tag{8.24b}$$

As limit cases, if $m = 0$ (or, alternatively, $\omega = 0$), we obtain the Eulerian buckling loads N_{c1} and N_{c2}, given by Equations 7.30a and b, respectively; whereas, if $N = 0$, then we obtain the natural frequencies of the unloaded system:

$$\omega_1 = \omega_2 = \sqrt{\frac{k}{m}} \tag{8.25}$$

As far as the eigenvectors are concerned, Equation 8.17 yields the **eigenvectors** corresponding, respectively, to the eigenfrequencies (Equations 8.24a and b), as functions of the axial force, N:

$$x_1 = \frac{N/l}{5N/l - 2k}x_2 \tag{8.26a}$$

$$x_1 = \frac{N/l}{3N/l - 2k}x_2 \tag{8.26b}$$

Dividing Equations 8.24a and b by ω_1^2 and ω_2^2, respectively, we derive two nondimensional relationships between the eigenvalues formally identical to Equations 8.22a and b. A graphical representation of the relationships between ω^2 and N for the present example is provided in Figure 8.3 in a nondimensional form. Also in this case, both the frequencies are decreasing functions of the compression axial load, and only the range $0 \le N/N_{c1} \le 1$ is of practical interest, unless suitable additional constraints are inserted into the system.

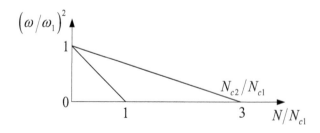

Figure 8.3

8.4 FLEXURAL OSCILLATIONS OF BEAMS SUBJECTED TO COMPRESSION AXIAL LOADS

Let us consider a slender elastic beam of constant cross section, inextensible and not deformable in shear, though deformable in bending; constrained at one end by a hinge and at the other by a roller support; loaded by an axial force, N (see Figure 7.6). In this case, with the purpose of analyzing the free flexural oscillations of the beam, the differential equation of the elastic line with second-order effects can be written by replacing the distributed load with the force of inertia:

$$EI\frac{\partial^4 v}{\partial z^4} + N\frac{\partial^2 v}{\partial z^2} = -\mu\frac{\partial^2 v}{\partial t^2} \tag{8.27}$$

where EI denotes the flexural rigidity of the beam and μ is its linear density (mass per unit length). Equation 8.27 can be rewritten in the following form:

$$\frac{\partial^4 v}{\partial z^4} + \alpha^2\frac{\partial^2 v}{\partial z^2} = -\frac{\mu}{EI}\frac{\partial^2 v}{\partial t^2} \tag{8.28}$$

where we have set $\alpha^2 = N/EI$, according to Equation 7.52.

Equation 8.28 is an equation with separable variables, the solution being represented as the product of two different functions, each one depending on a single variable as expressed by Equation 6.4.

Introducing Equation 6.4 into Equation 8.28 leads to

$$\frac{d^4\eta}{dz^4}f + \alpha^2\frac{d^2\eta}{dz^2}f + \frac{\mu}{EI}\eta\frac{d^2f}{dt^2} = 0 \tag{8.29}$$

Dividing Equation 8.29 by the product ηf, we find

$$-\frac{\dfrac{d^2f}{dt^2}}{f} = \frac{EI}{\mu}\frac{\dfrac{d^4\eta}{dz^4} + \alpha^2\dfrac{d^2\eta}{dz^2}}{\eta} = \omega^2 \tag{8.30}$$

where ω^2 represents a positive constant, the left- and the right-hand sides of Equation 8.30 being, at most, functions of the time t and of the coordinate z, respectively. From Equation 8.30, there follow two ordinary differential equations, the first of which is expressed by Equation 6.7a, whereas the other is given by

$$\frac{d^4\eta}{dz^4} + \alpha^2\frac{d^2\eta}{dz^2} - \beta^4\eta = 0 \tag{8.31}$$

where

$$\beta = \sqrt[4]{\frac{\mu\omega^2}{EI}} \tag{8.32}$$

Whereas Equation 6.7a is the equation of the harmonic oscillator, whose complete integral is given by Equation 6.9a, Equation 8.31 has the following complete integral:

$$\eta(z) = Ce^{\lambda_1 z} + De^{\lambda_2 z} + Ee^{-\lambda_1 z} + Fe^{-\lambda_2 z} \tag{8.33}$$

where λ_1 and λ_2 are functions of α and β:

$$\lambda_{1,2} = \sqrt{\frac{-\alpha^2 \pm \sqrt{\alpha^4 + 4\beta^4}}{2}} \tag{8.34}$$

As in modal analysis, the constants A and B can be determined on the basis of the initial conditions, while the constants C, D, E, and F can be determined by imposing the boundary conditions. As will be shown in the following, for a given value of α, the parameters β and ω can be determined by solving a generalized eigenvalue problem resulting from the imposition of the boundary conditions. From the mathematical point of view, this eigenvalue problem is analogous to that shown for the discrete systems. On the other hand, since we are considering a continuous mechanical system that has infinite degrees of freedom, we shall obtain an infinite number of eigenvalues β_i and ω_i, just as there will be an infinite number of eigenfunctions f_i and η_i. The complete integral of the differential Equation 8.27 may, therefore, be expressed in the form given by Equation 6.10, according to the principle of superposition, with $f_i(t)$ given by Equation 6.11a and $\eta_i(z)$ expressed as follows:

$$\eta_i(z) = C_i e^{\lambda_{1i} z} + D_i e^{\lambda_{2i} z} + E_i e^{-\lambda_{1i} z} + F_i e^{-\lambda_{2i} z} \tag{8.35}$$

It is possible to prove that the eigenfunctions η_i are still **orthonormal functions** when the beam is subjected to an axial load N, as in the classical modal analysis. We may, in fact, write Equation 8.31 for two different eigensolutions:

$$\eta_i^{IV} + \alpha^2 \eta_i'' = \beta_i^4 \eta_i \tag{8.36a}$$

$$\eta_k^{IV} + \alpha^2 \eta_k'' = \beta_k^4 \eta_k \tag{8.36b}$$

Multiplying Equation 8.36a by η_k and Equation 8.36b by η_j, and integrating over the beam length, we obtain

$$\int_0^l \eta_k \eta_i^{IV} dz + \alpha^2 \int_0^l \eta_k \eta_i'' dz = \beta_j^4 \int_0^l \eta_k \eta_j dz \tag{8.37a}$$

$$\int_0^l \eta_j \eta_k^{IV} dz + \alpha^2 \int_0^l \eta_j \eta_k'' dz = \beta_k^4 \int_0^l \eta_j \eta_k dz \tag{8.37b}$$

Integrating by parts the left-hand sides, the foregoing equations transform as follows:

$$\left[\eta_k \eta_i'''\right]_0^l - \left[\eta_k' \eta_i''\right]_0^l + \int_0^l \eta_k'' \eta_i'' dz + \alpha^2 \left[\eta_k \eta_i'\right]_0^l$$

$$- \alpha^2 \int_0^l \eta_k' \eta_i' dz = \beta_j^4 \int_0^l \eta_k \eta_j dz \tag{8.38a}$$

$$\left[\eta_j\eta_k'''\right]_0^l - \left[\eta_j'\eta_k''\right]_0^l + \int_0^l \eta_j''\eta_k''dz + \alpha^2\left[\eta_j\eta_k\right]_0^l$$

$$-\alpha^2\int_0^l \eta_j'\eta_k'dz = \beta_k^4\int_0^l \eta_j\eta_k dz \tag{8.38b}$$

When each of the two ends of the beam is constrained by a built-in support ($\eta = \eta' = 0$), or by a hinge ($\eta = \eta'' = 0$), the quantities within square brackets vanish. On the other hand, when the end at $z = 0$ is either unconstrained ($\eta''' = \eta'' = 0$) or constrained by a double rod ($\eta''' = \eta' = 0$), the remaining end of the beam has to be constrained either by a built-in support ($\eta = \eta' = 0$) or by a simple support ($\eta = \eta'' = 0$). For both configurations, only the terms $[\eta_j\eta_k]_0^l$ are different from zero.

In any case, subtracting member by member, these quantities are canceled and we have (see Equation 6.15)

$$\left(\beta_j^4 - \beta_k^4\right)\int_0^l \eta_j\eta_k = 0 \tag{8.39}$$

which leads to the orthonormality condition (Equation 6.16). Thus, when the eigenvalues are distinct, the integral of the product of the corresponding eigenfunctions vanishes. When, instead, the indices j and k coincide, the condition of normality reminds us that the eigenfunctions are defined but for a factor of proportionality.

The orthonormality of the eigenfunctions η_i permits us to determine the constants A_i and B_i in Equation 6.11a via the initial conditions (Equations 6.17a and b).

As regards the boundary conditions, let us consider as an example a beam of length l supported at both ends:

$$\begin{matrix} \eta(0) = 0, \\ \eta''(0) = 0, \\ \eta(l) = 0, \\ \eta''(l) = 0, \end{matrix} \Rightarrow \begin{bmatrix} 1 & 1 & 1 & 1 \\ \lambda_1^2 & \lambda_2^2 & \lambda_1^2 & \lambda_2^2 \\ e^{\lambda_1 l} & e^{\lambda_2 l} & e^{-\lambda_1 l} & e^{-\lambda_2 l} \\ \lambda_1^2 e^{\lambda_1 l} & \lambda_2^2 e^{\lambda_2 l} & \lambda_1^2 e^{-\lambda_1 l} & \lambda_2^2 e^{-\lambda_2 l} \end{bmatrix} \begin{bmatrix} C \\ D \\ E \\ F \end{bmatrix} = \begin{bmatrix} 0 \\ 0 \\ 0 \\ 0 \end{bmatrix} \tag{8.40}$$

For a nontrivial solution to the system in Equation 8.40, the determinant of the coefficient matrix must vanish. The resulting eigenequation permits us, for each given value of the parameter α, to determine the eigenvalues β_i of the system. Ultimately, the corresponding **natural frequencies** ω_i can be obtained by inverting Equation 8.32.

As an illustrative example, the first three nondimensional frequencies of the simply supported beam shown in Figure 7.6 are reported in Figure 8.4 as functions of the applied nondimensional axial force. Parameters ω_i and N_{ci} denote, respectively, the ith frequency of the system determined according to classic modal analysis (Equation 6.31) and the ith buckling load determined according to Euler's formula (Equation 7.65). In close analogy with the discrete mechanical systems, the curves in the $(\omega/\omega_1)^2$ versus N/N_{c1} plane are represented by straight lines (this is not generally true for other boundary conditions). Also in this case, the coordinates of the points along these lines provide the natural vibration frequencies as functions of the applied compression axial force. Lastly, we notice that axial loads larger than N_{c1} can only be experienced by providing suitable additional constraints to the beam.

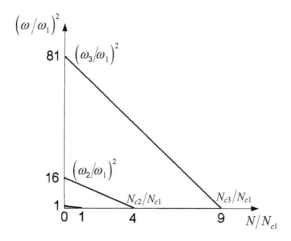

Figure 8.4

8.5 OSCILLATIONS AND LATERAL TORSIONAL BUCKLING OF DEEP BEAMS

Let us consider a beam of thin rectangular cross section, constrained at the ends so that rotation about the longitudinal axis Z is prevented. Let this beam be subjected to uniform bending by means of the application at the ends of two moments m contained in the plane YZ of maximum flexural rigidity (see Figure 7.19a).

Considering a deformed configuration of the beam, with deflection thereof in the XZ plane of minimum flexural rigidity and simultaneous torsion about the Z axis (see Figure 7.19b); flexural–torsional out-of-plane vibrations and buckling of the beam are described by the following partial differential equations:

$$EI_y \frac{\partial^4 u}{\partial z^4} + m \frac{\partial^2 \varphi_z}{\partial z^2} = -\mu \frac{\partial^2 u}{\partial t^2} \tag{8.41a}$$

$$-GI_t \frac{\partial^2 \varphi_z}{\partial z^2} + m \frac{\partial^2 u}{\partial z^2} = -\mu \rho^2 \frac{\partial^2 \varphi_z}{\partial t^2} \tag{8.41b}$$

where:

$u(z, t)$ and $\varphi_z(z, t)$ are, respectively, the out-of-plane deflection and the twist angle of the beam cross section

EI_y and GI_t are the flexural and torsional rigidities

μ is the mass of the beam per unit length

$\rho = \sqrt{I_P / A}$ is the polar radius of inertia of the beam cross section

A solution to Equations 8.41 can be found in the following separated-variable form:

$$u(z,t) = U(t)\eta(z) \tag{8.42a}$$

$$\varphi_z(z,t) = \Phi(t)\psi(z) \tag{8.42b}$$

where the functions $\eta(z)$ and $\psi(z)$ are such that the boundary conditions $\eta(0) = \eta(l) = \eta''(0) = \eta''(l) = 0$ and $\psi(0) = \psi(l) = 0$ are satisfied. By assuming $\eta(z) = \psi(z) = \sin(n\pi z/l)$, with n being a natural number, we obtain the following matrix equation:

$$\begin{bmatrix} \mu & 0 \\ 0 & \mu\rho^2 \end{bmatrix}\begin{bmatrix} \ddot{U} \\ \ddot{\Phi} \end{bmatrix} + \begin{bmatrix} EI_y\dfrac{n^4\pi^4}{l^4} & 0 \\ 0 & GI_t\dfrac{n^2\pi^2}{l^2} \end{bmatrix}\begin{bmatrix} U \\ \Phi \end{bmatrix} - m\begin{bmatrix} 0 & \dfrac{n^2\pi^2}{l^2} \\ \dfrac{n^2\pi^2}{l^2} & 0 \end{bmatrix}\begin{bmatrix} U \\ \Phi \end{bmatrix} = \begin{bmatrix} 0 \\ 0 \end{bmatrix} \tag{8.43}$$

which can be rewritten in compact form as

$$[M]\{\ddot{q}\} + [K]\{q\} - m[K_g]\{q\} = \{0\} \tag{8.44}$$

where $\{q\} = (U, \Phi)^{\mathrm{T}}$. The mass matrix $[M]$, the elastic stiffness matrix $[K]$, and the geometric stiffness matrix $[K_g]$ in Equation 8.44 can be defined in relation to Equation 8.43. Looking for a general solution in the form $\{q\} = \{q_0\}e^{i\omega t}$, we obtain

$$\left([K] - m[K_g] - \omega^2[M]\right)\{q_0\} = \{0\} \tag{8.45}$$

A nontrivial solution to Equation 8.45 exists if and only if the determinant of the resultant coefficient matrix of the vector $\{q_0\}$ vanishes. This yields the following **generalized eigenvalue problem**:

$$\det\left([K] - m[K_g] - \omega^2[M]\right) = 0 \tag{8.46}$$

where m and ω^2 are the **eigenvalues**.

As limit cases, if $\mu = 0$ (or, alternatively, $\omega = 0$), we then obtain the critical bending moments given by Prandtl's formula (compare with Equation 7.128):

$$m_{cn} = \frac{n\pi}{l}\sqrt{EI_y GI_t} \tag{8.47}$$

whereas, if $m = 0$, then we obtain the flexural and torsional eigenfrequencies of the unloaded beam:

$$\omega_n^{\text{flex}} = \left(\frac{n\pi}{l}\right)^2\sqrt{\frac{EI_y}{\mu}} \tag{8.48a}$$

$$\omega_n^{\text{tors}} = \frac{n\pi}{\rho l}\sqrt{\frac{GI_t}{\mu}} \tag{8.48b}$$

Considering a rectangular-section beam with a depth-to-span ratio of 1/3 and a thickness-to-depth ratio of 1/10, the variations of the first two flexural and torsional eigenfrequencies of the system are shown in Figure 8.5 as functions of the applied bending moment. In this

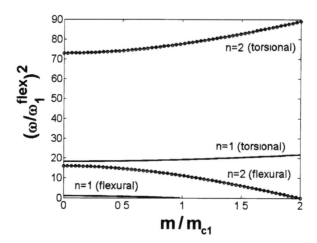

Figure 8.5

case, the curves in the nondimensional plane $(\omega/\omega_1^{flex})^2$ versus m/m_{c1} are no longer straight lines. Moreover, we observe that, as m increases from zero (condition of free vibrations of the unloaded beam), the resonance frequencies related to flexural oscillations progressively decrease from ω_n^{flex} down to zero in correspondence to the critical bending moments m_{cn} given by Prandtl's formula; whereas the resonance frequencies related to torsional oscillations increase indefinitely from ω_n^{tors}. From a physical point of view, this means that, if we take into account the geometric nonlinearities, the effect of the external moments m is to increase the torsional rigidity of the beam, decreasing its out-of-plane flexural rigidity at the same time. When the flexural rigidity of the fundamental lateral mode vanishes, the condition of neutral equilibrium with respect to the lateral torsional buckling of the beam is reached. In fact, even though instability originates in the flexural mode, it inevitably involves also the other mode because of the static coupling between bending and torsion (see Equations 8.41). Similarly to the previous cases, end moments m larger than m_{c1} cannot be applied without providing additional constraints to the beam.

8.6 FINITE ELEMENT FORMULATION FOR BEAMS, PLATES, AND SHELLS

When the mechanical system cannot be reduced to elementary schemes such as those previously analyzed, it is possible to apply the finite element method. According to this discretizing approach, the equations of motion for an elastic system with a finite number of degrees of freedom can be expressed in matrix form, also taking into account the effect of the geometric nonlinearity through the geometric stiffness matrix.

For a generic discretized elastic structure subjected to a given loading condition, we can obtain the following matrix equation for each finite element V_e (containing m nodal points, each one having g degrees of freedom):

$$[M_e]\{\ddot{\delta}_e\} + ([K_e] - [K_{ge}])\{\delta_e\} = \{0\} \tag{8.49}$$

where $[M_e]$, $[K_e]$, and $[K_{ge}]$ denote, respectively, the **local mass matrix**, the **local elastic stiffness matrix**, and the **local geometric stiffness matrix** of the finite element, and $\{\delta_e\}$ is the **nodal displacement vector**, which is a function of time t.

As usual, the local mass and elastic stiffness matrices are given by Equations 6.166 and 4.45, respectively.

The geometric stiffness matrix depends both on the initial load and on the geometry (topology) of the element (in fact, it is also called the **initial stress matrix** or **load-geometric matrix**). The initial stress state can be determined by a linear elastic static analysis. In the case of a compressive stress field, the geometric stiffness terms reduce the corresponding elements of the local elastic stiffness matrix, as shown previously for the discrete mechanical systems (see Sections 7.3 and 7.5). On the other hand, the information related to the finite element topology is simply included in the matrix $[\eta_e]$ that collects the shape functions. Some cases of particular relevance for practical applications will be discussed next.

The geometric stiffness matrix of a **beam in the plane** (two-node beam element) of constant cross section, loaded by a constant compression axial load, has been derived in Section 7.5 (Equation 7.86). To analyze the influence of axial loads on the free vibrations of **plane frames**, the geometric stiffness matrix of the ith element defined in Equation 7.86, of dimensions (4×4), must be expanded to dimensions (6×6) by adding zeros in the rows and columns corresponding to the axial degrees of freedom. Moreover, since, in general, the axial load N_i is not the same for all elements, it is convenient to include the initial axial load in the local geometric stiffness matrix of the ith element by multiplying the matrix in Equation 7.86 by N_i. The initial values N_i corresponding to a given loading condition can be determined from a geometrically linear analysis.

The previous formulation can be extended to the case of a **beam in space** (two-node 3-D beam element) of constant cross section, to account for the influence of a constant compression axial load on the flexural vibration frequencies of **space frames**. In this case, considering only bending deformation, we have two transverse displacements and two rotations per node. Analyzing the two beam deflections separately and adopting the principle of superposition, the local geometric stiffness matrix can be derived according to Equation 7.84b, where, for the shape functions $\{\eta_i\}$, we can assume the expressions reported in Figure 7.15 for both deflections. In this way, we obtain an (8×8) matrix that must be expanded to dimensions (12×12) by adding zeros in the rows and columns that correspond to the axial and torsional degrees of freedom. Ultimately, the local geometric stiffness matrix of the 3-D beam element appears as follows (N_i is positive if compression, negative otherwise; the elements are ordered in agreement with Equation 2.59):

$$\left[K_{gi}\right]_{(12\times12)} = \frac{N_i}{l_i} \begin{bmatrix} \left[K_{gi,11}\right] & \left[K_{gi,21}\right] \\ \left[K_{gi,12}\right] & \left[K_{gi,22}\right] \end{bmatrix} \tag{8.50}$$

where

$$\left[K_{gi,11}\right] = \begin{bmatrix} \dfrac{6}{5} & 0 & 0 & 0 & -\dfrac{1}{10} & 0 \\ 0 & \dfrac{6}{5} & 0 & -\dfrac{1}{10} & 0 & 0 \\ 0 & 0 & 0 & 0 & 0 & 0 \\ 0 & -\dfrac{1}{10} & 0 & \dfrac{2}{15} & 0 & 0 \\ -\dfrac{1}{10} & 0 & 0 & 0 & \dfrac{2}{15} & 0 \\ 0 & 0 & 0 & 0 & 0 & 0 \end{bmatrix} \tag{8.51a}$$

$$[K_{gi,12}] = [K_{gi,21}]^T = \begin{bmatrix} -\dfrac{6}{5} & 0 & 0 & 0 & \dfrac{1}{10} & 0 \\ 0 & -\dfrac{6}{5} & 0 & \dfrac{1}{10} & 0 & 0 \\ 0 & 0 & 0 & 0 & 0 & 0 \\ 0 & -\dfrac{1}{10} & 0 & -\dfrac{1}{30} & 0 & 0 \\ -\dfrac{1}{10} & 0 & 0 & 0 & -\dfrac{1}{30} & 0 \\ 0 & 0 & 0 & 0 & 0 & 0 \end{bmatrix} \tag{8.51b}$$

$$[K_{gi,22}] = \begin{bmatrix} \dfrac{6}{5} & 0 & 0 & 0 & \dfrac{1}{10} & 0 \\ 0 & \dfrac{6}{5} & 0 & \dfrac{1}{10} & 0 & 0 \\ 0 & 0 & 0 & 0 & 0 & 0 \\ 0 & \dfrac{1}{10} & 0 & \dfrac{2}{15} & 0 & 0 \\ \dfrac{1}{10} & 0 & 0 & 0 & \dfrac{2}{15} & 0 \\ 0 & 0 & 0 & 0 & 0 & 0 \end{bmatrix} \tag{8.51c}$$

It must be noted that, according to this formulation, the geometric stiffness matrix depends only on the axial load acting in the element. On the other hand, disregarding the presence of external torsion and assuming that, in a deformed configuration, the torsional rotation is described by a linear polynomial expression (uniform or Saint-Venant's torsion), it can be proved that the geometric stiffness matrix of a two-node 3-D beam element is independent of the external bending moments and axial loads. As a consequence, lateral–torsional and purely torsional buckling cannot be treated using the present discretization procedure. For these reasons, the present formulation allows us to deal with beams having compact cross section, as well as thin-walled beams with closed section. In the case of thin-walled beams with open section, we need to adopt enhanced formulations based on *ad hoc* beam models.

In addition to the case of one-dimensional structural elements, such as beams, of great practical importance is also the vibration of stretched or compressed bidimensional elements, that is, plates and shells. For a generic **plate element** of constant thickness h, the local geometric stiffness matrix can be obtained, analogously to the beam element, as follows:

$$[K_{ge}] = h \iint [G_e]^T [\sigma_e][G_e] dx dy \tag{8.52}$$

where the integration is taken over the element area. The matrix $[\sigma_e]$ is related to the **components of the stress field**:

$$[\sigma_e] = \begin{bmatrix} \sigma_x & 0 & \tau_{xy} & 0 \\ 0 & \sigma_x & 0 & \tau_{xy} \\ \tau_{xy} & 0 & \sigma_y & 0 \\ 0 & \tau_{xy} & 0 & \sigma_y \end{bmatrix} \tag{8.53}$$

while the matrix $[G_e]$, called the **slope matrix**, is related to the first derivatives of the shape functions through the differential operator $[\bar{\partial}]$:

$$[G_e] = [\bar{\partial}][\eta_e] \tag{8.54}$$

where

$$[\bar{\partial}] = \begin{bmatrix} \dfrac{\partial}{\partial x} & 0 \\[2ex] 0 & \dfrac{\partial}{\partial x} \\[2ex] \dfrac{\partial}{\partial y} & 0 \\[2ex] 0 & \dfrac{\partial}{\partial y} \end{bmatrix} \tag{8.55}$$

According to this formulation, we note that the geometric stiffness matrix is a function of the stress components through the matrix $[\sigma_e]$. In this way, knowing the shape functions of the adopted finite element, it is possible to compute $[K_{ge}]$ for any given in-plane system of forces.

The preceding procedure can also be extended to the calculation of the geometric stiffness matrix of **shell elements**. However, we must observe that, due to the high imperfection sensitivity exhibited by shells with respect to stability, linear elastic buckling analyses do not give reliable results in the case of external compressive loads. In this case, geometrically nonlinear step-by-step analyses have to be adopted to calculate the tangent stiffness matrix.

Independently of the type of structural element, by performing the usual operations of **rotation**, **expansion**, and **assemblage** of the local matrices and vectors, Equation 8.49 can be written in global form for the entire structure:

$$[M]\{\ddot{\delta}\} + ([K] - \lambda[K_g])\{\delta\} = \{0\} \tag{8.56}$$

where λ is the **loading multiplier**, that is, a factor that increases linearly the initial static characteristics calculated for some specific external loading condition (see, e.g., Equation 7.88).

Looking for the solution to Equation 8.56 in the general form $\{\delta\} = \{\delta_0\}e^{i\omega t}$, where ω is the **natural angular frequency** of the system, we obtain the following homogeneous system in matrix form:

$$([K] - \lambda[K_g] - \omega^2[M])\{\delta_0\} = \{0\} \tag{8.57}$$

which ultimately leads to the **generalized eigenproblem**:

$$\det([K] - \lambda[K_g] - \omega^2[M]) = 0 \tag{8.58}$$

Once the eigenfrequencies ω_i corresponding to a specific value of λ have been determined, the related **mode shapes** (eigenvectors) of the structure can be found by solving Equation 8.57. It must be noted that, unless $[K_g]$ has a structure similar to $[M]$, the buckling modes and the vibration modes will not be identical. In many cases, the modes are similar, but in general this will not be the case.

Therefore, the numerical procedure for the determination of the natural frequency versus loading multiplier diagrams for an elastic system discretized by finite elements consists of the following steps:

1. Compute the local elastic stiffness and mass matrices for the adopted finite element.
2. Perform the rotation, expansion, and assemblage operations to obtain the global matrices for the entire system.
3. For a given loading configuration, perform a geometrically linear static analysis to find the initial values of the internal static characteristics (or stresses) acting on each element (i.e., solve Equation 4.55).
4. Compute the local geometric stiffness matrix for each finite element considering the initial internal forces (or stresses) obtained at Step 3.
5. Perform the rotation, expansion, and assemblage operations to obtain the global geometric stiffness matrix.
6. For a fixed value of the loading multiplier λ, solve the generalized eigenvalue problem of Equation 8.58 to find the corresponding eigenfrequencies ω_i of the system, with $i = 1, 2, ..., (g \times n)$.
7. Repeat Step 6 for different values of λ.
8. Plot the $\omega_i - \lambda$ (or $\omega_i^2 - \lambda$) curves by using the results obtained in the previous steps.

8.7 NONCONSERVATIVE LOADING AND FLUTTER

Let us consider a slender elastic cantilever beam of constant cross section, inextensible and not deformable in shear, though deformable in bending, with a mass m placed in correspondence to the free end. We assume, for simplicity, that the distributed mass of the beam is negligible compared with the mass m concentrated at the tip. Let the beam be loaded by a **follower compression force**, that is, a force that after deformation rotates together with the end section of the beam and always remains tangential to the deformed axis (Figure 8.6). We shall leave open the question of how such a follower compression force is realized.

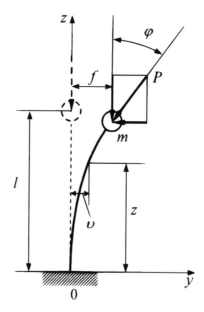

Figure 8.6

To explain why this problem is different from those analyzed so far, we shall start by applying the Euler method (i.e., the static or direct equilibrium method) to analyze the stability of the cantilever beam shown in Figure 8.6.

By virtue of the assumed smallness of the deflection $v(z)$, we have $P_z \cong P$ and $P_y \cong P\varphi$, where $\varphi \cong v'(l)$ is the rotation angle of the end section. Having chosen the coordinate axes as shown in Figure 8.6, the differential equation of the elastic line of the slightly bent column is given by

$$EI \frac{d^2 v}{dz^2} = P(f - v) - P\varphi(l - z) \tag{8.59}$$

where EI is the bending rigidity and f is the deflection at the free end of the beam. Introducing the definition:

$$\frac{P}{EI} = \alpha^2 \tag{8.60}$$

Equation 8.59 can be rewritten in the form:

$$\frac{d^2 v}{dz^2} + \alpha^2 v = \alpha^2 \left[f - \varphi(l - z) \right] \tag{8.61}$$

The general solution of Equation 8.61 is

$$v = C_1 \sin \alpha z + C_2 \cos \alpha z + f - \varphi(l - z) \tag{8.62}$$

To find the four constants C_1, C_2, f, and φ, we have the following four boundary conditions: $v(0) = 0$, $v'(0) = 0$, $v(l) = f$, $v'(l) = \varphi$. By imposing the boundary conditions on Equation 8.62, we obtain the following homogeneous linear system:

$$C_2 + f - \varphi l = 0 \tag{8.63a}$$

$$\alpha C_1 + \varphi = 0 \tag{8.63b}$$

$$C_1 \sin \alpha l + C_2 \cos \alpha l = 0 \tag{8.63c}$$

$$C_1 \cos \alpha l - C_2 \sin \alpha l = 0 \tag{8.63d}$$

The determinant of the coefficient matrix of Equations 8.63 is

$$\begin{vmatrix} 0 & 1 & 1 & -l \\ \alpha & 0 & 0 & 1 \\ \sin \alpha l & \cos \alpha l & 0 & 0 \\ \cos \alpha l & -\sin \alpha l & 0 & 0 \end{vmatrix} = -1 \tag{8.64}$$

which is nonzero for all values of P. This means that there are no values of P for which there can exist curvilinear and equilibrated forms of the beam that are close to the unperturbed (rectilinear) form. In other words, the only possible solution is the trivial one, that is, the undeformed straight configuration.

Based on the previous result, one could conclude that the beam shown in Figure 8.6 is stable for all values of P. As we will show hereafter, this conclusion is wrong; the reason is that the general concept of stability is not, in fact, implicit in the Euler approach, because stability is essentially a dynamic concept. In using the Euler method, in fact, we automatically exclude from our analysis any possible forms of motion. To encompass this possibility, we must consider the dynamic behavior of the system.

On the other hand, in this case, the **energy approach** also cannot be applied by investigating stability. In fact, it is not difficult to see that the follower force has no potential. The indication of whether a potential of the external forces exists is provided by the independence of the work done by these forces of the path followed to attain the final state. Figure 8.7 shows three different ways in which the cantilever beam can reach a state defined by a transverse deflection f and an angle of rotation φ of the end section. In case (a) (rotation through an angle φ with a subsequent displacement), the work done by the force P is obviously negative; in case (b) (displacement with a subsequent rotation), it is zero; in case (c) (rotation through an angle $-\varphi$, a displacement, and a final rotation through an angle 2φ), the work done is positive.

The method for investigating nonconservative problems in the theory of elastic stability is, therefore, the **dynamic method**. According to this approach, local stability is studied by applying a small perturbation to the equilibrium state to test, and then investigating the resulting motion.

We shall now apply the dynamic method to the problem of stability of the cantilever beam loaded by a follower force, represented in Figure 8.6. The equation of small oscillations of the beam about its position of equilibrium is

$$EI\frac{\partial^2 v}{\partial z^2} = P(f - v) - P\varphi(l - z) - m(l - z)\frac{d^2 f}{dt^2} \qquad (8.65)$$

where $v(z, t)$ is the dynamic deflection of the generic section, and $f(t)$ and $\varphi(t)$ are the deflection and the rotation at the free end. Equation 8.65 is satisfied by the following expressions:

$$v(z, t) = V(z)e^{i\omega t} \qquad (8.66a)$$

$$f(t) = F e^{i\omega t} \qquad (8.66b)$$

$$\varphi(t) = \Phi\, e^{i\omega t} \qquad (8.66c)$$

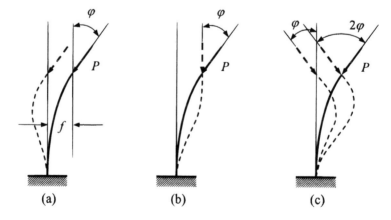

(a)   (b)   (c)

Figure 8.7

where ω is a constant to be found. If this constant is real, then the beam performs harmonic oscillations with angular frequency ω about its rectilinear configuration. If ω proves to be a complex number and its imaginary part is negative, then this would correspond to disturbances that become infinitely large with time. If we set $\omega = \omega_r + i\omega_i$, then

$$v(z, t) = V(z)e^{(i\omega_r - \omega_i)t} \tag{8.67}$$

and if $\omega_i < 0$ (i.e., $\text{Im}(\omega) < 0$), then an exponentially increasing term appears in the expression $v(z, t)$.

Substituting Equations 8.66 into Equation 8.65 and using the definition in Equation 8.60, we obtain

$$\frac{d^2 V}{dz^2} + \alpha^2 V = \alpha^2 F - \alpha^2 \Phi(l - z) + \frac{m\omega^2 F}{EI}(l - z) \tag{8.68}$$

The general solution of Equation 8.68 is

$$V(z) = C_1 \sin \alpha z + C_2 \cos \alpha z + F - \Phi(l - z) + \frac{m\omega^2 F}{\alpha^2 EI}(l - z) \tag{8.69}$$

where the constants C_1, C_2, F, and Φ can be determined by imposing the following four boundary conditions: $V(0) = 0$, $V'(0) = 0$, $V(l) = F$, $V'(l) = \Phi$. Making use of these conditions, we obtain a system of four linear algebraic equations in C_1, C_2, F, and Φ, and if we equate its determinant to zero, we obtain the characteristic equation:

$$\begin{vmatrix} 0 & 1 & 1 + \dfrac{m\omega^2 l}{\alpha^2 EI} & -l \\[2mm] \alpha & 0 & -\dfrac{m\omega^2}{\alpha^2 EI} & 1 \\[2mm] \sin \alpha l & \cos \alpha l & 0 & 0 \\[2mm] \alpha \cos \alpha l & -\alpha \sin \alpha l & -\dfrac{m\omega^2}{\alpha^2 EI} & 0 \end{vmatrix} = 0 \tag{8.70}$$

the solution of which gives the formula:

$$\omega = \pm \sqrt{\frac{P}{ml\left(\dfrac{\sin \alpha l}{\alpha l} - \cos \alpha l\right)}} \tag{8.71}$$

As the force P increases from zero, the natural frequencies increase in magnitude. When

$$\tan \alpha l = \alpha l \tag{8.72}$$

they become infinite, and, with a further increase in P, they become purely imaginary, their signs remaining opposite. The smallest root of Equation 8.72 is $\alpha l = 4.493$, which, from Equation 8.60, gives a **critical value** of the follower force:

$$P_c = \frac{20.19 \, EI}{l^2} \tag{8.73}$$

Thus, in the case of a cantilever beam under the action of a follower force, the Euler method leads to erroneous results: the rectilinear form of the beam, for sufficiently large values of the force P, proves to be unstable, even though in this particular problem there cannot exist curvilinear forms of equilibrium close to the initial rectilinear form, as previously pointed out. In other words, this means that neutral equilibrium does not exist at the critical load. The reason is that we are dealing with a different type of instability, called **oscillatory instability** or **flutter**, which is characterized by a divergent oscillatory motion. The critical load corresponding to this unstable behavior is called the **flutter load**. The present example is a limit case of flutter, in which the frequency tends to infinity at the loss of stability (the infinite value of the frequency is due to neglecting the distributed mass).

The features of the problem considered in this section become even more apparent if we consider a system with two degrees of freedom. For such a system, we can take as an example a cantilever beam with two concentrated masses.

Let m_1 and m_2 be the masses and $f_1(t)$ and $f_2(t)$ their displacements in the transverse direction (Figure 8.8). To obtain the characteristic equations, we choose a procedure slightly different from the one previously employed. On the basis of **D'Alembert's principle**, we establish the equations of the small oscillations of the beam about the unperturbed equilibrium configuration:

$$f_1 = -\delta_{11} m_1 \frac{d^2 f_1}{dt^2} - \delta_{12} m_2 \frac{d^2 f_2}{dt^2} \qquad (8.74a)$$

$$f_2 = -\delta_{21} m_1 \frac{d^2 f_1}{dt^2} - \delta_{22} m_2 \frac{d^2 f_2}{dt^2} \qquad (8.74b)$$

where $\delta_{11}(P)$, $\delta_{12}(P)$, $\delta_{21}(P)$, and $\delta_{22}(P)$ are the **flexibility influence coefficients** (e.g., δ_{12} is the displacement of the mass m_1 produced by a unit transverse force applied to the beam in correspondence to the mass m_2). These displacements are determined taking into account the

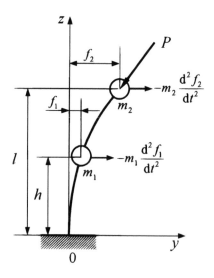

Figure 8.8

follower force P. For this purpose, we use the equation of the elastic line with second-order effects:

$$EI\frac{d^4v}{dz^4} + P\frac{d^2v}{dz^2} = 0 \tag{8.75}$$

which remains valid even in the case of a follower force, as can be checked by differentiating Equation 8.65 with respect to z twice (here, the dependence on time t is not relevant).

Equation 8.75 possesses the following general integral:

$$v(z) = C_1\sin\alpha z + C_2\cos\alpha z + C_3 z + C_4 \tag{8.76}$$

where, as before, the notation from Equation 8.60 is used. Considering, in succession, the cases of the effect of a unit force applied to the masses m_1 and m_2, and imposing the boundary conditions, we obtain

$$\delta_{11} = \frac{1}{\alpha P}(\sin\alpha h - \alpha h\cos\alpha h) \tag{8.77a}$$

$$\delta_{12} = \frac{1}{\alpha P}\left[\sin\alpha l - \alpha h\cos\alpha l - \sin\alpha(l-h)\right] \tag{8.77b}$$

$$\delta_{21} = \frac{1}{\alpha P}\left[\sin\alpha l - \alpha l\cos\alpha l + \sin\alpha(l-h)\right] \tag{8.77c}$$

$$\delta_{22} = \frac{1}{\alpha P}(\sin\alpha l - \alpha l\cos\alpha l) \tag{8.77d}$$

Let us observe that, in general, $\delta_{12} \neq \delta_{21}$; that is, the reciprocity principle for displacements is not valid (this is due to the nonconservativeness of the load). Analogously to Equation 8.66b, we set in this case:

$$f_1(t) = F_1 e^{i\omega t} \tag{8.78a}$$

$$f_2(t) = F_2 e^{i\omega t} \tag{8.78b}$$

As a result, we obtain the following condition:

$$\begin{vmatrix} m_1\delta_{11}\omega^2 - 1 & m_2\delta_{12}\omega^2 \\ m_1\delta_{21}\omega^2 & m_2\delta_{22}\omega^2 - 1 \end{vmatrix} = 0 \tag{8.79}$$

which gives the **characteristic equation**. Calculating the determinant and solving the equation, we obtain

$$\omega^2 = \frac{m_1\delta_{11} + m_2\delta_{22} \pm \sqrt{\left[(m_1\delta_{11} - m_2\delta_{22})^2 + 4m_1 m_2\delta_{12}\delta_{21}\right]}}{2m_1 m_2(\delta_{11}\delta_{22} - \delta_{12}\delta_{21})} \tag{8.80}$$

For sufficiently low values of the force P, Equation 8.80 gives real values for the frequency of vibration. However, as the force P reaches a certain critical value P_c, the expression underneath the radical becomes equal to zero and the characteristic equation will have a double root. With further increase in the force P, among the natural frequencies there will be a frequency that has a negative imaginary part. As may be seen from Equations 8.78a and b, this corresponds to an unbounded increase in perturbations with time, that is, instability of the rectilinear configuration of the beam.

It must be remarked that this kind of instability is characterized by oscillations of increasing amplitude (**dynamic instability**), since the real part ω_r (which represents the actual angular frequency of oscillation) of the complex frequency ω is nonzero when the imaginary part ω_i becomes negative. The critical value assumed by the frequency of motion ω_r at the flutter condition is called the **flutter frequency**.

The formula for the **critical flutter force** may be expressed in the form:

$$P_c = \beta \frac{EI}{l^2} \tag{8.81}$$

where β is a coefficient that depends on the location of the mass m_1 (ratio h/l) and on the mass ratio m_1/m_2. For example, if $m_1 = 0$, we obtain $\beta \cong 20.19$; that is, we are led to Equation 8.73. With $m_2 = 0$ and $h = l/2$, the critical value, as should have been expected, becomes exactly four times larger, whereas with $m_1 = m_2$ and $h = l/2$, we have $\beta \cong 1.48\pi^2$.

If the cantilever beam had been compressed by a force with a fixed line of action, it would have lost its stability in a static manner (i.e., without oscillations), and the classic Euler buckling would have occurred.

Therefore, according to the dynamic approach to the analysis of elastic stability, the static loss of stability (**divergence**) is characterized by the condition $\omega_r = \text{Re}(\omega) = 0$, $\omega_i = \text{Im}(\omega) < 0$ (Figure 8.9a), whereas the condition corresponding to the dynamic loss of stability (**flutter**) is $\omega_r = \text{Re}(\omega) \neq 0$, $\omega_i = \text{Im}(\omega) < 0$ (Figure 8.9b). In the former case, we have an aperiodic, exponentially growing motion (Figure 8.9c), whereas in the latter we have a periodic, exponentially growing motion (Figure 8.9d).

Usually, when applying the dynamic method, instead of the frequencies ω, we introduce the so-called **characteristic exponents** s, connected to the frequencies by the simple relation $s = i\omega$. In this case, the form of equilibrium under investigation becomes unstable if one of the characteristic exponents has a positive real part (i.e., $\text{Re}(s) > 0$), with the actual angular frequency of motion now represented by the imaginary part of s.

Closing this section, the main conclusion to draw is that the Euler method is applicable (to structures that behave linearly in the precritical regime) in cases when the external forces acting on the body are conservative. In this case, in fact, stability is always lost by divergence (i.e., buckling). Conversely, if the external forces are not conservative, although in a number of problems it may give the correct solution, the Euler method should be replaced by the more general dynamic method to avoid errors. When stability is lost by divergence, the Euler method and the dynamic method lead to the same results.

Another aspect that should be emphasized is that, so far, we have systematically neglected mechanical damping. When stability is lost by divergence, linear mechanical damping has no stabilizing (or destabilizing) influence. This is generally true for linear systems with conservative loads, although not always true for systems with nonconservative loads. In a nonconservative system, a small internal (viscous) damping can have a destabilizing influence; that is, it can lower the critical load.

Divergence

Flutter

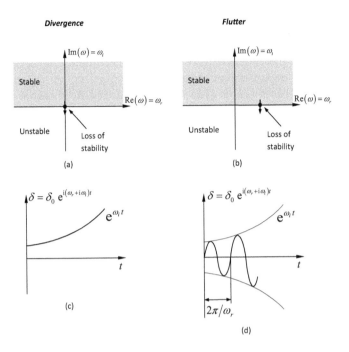

(a)

(b)

(c)

$\left|2\pi/\omega_r\right.$

(d)

Figure 8.9

8.8 WIND EFFECTS ON LONG-SPAN SUSPENSION OR CABLE-STAYED BRIDGES

Suspension bridges and cable-stayed bridges, due to their flexibility, are structures typically susceptible to wind-induced problems (Figure 8.10). In cable-stayed bridges, inclined cables (stays) sustain the deck and limit its deformation through their extensional stiffness, while in suspension bridges, larger deck deformations are allowed by a change of configuration in the main cables. Conversely, the axial load induced by the stays in the deck of cable-stayed bridges imposes an upper bound to their maximum span, which is, in fact, sensibly shorter than that of suspension bridges, where the deck is connected to the suspension cables by means of a series of vertical cables (hangers). Suspension bridges are, therefore, generally more flexible and prone to wind effects than cable-stayed bridges.

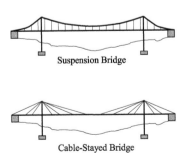

Suspension Bridge

Cable-Stayed Bridge

Figure 8.10

Wind may induce instability or excessive vibration in long-span bridges. Instability caused by the interaction between an air flow and an elastic structure is termed **aeroelastic instability**. Aeroelastic instability of bridges includes torsional divergence, galloping, and flutter. Normally, design against instability requires that the maximum wind speed expected at the bridge site is sufficiently lower than a critical value.

On the other hand, vibration is a cyclic motion induced by dynamic forcing and may cause fatigue or serviceability problems. Typical wind-induced vibrations consist of vortex shedding and buffeting. In this case, the design requires the analysis of the structural response in the presence of the dynamic forcing to check the reliability of the response itself.

The types of instability and vibration problems mentioned may occur alone or in combination. For example, a structure must experience vibration to some extent before flutter instability starts. Modern design of long-span bridges is normally conducted by combining computer analysis with experimental investigation through wind tunnel testing. Computer analyses rely on analytical or semianalytical models, as well as on numerical procedures, which are implemented in computer programs and usually need parameters that are experimentally determined.

The interaction between bridge vibration and wind flow is usually idealized as consisting of two kinds of forces: motion dependent and motion independent. The former vanish if the structures are rigidly constrained. The latter, being purely dependent on the wind characteristics and section geometry, exists whether or not the bridge is moving. According to this schematization, the equation of motion in the presence of the aerodynamic forces can be expressed in the following general form:

$$[M]\{\ddot{\delta}\} + [C]\{\dot{\delta}\} + [K]\{\delta\} = \left\{F\left(\delta,\dot{\delta}\right)\right\}_{md} + \{F\}_{mi} \tag{8.82}$$

where:

$[M]$ is the mass matrix
$[C]$ is the damping matrix
$[K]$ is the stiffness matrix
$\{\delta\}$ is the displacement vector
$\{F(\delta,\dot{\delta})\}_{md}$ is the motion-dependent aerodynamic force vector
$\{F\}_{mi}$ is the motion-independent wind force vector

While both motion-independent and motion-dependent forces cause deformation, aeroelastic instability is only due to the motion-dependent part. The difference between short-span and long-span bridges lies in the motion-dependent part. For short-span bridges, the motion-dependent part is insignificant and there is no concern about aeroelastic instability. In the following, we shall give an introductive description of the main forms of aeroelastic instability. For flexible structures such as long-span bridges, however, both instability and vibration need to be carefully investigated.

Aeroelastic phenomena are characterized by the peculiarity that the elastic structure and the air flow combine together to form a single dynamic system with proper features that differ from those of the two components taken separately. Some instability phenomena, such as torsional divergence and galloping, can be treated with a quasi-static approach. Conversely, flutter instability must be analyzed as a fully dynamic phenomenon.

8.9 TORSIONAL DIVERGENCE

Aeroelastic divergence is usually treated as a static instability, caused by the vanishing of the **total stiffness** (elastic plus aerodynamic) associated with one of the fundamental modes of the structure, typically the torsional one. As is well known, wind flowing against a structure

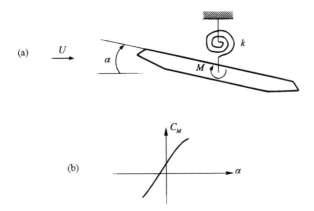

(a)

(b)

Figure 8.11

exerts a mean pressure proportional to the square of the wind velocity. Wind pressure generally induces both aerodynamic forces and moments in a structure. At a critical wind velocity, the edge-loaded bridge may buckle out of plane under the action of a drag force or torsionally diverge under a wind-induced moment that increases with the twist angle. In fact, the divergence of a bridge deck involves an inseparable combination of lateral (vertical) buckling and torsional divergence.

A simplified model to study **torsional divergence** that considers only a representative deck section immersed in a two-dimensional flow is the single-degree-of-freedom system shown in Figure 8.11a. Considering a small rotation angle α, the aerodynamic moment resulting from wind, M, is given by (Figure 8.11b)

$$M = \frac{1}{2}\rho U^2 B^2 C_M(\alpha)$$

$$= \frac{1}{2}\rho U^2 B^2 \left[C_{M0} + \left(\frac{dC_M}{d\alpha}\right)_{\alpha=0} \alpha \right] \tag{8.83}$$

where:

ρ is the air density
U is the wind speed
B is the deck width
$C_M(\alpha)$ is the aerodynamic moment coefficient, which is a function of the angle of attack α, that can be determined by wind tunnel tests
$C_{M0} = C_M(0)$ denotes the moment coefficient for the angle of attack equal to zero

When the aerodynamic moment caused by wind exceeds the resisting torsional capacity, the displacement of the bridge diverges. Equating the aerodynamic moment given by Equation 8.83 to the internal elastic moment $k\alpha$ gives

$$k\alpha - \frac{1}{2}\rho U^2 B^2 \left[C_{M0} + \left(\frac{dC_M}{d\alpha}\right)_{\alpha=0} \alpha \right] = 0 \tag{8.84}$$

where k is the spring constant of torsional stiffness. Equation 8.84 can be rewritten as follows:

$$\left[k - \frac{1}{2}\rho U^2 B^2 \left(\frac{dC_M}{d\alpha} \right)_{\alpha=0} \right] \alpha - \frac{1}{2}\rho U^2 B^2 C_{M0} = 0 \tag{8.85}$$

where the terms within square brackets represent the total (elastic plus aerodynamic) torsional stiffness of the system. Equating the total stiffness to zero (i.e., rotation α becomes infinitely large), we obtain the **critical wind speed for torsional divergence:**

$$U_D = \sqrt{\frac{2k}{\rho B^2 \left(\dfrac{dC_M}{d\alpha} \right)_{\alpha=0}}} \tag{8.86}$$

The critical speed given by Equation 8.86 should be sufficiently higher than the maximum wind speed at the bridge site. However, it should be observed that the previous formulation that highlights the essence of the phenomenon usually overestimates the torsional divergence speed with respect to more sophisticated models, in which the coupling between vertical bending and torsion is taken into account.

8.10 GALLOPING

Galloping is a dynamic aeroelastic instability typical of slender structures having special cross-sectional shapes such as, for example, rectangular or D-sections, or the effective sections of some ice-coated power line cables. Under certain conditions that will be defined later, these structures can exhibit large-amplitude oscillations in the direction normal to the flow, due to the vanishing of the **total damping** (mechanical plus aerodynamic) of the system.

The elementary model adopted for ordinary galloping consists of a damped oscillator, with a single degree of freedom, immersed in a two-dimensional flow. Experience has shown that knowledge of the mean lift and drag coefficients obtained under static conditions as functions of the angle of attack is sufficient for developing a satisfactory analytical description of the galloping phenomenon; that is, this kind of problem is governed primarily by quasi-steady forces.

Consider a bridge deck section that oscillates in the across-wind direction y in a flow with velocity U (Figure 8.12a). The motion of the structure implies that the wind direction varies relative to the structure. The case in which the deck section is moving downward with velocity $\dot{y}$ is kinematically equivalent to the case of Figure 8.12b, where the deck section is motionless and the wind blows upward with an angle of attack α:

$$\tan \alpha = -\frac{\dot{y}}{U} \tag{8.87}$$

If the wind-induced vertical force of this case is directed downward (which is equivalent to a negative aerodynamic damping), then the deck section will be pushed downward; this will result in a divergent vibration, or galloping, if the mechanical damping is insufficient. Otherwise, the vibration is stable. Galloping is therefore caused by the change in the effective attack angle due to vertical motion of the structure. A negative slope in the plot of the

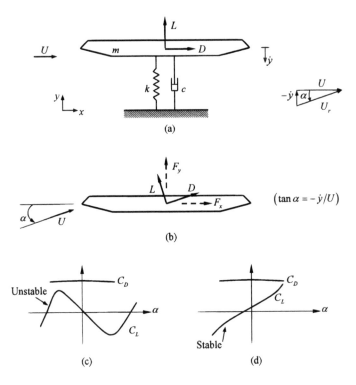

Figure 8.12

static lift coefficient versus the angle of attack, shown in Figure 8.12c, usually implies a tendency to galloping.

According to the elementary model, the equation of motion describing the phenomenon is

$$m\ddot{y}(t) + c\dot{y}(t) + ky(t) = F_y(t) \tag{8.88}$$

where:

   m, c, and k are the mass, the mechanical damping, and the elastic stiffness of the deck section, respectively

   $F_y(t)$ is the aerodynamic force acting on the body, which depends in general on the angle of attack

By assuming that the mean aerodynamic lift and drag coefficients $C_L(\alpha)$ and $C_D(\alpha)$ for the oscillating body and the fixed body are the same, $F_y(t)$ can be determined experimentally by testing the body in a wind tunnel under different values of the angle of attack (Figures 8.12c and d). This is equivalent to accepting that the expressions of the steady aerodynamic forces are valid instant by instant (quasi-steady theory). This assumption is valid when $\dot{y} \ll U$, which is the condition that usually applies for galloping to occur.

The static lift and drag forces—functions of the attack angle—are expressed as follows:

$$L = \frac{1}{2}\rho U^2 B C_L(\alpha) \tag{8.89a}$$

$$D = \frac{1}{2}\rho U^2 B C_D(\alpha) \qquad (8.89b)$$

For small values of the angle α, we have (Figure 8.12)

$$\alpha \cong -\dot{y}/U \qquad (8.90a)$$

$$C_L \cong \left(\frac{dC_L}{d\alpha}\right)_{\alpha=0} \alpha \qquad (8.90b)$$

$$C_D \cong C_D(0) \qquad (8.90c)$$

The aerodynamic force F_y, acting along the direction of the bridge deck motion, can be obtained from the lift and drag forces as (Figure 8.12b)

$$
\begin{aligned}
F_y &= L\cos\alpha + D\sin\alpha \cong L + D\alpha \\
&= \frac{1}{2}\rho U^2 B\left[\left(\frac{dC_L}{d\alpha}\right)_{\alpha=0}\alpha + C_D(0)\alpha\right] \\
&= -\frac{1}{2}\rho U B\left(\frac{dC_L}{d\alpha} + C_D\right)_{\alpha=0}\dot{y} \qquad (8.91)
\end{aligned}
$$

Introducing Equation 8.91 into Equation 8.88, the equation of motion eventually becomes

$$m\ddot{y} + c\dot{y} + ky = -\frac{1}{2}\rho U B\left(\frac{dC_L}{d\alpha} + C_D\right)_{\alpha=0}\dot{y} \qquad (8.92)$$

the right-hand side of which represents the aerodynamic damping. If the total damping is less than or equal to zero, that is,

$$c_{tot} = c + \frac{1}{2}\rho U B\left(\frac{dC_L}{d\alpha} + C_D\right)_{\alpha=0} \leq 0 \qquad (8.93)$$

the system tends toward instability (zero damping means a critical condition). Since the mechanical damping c is positive, this situation is possible if and only if the following **Glauert–Den Hartog criterion** is satisfied:

$$\left(\frac{dC_L}{d\alpha} + C_D\right)_{\alpha=0} \leq 0 \qquad (8.94)$$

Solving the equation corresponding to the inequality (Equation 8.93) gives the **critical wind speed for galloping:**

$$U_G = -\frac{2c}{\rho B\left(\frac{dC_L}{d\alpha} + C_D\right)_{\alpha=0}} \qquad (8.95)$$

From Equation 8.94, it is clear that stability against galloping depends exclusively on the aerodynamic coefficients; that is, on the geometry of the section and on the direction of the flow. Therefore, wind tunnel tests are usually conducted to check Equation 8.94 and to make necessary improvements of the section to eliminate negative effects of the possible wind speed at the bridge site (Figure 8.12d).

It should be noted that the **self-excited** nature of the galloping oscillation is in opposition to that of resonance phenomena, where the oscillation is caused by an external dynamic forcing that is independent of the motion of the structure. The right-hand side of the motion Equation 8.92 represents a forcing produced by the motion of the deck, which varies in time if the deck oscillates, and is steady (and therefore cannot induce oscillation) if the deck is motionless. Equation 8.92 represents, in fact, a homogeneous equation where the effect of the aerodynamic forcing appears in the effective damping term c_{tot}:

$$m\ddot{y} + c_{tot}\dot{y} + ky = 0 \tag{8.96}$$

The complete solution of Equation 8.96 is, in fact

$$y(t) = \exp\left(-\frac{c_{tot}}{2m}t\right)\cos(\omega_G t) \tag{8.97}$$

where

$$\omega_G = \sqrt{\frac{k}{m} - \frac{c_{tot}^2}{4m^2}} \tag{8.98}$$

represents the eigenvalue (the condition for which a nontrivial solution of Equation 8.96 is possible) and is called the **galloping frequency**. Note that if $U = U_G$ (the condition for which $c_{tot} = 0$), this frequency reduces to that of the motion of the undamped system. In other words, at the critical speed for galloping, the aerodynamic forces inject into the system exactly the same amount of power needed to compensate that dissipated by the mechanical damping, so that the system behaves as if it were undamped. For lower speeds, the said compensation is insufficient, whereas for higher speeds the power injection is exuberant and the oscillation is exalted (dynamic instability).

In agreement with what we have pointed out, Equation 8.95 shows that the galloping speed is zero for an undamped mechanical system ($c = 0$); structures with a low mechanical damping are therefore very prone to this kind of aeroelastic instability. Moreover, we observe that the critical wind speed for galloping given by Equation 8.95 is independent of the mass m of the body and of the stiffness k of its suspension. On those parameters depend both the frequency ω_G of the oscillation and its amplification rate $-c_{tot}/2m$, instead. Lastly, it is important to note that, in this simplified model, the frequency ω_G depends only on the physical characteristics (stiffness and mass) of the system and on those of the average wind contained in c_{tot}, whereas there is no reference to turbulence or to vortex shedding.

8.11 FLUTTER

Aeroelastic **flutter** is a dynamic instability phenomenon that originates from the mutual interaction of elastic, inertial, and **self-excited aerodynamic forces**, whereby at some critical wind speed the structure oscillates in a divergent, destructive manner. If a system immersed in a wind flow is given a small perturbation, its motion will either decay or diverge, depending

Figure 8.13

on whether the energy extracted from the flow is smaller or larger than the energy dissipated by mechanical damping. The theoretical line dividing decay from divergent motions
represents the critical condition. The corresponding wind speed is called the **flutter speed**,
at which the motion of the structure exhibits oscillations of increasing amplitude at a constant frequency, called the **flutter frequency**. Actually, the amplitude of the oscillation does
not grow indefinitely as predicted by linear theories, but stabilizes on a maximum value
(the **limit cycle**) due to structural and aerodynamic nonlinearities. Nevertheless, even if this
makes the phenomenon less catastrophic, its dangerousness is not reduced and the structure
can be destroyed even after a few cycles, due to damage of some key elements.

Flutter, originally studied in aeronautics, became of interest for bridge engineering after
the collapse of the first Tacoma Narrows bridge in 1940 (Figure 8.13). Bridge flutter analysis
combines both experimental and analytical procedures. Flutter design consists of analyzing a proposed bridge deck configuration for determination of the lowest wind speed that
initiates instability. The critical wind speed for flutter should be sufficiently higher than
meteorological possible wind speeds at the bridge site.

In a flutter analysis, only the onset instability condition is normally sought for the design
of bridge structures. The most relevant aspects can be discussed with reference to a bridge
deck section immersed in a two-dimensional flow. Therefore, under the assumption of small
oscillations perturbating the flow, the structure can be modeled as a damped linear oscillator with two degrees of freedom:

$$m\ddot{h}(t) + c_h \dot{h}(t) + k_h h(t) = L_{se}(t, K) \tag{8.99a}$$

$$I\ddot{\alpha}(t) + c_\alpha \dot{\alpha}(t) + k_\alpha \alpha(t) = M_{se}(t, K) \tag{8.99b}$$

where:

h and α are the heaving displacement (vertical bending) and the pitching rotation (torsion angle) (Figure 8.14a)

m and I are the mass and the polar mass moment of inertia per unit length

c_h and c_α are the mechanical damping coefficients

k_h and k_α are the stiffness coefficients in the heaving and pitching modes,
respectively

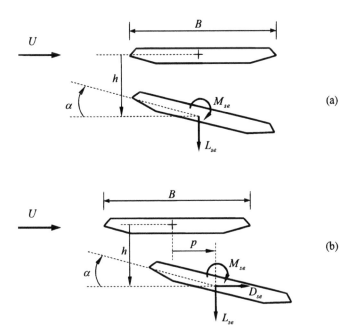

Figure 8.14

L_{se} and M_{se} are the **self-excited lift and moment** per unit length, depending on time t and on the deck oscillation through the nondimensional aeroelastic parameter $K = \omega B/U = 2\pi f B/U$, called the **reduced frequency of oscillation**

where:

B is the deck width
ω the angular frequency (f the frequency) of oscillation
U the undisturbed mean wind speed

A closed-form expression for the unsteady aerodynamic forces acting on oscillating bridge decks cannot be obtained. To solve the problem, Scanlan and Tomko extended the solution developed by Theodorsen for the thin airfoil to bridge sections. They proposed a semiempirical model, in which the aerodynamic forces are expressed in terms of some unsteady coefficients, the so-called **flutter (or aeroelastic) derivatives** that are experimentally determined in the wind tunnel. According to Scanlan and Tomko, the self-excited forces can be assumed as linear functions of structural displacements and velocities, parametrically dependent on the reduced frequency of oscillation:

$$L_{se}\left(t,K\right) = \frac{1}{2}\rho U^2 B\left[KH_1^*(K)\frac{\dot{h}(t)}{U} + KH_2^*(K)\frac{B\dot{\alpha}(t)}{U} \right.$$

$$\left. +K^2 H_3^*(K)\alpha(t) + K^2 H_4^*(K)\frac{h(t)}{B} \right] \qquad (8.100a)$$

$$M_{se}(t,K) = \frac{1}{2}\rho U^2 B^2 \left[KA_1^*(K)\frac{\dot{h}(t)}{U} + KA_2^*(K)\frac{B\dot{\alpha}(t)}{U} \right.$$

$$\left. + K^2 A_3^*(K)\alpha(t) + K^2 A_4^*(K)\frac{h(t)}{B} \right] \tag{8.100b}$$

where the coefficients H_i^* and A_i^* ($i = 1$–4) are the flutter derivatives, which are functions of the reduced frequency of oscillation K, as well as of the mean angle of attack. The coefficients that multiply generalized displacements are intended as **aerodynamic stiffness**, while those which multiply generalized velocities represent **aerodynamic damping**. We observe that Equations 8.100a and b do not explicitly include additional mass terms in h and $\ddot{\alpha}$, which are considered to be negligible in wind engineering applications. Moreover, notice that quantities such as α, $\dot{h}/U$, and $B\dot{\alpha}/U$ are effective angles of attack; this explains the reason why the coefficients H_i^* and A_i^* are called flutter derivatives (see the analogy with Equation 8.90b). The flutter derivatives are usually plotted as a function of the so-called **reduced velocity** $U_r = U/fB = 2\pi/K$; some examples are reported in Figure 8.15.

The previous model with two degrees of freedom can be extended to multimode analysis of the bridge structure. Nevertheless, in most cases the flutter speed can be accurately calculated by considering two modes only.

Basically, two main types of flutter exist. In **classical flutter** (also called coupled or stiffness-driven flutter), the two degrees of freedom couple together to originate a motion that introduces energy into the system, leading to divergent or large-amplitude oscillations (Figure 8.16). By looking for a harmonic solution to Equations 8.99 in the form

$$h(t) = h_0 e^{i\omega t} \tag{8.101a}$$

$$\alpha(t) = \alpha_0 e^{i\omega t} \tag{8.101b}$$

and causing the system complex determinant to vanish, after separating the real and the imaginary parts one obtains a fourth-degree and a third-degree polynomial equations with respect to the reduced frequency of oscillation, the common solution of which gives the critical reduced frequency for flutter K_F. The frequency ω_F that simultaneously satisfies both polynomial equations is the **flutter frequency**. Therefore, the **flutter speed** can be obtained as $U_F = B\omega_F/K_F$. If more than one intersection point is found in the selected range of K, the one with the lowest U_F will be the required solution. Alternatively, the flutter speed can be calculated by the eigenvalues $\omega_n = \omega_{r,n} + i\omega_{i,n}$ ($n = 1, 2, 3,...$) of the system, increasing the wind speed until one of those presents a negative imaginary part. The real part of that eigenvalue represents the flutter frequency. In practice, other solution strategies are also adopted.

Relevant for bridges is also **single degree of freedom flutter**, wherein negative damping in a single mode (typically the torsional one) can be attained without any coupling with other modes. An uncoupled flutter is called **damping-driven flutter**, since it is caused by zero damping. An examination of the flutter derivatives gives a preliminary judgment on the flutter behavior of the section. Necessary section modifications should be made to eliminate the positive flutter derivatives as shown in Figure 8.15, especially A_2^* and H_1^*. The parameter A_2^* controls torsional flutter and H_1^* controls vertical flutter. In bridges, however, torsional flutter is more common than vertical flutter. It should be noted that, for a coupled flutter, zero damping is a sufficient but not a necessary condition.

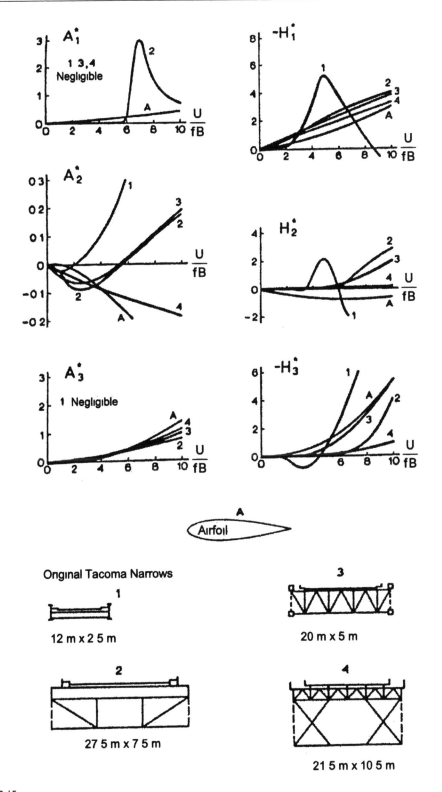

Figure 8.15

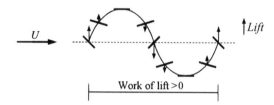

Figure 8.16

If, among the aerodynamic damping terms, only those associated with H_1^* and A_2^* are considered significant, the total structural plus aerodynamic damping can be written as

$$c_{h,tot} = c_h - \frac{1}{2}\rho UBKH_1^*(K) \tag{8.102a}$$

$$c_{\alpha,tot} = c_\alpha - \frac{1}{2}\rho UB^3KA_2^*(K) \tag{8.102b}$$

for vertical and torsional degrees of freedom, respectively.

For the airfoil, H_1^* and A_2^* are both negative for all values of K (see Figure 8.15). The total damping is, therefore, always positive for both h and α. It follows that, in an incompressible flow, the airfoil is not capable of experiencing single degree of freedom flutter in a vertical or torsional mode; that is, the critical mechanism always involves a coupling between these two modes (classic flutter instability). In contrast, such coupling is not always involved in the flutter of bridge decks. For example, the section of the original Tacoma Narrows bridge has negligible H_1^* values (see Figure 8.15), meaning that flutter in the vertical mode is not possible. However, A_2^* assumes positive values from $U/fB \cong 2$. Thus, it is easy to see that, for sufficiently high wind speeds, the total damping given by Equation 8.102b becomes negative, and flutter involving only the torsional mode occurs.

As an example of the application of the previous model with two degrees of freedom, let us consider the case study of the Golden Gate bridge (San Francisco, California), for which the data can be easily found in the literature. Referring to a deck portion of unit length, the parameters necessary for the flutter analysis are the following:

- Deck width, $B = 27.5$ m
- Air density, $\rho = 1.2$ kg/m³
- Deck mass, $m = 5,208.1$ kg
- Deck polar mass moment of inertia, $I = 3,680,000$ kg m²
- Vertical damping ratio, $\xi_h = c_h/2m\omega_h = 0.005$ (arbitrary choice), $(\omega_h^2 = k_h/m)$
- Torsional damping ratio, $\xi_\alpha = c_\alpha/2I\omega_\alpha = 0.005$ (arbitrary choice), $(\omega_\alpha^2 = k_\alpha/I)$
- Natural frequency of bending oscillation (6th mode, vert. symm.), $f_h = 0.164$ Hz
- Natural frequency of torsional oscillation (7th mode, tors. symm.), $f_\alpha = 0.192$ Hz
- Bending/torsion coupling coefficient, $C_{h\alpha} = 0.34$
- Torsion/bending coupling coefficient, $C_{\alpha h} = 0.32$

The flutter derivatives H_i^* and A_i^* ($i = 1,\dots, 4$) related to the case under investigation are reported in Figure 8.17.

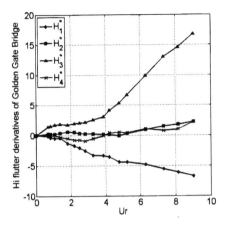
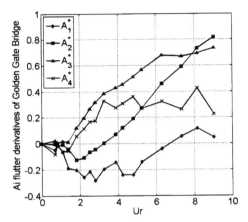

Figure 8.17

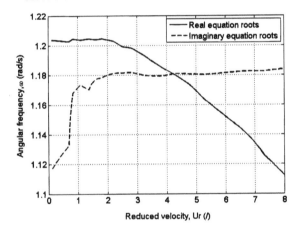

SOLUTION OF FLUTTER EQUATIONS FOR THE GOLDEN GATE BRIDGE

Figure 8.18

By solving the flutter Equations 8.99 for the Golden Gate bridge with the data listed, we obtain a critical speed $U_c = 21.83$ m/s (78.6 km/h) and a critical frequency $f_c = 0.188$ Hz ($\omega_c = 1.181$ rad/s), corresponding to the reduced frequency $K_c = 1.488$ (reduced velocity $U_{rc} = 4.22$). The graphical solution of the flutter equations is shown in Figure 8.18.

The critical parameters reported by Simiu and Scanlan for the same case study, obtained through a simplified analysis that considers a purely torsional instability (single degree of freedom flutter), are equal to a critical speed $U_c = 22.75$ m/s (81.9 km/h) and a critical frequency $f_c = f_\alpha = 0.192$ Hz ($\omega_c = 1.204$ rad/s), corresponding to a reduced velocity $U_{rc} = 4.32$. We note that the flutter speed given by the simplified single degree of freedom model is slightly higher than that predicted by the two degrees of freedom model.

In very long-span bridges, the lateral (sway) component of the motion may be relevant for flutter instability. Taking into account also the along-wind force and displacement

(Figure 8.14b), the general expression of the self-excited forces written in matrix form for a finite element analysis is

$$
\begin{Bmatrix} L_{se} \\ D_{se} \\ M_{se} \end{Bmatrix} = \frac{1}{2}\rho U^2 B \left(\begin{bmatrix} \dfrac{K^2 H_4^*}{B} & \dfrac{K^2 H_6^*}{B} & K^2 H_3^* \\ \dfrac{K^2 P_4^*}{B} & \dfrac{K^2 P_6^*}{B} & K^2 P_3^* \\ K^2 A_4^* & K^2 A_6^* & K^2 A_3^* B \end{bmatrix} \begin{Bmatrix} h \\ p \\ \alpha \end{Bmatrix} \right.
$$

$$
\left. + \begin{bmatrix} \dfrac{KH_1^*}{U} & \dfrac{KH_5^*}{U} & \dfrac{KH_2^* B}{U} \\ \dfrac{KP_1^*}{U} & \dfrac{KP_5^*}{U} & \dfrac{KP_2^* B}{U} \\ \dfrac{KA_1^* B}{U} & \dfrac{KA_5^* B}{U} & \dfrac{KA_2^* B^2}{U} \end{bmatrix} \begin{Bmatrix} \dot{h} \\ \dot{p} \\ \dot{\alpha} \end{Bmatrix} \right)
$$

$$
= \frac{1}{2}\rho U^2 \left([F_d]\{q\} + \frac{1}{U}[F_v]\{\dot{q}\} \right) \tag{8.103}
$$

where:

L_{se}, D_{se}, and M_{se} are the **self-excited lift force, drag force,** and **pitch moment,** respectively

h, p, and α are the displacements at the center of the deck section in the directions corresponding to L_{se}, D_{se}, and M_{se}, respectively

H_i^*, P_i^* and A_i^* ($i = 1$–6) are the **generalized flutter derivatives**

$[F_d]$ and $[F_v]$ are the **flutter derivative matrices** corresponding to displacement and velocity, respectively

In linear analysis, the general aeroelastic motion equations of bridge systems are expressed in terms of the generalized modal coordinate vector $\{\delta\}$:

$$
[M]\{\ddot{\delta}\} + \left([C] - \frac{1}{2}\rho U[C^*]\right)\{\dot{\delta}\} + \left([K] - \frac{1}{2}\rho U^2[K^*]\right)\{\delta\} = \{0\} \tag{8.104}
$$

where $[M]$, $[C]$, and $[K]$ are the generalized mass, damping, and stiffness matrices, respectively, while $[C^*]$ and $[K^*]$ are the generalized aerodynamic damping and aerodynamic stiffness matrices, respectively. Matrices $[M]$, $[C]$, and $[K]$ are derived in the same way as in the classical dynamic analysis. Matrices $[C^*]$ and $[K^*]$, corresponding to $[F_v]$ and $[F_d]$ in Equation 8.103, respectively, are assembled from local aerodynamic forces. Notice that even if the structural matrices $[K]$, $[M]$, and $[C]$ are uncoupled as regards the different modes, the motion equation is always coupled due to the coupling of the aerodynamic matrices $[K^*]$ and $[C^*]$.

By assuming harmonic oscillations in the form $\{\delta\} = \{\delta_0\}e^{i\omega t}$, we obtain the following characteristic problem:

$$
\left(-\omega^2 [M] + i\omega\left([C] - \frac{1}{2}\rho U[C^*]\right) + [K] - \frac{1}{2}\rho U^2[K^*] \right)\{\delta_0\} = \{0\} \tag{8.105}
$$

The **flutter speed** U_F and the **flutter frequency** ω_F can be obtained from the nontrivial solution of Equation 8.105, which is given by the following condition:

$$\det\left(-\omega^2[M]+i\omega\left([C]-\frac{1}{2}\rho U[C^*]\right)+[K]-\frac{1}{2}\rho U^2[K^*]\right)=0 \qquad (8.106)$$

The previous finite element formulation allows us to deal with any kind of bridge flutter, both coupled and uncoupled. *Ad hoc* numerical procedures, such as the pK-F method, have been developed to solve Equation 8.105 efficiently.

An uncoupled single degree of freedom flutter can be analyzed as follows. Pre- and post-multiplying Equation 8.105 by the eigenvector $\{\delta_{0,j}\}$ related to the jth mode, we have

$$\{\delta_{0,j}\}^{\mathrm{T}}\left(-\omega^2[M]+i\omega\left([C]-\frac{1}{2}\rho U[C^*]\right)+[K]-\frac{1}{2}\rho U^2[K^*]\right)\{\delta_{0,j}\}=\{0\} \qquad (8.107)$$

which reduces to the following uncoupled equation for mode j:

$$-\omega^2 m_j+i\omega\left(c_j-\frac{1}{2}\rho U c_j^*\right)+k_j-\frac{1}{2}\rho U^2 k_j^*=0 \qquad (8.108)$$

where the undamped natural frequency corresponding to the wind speed U can be obtained from the following definition:

$$\omega^2=\frac{k_j-\frac{1}{2}\rho U^2 k_j^*}{m_j} \qquad (8.109)$$

By setting the total damping in Equation 8.108 equal to zero, we obtain the flutter speed for the uncoupled mode j:

$$U_F=\frac{2}{\rho}\frac{c_j}{c_j^*} \qquad (8.110)$$

and, therefore, the flutter frequency from

$$\omega_F^2=\frac{k_j-\frac{1}{2}\rho U_F^2 k_j^*}{m_j} \qquad (8.111)$$

Turbulence is assumed to be beneficial for flutter stability and is usually ignored. Some studies include turbulence effects by treating the along-wind velocity U as mean velocity, $\bar{U}$, plus a turbulent component, $u(t)$. The random nature of $u(t)$ results in an equation of random damping and stiffness. Complicated mathematics, such as stochastic differentiation, need to be employed to solve the equation.

Time-history and nonlinear analyses can be conducted to investigate postflutter behavior and to include the effects of both geometric and material nonlinearities. However, this is not necessary for most practical applications.

Chapter 9

High-rise structures

Statics and dynamics

9.1 INTRODUCTION

Tall buildings represent, today, the symbol of economic and technological supremacy of certain geographical areas. Since its first appearance, this architectural typology has met approval in public opinion and, especially from a scientific point of view, has become an appealing challenge for structural designers.

Originally, high-rise structures were an American prerogative; nowadays, they represent a worldwide architectural option, even for those countries regarded as less advanced, but demonstrating fast industrial growth. Most of the last supertall buildings are located far from the United States: China, Korea, Saudi Arabia, Malaysia, and the United Arab Emirates. They were favored by considerable economic and technological capabilities, and represent evidence of these current trends. Nevertheless, even if the geographical location of such constructions has changed, the human attempt to overcome the limits already achieved is still the main reason that keeps alive the interest in this field. Figure 9.1 shows the Dubai city skyline, where the current world's tallest building, the Burj Khalifa (828 m), is clearly visible.

Historically, the appearance of tall buildings was due to the Industrial Revolution, at the end of the nineteenth century. In the construction field, technical evolution permitted the availability of advanced materials and equipment that were indispensable for the realization of tall structures. From this point of view, the invention of the elevator facilitated the evolution of these buildings as well as, in the same period, the use of new materials, which was a crucial issue. In particular, the presence of steel in the structural skeleton, with a high resistance-to-weight ratio, produced a series of benefits, such as decreased construction times, a variety of shapes, possible reuse, and a reduced degree of uncertainty about material properties that contributed to changes in the conceptual design of the constructions, especially regarding their slenderness.

At that time, the early reason for the growth in height was only a commercial one, having to compensate for the lack of space and natural light in densely populated urban areas. However, the higher the building, the more sensitive it becomes to lateral actions from wind and earthquakes. Without lateral stiffeners, the dimensions of the structural elements would increase, so that they would no longer be a satisfactory solution from an architectural point of view. For this purpose, the conventional load-bearing systems were replaced by new solutions that reduced the dimensions of the structural members, while at the same time guaranteeing the global stability of the building. The first solution was a framed steel structure that exploited the resistance properties of the material to reach an adequate stiffness, without compromising architectural demands. This first typology was followed by other systems designed to absorb and distribute the loads according to their own stiffness. At this new

Figure 9.1

stage, moment-resisting frames, braced frames, shear walls, and interactive frame-shear wall systems appeared.

Later, designers proposed that the building should be treated in a global manner and, therefore, analyzed as a three-dimensional body rather than as a series of parallel planar systems. This innovative vision gave rise to various different typologies that increased lateral resistance and stiffness without excessive use of structural materials. As a consequence, the traditional planar analyses were gradually replaced by three-dimensional global approaches. The global structure was considered as a unique vertical cantilever, or as a system of cantilevers clamped to the ground and transversally congruent.

According to the preceding considerations and depending on the height of the construction, several different solutions were realized, such as framed-tube and tube-in-tube systems. The latter refer to the basic idea of tube schemes that concentrate lateral stiffness on the external perimeter of the building. As significant examples, the John Hancock Center (Figure 9.2a) and the Sears Tower, now the Willis Tower (Figure 9.2b), are schematically represented.

Only recently has aesthetics acquired a leading role in the design of high-rise buildings. Changes in structural shape are suggested by emerging architectural trends and by developments in structural analysis, made possible by the advent of high-speed digital computers.

The most recent architectural trends are leading building design to new solutions, such as tapered, twisted, tilted, or even aerodynamic and free shapes.

9.2 PARALLEL-ARRANGED SYSTEM OF VERTICAL CANTILEVERS: GENERAL ALGORITHM

The elementary solution for the absorption of horizontal actions is the moment-resisting frame, in which global behavior is based on the flexural and shear stiffness of the network of beams and columns, being the corresponding joints designed as perfectly rigid. This solution proves to be adequate if horizontal forces are not predominant compared with vertical ones. Otherwise, an excessive increase in the dimensions of the structural components is expected. To overcome this difficulty, a different approach considers vertical behavior as

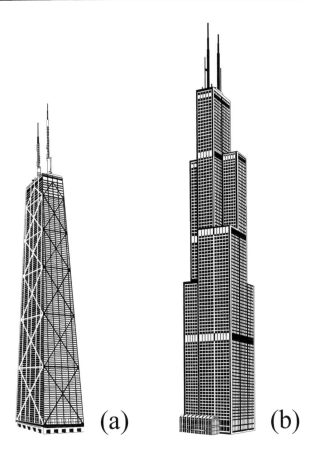

Figure 9.2

separate from horizontal behavior, and specific structural elements able to absorb the entire horizontal load can be utilized: this is the case for shear walls. These are cantilevers developing from the ground to the top of the building and, often, characterized by thin-walled open sections that allow the housing of stairwells and/or elevator shafts.

In fact, moment-resisting frames and shear walls can be adopted together to increase the global stiffness of the building and to reduce its lateral sway, which represents one of the most restrictive conditions coming from the standard codes of practice.

The analysis of this system, thus arranged, is primarily focused on the identification of the distribution of external forces among the single components. A planar configuration, in which the degrees of freedom are represented by the horizontal floor displacements, may be taken into account.

The effectiveness of this system is due to the different deformation that characterizes the frame with respect to the shear wall. In the presence of horizontal actions, the former is mainly subjected to shear deformations, whereas the latter is subjected to flexural ones. In this way, in the bottom part of the building, the shear wall sustains the frame, whereas, in the top part, the frame restrains the shear wall, thus mutually reducing the global deformation of the entire system (Figure 9.3).

From a structural point of view, to design each component of the scheme it is necessary to identify the amount of external load that is carried by each single vertical element. For this purpose, taking into account the model of Figure 9.3, a simplified approach assumes that

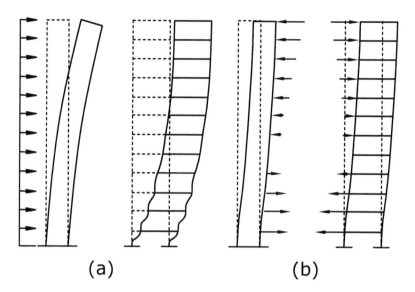

(a)   (b)

Figure 9.3

the connections between the two substructures can be represented by rigid rods, so that the congruence of the horizontal displacements at each floor is satisfied.

If $\{F\}$ represents the external load vector and $\{X\}$ the redundant unknown forces transmitted through the rods, due to the congruence conditions:

$$[C_1](\{F\}-\{X\}) = [C_2]\{X\} \tag{9.1}$$

where $[C_1]$ and $[C_2]$ are the compliance matrices of the frame and shear wall, respectively. Defining $[C]$ as the sum of matrices $[C_1]$ and $[C_2]$, the numerical solution of Equation 9.1 is

$$\{X\} = [C]^{-1}[C_1]\{F\} \tag{9.2}$$

which identifies the distribution of the redundant forces exchanged (Figure 9.3b).

The drawback of this analytical approach is that it can be applied to a limited number of cases, defined by simple structural combinations. It is characterized by a deep lack of generality that prevents the analysis of different and more complex structural solutions and that, above all, tends to reduce a three-dimensional problem to a planar one. In this way, it is evident that it is an inappropriate solution, especially in the case of very complex shapes, which cannot be grossly simplified.

In most buildings, the resistant system is composed of different vertical elements that constitute a three-dimensional structural skeleton. In this case, a more general semianalytical approach can be proposed, in which three degrees of freedom per story are taken into account. In this way, the bending and the torsional behaviors of the structure are studied at the same time. The approach proves to be general, since it is possible to consider any type of vertical bracing, from simple frames to free-shaped tubular elements, provided that their own stiffness matrix is known.

The formulation is based on the following fundamental hypotheses:

1. The structural material is homogeneous, isotropic, and obeys Hooke's law.
2. The floor slabs are rigid in their own plane but their out-of-plane rigidity is negligible.
3. In transversal analysis, the axial deformation of the structural elements due to gravity loads is negligible.

Based on the previously mentioned hypotheses, an N-story building is considered to have M vertical bracings, each defined by an arbitrary position in the floor plan. The right-handed system XYZ defines the global coordinate system. Since the slabs, which connect the bracings to each other, are considered to be infinitely rigid in their own planes, the degrees of freedom are represented by the transverse displacements of the single floors: two translations ξ and η in the directions X and Y, and the torsional rotation ϑ, for each story. In the same way, the external load applied to the origin of the reference system is expressed by a $3N$-vector $\{F\}$, in which $2N$ shearing forces $\{p_x\}$, $\{p_y\}$ and N torsional moments $\{m_z\}$ are included (Figure 9.4):

$$\{F_i\} = \begin{Bmatrix} p_i \\ m_{z,i} \end{Bmatrix} = \begin{Bmatrix} p_{x,i} \\ p_{y,i} \\ m_{z,i} \end{Bmatrix} \tag{9.3}$$

Within the right-handed system $X_i^* Y_i^* Z_i^*$, the local coordinate system of the ith bracing, the $3N$-load vector $\{F_i^*\}$, and the $3N$-displacement vector $\{\delta_i^*\}$ describe the amount of external load carried by the ith element and its transverse displacements, respectively.

The loading vector $\{F_i^*\}$ can be reduced to $\{F_i\}$, which refers to the global coordinate system XYZ, by means of the following expressions, valid for each bracing:

$$\{p_i^*\} = [N_i]\{p_i\} \tag{9.4}$$

$$m_{z,i}^* = m_{z,i} - \{\psi_i\} \wedge \{p_i\} \times \{u_z\} \tag{9.5}$$

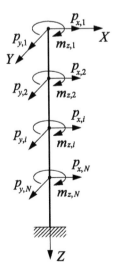

Figure 9.4

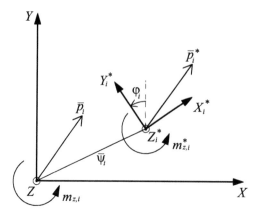

Figure 9.5

In matrix form:

$$\begin{Bmatrix} p^{*}_{x,i} \\ p^{*}_{y,i} \\ m^{*}_{z,i} \end{Bmatrix} = \begin{bmatrix} [N_i] & [0] \\ -\{u_z\} \wedge \{\psi_i\} & [1] \end{bmatrix} \begin{Bmatrix} p_{x,i} \\ p_{y,i} \\ m_{z,i} \end{Bmatrix}$$

(9.6)

where:

$[N_i]$ represents the orthogonal rotation matrix from system XY to system $X_i^* Y_i^*$
$\{\psi_i\}$ is the coordinate vector of the origin of the local system in the global one
$\{u_z\}$ is the unit vector associated with the direction Z
$[1]$ is the identity matrix
$[0]$ is the null matrix

The orthogonal matrix $[N_i]$, extended to consider all floors, can be represented by means of the angle φ_i between the Y and Y_i^* axes (Figure 9.5):

$$[N_i] = \begin{bmatrix} [\cos \varphi_i] & [\sin \varphi_i] \\ -[\sin \varphi_i] & [\cos \varphi_i] \end{bmatrix}$$

(9.7)

where each term is a diagonal $(N \times N)$ submatrix:

$$[\cos \varphi_i] = \begin{bmatrix} \cos \varphi_i & 0 & \cdots & 0 \\ 0 & \cos \varphi_i & \cdots & 0 \\ \vdots & \vdots & & \vdots \\ 0 & 0 & \cdots & \cos \varphi_i \end{bmatrix}$$

(9.8a)

$$[\sin \varphi_i] = \begin{bmatrix} \sin \varphi_i & 0 & \cdots & 0 \\ 0 & \sin \varphi_i & \cdots & 0 \\ \vdots & \vdots & & \vdots \\ 0 & 0 & \cdots & \sin \varphi_i \end{bmatrix}$$

(9.8b)

Taking into account all floors, Equation 9.6 can be rewritten in the following form:

$$\{F_i^*\} = [A_i]\{F_i\}$$

(9.9)

Matrix $[A_i]$ gathers the information regarding the reciprocal rotation between the local and global coordinate systems and the location of the ith bracing in the global system XY:

$$[A_i] = \begin{bmatrix} [N_i] & [0] \\ -\{u_z\} \wedge \{\psi_i\} & [1] \end{bmatrix}$$

(9.10)

The component $-\{u_z\} \wedge \{\psi_i\}$, valid for each floor, is obtained from Equation 9.5, exploiting the fact that the scalar triple product is invariant under any cyclic permutation of its factors. For the sake of simplicity, to take into account the N floors of the structure, this vector product can be written as a $(2N \times N)$ matrix $[C_i]$ composed of two diagonal submatrices containing the coordinates $(x_i; y_i)$ of the origin of the local system $X_i^* Y_i^*$:

$$-\{u_z\} \wedge \{\psi_i\} = -\begin{vmatrix} \bar{i} & \bar{j} & \bar{k} \\ 0 & 0 & 1 \\ x_i & y_i & 0 \end{vmatrix} = -[-y_i \quad x_i] = -[C_i]^T$$

(9.11)

Thus, the final expression for matrix $[A_i]$ is

$$[A_i] = \begin{bmatrix} [N_i] & [0] \\ -[C_i]^T & [1] \end{bmatrix}$$

(9.12)

In the same way, the vector $\{\delta_i^*\}$, constituted by $2N$ translations ξ_i^*, η_i^* and N rotations ϑ_i^*, can be connected to the corresponding $\{\delta_i\}$, which is referred to the global coordinate system

$$\{\delta_i^*\} = \begin{Bmatrix} \xi_i^* \\ \eta_i^* \\ \vartheta_i^* \end{Bmatrix}$$

(9.13)

The displacements $\{\delta_i\}$ in the global coordinate system XY are then connected to the displacements $\{\delta_i^*\}$ in the local coordinate system $X_i^* Y_i^*$ by the orthogonal matrix $[N_i]$:

$$\begin{Bmatrix} \xi_i^* \\ \eta_i^* \end{Bmatrix} = [N_i] \begin{Bmatrix} \xi_i \\ \eta_i \end{Bmatrix}$$

(9.14a)

$$\vartheta_i^* = \vartheta_i$$

(9.14b)

Taking into account all floors, Equations 9.14 can be rewritten in the following form by means of the compact $(3N \times 3N)$ matrix $[B_i]$:

$$\{\delta_i^*\} = [B_i]\{\delta_i\} \tag{9.15}$$

where matrix $[B_i]$ is similar to $[A_i]$, the term $[C_i]^T$ being reduced to a null matrix:

$$[B_i] = \begin{bmatrix} [N_i] & [0] \\ [0] & [1] \end{bmatrix} \tag{9.16}$$

A relation between $\{F_i^*\}$ and $\{\delta_i^*\}$ is considered known through the condensed stiffness matrix $[K_i^*]$, referred to the local coordinate system:

$$\{F_i^*\} = [K_i^*]\{\delta_i^*\} \tag{9.17}$$

Substituting Equations 9.9 and 9.15 into Equation 9.17, the load vector $\{F_i\}$ turns out to be connected to the displacement vector $\{\delta_i\}$ through a product of matrices, which identifies the stiffness matrix $[K_i]$ of the ith bracing in the global coordinate system XY:

$$\{F_i\} = \left([A_i]^{-1}[K_i^*][B_i]\right)\{\delta_i\} = [K_i]\{\delta_i\} \tag{9.18}$$

Due to the presence of in-plane rigid slabs connecting the vertical cantilevers, the transverse displacements of each element can be computed considering only three generalized displacements, ξ, η, and ϑ, per floor. This step, extended to consider all floors, is performed through the matrix $[T_i]$ that takes into account the location of each bracing in the plan by means of the coordinates $(x_i; y_i)$ and, therefore, the matrix $[C_i]$:

$$\{\delta_i\} = \begin{bmatrix} [1] & [C_i] \\ [0] & [1] \end{bmatrix}\{\delta\} = [T_i]\{\delta\} \tag{9.19}$$

where $\{\delta\}$ is the floor displacement vector, that is, the displacement vector associated with the origin of the global reference system.

The substitution of Equation 9.19 into Equation 9.18 allows the identification of the stiffness matrix of the ith bracing, in reference to the global coordinate system XYZ and to the generalized floor displacements ξ, η, and ϑ:

$$\{F_i\} = \left([K_i][T_i]\right)\{\delta\} = [\overline{K}_i]\{\delta\} \tag{9.20}$$

For global equilibrium, the external load $\{F\}$ applied to the structure is equal to the sum of the M vectors $\{F_i\}$. In this way, a relationship between the external load and the floor displacements is obtained and the global stiffness matrix of the structure is computed. By means of this matrix, once the external load is defined, the displacements of the structure are acquired, from which the information regarding each single bracing can be deduced:

$$\{F\} = \sum_{i=1}^{M} \{F_i\} = \left(\sum_{i=1}^{M} \left[\overline{K}_i\right] \right) \{\delta\} = \left[\overline{K}\right]\{\delta\} \qquad (9.21)$$

and therefore

$$\{\delta\} = \left[\overline{K}\right]^{-1} \{F\} \qquad (9.22)$$

Recalling Equation 9.20 and comparing it with Equation 9.22, an equation connecting the vectors $\{F\}$ and $\{F_i\}$ allows the definition of the amount of the external load carried by the ith vertical stiffening element:

$$\{\delta\} = \left[\overline{K}\right]^{-1} \{F\} = \left[\overline{K}_i\right]^{-1} \{F_i\} \qquad (9.23)$$

from which we obtain

$$\{F_i\} = \left[\overline{K}_i\right]\left[\overline{K}\right]^{-1} \{F\} = [R_i]\{F\} \qquad (9.24)$$

The load distribution matrix $[R_i]$, shown in Equation 9.24, demonstrates that each bracing is subjected to a load $\{F_i\}$ that is given by the external load $\{F\}$ premultiplied by the own stiffness matrix and the inverse of the global stiffness matrix.

Once the generalized displacement vector $\{\delta\}$ is known, recalling Equations 9.13, 9.17, and 9.19, the displacements and the forces related to the ith bracing in its local coordinate system can be computed. Consequently, since the loads applied to each element are clearly identified, a preliminary assessment can easily be performed.

Equation 9.24 solves the problem of the external load distribution between the resistant elements employed to stiffen a three-dimensional tall building. Such formulation proves to be general and can be adopted with any kind of structural element, provided that their own condensed stiffness matrix $[K_i^*]$ is known. Therefore, most of the common vertical stiffeners, such as frames, braced frames, shear walls, and tubular elements can be easily implemented in this static formulation.

Further benefits can be highlighted: first, an easy identification of the structural parameters, which govern the lateral behavior of the building, can be performed; secondly, the formulation proves to be extremely clear and concise, limiting, in this way, the risk of unexpected errors and guaranteeing, in the presence of very complex structures, relatively short times of modeling and analysis, if compared with finite element computations.

9.3 VLASOV'S THEORY OF THIN-WALLED OPEN-SECTION BEAMS IN TORSION

As described in the introduction, most of the resistant solutions employed in tall buildings are represented by vertical elements arranged as parallel cantilevers clamped at the base and designed to absorb the total horizontal force coming from earthquakes and winds. These members, commonly known as shear walls, can be freely located in the plan of the building and used with or without other vertical bracings to obtain an adequate stability. In the case of nonexcessive heights, they can be constituted by a simple plane element whose resistance is proportional to the maximum size of the section. For greater heights, they are designed to behave as three-dimensional elements, having an appropriate bending resistance in the two principal directions, as well as a good torsional stiffness, giving rise to thin-walled hollow or open-section walls. Beyond the mechanical function, these members allow the housing of stairwells and/or elevator shafts, which are indispensable in a tall building.

Unlike hollow sections, in the presence of torsional actions, thin-walled, open-section elements reveal a particular behavior that is far from Saint-Venant's results. Once the torsional deformation takes place, the section twists around its shear center but, at the same time, does not remain plane, since it undergoes different longitudinal extensions causing an out-of-plane distortion, or so-called **warping**, of the section. As a consequence, a further longitudinal stress, absent in the theory of primary torsion, develops in the thickness of the section.

Let us consider the case of a cantilever I-beam subjected to a concentrated load on one of its flanges (Figure 9.6). Based on the superposition principle, this load can be reduced to the sum of four different loading cases: one is purely axial, two are purely flexural, while the remaining is defined as flexural torsion, since the two flanges are forced to bend in opposite directions on their own planes. In the latter case, the section does not remain plane and additional normal stresses appear. These additional normal stresses give rise to a generalized action, called the **bimoment**, which is directly connected to the warping of the section and consists of two bending moments, each one acting on a single flange, having the same magnitude but opposite signs.

In the case of compact sections, this self-equilibrated action only has a local effect that rapidly falls off by increasing the distance from the beam end. On the contrary, in the case of thin-walled, open-section beams, the warping stresses fall off slowly as much as the walls are thin. The intensity of this stress state cannot be neglected for these profiles and the application of Saint-Venant's theory could lead to gross errors.

Two main geometrical hypotheses are at the basis of Vlasov's theory:

1. The section is considered rigid and, therefore, its shape is undeformable.
2. The shearing strains on the midline of the section are assumed to vanish.

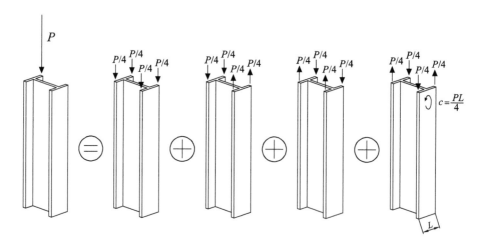

Figure 9.6

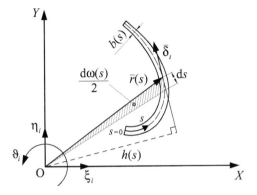

Figure 9.7

Let us consider a free-shaped, thin-walled, open-section beam, located in a generic coordinate system, in which the Z axis is parallel to the longitudinal axis of the beam. A specific cross section is defined at $z = constant$; X and Y axes complete the right-handed coordinate system XYZ. Each point of the midline can be identified by using the coordinates (x, y) or, along the midline, the curvilinear coordinate s (Figure 9.7).

With the aim of defining the equations that govern the structural behavior of thin-walled open profiles, it is assumed that the beam is subjected to some torsional deformations. As a result of these, each point of the section is characterized by a new position in the general coordinate system XYZ. According to the first geometrical hypothesis, the beam is deformable, although the shape of the section remains unchanged. Therefore, it behaves as a perfectly rigid body, whose position can be evaluated by means of three independent variables corresponding to the three generalized displacements of an arbitrarily chosen point: two translations ξ and η in the X and Y directions, respectively, and the rotation ϑ.

The transverse displacements ξ and η of any point belonging to the cross section can be determined through the well-known expressions:

$$u = \xi(z) - \vartheta(z)y \tag{9.25a}$$

$$v = \eta(z) + \vartheta(z)x \tag{9.25b}$$

The tangential displacement δ_t, related to the generic point of the section, can be computed by

$$\delta_t = \{\delta\}^{\mathrm{T}}\{u_t\} = u\frac{dx}{ds} + v\frac{dy}{ds} \tag{9.26}$$

and then

$$\delta_t = \xi\frac{dx}{ds} + \eta\frac{dy}{ds} + \vartheta\, h(s) \tag{9.27}$$

in which the term $h(s)$ represents the distance between the origin of the reference system and the tangent line to the section midline (Figure 9.7):

$$h(s) = \{r\}^\mathsf{T}\{u_n\} = x\frac{dy}{ds} - y\frac{dx}{ds} \tag{9.28}$$

The longitudinal displacement component w can be obtained by Vlasov's second hypothesis, according to which the shearing strains on the midline are considered negligible:

$$\gamma_{zs} = \frac{\partial w}{\partial s} + \frac{\partial \delta_t}{\partial z} = 0 \tag{9.29}$$

Taking into account the following relationship:

$$\omega(s) = \int_0^s h(s)\, ds \tag{9.30}$$

the analytical expression of w is derived by integration:

$$w = \zeta(z) - \int_0^s \frac{\partial \delta_t}{\partial z} ds = \zeta(z) - \xi'x - \eta'y - \vartheta'\omega \tag{9.31}$$

The term $\zeta(z)$ is an arbitrary function, depending only on z, which describes a longitudinal translation of the entire section; $\omega(s)$ is the **sectorial area,** that is, double the area swept by the radius vector $\{r\}$ from $s=0$ to the current point s of the section's midline. The points O and $s=0$ are the origin of an arbitrary reference system and the sectorial origin, respectively (Figure 9.7).

The longitudinal component w is composed of four terms: the first three are well-known and arise from extension and bending in the XZ and YZ planes. The component that describes the warping of the section is expressed by the fourth term and, in particular, ϑ' can be considered as an amplification factor, whereas ω is the shape of the warped section.

By differentiating w with respect to z, it is possible to obtain the expression of the longitudinal deformation ε_z:

$$\varepsilon_z = \frac{\partial w}{\partial z} = \zeta' - \xi''x - \eta''y - \vartheta''\omega \tag{9.32}$$

The fourth term of Equation 9.32 demonstrates that the hypothesis of primary torsion, according to which the unit angle of torsion should be constant, can, in general, be removed.

The general expression of the normal stresses is obtained by multiplying Equation 9.32 by the elastic modulus E:

$$\sigma_z = E(\zeta' - \xi''x - \eta''y - \vartheta''\omega) \tag{9.33}$$

In each section of the beam, the normal stress σ_z is the sum of two contributions:

$$\sigma_z = \sigma_z^{SV} + \sigma_z^{VL} \tag{9.34}$$

where

$$\sigma_z^{VL} = -E\vartheta''\omega \tag{9.35}$$

These expressions demonstrate that normal stresses can appear not only in the presence of uniform extension and bending of the beam, but also as a result of the nonuniform torsion of the cross section. On the other hand, this specific contribution is usually assumed to vanish in the theory of primary torsion.

Equation 9.33 allows the definition by integration of the internal actions related to the extensional and flexural behavior of the beam:

$$N = \int_A \sigma_z \, dA = E\left(A\zeta' - S_y\xi'' - S_x\eta'' - S_\omega\vartheta''\right) \tag{9.36a}$$

$$M_y = \int_A \sigma_z x \, dA = E\left(S_y\zeta' - I_{yy}\xi'' - I_{yx}\eta'' - I_{y\omega}\vartheta''\right) \tag{9.36b}$$

$$M_x = \int_A \sigma_z y \, dA = E\left(S_x\zeta' - I_{xy}\xi'' - I_{xx}\eta'' - I_{x\omega}\vartheta''\right) \tag{9.36c}$$

$$B = \int_A \sigma_z \omega \, dA = E\left(S_\omega\zeta' - I_{\omega y}\xi'' - I_{\omega x}\eta'' - I_{\omega\omega}\vartheta''\right) \tag{9.36d}$$

in which the sectorial characteristics of the section are expressed by the sectorial static moment S_ω, the sectorial moment of inertia $I_{\omega\omega}$, and the sectorial products of inertia $I_{\omega x}$ and $I_{\omega y}$, defined as follows:

$$S_y = \int_A x \, dA \tag{9.37a}$$

$$S_x = \int_A y \, dA \tag{9.37b}$$

$$S_\omega = \int_A \omega \, dA \tag{9.37c}$$

$$I_{yy} = \int_A x^2 \, dA \tag{9.38a}$$

$$I_{xx} = \int_A y^2 dA \tag{9.38b}$$

$$I_{\omega\omega} = \int_A \omega^2 dA \tag{9.38c}$$

$$I_{yx} = I_{xy} = \int_A xy dA \tag{9.39a}$$

$$I_{x\omega} = I_{\omega x} = \int_A \omega y dA \tag{9.39b}$$

$$I_{y\omega} = I_{\omega y} = \int_A \omega x dA \tag{9.39c}$$

Equation 9.36d defines the **bimoment** action, which represents a generalized self-balanced force system equivalent to two bending moments, having the same magnitude but opposite signs.

The tangential stresses τ_{zs}, supposed to be defined by a constant distribution on the thickness of the section, can be obtained considering the longitudinal equilibrium of an elementary portion of the beam, whose dimensions are length dz, width ds, and thickness b (Figure 9.8):

$$\left(\frac{\partial \tau_{zs}}{\partial s} ds \right) b dz + \left(\frac{\partial \sigma_z}{\partial z} dz \right) b ds = 0 \tag{9.40}$$

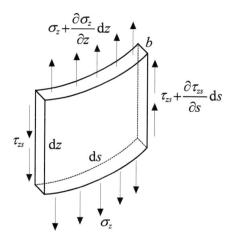

Figure 9.8

Dividing Equation 9.40 by $dsdz$, the following expression is obtained:

$$\frac{\partial(\tau_{zs}b)}{\partial s} + \frac{\partial(\sigma_z b)}{\partial z} = 0 \tag{9.41}$$

On the basis of Equation 9.41, three additional transverse internal actions, the shearing forces, and the secondary torsional moment, can be defined:

$$T_x = \int_A \tau_{zs} \frac{dx}{ds} dA \tag{9.42a}$$

$$T_y = \int_A \tau_{zs} \frac{dy}{ds} dA \tag{9.42b}$$

$$M_z^{VL} = \int_A \tau_{zs}b \, dA \tag{9.42c}$$

Integrating by parts and applying Equation 9.41, the following relations are obtained:

$$T_x = -\int_C \frac{\partial(\tau_{zs}b)}{\partial s} x ds = \int_C \frac{\partial(\sigma_z b)}{\partial z} x ds = \frac{d}{dz}\int_A \sigma_z x dA \tag{9.43a}$$

$$T_y = -\int_C \frac{\partial(\tau_{zs}b)}{\partial s} y ds = \int_C \frac{\partial(\sigma_z b)}{\partial z} y ds = \frac{d}{dz}\int_A \sigma_z y dA \tag{9.43b}$$

$$M_z^{VL} = -\int_C \frac{\partial(\tau_{zs}b)}{\partial s} \omega ds = \int_C \frac{\partial(\sigma_z b)}{\partial z} \omega ds = \frac{d}{dz}\int_A \sigma_z \omega dA \tag{9.43c}$$

Comparing Equations 9.43 with Equations 9.36, it is possible to recognize a fundamental differential relationship between the longitudinal and the transverse actions:

$$T_x = \frac{dM_y}{dz} = E\left(S_y\zeta'' - I_{yy}\xi''' - I_{yx}\eta''' - I_{y\omega}\vartheta'''\right) \tag{9.44a}$$

$$T_y = \frac{dM_x}{dz} = E\left(S_x\zeta'' - I_{xy}\xi''' - I_{xx}\eta''' - I_{x\omega}\vartheta'''\right) \tag{9.44b}$$

$$M_z^{VL} = \frac{dB}{dz} = E\left(S_\omega\zeta'' - I_{\omega y}\xi''' - I_{\omega x}\eta''' - I_{\omega\omega}\vartheta'''\right) \tag{9.44c}$$

The last equation highlights that, due to the warping of the section, an unexpected torsional moment M_z^{VL} is generated, it being the first derivative of the bimoment action.

The secondary torsional moment M_z^{VL} is generated by the τ_{zs} stresses due to the shearing actions T_x and T_y.

A further step of differentiation leads to the equilibrium equations that take into account the distributed external loads p_x, p_y, and m_z (known terms):

$$p_x = -\frac{dT_x}{dz} = E\left(-S_y\zeta''' + I_{yy}\xi^{IV} + I_{yx}\eta^{IV} + I_{y\omega}\vartheta^{IV}\right) \tag{9.45a}$$

$$p_y = -\frac{dT_y}{dz} = E\left(-S_x\zeta''' + I_{xy}\xi^{IV} + I_{xx}\eta^{IV} + I_{x\omega}\vartheta^{IV}\right) \tag{9.45b}$$

$$m_z^{VL} = -\frac{dM_z^{VL}}{dz} = E\left(-S_\omega\zeta''' + I_{\omega y}\xi^{IV} + I_{\omega x}\eta^{IV} + I_{\omega\omega}\vartheta^{IV}\right) \tag{9.45c}$$

In fact, the thin-walled open section is subjected to an external torsional moment, which can be subdivided into two portions: the first, according to Saint-Venant's theory, is due to a linear variation of tangential stresses through its thickness and is equal to zero on the midline; the second, according to Vlasov's theory, is due to a constant distribution of tangential stresses through the thickness, and is related to equilibrium, with the normal stresses coming from the differential warping of the section.

In each section of the beam, the torsional moment M_z is the sum of the two contributions

$$M_z = M_z^{SV} + M_z^{VL} = GI_t\vartheta' + E\left(S_\omega\zeta'' - I_{\omega y}\xi''' - I_{\omega x}\eta''' - I_{\omega\omega}\vartheta'''\right) \tag{9.46}$$

and, therefore, the global equilibrium Equation 9.45c becomes

$$m_z = m_z^{SV} + m_z^{VL} = -\frac{dM_z^{SV}}{dz} - \frac{dM_z^{VL}}{dz} = -GI_t\vartheta'' + E\left(-S_\omega\zeta''' + I_{\omega y}\xi^{IV} + I_{\omega x}\eta^{IV} + I_{\omega\omega}\vartheta^{IV}\right) \tag{9.47}$$

where G is the shear modulus and I_t is the torsional stiffness factor of the section.

Finally, an expression of the tangential stresses τ_{zs} can be obtained by substituting Equation 9.33 into Equation 9.41:

$$\frac{\partial(\tau_{zs}b)}{\partial s} + Eb(\zeta'' - \xi'''x - \eta'''y - \vartheta''\omega) = 0 \tag{9.48}$$

and integrating with respect to s:

$$\tau_{zs} = -\frac{E}{b}\left[\zeta''A(s) - \xi'''S_y(s) - \eta'''S_x(s) - \vartheta''S_\omega(s)\right] \tag{9.49}$$

where the following geometrical expressions are used:

$$A(s) = \int_0^s b\,ds \tag{9.50a}$$

$$S_y(s) = \int_0^s xb\,ds \tag{9.50b}$$

$$S_x(s) = \int_0^s yb\,ds \tag{9.50c}$$

$$S_\omega(s) = \int_0^s \omega b\,ds \tag{9.50d}$$

The system of differential equilibrium Equations 9.45 allows the computation of the unknown displacements.

9.4 CAPURSO'S METHOD: LATERAL LOADING DISTRIBUTION BETWEEN THE THIN-WALLED OPEN-SECTION VERTICAL CANTILEVERS OF A TALL BUILDING

Thin-walled, open-section shear walls are primarily employed in tall buildings to contribute to their horizontal resistance and stiffness. According to the previous hypotheses, the analytical formulation proposed by Vlasov can be adopted to evaluate the structural behavior of a tall building stiffened by a single thin-walled, open-section cantilever.

The computation of the sectorial terms is carried out considering the origin of the generic right-handed system XYZ. Transverse distributed actions p_x, p_y, and m_z are considered in the analysis.

Supposing the axial force in the vertical bracing to be null, the term ζ' in Equation 9.36a can be eliminated:

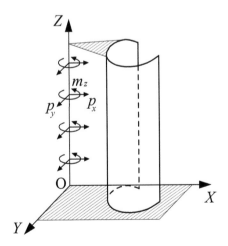

Figure 9.9

$$\zeta' = \frac{S_y}{A}\xi'' + \frac{S_x}{A}\eta'' + \frac{S_\omega}{A}\vartheta'' = x_G\xi'' + y_G\eta'' + \omega_0\vartheta'' \tag{9.51}$$

The substitution of Equation 9.51 into Equations 9.36b, c, and d permits the definition of new expressions for the longitudinal actions without the term ζ':

$$M_y = -E\left(J_{yy}\xi'' + J_{yx}\eta'' + J_{y\omega}\vartheta''\right) \tag{9.52a}$$

$$M_x = -E\left(J_{xy}\xi'' + J_{xx}\eta'' + J_{x\omega}\vartheta''\right) \tag{9.52b}$$

$$B = -E\left(J_{\omega y}\xi'' + J_{\omega x}\eta'' + J_{\omega\omega}\vartheta''\right) \tag{9.52c}$$

where

$$J_{yy} = I_{yy} - Ax_G^2 \tag{9.53a}$$

$$J_{xx} = I_{xx} - Ay_G^2 \tag{9.53b}$$

$$J_{xy} = I_{xy} - Ax_Gy_G \tag{9.53c}$$

$$J_{\omega\omega} = I_{\omega\omega} - A\omega_0^2 \tag{9.54a}$$

$$J_{\omega y} = I_{\omega y} - A\omega_0 x_G \tag{9.54b}$$

$$J_{\omega x} = I_{\omega x} - A\omega_0 y_G \tag{9.54c}$$

Equations 9.53 represent the implementation of the Huygens–Steiner theorem, which transfers the system XYZ from the generic origin to the centroid of the section. Similarly, Equations 9.54 express the sectorial properties with respect to the baricentric axes and to the sectorial centroid.

Disregarding the torsional rigidity GI_t, Equations 9.53 and 9.54 also affect the system of Equations 9.44, which become

$$T_x = -E\left(J_{yy}\xi''' + J_{yx}\eta''' + J_{y\omega}\vartheta'''\right) \tag{9.55a}$$

$$T_y = -E\left(J_{xy}\xi''' + J_{xx}\eta''' + J_{x\omega}\vartheta'''\right) \tag{9.55b}$$

$$M_z^{VL} = -E\left(J_{\omega y}\xi''' + J_{\omega x}\eta''' + J_{\omega\omega}\vartheta'''\right) \tag{9.55c}$$

and, similarly, the distributed external loads p_x, p_y, and m_z of the system of Equations 9.45 turn into

$$p_x = E\left(J_{yy}\xi^{IV} + J_{yx}\eta^{IV} + J_{y\omega}\vartheta^{IV}\right)$$ (9.56a)

$$p_y = E\left(J_{xy}\xi^{IV} + J_{xx}\eta^{IV} + J_{x\omega}\vartheta^{IV}\right)$$ (9.56b)

$$m_z^{VL} = E\left(J_{\omega y}\xi^{IV} + J_{\omega x}\eta^{IV} + J_{\omega\omega}\vartheta^{IV}\right)$$ (9.56c)

If the matrix of inertia $[J]$ and the vectors $\{\delta\}$, $\{M\}$, $\{T\}$, and $\{F\}$ are introduced, it is possible to write the systems of Equations 9.52, 9.55, and 9.56 in a compact form:

$$[J] = \begin{bmatrix} J_{yy} & J_{yx} & J_{y\omega} \\ J_{xy} & J_{xx} & J_{x\omega} \\ J_{\omega y} & J_{\omega x} & J_{\omega\omega} \end{bmatrix}$$ (9.57)

$$\{\delta\} = \begin{Bmatrix} \xi \\ \eta \\ \vartheta \end{Bmatrix}$$ (9.58a)

$$\{M\} = \begin{Bmatrix} M_y \\ M_x \\ B \end{Bmatrix}$$ (9.58b)

$$\{T\} = \begin{Bmatrix} T_x \\ T_y \\ M_z^{VL} \end{Bmatrix}$$ (9.58c)

$$\{F\} = \begin{Bmatrix} p_x \\ p_y \\ m_z \end{Bmatrix}$$ (9.58d)

$$\{M\} = -E[J]\{\delta''\}$$ (9.59a)

$$\{T\} = -E[J]\{\delta'''\}$$ (9.59b)

$$\{F\} = E[J]\{\delta^{IV}\}$$ (9.59c)

Since the matrix of inertia is symmetrical and positive definite until the geometry of the section is such that the determinant of $[J]$ is not zero, it can be inverted to obtain a relationship between the fourth derivatives of the displacements and the external distributed actions:

$$\{\delta^{IV}\} = \frac{1}{E}[J]^{-1}\{F\} \tag{9.60}$$

The transverse displacements of the section are obtained by integrating Equation 9.60 and applying the boundary conditions at the base and at the top of the cantilever.

At the constrained end:

$$\{\delta\} = \{0\} \tag{9.61a}$$

$$\{\delta'\} = \{0\} \tag{9.61b}$$

for $z = 0$

whereas, at the top:

$$\{\delta''\} = \{0\} \tag{9.62a}$$

$$\{\delta'''\} = \{0\} \tag{9.62b}$$

for $z = l$.

Once ξ, η, and ϑ are known, the application of Equation 9.51 yields the uniform axial displacement ζ with the corresponding boundary condition:

$$\zeta(z = 0) = 0 \tag{9.63}$$

Eventually, the displacement components δ_t and w can be easily derived from Equations 9.27 and 9.31, as well as the internal stress state given by Equation 9.33, to which the effect of the primary torsion has to be added.

This analytical formulation cannot be applied in the presence of specific sections for which the matrix $[J]$ is singular. These are the cases of shear walls constituted by a single thin rectangular plate or different thin plates converging to a single point, as shown in Figure 9.10. In these cases, the warping function vanishes.

The previous formulation can be extended to consider the case of M vertical cantilevers that represent the resistant skeleton of a tall building loaded by transverse actions applied to the floors with reference to the global coordinate system XYZ. The vertical bracings are connected to each other by means of in-plane rigid slabs, whose out-of-plane rigidity can be considered negligible.

The unknown variables of the problem are the floor displacements, identified by the translations ξ and η in the X and Y directions, respectively, and the torsional rotation ϑ. If $\{F_i\}$

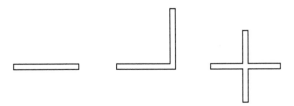

Figure 9.10

indicates the vector of the transverse actions transmitted to the ith cantilever, by virtue of Equation 9.59c we have

$$\{F_i\} = E[J_i]\{\delta^{IV}\} \tag{9.64}$$

where matrix $[J_i]$ contains the moments of inertia with reference to the centroid of the section and to the sectorial centroid, whereas the vector $\{\delta^{IV}\}$ gathers fourth order derivatives of the floor displacements ξ, η, and ϑ.

If $\{F\}$ is the vector of the external loads, the equilibrium condition imposes

$$\{F\} = \sum_{i=1}^{M}\{F_i\} = E\left(\sum_{i=1}^{M}[J_i]\right)\{\delta^{IV}\} = E[J]\{\delta^{IV}\} \tag{9.65}$$

Therefore, the combination of M cantilevers behaves as a single cantilever whose matrix of inertia is given by the sum of the M matrices related to the single cantilevers:

$$[J] = \sum_{i=1}^{M}[J_i] \tag{9.66}$$

Equation 9.65 can be solved following the procedure previously described in the case of a single vertical bracing. Once the floor displacements are known, the displacements of each cantilever can be deduced and information regarding the stress state can also be obtained. Finally, it is interesting to observe, from the relation between the vector $\{F_i\}$ of the ith cantilever and the global vector $\{F\}$, that each bracing is subjected to an external load vector produced by the product of its own inertia matrix and the inverse of the global one, analogously to what emerges in the general algorithm of Section 9.2:

$$\{F_i\} = [J_i][J]^{-1}\{F\} \tag{9.67}$$

In the case of a discrete distribution of transverse forces corresponding to the different floors, the (3×3) matrix $[J]$, which is a function of the longitudinal coordinate z, can be expanded to a $(3N \times 3N)$ stiffness matrix to be inserted in the general algorithm.

9.5 DIAGONALIZATION OF VLASOV'S EQUATIONS

The system of Equations 9.45 can be strongly simplified by making certain choices. In fact, if a centroidal coordinate system is considered, the following conditions are all immediately satisfied:

$$S_y = \int_A x dA = 0 \tag{9.68a}$$

$$S_x = \int_A y dA = 0 \tag{9.68b}$$

In addition, if the reference system is also principal, the product of inertia is null:

$$I_{xy} = I_{yx} = \int_A xy dA = 0 \tag{9.69}$$

On the other hand, if the sectorial pole coincides with the shear center of the section, it can be shown that

$$I_{\omega y} = I_{y\omega} = \int_A \omega x dA = 0 \tag{9.70a}$$

$$I_{\omega x} = I_{x\omega} = \int_A \omega y dA = 0 \tag{9.70b}$$

In addition, if the sectorial origin is in the sectorial centroid, by definition it follows that the sectorial static moment is also null:

$$S_\omega = \int_A \omega dA = 0 \tag{9.71}$$

Taking into account Equations 9.52 and considering the principal reference system with its origin in the shear center, the internal actions can be defined as

$$M_y = -EJ_{yy}\xi'' \tag{9.72a}$$

$$M_x = -EJ_{xx}\eta'' \tag{9.72b}$$

$$B = -EJ_{\omega\omega}\vartheta'' \tag{9.72c}$$

When the centroid and shear center do not coincide, the diagonalization of Vlasov's equations, is possible only in the case $N = 0$.

The substitution of Equations 9.72 into Equation 9.33 gives an expression of the normal stress based on the corresponding internal actions:

$$\sigma_z = \frac{M_y}{J_{yy}} x + \frac{M_x}{J_{xx}} y + \frac{B}{J_{\omega\omega}} \omega \tag{9.73}$$

The first two contributions derive from the well-known Saint-Venant's theory and are based on the hypothesis of plane sections; the third describes the normal stresses due to the out-of-plane warping of the profile.

The internal actions producing tangential stresses are also diagonalized:

$$T_x = -EJ_{yy}\xi''' \tag{9.74a}$$

$$T_y = -EJ_{xx}\eta''' \tag{9.74b}$$

$$M_z^{VL} = -EJ_{\omega\omega}\vartheta''' \tag{9.74c}$$

This means that the system of Equations 9.56 is reduced to the following decoupled equilibrium equations:

$$p_x = EJ_{yy}\xi^{IV} \tag{9.75a}$$

$$p_y = EJ_{xx}\eta^{IV} \tag{9.75b}$$

$$m_z = EJ_{\omega\omega}\vartheta^{IV} - GI_t\vartheta'' \tag{9.75c}$$

Imposing the boundary conditions, the system can be solved and functions ξ, η, and ϑ can be determined together with the normal and tangential stresses.

It is interesting to observe that Equation 9.75c is formally the same as the equation of the elastic line with effects of the second order, due to a tensile axial load N:

$$q(z) = EIv^{IV} - Nv'' \tag{9.76}$$

The substitution of Equations 9.74 into Equation 9.49 gives an expression for the tangential stresses:

$$\tau_{zs} = \frac{1}{b}\left[\frac{T_x}{J_{yy}} S_y(s) + \frac{T_y}{J_{xx}} S_x(s) + \frac{M_z^{VL}}{J_{\omega\omega}} S_\omega(s) \right] \tag{9.77}$$

The first two terms derive from Jourawski's theory, whereas the last derives from Vlasov's theory.

It is worthwhile to emphasize the formal analogy between the well-known equations of the elastic line describing the flexural behavior of a beam, and the diagonalized differential equations describing the torsional behavior of thin-walled, open-section beams.

As in the case of flexural curvature, in torsional behavior the term ϑ'' vanishes where the bimoment is null, or, in other words, the bimoment is zero where the line describing the rotations of the beam shows an inflection point.

If the contribution related to primary torsion $GI_t\vartheta''$ is negligible, Equation 9.75c can be more easily integrated.

9.6 DYNAMIC ANALYSIS OF TALL BUILDINGS

It is well-known that the higher the building, the more sensitive it becomes to dynamic actions from wind and earthquakes. In the stage of conceptual design, the need for a preliminary assessment of the free-vibration frequencies is essential.

Since only mode shapes and natural frequencies will be evaluated, external actions are not taken into account; nor is forced ground motion included in the analysis.

Due to D'Alembert's principle, the inertial forces of the structure can be reduced to static forces and, therefore, be included in Equation 9.21. In particular, the masses of the building floors, together with the corresponding accelerations, appear in the global equilibrium equations. On the other hand, the mass corresponding to the vertical elements is considered negligible and its effect is omitted. As a consequence, the load vector is represented in this case by the product of a mass matrix by a vector containing the inertial accelerations of the building stories.

The inertial forces are (Figure 9.11)

$$p_x = -\rho A\ddot{\xi}_G \tag{9.78a}$$

$$p_y = -\rho A\ddot{\eta}_G \tag{9.78b}$$

Let the shear center C be the origin of the local coordinate system; the transverse displacements of the centroid can be written in terms of the global floor displacements ξ, η, and ϑ through the following expressions:

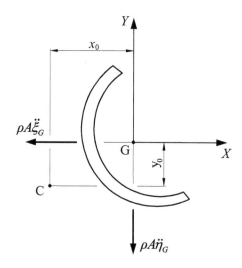

Figure 9.11

$$\ddot{\xi}_G = \frac{d^2}{dt^2}(\xi - y_0 \vartheta) \tag{9.79a}$$

$$\ddot{\eta}_G = \frac{d^2}{dt^2}(\eta + x_0 \vartheta) \tag{9.79b}$$

where x_0 and y_0 define the position of the centroid with respect to the shear center.

The actions described by Equations 9.78, applied to the centroid of the section, produce a torsional moment with respect to the shear center:

$$m_z = -\rho J_p \frac{d^2 \vartheta}{dt^2} + \left[\rho A \frac{d^2}{dt^2}(\xi - y_0 \vartheta)\right] y_0 - \left[\rho A \frac{d^2}{dt^2}(\eta + x_0 \vartheta)\right] x_0 \tag{9.80}$$

where J_p is the polar moment of inertia of the section referred to the centroid of the section.

Substituting Equations 9.78 and 9.80 into Equations 9.75 yields

$$EJ_{yy} \frac{\partial^4 \xi}{\partial z^4} + \rho A \frac{\partial^2}{\partial t^2}(\xi - y_0 \vartheta) = 0 \tag{9.81a}$$

$$EJ_{xx} \frac{\partial^4 \eta}{\partial z^4} + \rho A \frac{\partial^2}{\partial t^2}(\eta + x_0 \vartheta) = 0 \tag{9.81b}$$

$$EJ_{\omega\omega} \frac{\partial^4 \vartheta}{\partial z^4} - GI_t \frac{\partial^2 \vartheta}{\partial z^2} + \rho J_p \frac{\partial^2 \vartheta}{\partial t^2} - \rho A y_0 \frac{\partial^2 \xi}{\partial t^2} + \rho A y_0^2 \frac{\partial^2 \vartheta}{\partial t^2} + \rho A x_0 \frac{\partial^2 \eta}{\partial t^2} + \rho A x_0^2 \frac{\partial^2 \vartheta}{\partial t^2} = 0 \tag{9.81c}$$

Using the relationship between the polar moment of inertia in reference to the shear center, I_0, and to the center of gravity, J_p:

$$J_p = I_0 - A y_0^2 - A x_0^2 \tag{9.82}$$

and substituting this equation into Equation 9.81c, we have:

$$EJ_{\omega\omega} \frac{\partial^4 \vartheta}{\partial z^4} - GI_t \frac{\partial^2 \vartheta}{\partial z^2} + \rho I_0 \frac{\partial^2 \vartheta}{\partial t^2} - \rho A y_0 \frac{\partial^2 \xi}{\partial t^2} + \rho A x_0 \frac{\partial^2 \eta}{\partial t^2} = 0 \tag{9.83}$$

In general, the three equations are coupled to each other. Only in the case of double symmetry is the bending problem decoupled from the torsion problem.

It is possible to separate the spatial problem from the temporal, expressing the unknowns ξ, η, and ϑ as the product of a spatial function $Z(z)$ and a time function $T(t)$:

$$\xi = U(z)T(t) \tag{9.84a}$$

$$\eta = V(z)T(t) \tag{9.84b}$$

$$\vartheta = \Theta(z)T(t) \tag{9.84c}$$

Substituting Equations 9.84 into Equations 9.81a and b and into Equation 9.83 yields

$$\frac{EJ_{yy}U^{IV}}{(-\rho AU + \rho Ay_0\Theta)} = \frac{\ddot{T}}{T} = -\omega_n^2 \tag{9.85a}$$

$$\frac{EJ_{xx}V^{IV}}{(\rho AV + \rho Ax_0\Theta)} = \frac{\ddot{T}}{T} = -\omega_n^2 \tag{9.85b}$$

$$\frac{EJ_{\omega\omega}\Theta^{IV} - GI_t\Theta''}{(-\rho I_0\Theta + \rho Ay_0U - \rho Ax_0V)} = \frac{\ddot{T}}{T} = -\omega_n^2 \tag{9.85c}$$

where ω_n^2 is the square of the angular frequency.

From the system of Equations 9.85, it is possible to obtain the time-dependent differential equation:

$$\ddot{T} + \omega_n^2 T = 0 \tag{9.86}$$

The general integral of this equation is given by

$$T(t) = A_n \cos\omega_n t + B_n \sin\omega_n t \tag{9.87}$$

The coefficients A_n and B_n can be obtained from the initial conditions of the problem.

Any vibrational motion of the beam can be described as a superposition effect of the mode shapes:

$$\xi = \sum_{n=1}^{\infty} U_n(z)T_n(t) \tag{9.88a}$$

$$\eta = \sum_{n=1}^{\infty} V_n(z)T_n(t) \tag{9.88b}$$

$$\vartheta = \sum_{n=1}^{\infty} \Theta_n(z)T_n(t) \tag{9.88c}$$

The boundary conditions at the constraint $(z=0)$ are

$$U = V = \Theta = U' = V' = \Theta' = 0 \tag{9.89}$$

while, at the top:

$$U'' = V'' = \Theta'' = U''' = V''' = GI_t\Theta' - EJ_{\omega\omega}\Theta''' = 0 \tag{9.90}$$

In the case of an N-story building, the dead load of the floors is prevalent with respect to the mass of the bracings. Consequently the inertial forces are evaluated as the product of the floors' mass matrix by the vector containing the accelerations of the same floors in the directions X and Y. If the viscous damping forces are neglected, after the expansion and assemblage procedures of the mass matrix and the stiffness matrix, the equation of motion can be expanded to a $3N \times 3N$ matrix relation. Following the procedure explained in Chapters 5 and 6, a vibrating system with $3N$ degrees of freedom can be deduced. Once the floors' displacements eigenvector $\{\delta\}$ have been obtained, applying the General Algorithm, it is possible to calculate the displacements of the ith element and, therefore, the stresses acting in it.

9.7 NUMERICAL EXAMPLE

A numerical example, referring to the case of thin-walled, closed- and open-section shear walls, is considered here. The open sections are analyzed according to the theory of sectorial areas proposed by Vlasov.

The matrix formulation proposed in Section 9.2 for three-dimensional problems proves to be a general approach that can be easily adapted to every case concerning a system of in-parallel members subjected to transverse actions. Indeed, if the stiffness matrix $[K_i^*]$ of the ith element in its own coordinate system is known (Equation 9.17), it is possible to implement the formulation in the presence of any type of bracing.

With the aim of highlighting the potentiality of the method, a specific numerical example is proposed regarding a 50-story building that is 200 m tall and subjected to horizontal forces $F_x = 80$ kN, applied at each floor level in the origin of the global coordinate system (Figure 9.12). The cantilevers are made of concrete, whose elastic modulus and Poisson ratio are 3×10^4 MPa and 0.18, respectively.

The innermost element exhibits a square hollow section, whereas the perimeter elements are L- or C-shaped open sections. The geometry of the system has no symmetry axis; therefore, torsional actions are expected to occur. The static analysis results are shown in Figures 9.13 through 9.15, whereas the dynamic analysis outcomes are represented in Figure 9.16.

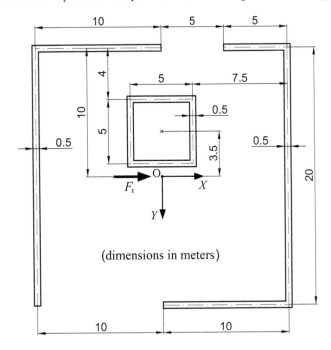

(dimensions in meters)

Figure 9.12

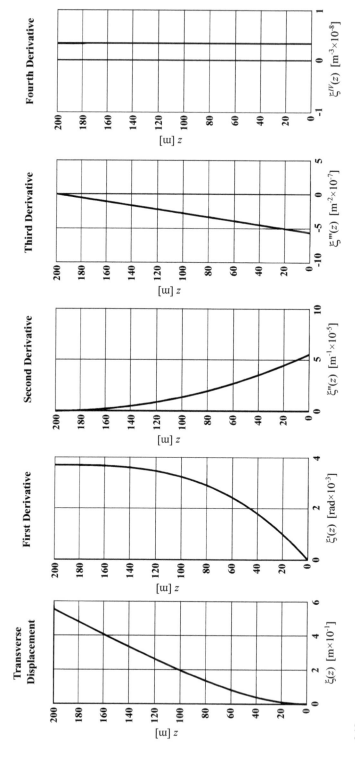

Figure 9.13a

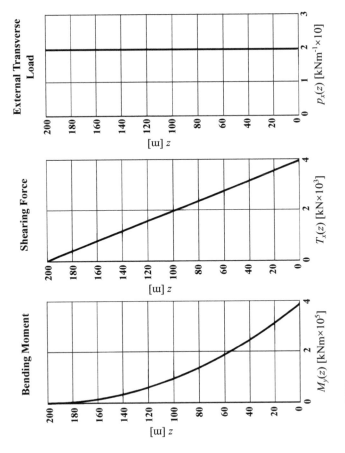

Figure 9.13b

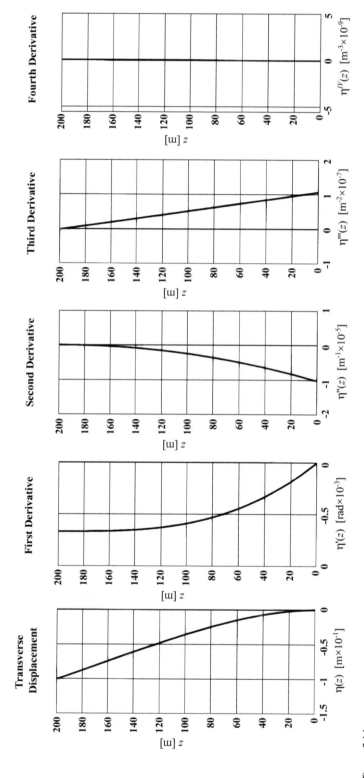

Figure 9.14a

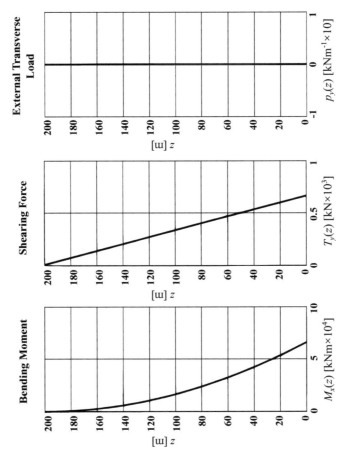

Figure 9.14b

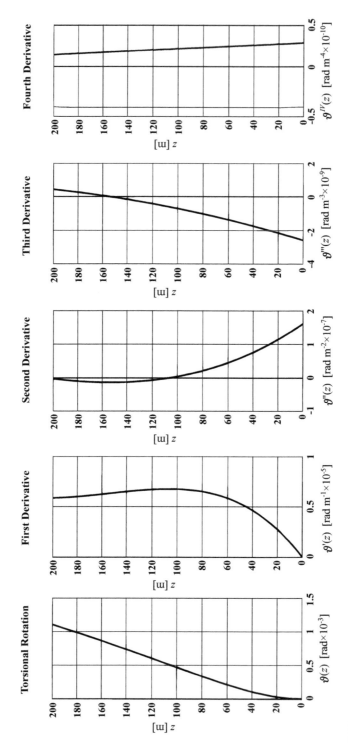

Figure 9.15a

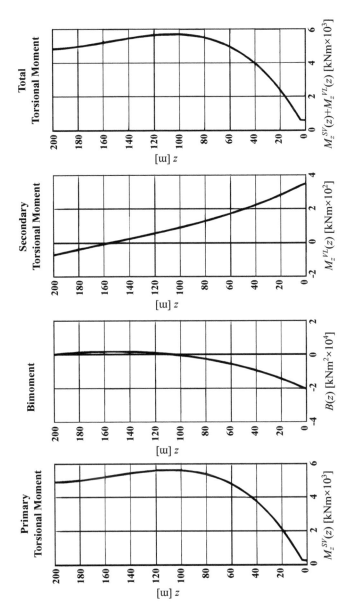

Figure 9.15b

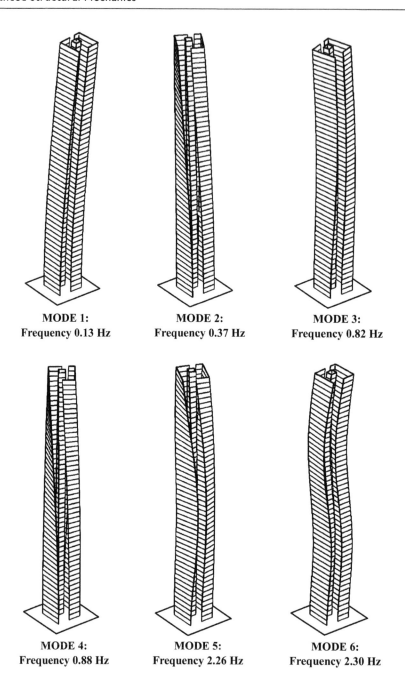

MODE 1:
Frequency 0.13 Hz

MODE 2:
Frequency 0.37 Hz

MODE 3:
Frequency 0.82 Hz

MODE 4:
Frequency 0.88 Hz

MODE 5:
Frequency 2.26 Hz

MODE 6:
Frequency 2.30 Hz

Figure 9.16

In particular, Figure 9.13a shows the transverse displacement $\xi(z)$ in the X direction and its successive derivatives. While the $\xi'(z)$ function represents the slope of the elastic line, the derivatives $\xi''(z)$, $\xi'''(z)$, and $\xi^{IV}(z)$, as shown in Equations 9.72a, 9.74a, and 9.75a, are proportional to the bending moment, the shearing force, and the distributed load, respectively. On the other hand, the values of the global internal actions, $M_y(z)$ and $T_x(z)$, and those of the uniformly distributed load $p_x(z)$, are represented in Figure 9.13b.

The transverse displacement in the Y direction, $\eta(z)$, with its subsequent derivatives, is shown in Figure 9.14a. Furthermore, the values of the global internal actions, $M_x(z)$ and $T_y(z)$, and those of the uniformly distributed load $p_y(z)$, are represented in Figure 9.14b. It is worth noting that, in Figure 9.14b, the uniformly distributed load $p_y(z)$ is equal to zero, in agreement with the $\eta^{IV}(z)$ value that is equal to zero along the entire structure.

Figure 9.15a shows the torsional rotation $\vartheta(z)$ of each floor, and its successive derivatives. Considering the trends of the diagrams, it can be seen that the inflection point in the $\vartheta(z)$ function for $z \cong 105$ m corresponds to a stationary point in the $\vartheta'(z)$ function, and to a zero value of the $\vartheta''(z)$ function at the same level. Similarly, the $\vartheta'(z)$ function presents an inflection point for $z \cong 150$ m, corresponding to a stationary point in the $\vartheta''(z)$ function, and to a zero value of the $\vartheta'''(z)$ function. The $\vartheta'(z)$ function also turns out to be null for $z = 0$. As a matter of fact, the structure is constrained at the base and, in agreement with Equation 9.31, warping is prevented. In the first three diagrams of Figure 9.15b, the values of the primary torsional moment $M_z^{SV}(z)$, the bimoment action $B(z)$, and the secondary torsional moment $M_z^{VL}(z)$ are represented. As shown in Equation 9.46, the primary torsional moment $M_z^{SV}(z)$ and the secondary torsional moment $M_z^{VL}(z)$ turn out to be proportional to $\vartheta'(z)$ and $\vartheta'''(z)$, respectively, while the bimoment $B(z)$ is proportional to $\vartheta''(z)$, as proved by Equation 9.46. In the fourth diagram of Figure 9.15b, the sum of the moments $M_z^{SV}(z)$ and $M_z^{VL}(z)$ is reported. It is to be noted that the secondary torsional moment is one order of magnitude lower than the primary torsional moment.

As regards the dynamic analysis, the first six mode shapes and the corresponding natural frequencies are shown in Figure 9.16. Generally, in such irregular buildings, the vibration modes are characterized by coupled bending and twisting motions.

In the first three vibration modes, nodal sections in deformed shapes do not appear, whereas, in the following three vibration modes, they manifest themselves. In particular, as can be observed in Figure 9.16, the first modal shape is a prevailing bending along the X axis; the second one is a prevailing bending along the Y axis; while the third mode is predominantly torsional. The fourth and fifth shapes are coupled bending–torsional modes, and present one nodal section. In the former case, the bending occurs mainly along the X axis, whereas in the latter it occurs mostly in the Y direction. Finally, the sixth mode shape is a bending along the X axis coupled with a simultaneous torsion, and presents two nodal sections, as shown in Figure 9.16.

The diagrams shown in this example demonstrate the ability of the present formulation to capture the main information regarding the static and dynamic behavior of a tall building subjected to lateral forces. The transverse displacements, as well as the distribution of the external actions between the vertical members constituting the whole resistant system, are, in fact, crucial for the necessary preliminary analysis. In addition, the simplicity of the approach prevents the use of time-consuming tools, such as the finite element programs, which can be adopted at a later stage for more detailed and executive computations.

Chapter 10

Theory of plasticity

10.1 INTRODUCTION

On the basis of the hypotheses of the linear elastic behavior of a material and of small displacements, the problem of elastic plates and shells, and more generally of elastic solids, may be resolved, as we have found in Chapter 3, by means of Lamé's equation, where the operator $[\mathcal{L}]$ is always **linear**. If, that is, $\{\mathcal{F}\}$ is the vector of the external forces and $\{\eta\}$ is the corresponding displacement vector, obtained by resolving Lamé's equation, the loads should be multiplied by a constant c, also the displacements, and hence the deformations and the static characteristics will be multiplied by the same constant:

$$[\mathcal{L}]\{c\eta\} = -\{c\mathcal{F}\}$$ (10.1)

Furthermore, if $\{\mathcal{F}_a\}$, $\{\mathcal{F}_b\}$ are two different vectors of the external forces and $\{\eta_a\}$, $\{\eta_b\}$ the corresponding displacement fields, in the case of the superposition of the forces, the principle of superposition will also hold good for displacements:

$$[\mathcal{L}]\{\eta_a + \eta_b\} = -\{\mathcal{F}_a + \mathcal{F}_b\}$$ (10.2)

A first case of nonlinearity was examined in Chapter 7, where it was shown how the external loads do not always increase proportionally to the induced displacements (Figures 7.1b, 7.2b, 7.8b, and 7.9b) in cases where such displacements cannot be considered small. A second fundamental case of nonlinearity will be examined in this chapter, where the ductile behavior of a material will be considered. The first is the case of **geometrical nonlinearity**, while the second is the case of **constitutive nonlinearity of the material**.

A first simple example of nonlinear structural behavior may be offered by the system of parallel connecting rods in Figure 10.1a, if we assume that they follow the law of elastic–perfectly plastic behavior (Figure 10.1b). This case has already been considered in the elastic regime in Section 2.2, where the reactions of the individual connecting rods have been determined, once the behavior of the cross member is assumed to be rigid. Applying Equations 2.3 and 2.4 and considering each of the two lateral connecting rods as having a cross section equal to half of the cross section of the central connecting rod, the two reactions in the elastic field are obtained:

$$X_I = F \frac{\dfrac{A}{l_I}}{\dfrac{A}{l_I} + \dfrac{A}{l_{II}}} = F \frac{l_{II}}{l_I + l_{II}}$$ (10.3a)

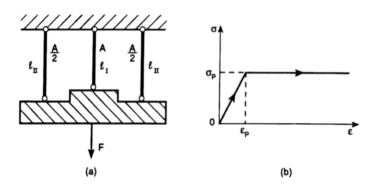

Figure 10.1

$$X_{II} = F \frac{\dfrac{A/2}{l_{II}}}{\dfrac{A}{l_I} + \dfrac{A}{l_{II}}} = \frac{F}{2} \frac{l_I}{l_I + l_{II}} \tag{10.3b}$$

If $l_I < l_{II}$, the higher tension is developed in the central connecting rod; therefore, if the external force F is increased, this element is the one that first undergoes plastic deformation. Yielding of the central connecting rod occurs for

$$F_1 = \sigma_P A \left(1 + \frac{l_I}{l_{II}} \right) \tag{10.4a}$$

$$\delta_1 = \frac{\sigma_P l_I}{E} \tag{10.4b}$$

where subscript 1 denotes the characteristics of first yielding (force applied to the cross member and vertical displacement thereof).

Yielding of the lateral connecting rods, on the other hand, occurs for

$$F_2 = 2\sigma_P A \tag{10.5a}$$

$$\delta_2 = \frac{\sigma_P l_{II}}{E} \tag{10.5b}$$

where subscript 2 indicates the characteristics of the second and ultimate yielding. For $\delta > \delta_2$, in fact, the total reaction of the connecting rods cannot increase and remains stationary at the value of ultimate plasticity F_2.

Hence, recapitulating, we obtain a global elastic behavior for $0 < \delta < \delta_1$:

$$F = EA \left(\frac{1}{l_I} + \frac{1}{l_{II}} \right) \delta \tag{10.6a}$$

a globally strain-hardening behavior for $\delta_1 < \delta < \delta_2$:

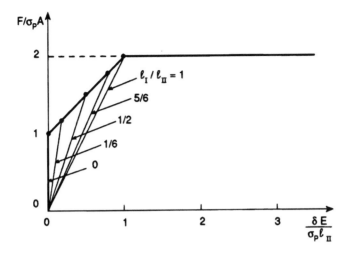

Figure 10.2

$$F = \sigma_P A + \frac{EA}{l_{II}} \delta \tag{10.6b}$$

and a perfectly plastic behavior (plastic flow) for $\delta > \delta_2$:

$$F = 2\sigma_P A \tag{10.6c}$$

Figure 10.2 presents, in nondimensional form, the force-displacement curves for different values of the ratio l_I/l_{II}. For $l_I \to 0$, the force is sustained in the elastic phase entirely by the central connecting rod, which then yields for $F = \sigma_P A$. When $l_I = l_{II}$, on the other hand, the hardening phase is not present because all three connecting rods yield at the same time. It may be noted that the straight line to which the hardening portion of the diagram in Figure 10.2 belongs does not depend on the ratio l_I/l_{II}. This is because, once the central rod has yielded, its length l_I no longer enters into the analysis.

Although the example just considered is particularly simple, since it contains only three elements subjected to axial force, it reflects conceptually the mechanical behavior of the more complex beam systems, where the prevalent characteristic is the bending moment. In such cases, the local plastic flow will be represented by a localized relative rotation and, as the number of such rotations increases, the degree of redundancy of the frame will diminish simultaneously.

In the following, various examples of **incremental plastic analysis** of beam systems will be shown. In these will be determined, step by step and as the external load increases, the position of the cross sections in which the localized plastic rotation occurs. To this type of incremental analysis of the process of plastic deformation, there corresponds the so-called **plastic limit analysis**, which, on the basis of two specific theorems, directly identifies an interval (one that is generally restricted) within which the ultimate load of global plastic flow must necessarily fall. This load is called the **load of plastic collapse**. Once this load is reached, the structure is reduced to a mechanism; that is, it is hypostatic, even though it is in equilibrium on account of the particular load condition, and is not able to sustain further increments of load. On the basis of the aforementioned theorems, it is also possible to identify the mechanism of collapse; that is, the positions of the centers of relative plastic rotation.

More particularly, beam systems loaded by concentrated forces will be distinguished from those loaded by distributed forces. In the latter, in fact, the identification of the mechanism of collapse is generally more complex. The closing sections of this chapter will deal briefly with problems of nonproportional loading and problems of repeated loading (**shake-down**), as well as with the problem of the plastic collapse of deflected plates.

10.2 ELASTIC–PLASTIC FLEXURE

Let us consider the rectangular cross section of base b and depth h (Figure 10.3) of a beam made of elastic–perfectly plastic material, with equal elastic modulus E and yield stress σ_P both in tension and in compression (Figure 10.1b).

Let us assume that as the applied bending moment increases, the cross section of the beam remains plane, even though part of the beam undergoes plastic deformation. This is equivalent to considering linear variations of the axial dilation ε_z through the depth of the beam (Figure 10.3a). The axial stress σ_z, on the other hand, will not be able to exceed its limit value σ_P and, once the moment of first plastic deformation M_e has been overstepped, will hence present a linear variation in the central part of the cross section and two plateaus in the outermost parts (Figures 10.3b and c). The diagrams in Figure 10.3 present the succession of patterns that both ε_z and σ_z follow through the depth of the beam, rendering the scales uniform with the values at the yielding points ε_P and σ_P, respectively. Therefore, it clearly emerges from these diagrams that the maximum dilation ε_{max} that is reached at the outermost edges of the beam exceeds the dilation ε_P, which corresponds to yielding. When $\varepsilon_{max} \to \infty$, and hence when plastic flow has occurred, the variation of the stress is birectangular, while the extension $2d$ of the elastic core of the beam vanishes (Figure 10.3d).

The **moment of first plastic deformation** (the maximum moment in the elastic regime) is readily obtainable as

$$M_e = \sigma_P \frac{I_x}{h/2} = \sigma_P b \frac{h^2}{6} \tag{10.7a}$$

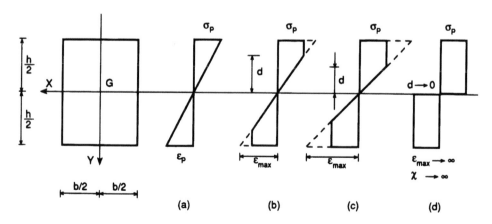

(a)   (b)   (c)   (d)

Figure 10.3

while the **moment of ultimate plastic deformation** or **plastic moment** may be evaluated using the diagram in Figure 10.3d:

$$M_P = \sigma_P\left(b\frac{h}{2}\right)\frac{h}{2} = \sigma_P b\frac{h^2}{4} \qquad (10.7b)$$

this being equal to that of a couple of forces $\sigma_P(bh/2)$ with moment arm $h/2$. The plastic moment M_P is therefore equal to $3/2\ M_e$, a fact that enables further exploitation of the load-bearing capacity of metal materials, with loads substantially higher than those that meet the **criterion of admissible stresses**.

It is possible to put the axial dilations ε_z in relation with the curvature χ_x of the beam element straddling the section under consideration:

$$\varepsilon_z = \frac{M_x}{EI_x}y = \chi_x y \qquad (10.8)$$

from which we obtain (Figure 10.3)

$$\chi_x = \frac{\varepsilon_{max}}{h/2} = \frac{\varepsilon_P}{d} \qquad (10.9)$$

Equation 10.9 warns us that, for $\varepsilon_{max}\to\infty$, or for $d\to 0$, the beam curvature tends to infinity, giving rise to a localized relative rotation with center in the cross section under examination.

We now intend to determine the moment–curvature law, M_x versus χ_x, corresponding to the plastic evolution of the cross section (Figure 10.3). At each step of this evolution, the applied moment may be evaluated on the basis of the known distribution of the forces:

$$M_x = 2\int_0^d\left(\sigma_P\frac{y}{d}\right)yb\,dy + 2\int_d^{h/2}\sigma_P yb\,dy \qquad (10.10)$$

Substituting the half-depth d of the elastic zone with the expression deriving from Equation 10.9, we have

$$M_x = 2\sigma_P b\left[\frac{\chi_x}{\varepsilon_P}\int_0^{\varepsilon_P/\chi_x}y^2\,dy + \int_{\varepsilon_P/\chi_x}^{h/2}y\,dy\right] \qquad (10.11)$$

from which, evaluating the integrals, we obtain

$$M_x = \sigma_P b\frac{h^2}{4}\left[1 - \frac{1}{3}\frac{\left(\varepsilon_P/\frac{h}{2}\right)^2}{\chi_x^2}\right] \qquad (10.12)$$

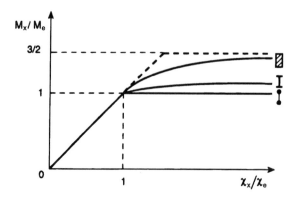

Figure 10.4

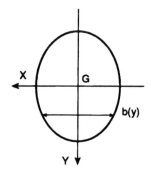

Figure 10.5

Via Equation 10.7a, the foregoing function can be cast in a particularly expressive form:

$$\frac{M_x}{M_e} = \frac{3}{2} - \frac{1}{2}\left(\frac{\chi_e}{\chi_x}\right)^2$$

(10.13)

where χ_e denotes the beam curvature when the first plastic deformation occurs. The diagram in Figure 10.4 thus represents a linear law for $\chi_x < \chi_e$, or $M_x < M_e$, and the hyperbolic law of Equation 10.13 for $\chi_x > \chi_e$, or $M_x > M_e$. This strain-hardening law is replaced in practice by the elastic–perfectly plastic law represented by the dashed line in the same figure.

When the cross section of the beam presents double symmetry (Figure 10.5), while at the same time not being rectangular, the foregoing reasoning still applies to a large extent. In particular, Equation 10.11 must present the width $b(y)$ that, in general, is a function of y, under the integral sign:

$$M_x = 2\sigma_P\left[\frac{\chi_x}{\varepsilon_P}\int_0^{\varepsilon_P/\chi_x} y^2 b(y)\,dy + \int_{\varepsilon_P/\chi_x}^{b/2} yb(y)\,dy\right]$$

(10.14)

The plastic moment is, therefore

$$M_P = \lim_{\chi_x \to \infty} M_x = 2\sigma_P\int_0^{b/2} yb(y)\,dy$$

(10.15)

where the integral represents the half-section static moment $S_x^{A/2}$ with respect to the X axis. The ratio

$$\frac{M_P}{M_e} = \frac{2S_x^{A/2}}{\left(I_x \bigg/ \dfrac{b}{2}\right)} \tag{10.16}$$

is, as has been stated, equal to 1.5 in the case of a rectangular section, while, in the limit case of a section consisting of two concentrated areas set apart at a distance b, it is equal to unity. This means that, in such a case, the moments of first and ultimate plastic deformation coincide. In the technically highly recurrent case of an I-section, there is no great departure from the limit case just considered, and the ratio given by Equation 10.16 is approximately equal to 1.15 (Figure 10.4). The I-sections are, thus, the most convenient in the elastic regime, while, in the plastic regime, they reveal poor reserves of flexural bearing capacity.

Consider a cross section with a single axis of symmetry that coincides with the axis of flexure (Figure 10.6). The neutral axis remains orthogonal to the axis of symmetry, even though its position may vary during the entire loading process. In the condition of full plastic deformation (Figure 10.6d), we have

$$\sigma_P A_1 = \sigma_P A_2 \tag{10.17}$$

where $A_1 = A_2 = A/2$ are the areas of the portions of cross section that remain, respectively, above and below the plastic neutral axis n_P. Consequently, the plastic moment is

$$M_P = \sigma_P \frac{A}{2}(d_1 + d_2) \tag{10.18}$$

where d_1 and d_2 are the distances of the plastic neutral axis n_P from the centroids of the two half-sections. When $M_e < M < M_P$, the neutral axis is between n_e and n_P, as shown in Figures 10.6b and c. Whereas, the elastic neutral axis renders the static moments of the two portions into which it divides the section equal in absolute value, the plastic neutral axis makes the areas equal.

As regards the other static characteristics, the twisting moment applied to a circular section presents a behavior altogether similar to that described previously for the bending moment. The twisting moment of first plastic deformation is

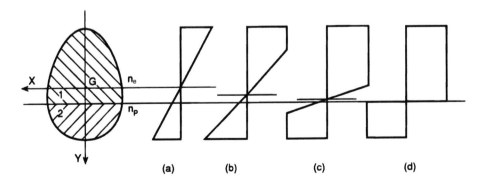

(a)   (b)   (c)   (d)

Figure 10.6

$$M_{ze} = \frac{I_p}{R} \tau_P = \frac{\pi}{2} R^3 \tau_P \tag{10.19a}$$

where, according to Tresca's criterion, $\tau_P = 1/2\ \sigma_P$. The plastic twisting moment is, on the other hand, equal to the product of the yielding stress τ_P by the polar static moment of the cross section:

$$M_{zP} = \frac{2\pi}{3} R^3 \tau_P \tag{10.19b}$$

The ratio M_{zP}/M_{ze} is hence 4/3.

The case of centered axial force is trivial and has already been dealt with in advance in Section 10.1. Clearly, the plastic axial force is

$$N_P = \sigma_P A \tag{10.20}$$

while the case of shearing force should be considered together with that of bending, although, generally speaking, in the framework of plastic calculation this characteristic has negligible influence.

As regards the combined loading conditions, the case of eccentric axial force is noteworthy. For a rectangular cross section loaded by an axial force N, applied on the Y axis with eccentricity e (Figure 10.7), four different phases succeed one another as N increases. These phases are relative to the conditions represented in Figure 10.7:

1. Elastic
2. Elastic–plastic, with yielding only at one edge
3. Elastic–plastic, with yielding at both edges
4. Fully plastic

The diagram in Figure 10.7d can be split into two, as shown in Figure 10.8: (a) representing the resultant axial force N:

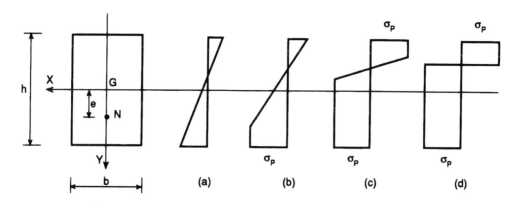

Figure 10.7

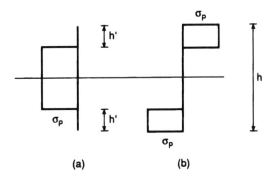

Figure 10.8

$$N = \sigma_P b\left(h - 2h'\right) \tag{10.21a}$$

and (b) representing the resultant bending moment $M = Ne$:

$$M = \sigma_P b h'\left(h - h'\right) \tag{10.21b}$$

On the basis of the plastic loadings:

$$N_P = \sigma_P bh \tag{10.22a}$$

$$M_P = \sigma_P b \frac{h^2}{4} \tag{10.22b}$$

it is possible to define the following nondimensional ratios:

$$\frac{N}{N_P} = 1 - 2\left(\frac{h'}{h}\right) \tag{10.23a}$$

$$\frac{M}{M_P} = 4\left(\frac{h'}{h}\right)\left(1 - \frac{h'}{h}\right) \tag{10.23b}$$

such that

$$\frac{M}{M_P} = 1 - \left(\frac{N}{N_P}\right)^2 \tag{10.24}$$

The plastic limit in the M–N plane is given by the closed curve in Figure 10.9, which is also called the **curve of interaction**. The couples M–N which are internal to the domain represent elastic–plastic conditions, while the couples that are on the boundary represent the ultimate conditions of full plastic deformation (plastic flow of the cross section). As will be

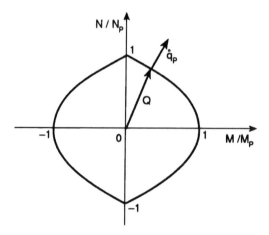

Figure 10.9

seen in the following, these occur with a localized rotation of the beam to which a localized axial dilation is added.

10.3 INCREMENTAL PLASTIC ANALYSIS OF BEAM SYSTEMS

Let us consider a cantilever beam of length l, loaded by an orthogonal force F at the free end (Figure 10.10a). As the force increases, the plastic collapse of the cantilever beam is reached as soon as the fixed-end moment equals the plastic moment:

$$F_P l = M_P \tag{10.25}$$

and hence we obtain $F_P = M_P/l$. At that point, a localized absolute rotation is produced in the fixed-end cross section, while the fixed-end moment cannot continue to grow and remains stationary at its limit value M_P. This situation is usually represented by inserting a hinge instead of the built-in constraint and by applying a resisting moment M_P in the neighborhood of the hinge (Figure 10.10b). The hinge allows localized rotations, while the moment M_P represents the rotational reaction exerted by the fixed-end cross section. The system has thus become hypostatic, but is in equilibrium on account of the particular loading condition. It should be noted that we have

$$F_P = \frac{3}{2} F_e \tag{10.26}$$

where F_e denotes the maximum force applicable in the framework of the criterion of admissible stresses. The ratio 3/2 thus represents a sort of safety factor in the framework of the criterion of admissible stresses in regard to the ultimate plastic condition.

As a second elementary case, let us take that of a simply supported beam with a force applied in the center (Figure 10.11a). As the force increases, plastic collapse is reached as soon as the moment in the center equals the plastic moment:

$$\frac{1}{4} F_P l = M_P \tag{10.27}$$

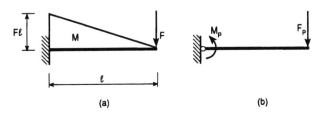

Figure 10.10

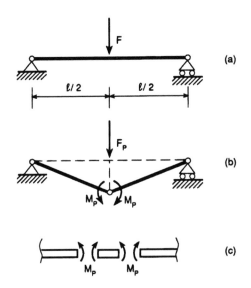

Figure 10.11

from which we obtain the collapse load $F_P = 4M_P/l$. At that point, a **plastic hinge** is created in the center; that is, a hinge with a constant resisting action equal to M_P. Note that the moments M_P, acting in the scheme of Figure 10.11b, tend to oppose the action of the external load and to cause the two arms to rotate in the direction opposite to that of the collapse mechanism. Also in this case, the mechanism is in equilibrium by virtue of the particular load condition. This equilibrium condition is neutral in the case of small displacements. If, on the other hand, the plastically deformed beam element is isolated, as shown in Figure 10.11c, the moments M_P acting on the element and on the two arms of the beam, in all cases, stretch the lower longitudinal fibers. Here again, Equation 10.26 applies, along with the observations deriving therefrom.

The safety factor, defined in accordance with Equation 10.26, is 3/2 for all the statically determinate systems of deflected beams of rectangular cross section, once the contributions of the axial force and shearing force are neglected. The formation of a single plastic hinge, in fact, directly leads to the collapse of the system (Figure 10.12). In statically determinate trusses, or, at any rate, in systems made up of bars and hence subjected to axial force alone, this factor is obviously equal to unity.

In the case of statically indeterminate systems of deflected beams, the safety factor is generally greater than 3/2. The formation of the first plastic hinge, in fact, does not bring about the collapse of the structure. In general, it may be stated that in a frame with n degrees of

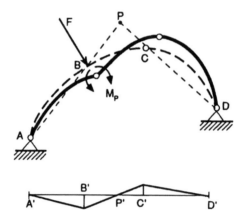

Figure 10.12

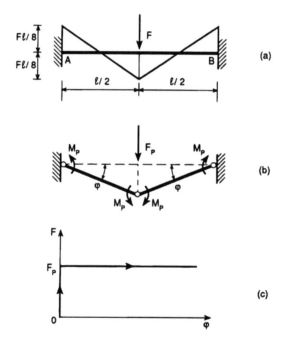

Figure 10.13

redundancy, the number of plastic hinges that are activated for collapse to occur is less than or equal to $(n+1)$.

Let us consider, for instance, the beam built in at both ends of Figure 10.13a. The fixed-end moments and the moment in the center are equal to $Fl/8$, and stretch the upper and lower fibers, respectively. The fixed-end cross sections and the cross section at the center hence reach full plasticity simultaneously:

$$\frac{1}{8}F_p l = M_p \tag{10.28}$$

from which we obtain the collapse load $F_p = 8M_p/l$. Applying the principle of virtual work to the collapse mechanism of Figure 10.13b, it is possible to reobtain the previous value:

$$F_p \varphi \frac{l}{2} - 4M_p \varphi = 0 \tag{10.29}$$

The absence of a strain-hardening phase in the loading process (Figure 10.13c) is due to the substantial statical determinacy of the structure.

In the case of the closed framework of Figure 1.26a, the maximum bending moment in the elastic phase is that found in the loaded cross sections. This equals 3/16 Fl, whereby the load producing the first two plastic hinges (Figure 10.14a) is

$$F_1 = \frac{16}{3} \frac{M_P}{l} \tag{10.30}$$

The scheme of Figure 10.14b describes this situation, taking into account the double symmetry of the framework, while Figure 10.14c gives the corresponding moment diagram. The subsequent four hinges are formed simultaneously at A, B, C, D, when the moment in the nodes also attains the value M_P, as the external forces F increase:

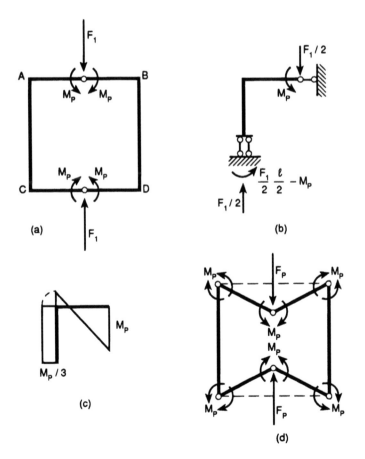

Figure 10.14

$$M_A = \frac{Fl}{4} - M_P = M_P \qquad (10.31)$$

whence we find the collapse load:

$$F_2 = F_P = \frac{8M_P}{l} \qquad (10.32)$$

It should be noted that this load is the same as that obtained in the case of the beam built in at both ends. This is due to the substantial identity of the corresponding mechanisms of collapse (Figures 10.13b and 10.14d). The safety factor, in the framework of the criterion of admissible stresses and in regard to plastic collapse, is, in this case:

$$\frac{F_P}{F_e} = \frac{F_2}{\frac{2}{3}F_1} = \frac{3}{2}\frac{(8M_P/l)}{(16M_P/3l)} = \frac{9}{4} \qquad (10.33)$$

In the case of the asymmetrical portal frame of Figure 1.30a, the force that causes the formation of the first plastic hinge is (Figure 10.15a)

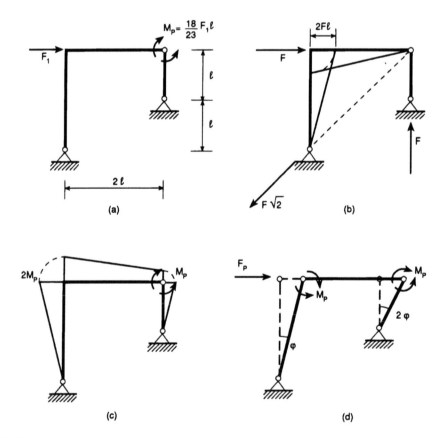

Figure 10.15

$$F_1 = \frac{23}{18}\frac{M_P}{l} \tag{10.34}$$

For $F > F_1$, the portal frame is transformed into a three-hinged arch, loaded by the external force F and the two plastic moments M_P. The partial moment diagrams for the two loads are given in Figures 10.15b and c. The diagram corresponding to the plastic moments M_P is virtual, since it shows cross member values $M > M_P$. A second plastic hinge is formed when the global moment in the left-hand node becomes equal to M_P (stretching the internal fibers):

$$2F_2 l - 2M_P = M_P \tag{10.35}$$

from which there follows

$$F_2 = F_P = \frac{3}{2}\frac{M_P}{l} \tag{10.36}$$

The value of F_P may, alternatively, be found by applying the principle of virtual work to the collapse mechanism of Figure 10.15d:

$$F_P(2l\varphi) - M_P\varphi - M_P(2\varphi) = 0 \tag{10.37}$$

In this case, the safety factor is

$$\frac{F_P}{F_e} = \frac{F_2}{\frac{2}{3}F_1} = \frac{3}{2}\frac{(3M_P/2l)}{(23M_P/18l)} = \frac{81}{46} \tag{10.38}$$

So far, only concentrated loads have been considered. A first simple example of a distributed load is furnished by the scheme of the beam built in at both ends shown in Figure 10.16a.

It is known that the maximum moment in the elastic regime is that of the built-in constraints, which is equal to $ql^2/12$ (Figure 10.16c).

Therefore, when the external load reaches the value

$$q_1 = 12\frac{M_P}{l^2} \tag{10.39}$$

two plastic hinges are formed at the built-in constraints. In line with the scheme of Figure 10.16b, the third plastic hinge in the center forms when

$$q_2\frac{l^2}{8} - M_P = M_P \tag{10.40}$$

whence we obtain the collapse load:

$$q_2 = q_P = 16\frac{M_P}{l^2} \tag{10.41}$$

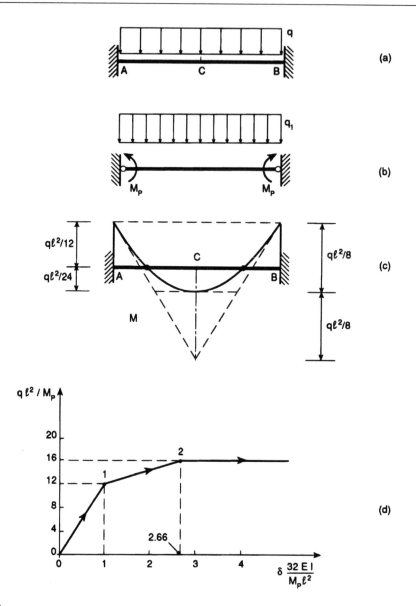

Figure 10.16

It is also possible to arrive at this collapse load by a simple application of the principle of virtual work. The safety factor in this case is

$$\frac{q_P}{q_e} = \frac{q_2}{\frac{2}{3}q_1} = \frac{3}{2}\frac{\left(16M_P/l^2\right)}{\left(12M_P/l^2\right)} = 2 \tag{10.42}$$

The deflection in the center for $q = q_1$ (Figure 10.16b) is given by the contributions of the external load q_1 and of the plastic moments M_P, respectively:

$$\delta_1 = \frac{5}{384} \frac{q_1 l^4}{EI} - \frac{M_p l^2}{8EI} \qquad (10.43)$$

which, via Equation 10.39, becomes

$$\delta_1 = \frac{M_p l^2}{32EI} \qquad (10.44)$$

On the other hand, the deflection in the center for $q = q_2$ likewise is

$$\delta_2 = \frac{5}{384} \frac{q_2 l^4}{EI} - \frac{M_p l^2}{8EI} \qquad (10.45)$$

and hence, inserting Equation 10.41:

$$\delta_2 = \frac{M_p l^2}{12EI} \qquad (10.46)$$

Plotting the points (δ_1, q_1) and (δ_2, q_2) on the nondimensional plane of Figure 10.16d, we immediately find the curve $\delta(q)$, that is, the structural response as the external load increases. It may be noted that this response is elastic between points 0 and 1, strain-hardening between points 1 and 2, and finally perfectly plastic for $\delta > \delta_2$.

Let us now examine the case of the beam built in at one end and supported at the other, illustrated in Figure 10.17a.

In the elastic regime, the maximum bending moment occurs at the built-in constraint and is equal to $q l^2 / 8$. Hence, a first plastic hinge is formed at the built-in constraint for $q_1 = 8 M_p / l^2$. At this point, the structure becomes statically determinate and presents a globally strain-hardening behavior, until the second and last plastic hinge is formed (Figure 10.17b). Whereas, in the case of concentrated loadings, it is straightforward to identify the location of the subsequent plastic hinges, in the case of distributed loads such identification is not usually immediate. In the case in point, for instance, which does not even present particular symmetries, it is necessary to calculate the maximum of the moment function in the hardening phase and to determine the value q_2 that makes this maximum equal to the plastic moment M_p. Expressed in formulas, this is

$$M(z) = -\frac{M_p}{l} z + \left(\frac{1}{2} q l z - \frac{1}{2} q z^2 \right) \qquad (10.47a)$$

$$T(z) = \frac{dM}{dz} = -\frac{M_p}{l} + \frac{1}{2} q l - q z \qquad (10.47b)$$

The shear vanishes for

$$z = \frac{l}{2} - \frac{M_p}{q l} \qquad (10.48)$$

whence we obtain

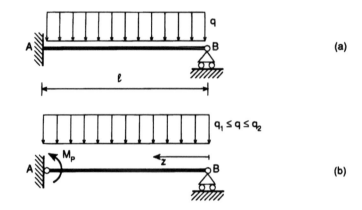

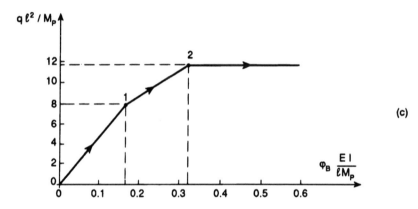

Figure 10.17

$$M_{max} = \frac{1}{8} q l^2 \left(1 - \frac{2M_P}{q l^2} \right)^2 \tag{10.49}$$

Setting $M_{max} = M_P$, we obtain a second-degree algebraic equation in the unknown q_2:

$$\left(1 - \frac{2M_P}{q_2 l^2} \right)^2 = 4 \frac{2M_P}{q_2 l^2} \tag{10.50}$$

which, resolved, yields the two roots

$$\frac{2M_P}{q_2 l^2} = 3 \pm 2\sqrt{2} \tag{10.51}$$

Whereas the first root must be rejected, since it would imply $q_2 < q_1$, the second yields the collapse load:

$$q_2 = q_P = \frac{M_P}{l^2} \frac{2}{3 - 2\sqrt{2}} \tag{10.52}$$

or

$$q_2 = q_P \simeq 11.6568 \frac{M_P}{l^2} \qquad (10.53)$$

The safety factor is, therefore:

$$\frac{q_P}{q_e} = \frac{q_2}{\frac{2}{3}q_1} = \frac{3}{8(3-2\sqrt{2})} \simeq 2.19 \qquad (10.54)$$

To represent the structural response with the increase in the load q, it is necessary to choose a suitable kinematic parameter; for example, the rotation of the end cross section B. In the elastic phase $(0 \leq q \leq q_1)$, we have

$$\varphi_B = \frac{ql^3}{24EI} - \left(q\frac{l^2}{8}\right)\frac{l}{6EI} = \frac{ql^3}{48EI} \qquad (10.55a)$$

while in the hardening phase $(q_1 \leq q \leq q_2)$, we obtain

$$\varphi_B = \frac{ql^3}{24EI} - M_P\frac{l}{6EI} = \frac{l^3}{24EI}\left(q - 4\frac{M_P}{l^2}\right) \qquad (10.55b)$$

Therefore, the rotations for the notable loads q_1 and q_2 are, respectively

$$\varphi_{B1} = \frac{1}{6}\frac{M_P l}{EI} \qquad (10.56a)$$

$$\varphi_{B2} = \frac{11.6568 - 4}{24}\frac{M_P l}{EI} \simeq 0.319\frac{M_P l}{EI} \qquad (10.56b)$$

The diagram $q(\varphi_B)$ is given in nondimensional form in Figure 10.17c. Between points 0 and 1, the global behavior of the structure is elastic; between points 1 and 2 it is hardening; while beyond point 2, it is perfectly plastic.

In the case of the continuous beam of Figure 10.18a, the maximum bending moment is reached in the right-hand span, and hence it will be at this cross section that the first plastic hinge will be formed. Isolating the right-hand span (Figure 10.18b), we obtain

$$M(z) = \frac{3}{7}qlz - \frac{1}{2}qz^2 \qquad (10.57a)$$

$$T(z) = \frac{dM}{dz} = \frac{3}{7}ql - qz \qquad (10.57b)$$

The shear vanishes for $z = 3/7\ l$, where we have

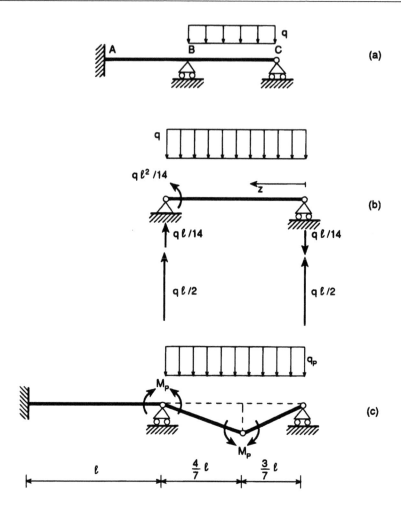

Figure 10.18

$$M_{\text{max}} = \frac{9}{98} q l^2 \tag{10.58}$$

Setting $M_{\text{max}} = M_P$, we obtain the load that produces the first plastic hinge:

$$q_1 = \frac{98}{9} \frac{M_P}{l^2} \approx 10.89 \frac{M_P}{l^2} \tag{10.59}$$

At this point, it is easy to understand that the second and last plastic hinge will be formed at the central support (Figure 10.18c). Applying the principle of virtual work to the collapse mechanism yields the equation:

$$-M_P\left(\frac{3}{7}\varphi\right) - M_P\varphi + \left(\frac{4}{7}ql\right)\left(\frac{3}{7}\varphi\right)\left(\frac{2}{7}l\right) \tag{10.60}$$

$$+\left(\frac{3}{7}ql\right)\left(\frac{4}{7}\varphi\right)\left(\frac{3}{14}l\right)=0$$

whence, performing the computation, we find

$$q_2 = q_P = \frac{70}{6}\frac{M_P}{l^2} \simeq 11.6666\frac{M_P}{l^2} \qquad (10.61)$$

Note that the collapse load of Equation 10.61 is greater than that of Equation 10.53, corresponding to the previously considered scheme, by what, from an engineering standpoint, is a negligible amount (~1‰). The safety factor is, however, less, even though the structure in this case has two degrees of static indeterminacy:

$$\frac{q_P}{q_e} = \frac{q_2}{\frac{2}{3}q_1} = \frac{3}{2}\frac{\left(\frac{70}{6}\right)}{\left(\frac{98}{9}\right)} \simeq 1.607 \qquad (10.62)$$

This is because, to bring about the collapse mechanism, the formation of two plastic hinges is sufficient, rather than three, as the global static determinacy would require in the proximity of collapse. Cases of this sort are referred to as instances of **partial collapse**, as opposed to complete collapse, whereby the entire structure becomes hypostatic.

As a final example, let us consider the portal frame with inclined stanchion of Figure 1.39a.

The maximum bending moment is reached, in the elastic phase, at the left-hand fixed-joint node (Figure 1.40a), whereby the load that produces the first plastic hinge in the same node is obtained as

$$q_1 = \frac{112}{16+3\sqrt{2}}\frac{M_P}{l^2} \simeq 5.53\frac{M_P}{l^2} \qquad (10.63)$$

Application of the principle of virtual work yields the moment X_2 in the right-hand fixed-joint node for $q>q_1$ (Figure 10.19a):

$$2M_P\varphi + 2X_2\varphi - ql\left(\frac{l}{2}\varphi\right)=0 \qquad (10.64)$$

whence we obtain

$$X_2 = \frac{1}{4}ql^2 - M_P \qquad (10.65)$$

Equilibrium with regard to the rotation of the cross member about the right-hand node yields the shear V, transmitted to the cross member by the left-hand stanchion (Figure 10.19b):

$$Vl - \frac{1}{4}ql^2 - \frac{1}{2}ql^2 = 0 \qquad (10.66)$$

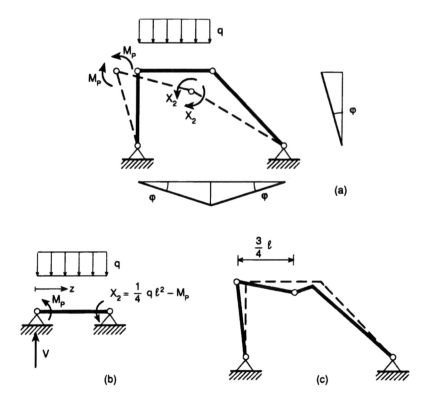

Figure 10.19

whence we obtain

$$V = \frac{3}{4}ql \tag{10.67}$$

The shear on the cross member vanishes and, therefore, the moment is at its maximum for $V = qz$, and hence for $z = 3/4\ l$:

$$M_{max} = M\left(\frac{3}{4}l\right) = \frac{3}{4}Vl - M_P - \frac{1}{2}q\left(\frac{3}{4}l\right)^2 \tag{10.68}$$

or

$$M_{max} = \frac{9}{32}ql^2 - M_P \tag{10.69}$$

The condition of second and ultimate plastic hinge formation is $M_{max} = M_P$, from which we find the collapse load:

$$q_2 = q_P = \frac{64}{9}\frac{M_P}{l^2} \approx 7.11\frac{M_P}{l^2} \tag{10.70}$$

while the mechanism of collapse consists of the articulated parallelogram of Figure 10.19c. The safety factor is

$$\frac{q_P}{q_e} = \frac{q_2}{\frac{2}{3}q_1} \simeq 1.93 \tag{10.71}$$

10.4 LAW OF NORMALITY OF INCREMENTAL PLASTIC DEFORMATION AND OF CONVEXITY OF PLASTIC LIMIT SURFACE

As will be illustrated in this and the ensuing sections, it is possible to avoid the unwieldy incremental plastic calculation and focus attention on the ultimate condition of collapse, when the entire structure, or part thereof, undergoes large increments of displacement resulting from small increments of load. This can be achieved by means of the theorems of plastic limit analysis, which will be demonstrated in the next section.

The present section will provide a preliminary demonstration of the two fundamental properties possessed, respectively, by the surface of plastic deformation in the space of principal stresses:

$$F(\sigma_1, \sigma_2, \sigma_3) = 0 \tag{10.72}$$

and the incremental plastic deformation.

As in the uniaxial condition, where the element of material is in the elastic state for $|\sigma| < \sigma_P$, likewise, in the biaxial (plane stress) condition, the element of material is in the elastic state for $F(\sigma_1, \sigma_2) < 0$. The function F, according to Von Mises' criterion, leads to

$$F_M(\sigma_1, \sigma_2) = \left(\sigma_1^2 + \sigma_2^2 - \sigma_1\sigma_2\right) - \sigma_P^2 \tag{10.73}$$

or, according to Tresca's criterion:

$$F_T(\sigma_1, \sigma_2) = \max\{|\sigma_1|, |\sigma_2|, |\sigma_1 - \sigma_2|\} - \sigma_P \tag{10.74}$$

Whereas, in the uniaxial condition, the characteristics of plastic flow are evident (i.e., there is a dilation collinear to stress), in multiaxial conditions it is difficult to make out the mechanics of the deformation.

Let us consider an element of a two-dimensional solid subjected to the stress condition (Figure 10.20):

$$\{\sigma_0\} = [\sigma_1, \sigma_2]^T \tag{10.75}$$

Suppose that an increment $\{\sigma\}-\{\sigma_0\}$ is applied to the same element, and that subsequently this increment is removed in a quasi-static manner. **Drucker's postulate** states that the material may be defined as stable when the work performed in the loading cycle is nonnegative. For a stress condition $\{\sigma\}$ lying on the surface of plastic deformation $F(\{\sigma\}) = 0$, and for each stress

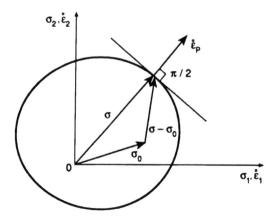

Figure 10.20

condition $\{\sigma_0\}$ that is admissible, and thus contained within the elastic domain or lying on the boundary, we must have

$$\left(\{\sigma\} - \{\sigma_0\}\right)^{\mathrm{T}} \{\dot{\varepsilon}_P\} \geq 0 \tag{10.76}$$

where $\{\dot{\varepsilon}_P\}$ is the incremental plastic strain that occurs when the stress reaches $\{\sigma\}$. It is possible to give a highly significant geometrical interpretation, superposing the spaces $\{\sigma\}$ and $\{\dot{\varepsilon}_P\}$ (Figure 10.20): the scalar product of Equation 10.76 is always positive or at least zero. It thus follows that

1. In each regular point of the surface of plasticity (single tangent plane), the incremental plastic strain $\{\dot{\varepsilon}_P\}$ is **normal** to the surface itself.
2. The surface of plasticity is **convex**.

In the cusps of the surface of plasticity (Figure 10.21a), $\{\dot{\varepsilon}_P\}$ cannot be external to the cone defined by the normals to the infinite tangent planes. In this case, more than one vector $\{\dot{\varepsilon}_P\}$ may correspond to a single vector $\{\sigma\}$. On the other hand, in the portions where the surface of plasticity is linear (i.e., not strictly convex), more than one vector $\{\sigma\}$ corresponds to a single vector $\{\dot{\varepsilon}_P\}$ (Figure 10.21b). These two conditions are both present in Tresca's hexagon.

The elastic domain includes the origin, and hence the inequality of Equation 10.76, when $\{\sigma_0\} = \{0\}$, becomes

$$\{\sigma\}^{\mathrm{T}} \{\dot{\varepsilon}_P\} = \dot{\Phi}\left(\{\dot{\varepsilon}_P\}\right) > 0 \tag{10.77}$$

where $\dot{\Phi}$ represents the energy dissipated in the unit of volume and is a function solely of the incremental plastic strain. This consideration remains valid also when the surface of plastic deformation presents cusps and linear portions. Consequently, the following statement is equivalent to Drucker's postulate: the energy dissipated in the unit of volume is a function only of the incremental plastic strain. Also, from this, it is possible to deduce the **law of normality** and the **convexity of the surface of plastic deformation.** Equation 10.77 shows, in fact, that each stress condition $\{\sigma\}$ capable of producing the incremental plastic strain $\{\dot{\varepsilon}_P\}$ must be on the plane normal to $\{\dot{\varepsilon}_P\}$ and distant $\dot{\Phi}\left(\{\dot{\varepsilon}_P\}\right)$ from the origin

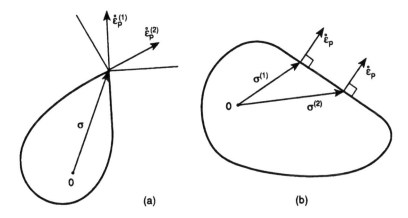

Figure 10.21

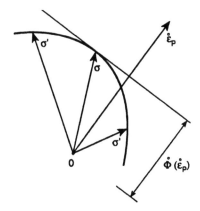

Figure 10.22

(Figure 10.22). As $\{\dot{\varepsilon}_P\}$ is made to turn about the origin, all these planes envelop the surface of plastic deformation, which is thus convex.

If $\{\dot{\sigma}\}$ is the incremental stress vector corresponding to the incremental plastic strain vector $\{\dot{\varepsilon}_P\}$, we have

$$\{\dot{\sigma}\}^T \{\dot{\varepsilon}_P\} \geq 0 \tag{10.78}$$

assuming $\{\sigma\}$ to be the initial stress condition and applying Equation 10.76. For an elastic–perfectly plastic material, we have, in particular:

$$\{\dot{\sigma}\}^T \{\dot{\varepsilon}_P\} = 0 \tag{10.79}$$

whereas, for a strain-softening material, we have

$$\{\dot{\sigma}\}^T \{\dot{\varepsilon}_P\} < 0 \tag{10.80}$$

and Drucker's postulate is violated.

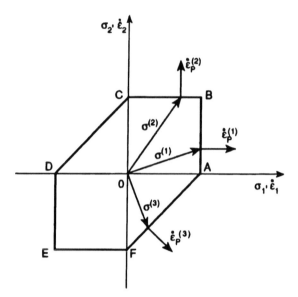

Figure 10.23

Figure 10.23 represents Tresca's criterion in two dimensions and the corresponding mechanisms of plastic flow (incremental plastic strain). Along the sides AB, BC, DE, EF only one of the principal dilations $\dot{\varepsilon}_1$, $\dot{\varepsilon}_2$ is activated, while, along the sides CD and FA, one dilation is positive and the other is negative; these are activated simultaneously and with equal intensity.

In the present section, we have faithfully reported the demonstration provided by Drucker. In the author's opinion, it is more correct to consider, as regards the scalar product in the inequality of Equation 10.76, the actual stress vector $\{\sigma\}$, rather than $\{\sigma\}-\{\sigma_0\}$. In fact, it is the actual stress vector $\{\sigma\}$ that produces work. Therefore, Drucker's postulate assumes only a geometrical character.

10.5 THEOREMS OF PLASTIC LIMIT ANALYSIS

Let us consider a rigid–perfectly plastic solid, subjected to a condition of proportional loading, measured by the multiplier λ (Figure 10.24). A stress field is said to be **statically admissible** when it is in equilibrium with the external load λ and at each point of the solid we have $F \leq 0$. On the other hand, a collapse mechanism is said to be **kinematically admissible** when the external constraints are respected and the corresponding dissipated energy is positive.

10.5.1 Theorem of maximum dissipated energy

Given an incremental plastic strain $\{\dot{\varepsilon}_P\}$, the energy dissipated by the stress $\{\sigma\}$ corresponding to this strain (Figure 10.22) is greater than or equal to the energy dissipated by any other possible stress $\{\sigma'\}$:

$$\{\sigma\}^{\mathrm{T}}\{\dot{\varepsilon}_P\} \geq \{\sigma'\}^{\mathrm{T}}\{\dot{\varepsilon}_P\} \tag{10.81}$$

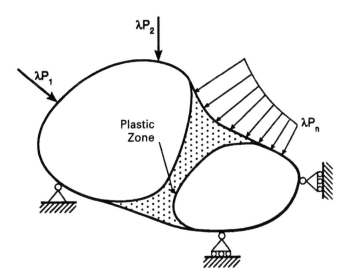

Figure 10.24

The foregoing inequality is valid at each point of the solid and, therefore, on the basis of Equation 10.77, it is possible to write

$$\int_V \dot{\Phi}\big(\{\dot{\varepsilon}_P\}\big)\mathrm{d}V \geq \int_V \{\sigma'\}^{\mathrm{T}}\{\dot{\varepsilon}_P\}\mathrm{d}V \tag{10.82}$$

10.5.2 Static Theorem (Upper Bound Theorem)

The multiplier of the loads λ^- corresponding to any statically admissible stress field is less than or equal to the multiplier of actual collapse λ_P.

Let $\{\sigma^-\}$ be a statically admissible stress field and λ^- the corresponding multiplier of the external loads. Then, let $\{\sigma\}$ be the stress field of collapse and $\{\dot{\eta}\}$, $\{\dot{\varepsilon}_P\}$ the incremental fields of displacement and of plastic strain, respectively, at the moment of collapse. Application of the principle of virtual work yields the following relations:

$$\int_V \{\sigma^-\}^{\mathrm{T}}\{\dot{\varepsilon}_P\}\mathrm{d}V = \sum_i \lambda^- P_i \dot{\eta}_i \tag{10.83a}$$

$$\int_V \{\sigma\}^{\mathrm{T}}\{\dot{\varepsilon}_P\}\mathrm{d}V = \sum_i \lambda_P P_i \dot{\eta}_i \tag{10.83b}$$

where P_i, $i = 1, 2, \ldots, n$ indicate the external loads applied to the solid. Recalling the inequality of Equation 10.81, we obtain

$$\int_V \big(\{\sigma\}-\{\sigma^-\}\big)^{\mathrm{T}}\{\dot{\varepsilon}_P\}\mathrm{d}V \geq 0 \tag{10.84}$$

and hence

$$\lambda_P \geq \lambda^-$$ (10.85)

10.5.3 Kinematic Theorem (Lower Bound Theorem)

The multiplier of the loads λ^+ corresponding to any kinematically admissible collapse mechanism is greater than or equal to the multiplier of actual collapse λ_P.

Let $\{\dot{\eta}^+\}, \{\dot{\varepsilon}^+\}$ be the incremental fields, respectively, of displacement and of plastic strain, corresponding to a kinematically admissible mechanism of collapse. Then, let $\{\sigma\}$ be the stress field of actual collapse. The multiplier of the external loads λ^+ corresponding to the kinematically admissible collapse mechanism is given by the following energy balance:

$$\int_V \{\sigma^+\}^T \{\dot{\varepsilon}^+\} dV = \sum_i \lambda^+ P_i \dot{\eta}_i^+$$ (10.86)

Application of the principle of virtual work to the stress field of actual collapse $\{\sigma\}$ and to the kinematically admissible collapse mechanism $\{\dot{\varepsilon}^+\}$ yields

$$\int_V \{\sigma\}^T \{\dot{\varepsilon}^+\} dV = \sum_i \lambda_P P_i \dot{\eta}_i^+$$ (10.87)

On the other hand, for the inequality of Equation 10.82, we have

$$\int_V \{\sigma^+\}^T \{\dot{\varepsilon}^+\} dV \geq \int_V \{\sigma\}^T \{\dot{\varepsilon}^+\} dV$$ (10.88)

From Equations 10.86 through 10.88, there follows

$$\lambda^+ \geq \lambda_P$$ (10.89)

10.5.4 Mixed Theorem

If the multiplier of the external loads λ corresponds to a statically admissible stress field and, at the same time, to a kinematically admissible collapse mechanism, then we have

$$\lambda = \lambda_P$$ (10.90)

This statement follows immediately from the two theorems demonstrated previously, since the **actual collapse mechanism** represents a kinematically admissible mechanism and, at the same time, presupposes a statically admissible stress field.

10.5.5 Theorem of Addition of Material

A dimensional increment of a perfectly plastic solid cannot produce a decrement in the collapse load.

In fact, the sum of the stress field of collapse in the original solid and of an identically zero stress field in the portion of additional material constitutes a statically admissible stress field. This means that the collapse load of the new solid is greater than or equal to that of the original solid, and certainly not less than it.

The properties of **convexity** of the elastic domain and of **normality** of the incremental plastic deformation, as well as the **theorems of limit analysis**, just demonstrated for the three-dimensional solids, can be readily extended to two-dimensional solids (plates) and one-dimensional solids (beams), replacing the stress vector $\{\sigma\}$ with the static characteristics vector $\{Q\}$, and the vector of incremental plastic strains $\{\dot{\varepsilon}_P\}$ with the vector of incremental plastic deformation characteristics $\{\dot{q}_P\}$. As an example of convexity of the elastic domain and normality of the plastic flow, consider the plastic limit of the bending moment versus axial force interaction in Figure 10.9. In the case where, as usually occurs, only the bending moment M is considered as the active characteristic, rather than the incremental vector $\{\dot{q}_P\}$, it is sufficient to consider the plastic increment of the curvature, $\dot{\chi}_P$ or, more simply, the relative rotation φ.

10.6 BEAM SYSTEMS LOADED PROPORTIONALLY BY CONCENTRATED FORCES

In the case of statically indeterminate systems of beams loaded proportionally by concentrated forces, application of the static theorem reduces the solution to that of a problem of **linear programming**. As an illustration of this, let us consider the continuous beam in Figure 10.25a, constrained by three supports and a built-in constraint, and loaded proportionally by two forces concentrated in the first and third spans. This structure has three degrees of redundancy, as is noted from an examination of the equivalent statically determinate system of Figure 10.25b, and presents five critical cross sections for the formation of the plastic hinges; namely, the two central supports, the built-in constraint, and the two sections in which the external forces are applied.

The total bending moment is expressible as the sum of four contributions, due to the external forces and the redundant moments (Figure 10.25b):

$$M(z) = \lambda M^{(0)} + \sum_{j=1}^{n} X_j M^{(j)} \tag{10.91}$$

with $n = 3$. In the $m = 5$ critical cross sections, we thus have

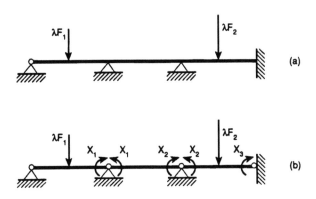

Figure 10.25

$$M_i = \lambda M_i^{(0)} + \sum_{j=1}^{n} X_j M_i^{(j)}, \text{ for } i = 1, 2, \dots, m \tag{10.92}$$

The static theorem states that the plastic collapse load is represented by the maximum value that the target function λ can assume in respect of the following $2m$ constraints:

$$-M_P \leq \lambda M_i^{(0)} + \sum_{j=1}^{n} X_j M_i^{(j)} \leq M_P, \text{ for } i = 1, 2, \dots, m \tag{10.93}$$

In the case of structures with many degrees of redundancy, this problem is resolvable with procedures of automatic calculation. It is, in fact, a problem of **linear programming** in the variables λ; $X_1, X_2, \dots, X_n$.

As an example, let us reconsider the elementary case of a beam built in at either end and subjected to the vertical load λF in the center (Figure 10.26a), whereby $n = 1$, $m = 2$. Designating the built-in constraint moment as X (Figure 10.26b), we have

$$M_1 = -X \tag{10.94a}$$

$$M_2 = -X + \frac{1}{4}\lambda Fl \tag{10.94b}$$

and, hence, the inequalities of Equation 10.93 in this case take on the following form:

$$-M_P \leq -X \leq M_P \tag{10.95a}$$

$$-M_P \leq \left(-X + \frac{1}{4}\lambda Fl\right) \leq M_P \tag{10.95b}$$

From Equations 10.95, we deduce the following four inequalities:

$$X \geq -M_P \tag{10.96a}$$

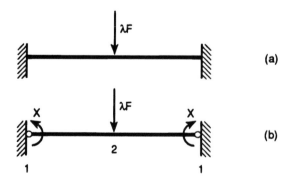

Figure 10.26

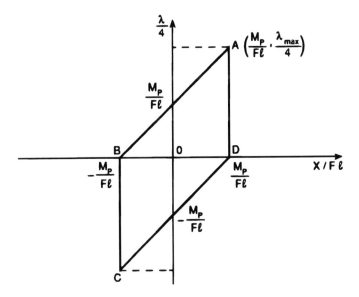

Figure 10.27

$$X \leq M_P \tag{10.96b}$$

$$X \leq M_P + \frac{1}{4}\lambda Fl \tag{10.96c}$$

$$X \geq -M_P + \frac{1}{4}\lambda Fl \tag{10.96d}$$

which, on the plane X-λ, define the parallelogram represented in Figure 10.27. The maximum value of λ on this domain is given by the ordinate of point A:

$$\lambda_{\max} = 8\frac{M_P}{Fl} \tag{10.97}$$

from which we find the collapse load:

$$F_P = \lambda_{\max}F = 8\frac{M_P}{l} \tag{10.98}$$

An alternative method for solving beam systems loaded proportionally by concentrated forces is that proposed by Neal and Symonds, which is also called the **method of combining mechanisms**. According to this method, each mechanism of collapse can be considered as the combination of a certain number of independent mechanisms. To each mechanism of collapse, it is possible to apply the principle of virtual work, so as to determine the corresponding multiplier of the loads λ. The actual mechanism of collapse is distinguished from among all the virtual mechanisms because it presents the minimum value of the multiplier λ due to the kinematic theorem. It is then a matter of examining the independent mechanisms with low values of the multiplier λ, and seeking to combine them to form mechanisms with even lower values of λ. To verify the validity of the result, it is then necessary to check its static admissibility.

The method proposed by Neal and Symonds will now be illustrated with reference to a simple portal frame, subjected to two equal forces, one horizontal and the other vertical (Figure 10.28a). Since the degrees of redundancy are $n=3$, and the number of critical cross sections is $m=5$, the number of supplementary equations of equilibrium and, hence, the number of independent mechanisms must be $m-n=2$. As independent

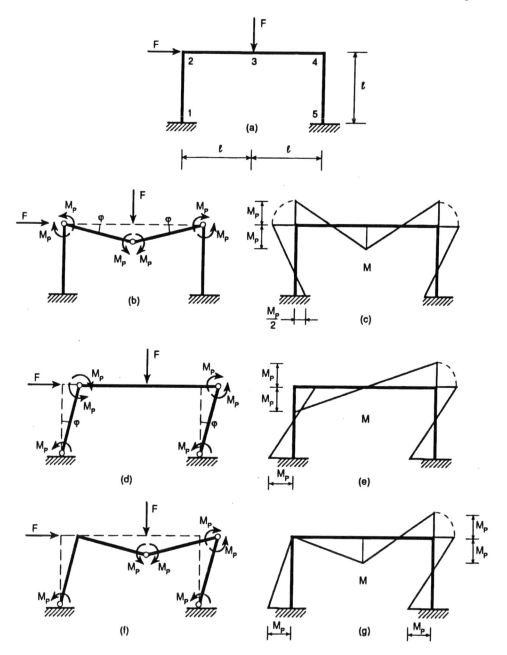

Figure 10.28

mechanisms, let the two represented in Figures 10.28b and d be chosen. These cause the horizontal force or the vertical force alternately to perform work. On the other hand, it may be demonstrated that both imply moment diagrams that are statically not admissible (Figures 10.28c and e) since the horizontal and vertical translation equilibrium equations of the cross member are, respectively, not satisfied, once both the applied forces are taken into consideration. Applying the principle of virtual work to the beam mechanism (Figure 10.28b) produces the equation:

$$Fl\varphi - 4M_P\varphi = 0 \tag{10.99}$$

from which we obtain

$$F = 4\frac{M_P}{l} \tag{10.100}$$

Applying the principle of virtual work to the sidesway mechanism (Figure 10.28d) produces, on the other hand, the same result.

If we now proceed to sum up algebraically (or combine) the two abovementioned mechanisms, we shall obtain the combined mechanism of Figure 10.28f, with four plastic hinges in the cross sections 1, 3, 4, and 5. The corresponding moment diagram (Figure 10.28g) proves to be statically admissible and, hence, we may conclude that the collapse mechanism of Figure 10.28f is the correct one. On the other hand, application of the principle of virtual work yields a collapse load smaller than the one corresponding to each of the two elementary mechanisms:

$$2Fl\varphi - 6M_P\varphi = 0 \tag{10.101}$$

whence we obtain

$$F_P = 3\frac{M_P}{l} \tag{10.102}$$

In the application of Neal and Symonds' method, both the theorems of limit analysis have been employed.

Now, let us consider the frame of Figure 10.28a, with the height equal to h instead of l, and let us determine the actual mechanism of collapse, varying the ratio l/h. Applying the principle of virtual work to the three abovementioned mechanisms (Figures 10.28b, d, and f), we have, respectively:

$$Fl\varphi - 4M_P\varphi = 0 \quad \rightarrow \quad \frac{F_P l}{M_P} = 4 \tag{10.103a}$$

$$Fh\varphi - 4M_P\varphi = 0 \quad \rightarrow \quad \frac{F_P l}{M_P} = 4\frac{l}{h} \tag{10.103b}$$

$$Fh\varphi + Fl\varphi - 6M_P\varphi = 0 \quad \rightarrow \quad \frac{F_P l}{M_P} = \frac{6}{\frac{h}{l}+1} \tag{10.103c}$$

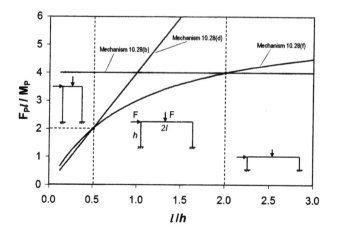

Figure 10.29

The ratio $F_p l/M_p$ for the three different mechanisms is graphically represented in Figure 10.29, where, for $0 < l/h \leq 0.5$, the lateral mechanism in Figure 10.28d prevails; for $0.5 \leq l/h \leq 2$, the combined mechanism in Figure 10.28f develops; whereas, for $l/h > 2$, the vertical mechanism in Figure 10.28b is activated.

10.7 BEAM SYSTEMS LOADED PROPORTIONALLY BY DISTRIBUTED FORCES

The solution of statically indeterminate beam systems also loaded proportionally by distributed forces presents greater difficulties than does the solution of systems involving concentrated forces. This is due to the impossibility of identifying a finite number of critical cross sections from the outset. Since, therefore, there does not exist any systematic method, the procedure is one of trial and error, applying the kinematic and static theorems alternately.

Let us consider, for instance, the portal frame of Figure 10.30a, subjected to a load uniformly distributed over the cross member and to a concentrated horizontal force of equal intensity. Let us take as the collapse mechanism the actual one from the concentrated force scheme (Figure 10.28a) and apply the principle of virtual work (Figure 10.30b):

$$2ql\left(\frac{l}{2}\varphi\right) + 2ql\left(l\varphi\right) - 6M_p\varphi = 0 \tag{10.104}$$

from which we obtain the load:

$$q = 2\frac{M_p}{l^2} \tag{10.105}$$

The constraint reactions at cross section 5 are obtained by assuming that, also in sections 3 and 4, the bending moment is equal to its plastic value M_p (Figure 10.30c):

$$M_4 = -M_p + Hl = M_p \tag{10.106a}$$

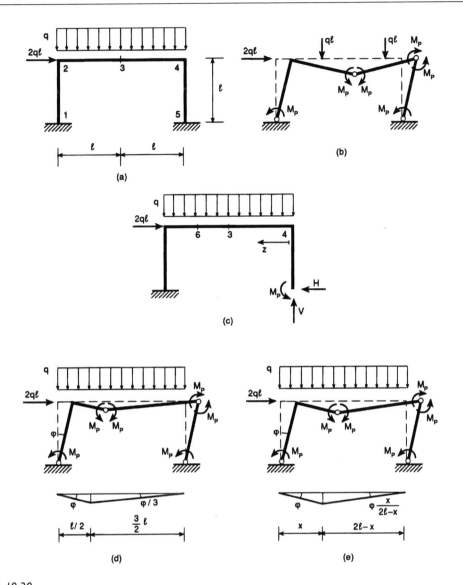

Figure 10.30

$$M_3 = M_P - Hl + Vl - \frac{1}{2}ql^2 = M_P \qquad (10.106b)$$

whence we obtain

$$H = 2\frac{M_P}{l} \qquad (10.107a)$$

$$V = 3\frac{M_P}{l} \qquad (10.107b)$$

The moment function on the cross member is given by the sum of four contributions:

$$M(z) = M_P + 3\frac{M_P}{l}z - 2\frac{M_P}{l}l - \frac{1}{2}\left(2\frac{M_P}{l^2}\right)z^2 = -M_P + 3\frac{M_P}{l}z - \frac{M_P}{l^2}z^2 \qquad (10.108)$$

while the shear is given by two contributions:

$$T(z) = \frac{dM}{dz} = 3\frac{M_P}{l} - 2\frac{M_P}{l^2}z \qquad (10.109)$$

and vanishes for $z = 3/2\ l$. The maximum moment is thus

$$M_{max} = M\left(\frac{3}{2}l\right) = \frac{5}{4}M_P \qquad (10.110)$$

and, since it is greater than M_P, it reveals the static inadmissibility of the mechanism of Figure 10.30b. On the other hand, dividing the load resulting from Equation 10.105 by 5/4, we obtain a statically admissible scheme and, hence, an application of the static and kinematic theorems leads to the conclusion that the actual collapse load must fall within the following interval:

$$1.6\frac{M_P}{l^2} < q_P < 2\frac{M_P}{l^2} \qquad (10.111)$$

Since the interval given by Equation 10.111 is still not sufficiently narrow, we assume that the collapse mechanism of second approximation presents three plastic hinges, again in sections 1, 4, 5, and the fourth hinge in the section that, in the previous scheme, was subjected to the maximum moment M_{max} (Figure 10.30d). Applying the principle of virtual work yields the equation:

$$q\frac{l}{2}\left(\frac{l}{4}\varphi\right) + \frac{3}{2}ql\left(\frac{3}{4}l\right)\left(\frac{1}{3}\varphi\right) + 2ql(l\varphi) - 4M_P\varphi - 2M_P\left(\frac{1}{3}\varphi\right) = 0 \qquad (10.112)$$

which presents the solution:

$$q = \frac{28}{15}\frac{M_P}{l^2} \qquad (10.113)$$

The constraint reactions at section 5 are obtained by assuming that, in cross section 4 and 6 (Figure 10.30c), the bending moment is also equal to its plastic value M_P:

$$M_4 = -M_P + Hl = M_P \qquad (10.114a)$$

$$M_6 = M_P - Hl + V\frac{3}{2}l - \frac{1}{2}q\left(\frac{3}{2}l\right)^2 = M_P \qquad (10.114b)$$

whence we obtain

$$H = 2\frac{M_P}{l} \tag{10.115a}$$

$$V = \frac{41}{15}\frac{M_P}{l} \tag{10.115b}$$

The bending moment and shearing force on the horizontal beam are therefore represented by the following functions:

$$M(z) = M_P + \left(\frac{41}{15}\frac{M_P}{l}\right)z - 2\frac{M_P}{l}l - \frac{1}{2}\left(\frac{28}{15}\frac{M_P}{l^2}\right)z^2 \tag{10.116a}$$

$$T(z) = \frac{dM}{dz} = \frac{41}{15}\frac{M_P}{l} - \frac{28}{15}\frac{M_P}{l^2}z \tag{10.116b}$$

The shear vanishes for $z = 41/28\ l$, and the maximum bending moment is thus

$$M_{\max} = M\left(\frac{41}{28}l\right) = \frac{841}{840}M_P \tag{10.117}$$

On the other hand, dividing the load obtained from Equation 10.113 by 841/840, we obtain a statically admissible scheme, and thus an application of the static and kinematic theorems yields the following interval for the actual collapse load:

$$\frac{840}{841}\times\frac{28}{15}\frac{M_P}{l^2} < q_P < \frac{28}{15}\frac{M_P}{l^2} \tag{10.118}$$

This interval is extremely narrow and, for engineering purposes, yields the actual collapse load with sufficient approximation (~1‰).

To improve this approximation still further, it would suffice to consider a third mechanism with the hinge in $z = 41/28\ l$, but this is not necessary, since it is possible to identify the actual collapse mechanism by minimizing the load q as the position of the plastic hinge on the horizontal beam varies (kinematic theorem).

Consider the mechanism of Figure 10.30e, with the plastic hinge in an intermediate position on the cross member, at a distance x from the left-hand fixed-joint node. As the diagram of vertical displacements shows, the left-hand portion turns clockwise by the angle φ, while the right-hand portion turns counterclockwise by the angle:

$$\vartheta = \varphi\frac{x}{2l - x} \tag{10.119}$$

Application of the principle of virtual work provides the following equation:

$$qx\left(\frac{x}{2}\varphi\right)+\frac{1}{2}q(2l-x)^2\,\varphi\frac{x}{2l-x}+2ql(l\varphi)$$

$$-4M_P\varphi-2M_P\varphi\frac{x}{2l-x}=0 \qquad (10.120)$$

from which we obtain the load:

$$q(x)=2\frac{M_P}{l}\frac{4l-x}{4l^2-x^2} \qquad (10.121)$$

The derivative of this load with respect to the coordinate x:

$$\frac{dq}{dx}=2\frac{M_P}{l}\frac{-x^2+8lx-4l^2}{\left(4l^2-x^2\right)^2} \qquad (10.122)$$

vanishes for

$$x=2l\left(2\pm\sqrt{3}\right) \qquad (10.123)$$

The larger root is to be rejected, whereas if we substitute the value $x=2l\left(2-\sqrt{3}\right)$ in Equation 10.121, we obtain the actual collapse load:

$$q_P=\frac{M_P}{l^2}\frac{3}{12-6\sqrt{3}} \qquad (10.124)$$

Rationalizing the ratio in Equation 10.123 up to the seventh decimal place, we have

$$q_P=1.8660254\frac{M_P}{l^2} \qquad (10.125)$$

so that the inequalities of Equation 10.118 are verified:

$$1.8644471<1.8660254<1.8666667$$

Ultimately, it should be noted that, if we set $2ql=F$ to make a comparison with the case where also the vertical load on the cross member is concentrated (Figure 10.28a), the collapse load given by Equation 10.125 can be expressed in the form

$$F_P=2q_Pl\simeq3.73\frac{M_P}{l} \qquad (10.126)$$

and is thus greater than the collapse load given by Equation 10.102 for the concentrated force in the center.

The case of the portal frame with inclined stanchion in Figure 1.39a has thus been solved in Section 10.3 by means of an incremental analysis. If, instead, we had chosen to proceed by trial and error, applying the upper bound and lower bound theorems, we could have assumed the first-approximation mechanism illustrated in Figure 10.19a, with two plastic hinges located in the fixed-joint nodes. Application of the principle of virtual work yields the load

$$q = 8\frac{M_P}{l^2} \tag{10.127}$$

which turns out to be greater than the collapse load given by Equation 10.70 by virtue of the kinematic theorem. The fact imposing the two plastic moments M_P at the ends of the cross member means that the bending moment in the right-hand part of the cross member exceeds the value M_P (Figure 10.31a). This is, on the other hand, a statically inadmissible situation.

To determine the maximum value of the moment, let us isolate the cross member (Figure 10.31b) and identify the cross section in which the shear vanishes:

$$T(z) = 2\frac{M_P}{l} + \frac{1}{2}ql - qz = 0 \tag{10.128a}$$

for

$$z = \frac{l}{2} + 2\frac{M_P}{ql} = \frac{3}{4}l \tag{10.128b}$$

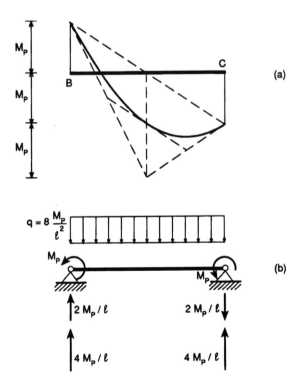

Figure 10.31

The maximum bending moment is thus

$$M_{max} = \left(6\frac{M_P}{l}\right)\left(\frac{3}{4}l\right) - M_P - \frac{1}{2}\left(8\frac{M_P}{l^2}\right)\left(\frac{3}{4}l\right)^2 = \frac{5}{4}M_P \tag{10.129}$$

Then, dividing the load given by Equation 10.127 by 5/4, the system is reduced to a statically admissible one, whereby the following inequalities result:

$$6.4\frac{M_P}{l^2} < q_P < 8\frac{M_P}{l^2} \tag{10.130}$$

These inequalities are confirmed by solution in Equation 10.70. Finally, it should be noted that the second-approximation mechanism, with the hinge at a distance $z = 3/4\ l$ from the left-hand node of the beam, is identified with the actual collapse mechanism (Figure 10.19d).

As our last example of the application of the limit analysis theorems, let us examine the case of the portal frame with a strut, illustrated in Figure 1.27a.

The moment and the shear on the portion BF are described by the following functions (Figure 10.32a):

$$M(z) = \frac{29}{124}qlz - \frac{3}{496}ql^2 - \frac{1}{2}qz^2 \tag{10.131a}$$

$$T(z) = \frac{dM}{dz} = \frac{29}{124}ql - qz \tag{10.131b}$$

The point of zero shear is given by

$$z = \frac{29}{124}l \tag{10.132}$$

and, hence, the maximum bending moment is

$$M_{max} = M\left(\frac{29}{124}l\right) = \frac{655}{30752}ql^2 \tag{10.133}$$

The load that produces the first plastic hinge is, therefore:

$$q_1 = \frac{30752}{655}\frac{M_P}{l^2} \simeq 46.95\frac{M_P}{l^2} \tag{10.134}$$

If we consider the mechanism of Figure 10.32b, the principle of virtual work yields the equation:

$$\frac{1}{2}q\left(\frac{33}{124}l\right)^2\left(\frac{29}{62}\varphi\right) + \frac{1}{2}q\left(\frac{29}{124}l\right)^2\left(\frac{33}{62}\varphi\right) - 2M_P\varphi = 0 \tag{10.135}$$

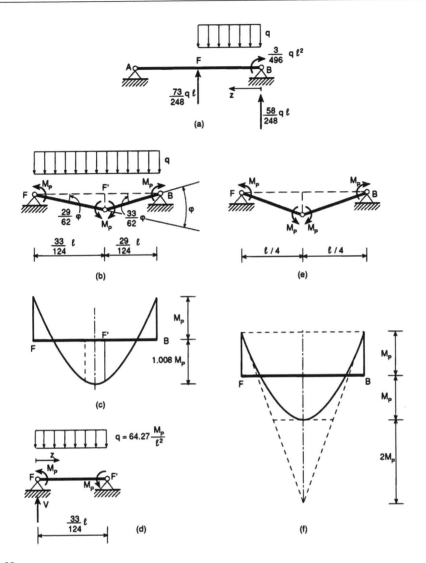

Figure 10.32

from which we obtain the load:

$$q\left(\frac{33^2 \times 29 + 33 \times 29^2}{124^2 \times 62}\right) = 4\frac{M_P}{l^2}$$ (10.136)

or

$$q \simeq 64.27\frac{M_P}{l^2}$$ (10.137)

On the other hand, the scheme of Figure 10.32b proves not to be statically admissible, since, in an intermediate portion of the beam BF, the moment assumes values greater than

M_P (Figure 10.32c). In fact, isolating the portion of length 33/124 l, contained between the plastic hinges F and F' (Figure 10.32d), we find the shear at F:

$$V \simeq \frac{2M_P}{(33l/124)} + 64.27 \frac{M_P}{l^2} \left(\frac{33}{248} l \right) \simeq 16.067 \frac{M_P}{l} \tag{10.138}$$

The shear between F and F':

$$T(z) = V - qz \tag{10.139}$$

vanishes for

$$z = \frac{V}{q} \simeq \frac{16.067}{64.27} l \simeq 0.25 l \tag{10.140}$$

from which we obtain the maximum bending moment:

$$M_{\max} = M\left(\frac{l}{4} \right)$$

$$\simeq 16.067 \frac{M_P}{l} \left(\frac{l}{4} \right) - M_P - \frac{1}{2} \left(64.27 \frac{M_P}{l^2} \right) \left(\frac{l}{4} \right)^2$$

$$\simeq 1.008 M_P \tag{10.141}$$

which is not statically admissible.

Dividing the load given by Equation 10.137 by 1.008, we obtain, on the other hand, a statically admissible system. The actual collapse load q_P must then be greater than

$$q = \frac{64.27}{1.008} \frac{M_P}{l^2} \tag{10.142}$$

and, consequently, contained within the following interval:

$$63.76 \frac{M_P}{l^2} < q_P < 64.27 \frac{M_P}{l^2} \tag{10.143}$$

Let us assume that our second-approximation mechanism presents the plastic hinges at B, F, and in the center of the portion BF, which proved to be the location of the maximum bending moment in the previous step (Figure 10.32e). The principle of virtual work yields the load:

$$q_P = 64 \frac{M_P}{l^2} \tag{10.144}$$

As the scheme in Figure 10.32f shows, the corresponding bending moment diagram is statically admissible, whence, by virtue of the mixed theorem, a mechanism that is both kinematically and statically admissible coincides with the actual collapse mechanism. The inequalities of Equation 10.143 are, in fact, verified by the collapse load given by Equation 10.144.

10.8 NONPROPORTIONALLY LOADED BEAM SYSTEMS

In the case of statically indeterminate beam systems, loaded nonproportionally by two or more concentrated forces, applying the kinematic theorem makes it possible to define the limit of collapse in the space of these forces.

In the case, for example, of the continuous beam of Figure 10.33a, loaded by two non-proportional forces F_1 and F_2, the introduction of five plastic hinges in the critical sections makes it possible to define two different collapse mechanisms (Figure 10.33b). Applying the principle of virtual work to each of them (and to its converse) yields the following equations:

$$-3M_P\varphi \pm F_1 \frac{l}{2}\varphi = 0 \qquad (10.145a)$$

$$-4M_P\varphi \pm F_2 \frac{l}{2}\varphi = 0 \qquad (10.145b)$$

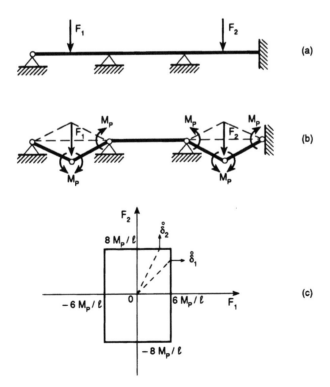

Figure 10.33

whence we obtain

$$F_1 = \pm 6 \frac{M_P}{l} \tag{10.146a}$$

$$F_2 = \pm 8 \frac{M_P}{l} \tag{10.146b}$$

The limit of collapse in the plane $F_1 - F_2$ is thus represented by the rectangle of Figure 10.33c. In the case where

$$-\frac{4}{3} < \frac{F_2}{F_1} < \frac{4}{3} \tag{10.147}$$

we have the activation of the first mechanism; otherwise, we have the activation of the second mechanism. The properties of convexity of the surface of plastic deformation and of normality of the incremental plastic deformation also apply, in fact, in the case of beam systems loaded by two or more concentrated forces.

A second example may be represented by the portal frame, already examined in Sections 10.6 and 10.7, loaded in this case by two independent concentrated forces, H and V (Figure 10.34). The introduction of five plastic hinges in the critical sections enables the four different collapse mechanisms, 1, 2, 3, 4, to be defined (Figure 10.35). Applying the principle of virtual work to these four collapse mechanisms and to their respective opposites, 5, 6, 7, 8, yields the following equations:

$$-4M_P\varphi \pm Hl\varphi = 0 \tag{10.148a}$$

$$-6M_P\varphi \pm (Vl\varphi + Hl\varphi) = 0 \tag{10.148b}$$

$$-4M_P\varphi \pm Vl\varphi = 0 \tag{10.148c}$$

$$-6M_P\varphi \pm (Vl\varphi - Hl\varphi) = 0 \tag{10.148d}$$

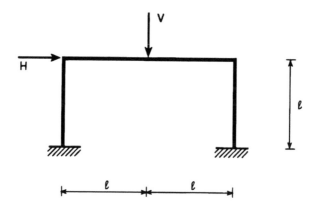

Figure 10.34

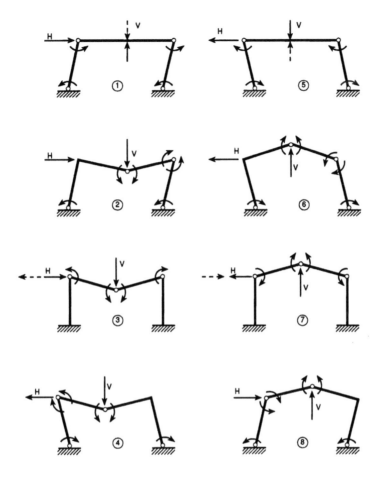

Figure 10.35

Once the factor φ, which does not affect the problem, has been canceled out, we then have the equations of the eight straight lines in the plane H–V, to which the respective sides of the boundary of collapse belong (Figure 10.36). The activations for each of the four pairs of mechanisms are given, respectively, by

$$\left|\frac{V}{H}\right| < \frac{1}{2} \tag{10.149a}$$

$$\frac{1}{2} < \frac{V}{H} < 2 \tag{10.149b}$$

$$\left|\frac{V}{H}\right| > 2 \tag{10.149c}$$

$$-2 < \frac{V}{H} < -\frac{1}{2} \tag{10.149d}$$

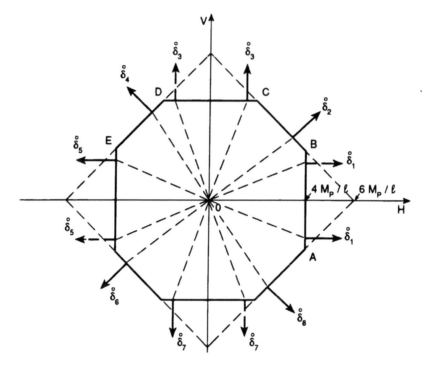

Figure 10.36

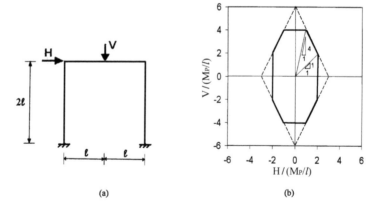

Figure 10.37

Also in this case, the properties of convexity of the plastic limit and normality of the incremental deformation are both verified. In fact, when the ratio $|V/H|$ is sufficiently small, the sidesway mechanism is activated (1,5). For intermediate values of $|V/H|$, the combined mechanisms are activated (2, 4, 6, 8). For sufficiently high values of $|V/H|$, the beam mechanism is activated (3, 7).

The limit of plastic collapse for a portal frame with $h/l=2$ (Figure 10.37a) is represented in Figure 10.37b, whereas for a portal frame with $h/l=1/2$ (Figure 10.38a) it is represented in Figure 10.38b.

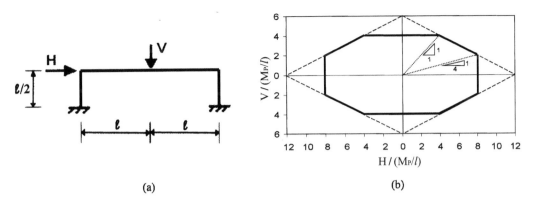

Figure 10.38

10.9 CYCLIC LOADING AND SHAKE-DOWN

Let us consider again the system of parallel connecting rods of Figure 10.1, and let us suppose that the rigid cross member is loaded repeatedly with a pulsating force (Figure 10.39). As was seen in Section 10.1, for

$$0 \le F \le \sigma_P A \left(1 + \frac{l_I}{l_{II}} \right) \tag{10.150}$$

the behavior of the system is elastic, and hence both loading and unloading occur along the segment O1 in Figure 10.40.

For greater values of the maximum load F, the central rod yields, so that, on unloading, this rod is found to be compressed by the lateral ones, which are assumed to obey an elastic constitutive law, devoid of yielding. On these hypotheses, a value F_{SD} of the load is shown to exist, such that the central rod also yields in compression.

When the external force is maximum, we have

$$X_I(\text{max}) + 2X_{II}(\text{max}) = F \tag{10.151a}$$

whereas, when the external force is zero, we have

$$X_I(\text{min}) + 2X_{II}(\text{min}) = 0 \tag{10.151b}$$

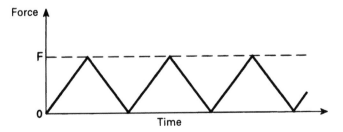

Figure 10.39

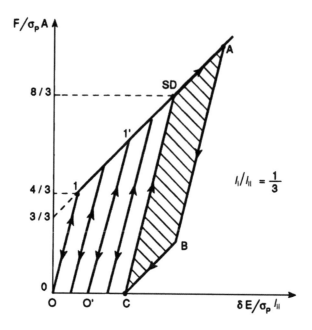

Figure 10.40

When the external force is maximum, the dilation of the central rod is

$$\varepsilon_I (\max) = \frac{l_{II}}{l_I} \varepsilon_{II} (\max) \tag{10.152}$$

and hence

$$\varepsilon_I (\max) = \frac{l_{II}}{l_I} \frac{X_{II} (\max)}{(EA/2)} \tag{10.153}$$

since the lateral rods are in elastic conditions.

On the other hand, when the external force is zero and in the case where inverse plastic deformation occurs, the dilation of the central rod is

$$\varepsilon_I (\min) = \varepsilon_I (\max) - \frac{2\sigma_P}{E} \tag{10.154}$$

Since it is also found that

$$\varepsilon_I (\min) = \frac{l_{II}}{l_I} \varepsilon_{II} (\min) \tag{10.155}$$

by virtue of the transitive law, we obtain

$$\frac{l_{II}}{l_I} \frac{X_{II} (\min)}{(EA/2)} = \frac{l_{II}}{l_I} \frac{X_{II} (\max)}{(EA/2)} - \frac{2\sigma_P}{E} \tag{10.156}$$

From Equation 10.156, we deduce the reaction of the lateral rods at the maximum load:

$$X_{II}(\max) = X_{II}(\min) + \sigma_P A \frac{l_I}{l_{II}} \tag{10.157}$$

On account of the equilibrium of the rigid cross member, Equation 10.151b yields

$$X_{II}(\min) = -\frac{1}{2} X_I(\min) \tag{10.158}$$

The hypothesis of inverse plastic deformation of the central rod gives

$$X_I(\min) = -\sigma_P A \tag{10.159}$$

so that

$$X_{II}(\min) = \sigma_P \frac{A}{2} \tag{10.160}$$

From Equation 10.157 we thus obtain

$$X_{II}(\max) = \sigma_P \frac{A}{2} + \sigma_P A \frac{l_I}{l_{II}} \tag{10.161}$$

Since at maximum load the central rod has yielded, we have

$$X_I(\max) = \sigma_P A \tag{10.162}$$

Finally, substitution of Equations 10.161 and 10.162 into Equation 10.151a yields the force of inverse plastic deformation:

$$F_{SD} = \sigma_P A + 2\left(\sigma_P \frac{A}{2} + \sigma_P A \frac{l_I}{l_{II}}\right) \tag{10.163}$$

or

$$F_{SD} = 2\sigma_P A\left(1 + \frac{l_I}{l_{II}}\right) \tag{10.164}$$

Only for

$$F > F_{SD} \tag{10.165}$$

does inverse plastic deformation of the central rod occur. Note that this threshold load F_{SD} is exactly twice the load of the first plastic deformation F_1 given by Equation 10.4a.

Summarizing, it is possible to state:

1. For $0 \leq F \leq F_1$, the system behaves elastically and its representative point in the plane F–δ (Figure 10.40) oscillates on the segment O1.
2. For $F_1 < F \leq F_{SD}$, the central rod yields only in tension and the representative point of the system oscillates on the corresponding segment O'1' (**shake-down**).
3. For $F > F_{SD}$, the central rod yields both in tension (loading) and in compression (unloading), and the representative point of the system traverses the parallelogram SD–A–B–C cyclically. More precisely, at the first loading the point moves to A; on subsequent elastic unloading, the point moves to B, where the plastic flow in compression of the central rod starts. This flow ceases at C. From here, the second loading cycle begins, which develops first elastically along the segment C–SD, and then plastically along the segment SD–A, and so forth. The energy that is dissipated in each **hysteresis cycle** is equal to the area of the parallelogram SD–A–B–C. The phenomenon of plastic dissipation just described is called **alternating plastic deformation**.

In the case where the cyclic load, instead of pulsating (Figure 10.39), is alternating and symmetrical (Figure 10.41), it is possible to demonstrate how the hysteresis cycle assumes the appearance represented in Figures 10.42a and b for $F < F_{SD}$ and $F > F_{SD}$,

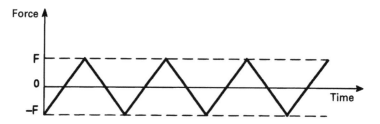

Figure 10.41

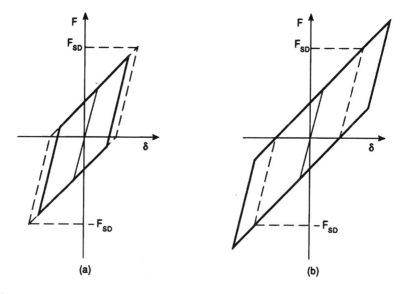

(a)   (b)

Figure 10.42

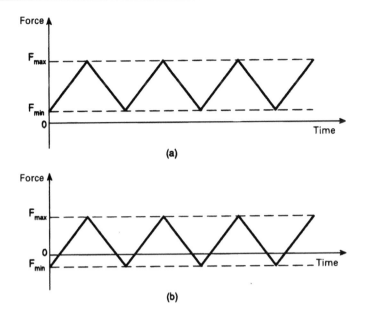

Figure 10.43

respectively. In the cases of cyclic loads pulsating from nonzero (Figure 10.43a) or of alternating nonsymmetrical cyclic loads (Figure 10.43b), the hysteresis cycle will again take on the appearance of a parallelogram, with the sides parallel to those of the alternating and symmetrical cycles. On the other hand, it will not be symmetrical with respect to the origin (Figures 10.44a and b).

Finally, it should be noted that in the cases where there is more than one hardening portion (Figure 10.45a), or in the cases where the plastic collapse of the system is reached (Figure 10.45b), the alternating plastic deformation develops through polygonal cycles that are polar-symmetrical with respect to specific points of the plane F–δ.

10.10 DEFLECTED CIRCULAR PLATES

Let us consider the case of a deflected circular plate, consisting of elastic–perfectly plastic material. The indefinite equations of equilibrium can be written in explicit form as follows:

$$\frac{\mathrm{d}}{\mathrm{d}r}(rT_r) = qr \tag{10.166a}$$

$$\frac{\mathrm{d}}{\mathrm{d}r}(rM_r) - M_\vartheta - rT_r = 0 \tag{10.166b}$$

Differentiating Equation 10.166b and using Equation 10.166a, we obtain the differential equation

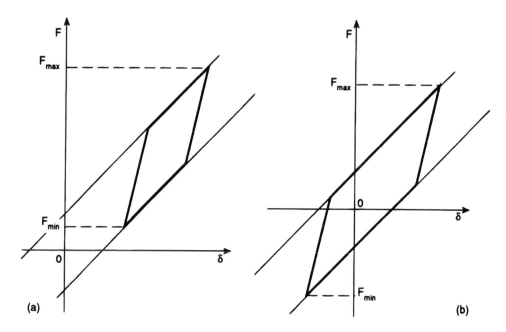

Figure 10.44

$$\frac{d^2}{dr^2}(rM_r) - \frac{dM_\vartheta}{dr} - qr = 0 \qquad (10.167a)$$

which is valid for $0 \le r \le R$, if R is the radius of the plate, with the boundary condition

$$M_r(R) = 0 \qquad (10.167b)$$

if the plate is simply supported at the boundary. On account of the isotropy of the stress condition in the center of the plate, we also have

$$M_r(0) = M_\vartheta(0) \qquad (10.167c)$$

Let us assume, as a statically admissible regime, a condition of complete plastic deformation of the plate. According to Equation 10.167c, the representative point of the characteristics that develop in the center of the plate must fall at the vertex B of Tresca's hexagon illustrated in Figure 10.46. Moreover, since the condition in Equation 10.167b holds good, it is legitimate to assume that the static regime is always represented by points belonging to the side BC of the same hexagon. Since, on this side, we have $M_\vartheta = M_P$, the static Equation 10.167a is transformed as follows:

$$\frac{d^2}{dr^2}(rM_r) - qr = 0 \qquad (10.168)$$

and, on integration, yields

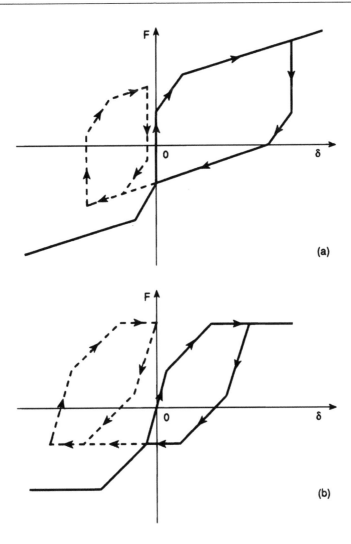

Figure 10.45

$$M_r = q\frac{r^2}{6} + C_1 + \frac{C_2}{r} \tag{10.169}$$

Applying the boundary conditions:

$$M_r(0) = M_P \tag{10.170a}$$

$$M_r(R) = 0 \tag{10.170b}$$

the two constants of integration are determined:

$$C_1 = M_P \tag{10.171a}$$

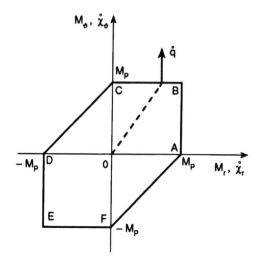

Figure 10.46

$$C_2 = 0 \tag{10.171b}$$

as well as the load:

$$q = -6\frac{M_P}{R^2} \tag{10.172}$$

The radial bending moment thus becomes

$$M_r(r) = -M_P\left[\left(\frac{r}{R}\right)^2 - 1\right] \tag{10.173a}$$

while the circumferential bending moment has been assumed to be constant and equal to the plastic moment:

$$M_\vartheta = M_P \tag{10.173b}$$

By virtue of the static theorem, the load given by Equation 10.172 represents a lower limit for the actual collapse load. On the other hand, on account of the property of normality of the incremental plastic deformation (Figure 10.46), the variations of the radial and circumferential curvatures must satisfy the conditions:

$$\dot{\chi}_r = 0, \quad \dot{\chi}_\vartheta \geq 0 \tag{10.174}$$

Via the kinematic Equation 3.72, the corresponding conditions on the deflection are obtained:

$$\frac{d^2\dot{w}}{dr^2} = 0, \qquad -\frac{1}{r}\frac{d\dot{w}}{dr} \geq 0 \tag{10.175}$$

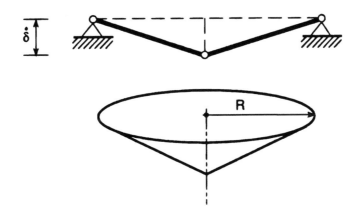

Figure 10.47

It is possible to verify immediately that the mechanism of Figure 10.47:

$$\dot{w} = \dot{\delta}\left(1 - \frac{r}{R}\right) \tag{10.176}$$

where $\dot{\delta}$ is the incremental plastic deflection of the central point, satisfies Equation 10.175 and may thus be associated with the static regime expressed by Equations 10.173.

Since the load given by Equation 10.172 corresponds both to a statically admissible condition and a kinematically admissible mechanism, by virtue of the mixed theorem it is possible to state that this load represents the actual collapse load of the plate.

10.11 DEFLECTED RECTANGULAR PLATES

Let us consider the case of a deflected rectangular plate, simply supported on its four edges, consisting of elastic–perfectly plastic material (Figure 10.48). Considering the plastic limit domain according to Von Mises:

$$F\left(M_x, M_y, M_{xy}\right) = M_x^2 - M_x M_y + M_y^2 + 3M_{xy}^2 \tag{10.177}$$

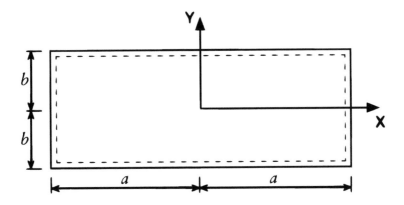

Figure 10.48

the law of normality yields the following equations:

$$\dot{\chi}_x = \frac{\partial F}{\partial M_x} \dot{\lambda} = \dot{\lambda}(2M_x - M_y) \tag{10.178a}$$

$$\dot{\chi}_y = \frac{\partial F}{\partial M_y} \dot{\lambda} = \dot{\lambda}(2M_y - M_x) \tag{10.178b}$$

$$\dot{\chi}_{xy} = \frac{\partial F}{\partial M_{xy}} \dot{\lambda} = 6\dot{\lambda}M_{xy} \tag{10.178c}$$

assuming $\dot{\lambda}$ as a positive factor.

Thus, we have

$$M_x = \frac{1}{3\dot{\lambda}}(2\dot{\chi}_x + \dot{\chi}_y) \tag{10.179a}$$

$$M_y = \frac{1}{3\dot{\lambda}}(2\dot{\chi}_y + \dot{\chi}_x) \tag{10.179b}$$

$$M_{xy} = \frac{1}{6\dot{\lambda}}\dot{\chi}_{xy} \tag{10.179c}$$

Substituting Equations 10.179 into Equation 10.177, for the limit condition $F(M_x, M_y, M_{xy}) = 1$, we obtain

$$\dot{\lambda} = \frac{1}{\sqrt{3}}\left(\dot{\chi}_x^2 + \dot{\chi}_x\dot{\chi}_y + \dot{\chi}_y^2 + \frac{\dot{\chi}_{xy}}{4}\right)^{1/2} \tag{10.180}$$

Therefore, the dissipated energy per unit area, depending on the incremental deformation, is

$$D = \frac{2M_P}{H}\dot{\lambda} = \frac{2M_P}{H\sqrt{3}}\left(\dot{\chi}_x^2 + \dot{\chi}_x\dot{\chi}_y + \dot{\chi}_y^2 + \frac{\dot{\chi}_{xy}}{4}\right)^{1/2} \tag{10.181}$$

where

$$H = \sqrt{\frac{M_x}{\sigma_x}} = \sqrt{\frac{M_y}{\sigma_y}} = \sqrt{\frac{M_{xy}}{\tau_{xy}}} \tag{10.182}$$

Since we know that

$$\dot{\chi}_x = -h^2 \frac{\partial^2 \dot{w}}{\partial x^2} \tag{10.183a}$$

$$\dot{\chi}_y = -h^2 \frac{\partial^2 \dot{w}}{\partial y^2} \tag{10.183b}$$

$$\dot{\chi}_{xy} = -2h^2 \frac{\partial^2 \dot{w}}{\partial x \partial y} \tag{10.183c}$$

where $\dot{w}$ represents the mechanism, and $h = H/a$, where a is the half-length of the plate side, Equation 10.181 can be rewritten as follows:

$$D = \frac{2M_p h}{a\sqrt{3}} \Lambda_w \tag{10.184a}$$

where

$$\Lambda_w = \left\{ \left(\frac{\partial^2 \dot{w}}{\partial x^2} \right)^2 + \frac{\partial^2 \dot{w}}{\partial x^2} \frac{\partial^2 \dot{w}}{\partial y^2} + \left(\frac{\partial^2 \dot{w}}{\partial y^2} \right)^2 + \left(\frac{\partial^2 \dot{w}}{\partial x \partial y} \right)^2 \right\}^{\frac{1}{2}} \tag{10.184b}$$

The total dissipated energy is thus

$$\dot{L}_i = \frac{2M_p ah}{\sqrt{3}} \int_{-1}^{1} \int_{-\beta}^{\beta} \Lambda_w \, dx \, dy \tag{10.185}$$

where $\beta = b/a$.

On the other hand, the external work is

$$\dot{L}_e = M_p ah \int_{-1}^{1} \int_{-\beta}^{\beta} p \dot{w} \, dx \, dy \tag{10.186}$$

assuming $p = qa^2/M_p$ and q as the uniformly distributed load on the plate.

The application of the principle of virtual work to Equations 10.185 and 10.186 leads to the general expression of the collapse load for the plate, corresponding to a kinematically admissible mechanism:

$$p^+ = \frac{2}{\sqrt{3}} \frac{\int_{-1}^{1} \int_{-\beta}^{\beta} \Lambda_w \, dx \, dy}{\int_{-1}^{1} \int_{-\beta}^{\beta} \dot{w} \, dx \, dy} \tag{10.187}$$

The integrals in Equation 10.187 are not easy to solve. The problem of calculating the limit load through the application of the kinematic theorem is simplified by referring to a collapse mechanism with the plastic deformation concentrated along yield lines, that is, cylindrical plastic hinges (Figure 10.49).

The integrals in Equation 10.187 can be rewritten as follows:

$$\int_{-1}^{1} \int_{-\beta}^{\beta} \Lambda_w \, dx \, dy = \sum_i \dot{\vartheta}_i \, l_i \tag{10.188a}$$

$$\int_{-1}^{1} \int_{-\beta}^{\beta} \dot{w} \, dx \, dy = \dot{V} \tag{10.188b}$$

where:
l_i　indicates the length of the plastic hinges
$\dot{\vartheta}_i$　is the incremental rotation of plate elements around the plastic hinges
$\dot{V}$　represents the volume generated by the collapse mechanism

Referring to the kinematically admissible mechanism of Figure 10.50, the total dissipated energy is

$$\dot{L}_i = \frac{2M_p a}{\sqrt{3}} h \left(2 \, \dot{\vartheta}_{2-4} \, \mu + 4 \, \dot{\vartheta}_{1-2} \, l_b \right) \tag{10.189}$$

where

$$l_b = \sqrt{(1-\mu)^2 + \beta^2} \tag{10.190a}$$

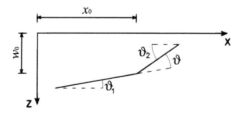

Figure 10.49

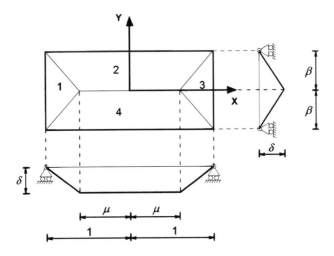

Figure 10.50

$$\dot{\vartheta}_{1-2} = \dot{\delta}\frac{\sqrt{(1-\mu)^2 + \beta^2}}{(1-\mu)\beta} \tag{10.190b}$$

$$\dot{\vartheta}_{2-4} = 2\frac{\dot{\delta}}{\beta} \tag{10.190c}$$

The external work is a function of the volume generated by the collapse mechanism:

$$\dot{L}_e = M_P ab \frac{2}{3}\beta(2+\mu)\dot{\delta} \tag{10.191}$$

The application of the principle of virtual work to Equations 10.189 and 10.191 leads to

$$p^+ = \frac{4\sqrt{3}}{\beta^2}\frac{1-\mu+\beta^2}{2-\mu-\mu^2} \tag{10.192}$$

The collapse load of the rectangular plate simply supported on its four edges, corresponding to a kinematically admissible mechanism, is thus proportional to the minimum of Equation 10.192 in the interval $0 \leq \mu \leq 1$.

As a second example, let us consider a deflected rectangular plate of elastic–perfectly plastic material, simply supported on three sides, and free at the fourth.

Referring to the yield lines scheme of Figure 10.51, and considering a unitary incremental displacement, the internal work of the plate is equal to

$$L_i = 2M_P\left(\frac{xa}{b} + \frac{b}{xa}\right) \tag{10.193}$$

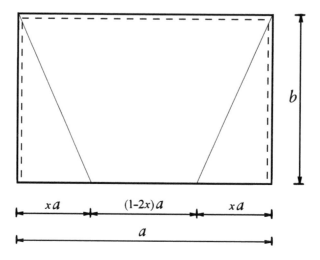

Figure 10.51

where x represents the projection of the cylindrical hinge on the plate edge of length a. The external work, assuming q as the uniformly distributed load on the plate, is

$$L_e = qab\frac{(3-2x)}{6}$$ (10.194)

By the application of the principle of virtual work to Equations 10.193 and 10.194 we have

$$q^+ = 12M_P\frac{\dfrac{xa}{b}+\dfrac{b}{xa}}{ab(3-2x)}$$ (10.195)

The minimum of Equation 10.195 with respect to the variable x leads to

$$x = \frac{b^2}{a^2}\left(\sqrt{\frac{a^2}{b^2}+\frac{4}{9}}-\frac{2}{3}\right)$$ (10.196)

Thus, the collapse load per unit width of the plate, simply supported on three sides, and free at the fourth, corresponding to the kinematically admissible mechanism in Figure 10.51, is

$$q^+ = \frac{24M_P}{b^2\left(\sqrt{9\dfrac{a^2}{b^2}+4}-2\right)}$$ (10.197)

Chapter 11

Plane stress and plane strain conditions

11.1 INTRODUCTION

This chapter will deal with two-dimensional elastic problems in the plane stress condition or the plane strain condition. The three-dimensional stress condition at a single point is said to be plane if the stress vector belongs to the same plane, independently of the cross section chosen. Likewise, the strain condition is said to be plane if the displacement vector belongs to the same plane, regardless of the direction chosen. A necessary and sufficient condition for a state of stress or strain to be defined as plane at a point is that one of the three principal values (of stress on the one hand and of strain on the other) should be equal to zero.

While it is possible for stress conditions to exist that are plane at a point, but not globally—such as in the case of the Saint-Venant solid, where the individual stress planes for the different points of the solid are not necessarily parallel—in what follows, we shall deal only with cases that are globally plane, that is, ones having planes of stresses or strains that are all parallel.

Having first introduced a mathematical method that is generally suited to the situation of plane problems and is based on the **Airy stress function**, we shall consider a number of notable cases, such as the deep beam, the thick-walled cylinder, the circular hole in a plate in tension, and the concentrated force acting on an elastic half-plane. The subsequent introduction of an additional method, based on the theory of complex functions of a complex variable (Muskhelishvili's method) will make possible the treatment of the elliptical hole in a plate in tension. This latter topic will, in turn, serve as an introduction to the problems of stress concentration and fracture mechanics, which will form the subject matter of the closing chapter. In the appendices, the cases of a circular disk subjected to inertial forces (Appendix VI) and to thermal stress (Appendix VII) will be illustrated.

11.2 PLANE STRESS CONDITION

The plane stress condition tends to occur in thin plates, loaded by forces contained in their own middle plane (Figure 11.1). There are originally five unknowns in plane stress problems: the three components of stress σ_x, σ_y, τ_{xy}, and the two components of displacement u and v. There are, likewise, five resolving equations, which are the two indefinite equations of equilibrium:

$$\frac{\partial \sigma_x}{\partial x} + \frac{\partial \tau_{xy}}{\partial y} + \mathcal{F}_x = 0 \tag{11.1a}$$

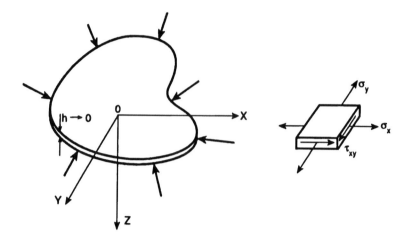

Figure 11.1

$$\frac{\partial \tau_{xy}}{\partial x} + \frac{\partial \sigma_y}{\partial y} + \mathcal{F}_y = 0 \tag{11.1b}$$

and the three elastic constitutive equations:

$$\varepsilon_x = \frac{1}{E}\left(\sigma_x - \nu\sigma_y\right) \tag{11.2a}$$

$$\varepsilon_y = \frac{1}{E}\left(\sigma_y - \nu\sigma_x\right) \tag{11.2b}$$

$$\gamma_{xy} = \frac{2(1+\nu)}{E}\tau_{xy} \tag{11.2c}$$

The equation of compatibility:

$$\frac{\partial^2 \varepsilon_x}{\partial y^2} + \frac{\partial^2 \varepsilon_y}{\partial x^2} = \frac{\partial^2 \gamma_{xy}}{\partial x \partial y} \tag{11.3}$$

on the basis of Equations 11.2, may be expressed in stress terms:

$$\frac{\partial^2}{\partial y^2}\left(\sigma_x - \nu\sigma_y\right) + \frac{\partial^2}{\partial x^2}\left(\sigma_y - \nu\sigma_x\right) = 2(1+\nu)\frac{\partial^2 \tau_{xy}}{\partial x \partial y} \tag{11.4}$$

On the other hand, differentiating the indefinite equilibrium Equations 11.1 in order with respect to x and y, and adding together the resulting equations, we obtain

$$-\frac{\partial^2 \sigma_x}{\partial x^2} - \frac{\partial^2 \sigma_y}{\partial y^2} - \frac{\partial \mathcal{F}_x}{\partial x} - \frac{\partial \mathcal{F}_y}{\partial y} = 2\frac{\partial^2 \tau_{xy}}{\partial x \partial y} \tag{11.5}$$

If we multiply both sides of Equation 11.5 by $(1+\nu)$, the right-hand side becomes the same as that of Equation 11.4. By the transitive law, we have

$$\frac{\partial^2}{\partial y^2}(\sigma_x - \nu\sigma_y) + \frac{\partial^2}{\partial x^2}(\sigma_y - \nu\sigma_x)$$
$$= -(1+\nu)\left[\frac{\partial^2 \sigma_x}{\partial x^2} + \frac{\partial^2 \sigma_y}{\partial y^2} + \frac{\partial \mathcal{F}_x}{\partial x} + \frac{\partial \mathcal{F}_y}{\partial y}\right] \tag{11.6}$$

Finally, collecting terms, we obtain

$$\nabla^2(\sigma_x + \sigma_y) = -(1+\nu)\left[\frac{\partial \mathcal{F}_x}{\partial x} + \frac{\partial \mathcal{F}_y}{\partial y}\right] \tag{11.7}$$

If the body forces are zero, from Equations 11.1 and 11.7 we obtain the following system of three differential equations in the three unknown functions σ_x, σ_y, τ_{xy}:

$$\frac{\partial \sigma_x}{\partial x} + \frac{\partial \tau_{xy}}{\partial y} = 0 \tag{11.8a}$$

$$\frac{\partial \tau_{xy}}{\partial x} + \frac{\partial \sigma_y}{\partial y} = 0 \tag{11.8b}$$

$$\nabla^2(\sigma_x + \sigma_y) = 0 \tag{11.8c}$$

The elastic characteristics E, ν of the material do not appear in the resolving Equations 11.8. It follows, therefore, that the plane stress field does not depend in any way on the material, but only on the boundary conditions. Of course, it is not possible to say the same of the strain field expressed in Equation 11.2 and, hence, of the elastic displacements induced by it.

Let us assume that the components of the stress field are obtainable by derivation of an unknown function Φ, called the **Airy stress function**:

$$\sigma_x = \frac{\partial^2 \Phi}{\partial y^2} \tag{11.9a}$$

$$\sigma_y = \frac{\partial^2 \Phi}{\partial x^2} \tag{11.9b}$$

$$\tau_{xy} = -\frac{\partial^2 \Phi}{\partial x \partial y} \tag{11.9c}$$

In this case, the indefinite equilibrium Equations 11.8a and b are identically satisfied, while compatibility Equation 11.8c is satisfied if and only if

$$\nabla^2 \nabla^2 \Phi = 0 \tag{11.10}$$

or

$$\nabla^4 \Phi = 0 \tag{11.11}$$

A function Φ that satisfies Equation 11.11 is said to be **biharmonic**.

11.3 PLANE STRAIN CONDITION

The plane strain condition tends to occur in cylindrical or prismatic solids of large thickness, loaded by orthogonal forces on the generatrices, the forces having constant distribution along these (Figure 11.2). If the end planes are considered as being constrained axially, it is legitimate to assume a zero dilation in the axial direction, $\varepsilon_z = 0$, and a strain condition that is repeated section by section.

The condition of annihilation of the axial dilation:

$$\varepsilon_z = \frac{1}{E}\left(\sigma_z - \nu\sigma_x - \nu\sigma_y\right) = 0 \tag{11.12}$$

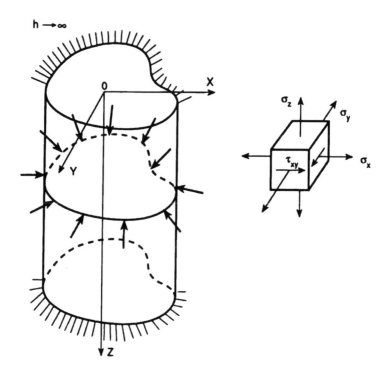

Figure 11.2

yields the axial stress as a function of the other two normal stresses:

$$\sigma_z = \nu\left(\sigma_x + \sigma_y\right) \tag{11.13}$$

Substituting Equation 11.13 in the elastic constitutive equations, we find

$$\varepsilon_x = \frac{1}{E}\left[\left(1-\nu^2\right)\sigma_x - \nu\left(1+\nu\right)\sigma_y\right] \tag{11.14a}$$

$$\varepsilon_y = \frac{1}{E}\left[\left(1-\nu^2\right)\sigma_y - \nu\left(1+\nu\right)\sigma_x\right] \tag{11.14b}$$

$$\gamma_{xy} = \frac{2\left(1+\nu\right)}{E}\tau_{xy} \tag{11.14c}$$

If we consider the so-called **stiffened elastic characteristics**:

$$E' = \frac{E}{1-\nu^2} \tag{11.15a}$$

$$\nu' = \frac{\nu}{1-\nu} \tag{11.15b}$$

Equations 11.14 may assume the following form:

$$\varepsilon_x = \frac{1}{E'}\left(\sigma_x - \nu'\sigma_y\right) \tag{11.16a}$$

$$\varepsilon_y = \frac{1}{E'}\left(\sigma_y - \nu'\sigma_x\right) \tag{11.16b}$$

$$\gamma_{xy} = \frac{2\left(1+\nu'\right)}{E'}\tau_{xy} \tag{11.16c}$$

which corresponds to that of Equations 11.2.

The equation of compatibility, analogously to Equation 11.7, can therefore be expressed as follows:

$$\nabla^2\left(\sigma_x + \sigma_y\right) = -\left(1+\nu'\right)\left[\frac{\partial \mathcal{F}_x}{\partial x} + \frac{\partial \mathcal{F}_y}{\partial y}\right] \tag{11.17}$$

and, hence, on the basis of Equation 11.15b:

$$\nabla^2\left(\sigma_x + \sigma_y\right) = -\frac{1}{1-\nu}\left[\frac{\partial \mathcal{F}_x}{\partial x} + \frac{\partial \mathcal{F}_y}{\partial y}\right] \tag{11.18}$$

If the body forces are zero, we again obtain the three Equations 11.8 that resolve the plane stress problem, plus a fourth equation:

$$\frac{\partial \sigma_z}{\partial z} = 0 \tag{11.19}$$

which implies the constancy of axial stress along the thickness.

11.4 DEEP BEAM

Let us consider a deep beam of rectangular cross section and unit base, in which the ratio of length l to depth h is not so high as to enable application of the elementary beam theory of Saint-Venant (Figure 11.3). Let this beam be supported at the ends and loaded by a constant distribution of vertical forces q. The boundary conditions on the upper and lower edges are

$$\tau_{xy}\left(y = \pm \frac{h}{2}\right) = 0 \tag{11.20a}$$

$$\sigma_y\left(y = \frac{h}{2}\right) = 0 \tag{11.20b}$$

$$\sigma_y\left(y = -\frac{h}{2}\right) = -q \tag{11.20c}$$

The conditions at the ends $x = \pm l/2$ are

$$\int_{-h/2}^{h/2} \tau_{xy}dy = \mp q\frac{l}{2} \tag{11.21a}$$

$$\int_{-h/2}^{h/2} \sigma_x dy = 0 \tag{11.21b}$$

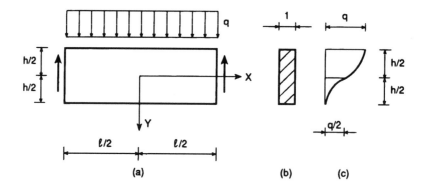

(a)   (b)   (c)

Figure 11.3

$$\int_{-b/2}^{b/2} \sigma_x y \, dy = 0 \tag{11.21c}$$

Equations 11.21b and c impose the disappearance of the axial force and the bending moment on the end sections.

Let us assume, for the stress components, polynomial expressions with unknown coefficients:

$$\sigma_x = a\left(x^2 y - \frac{2}{3} y^3 \right) \tag{11.22a}$$

$$\sigma_y = \frac{1}{3} a y^3 + by + c \tag{11.22b}$$

$$\tau_{xy} = -axy^2 - bx \tag{11.22c}$$

that satisfy Equations 11.8a and b.

From the conditions expressed by Equations 11.20, we obtain the system:

$$-a\frac{b^2}{4} - b = 0 \tag{11.23a}$$

$$\frac{1}{3} a \frac{b^3}{8} + b \frac{b}{2} + c = 0 \tag{11.23b}$$

$$-\frac{1}{3} a \frac{b^3}{8} - b \frac{b}{2} + c = -q \tag{11.23c}$$

which, on resolution, yields

$$a = -6\frac{q}{b^3} \tag{11.24a}$$

$$b = \frac{3}{2} \frac{q}{b} \tag{11.24b}$$

$$c = -\frac{q}{2} \tag{11.24c}$$

Noting that the moment of inertia of the rectangular cross section of unit base is $I = b^3/12$, we obtain $a = -q/2I$, and therefore Equations 11.22 offers the following expressions:

$$\sigma_x = -\frac{q}{2I}\left(x^2 y - \frac{2}{3} y^3 \right) \tag{11.25a}$$

$$\sigma_y = -\frac{q}{2I}\left(\frac{1}{3}y^3 - \frac{h^2}{4}y + \frac{h^3}{12}\right)$$ (11.25b)

$$\tau_{xy} = -\frac{q}{2I}\left(\frac{h^2}{4} - y^2\right)x$$ (11.25c)

These three components of stress also satisfy the conditions set by Equations 11.21a and b. For the moments on the end cross sections also to become zero, it is necessary to superpose on the solution represented by Equations 11.25 a stress field of pure bending, $\sigma_x = yd$, $\sigma_y = \tau_{xy} = 0$, so as to determine the constant d via Equation 11.21c for $x = \pm l/2$:

$$\int_{-h/2}^{h/2} \sigma_x y\, dy = \int_{-h/2}^{h/2}\left[-6\frac{q}{h^3}\left(\frac{l^2}{4}y - \frac{2}{3}y^3\right) + yd\right] y\, dy = 0$$ (11.26)

from which we find

$$d = \frac{3}{2}\frac{q}{h}\left(\frac{l^2}{h^2} - \frac{2}{5}\right)$$ (11.27)

whereby, finally, we have

$$\sigma_x = \frac{q}{2I}\left(\frac{l^2}{4} - x^2\right)y + \frac{q}{2I}\left(\frac{2}{3}y^3 - \frac{1}{10}h^2 y\right)$$ (11.28)

The first term on the right-hand side of Equation 11.28 represents the stress given by the usual elementary theory of bending, whereas the second term represents its correction. This term does not depend on the abscissa x and is negligible only in the cases where the span of the beam is large compared with its depth. Note that Equation 11.28 represents an exact solution only if the axial stresses at the ends are distributed according to the following law:

$$\sigma_x\left(x = \pm\frac{l}{2}\right) = 6\frac{q}{h^3}\left(\frac{2}{3}y^3 - \frac{1}{10}h^2 y\right)$$ (11.29)

These stresses present both the resultant force and the resultant moment as zero. Consequently, by virtue of Saint-Venant's principle, it is possible to deduce that their effect, at distances from the ends greater than the depth h of the beam, diminishes gradually until it vanishes altogether.

From Equation 11.25b, we detect the existence of compressive stresses σ_y that are absent in the elementary theory. The distribution of these compressive stresses over the depth of the beam is shown in Figure 11.3c.

The distribution of the shearing stresses τ_{xy} given by Equation 11.25c coincides instead with that furnished by the usual elementary theory.

Using Equations 11.2, 11.25b and c, and 11.28, we obtain the components of strain:

$$\varepsilon_x = \frac{q}{2EI}\left\{\left[\left(\frac{l^2}{4}-x^2\right)y+\left(\frac{2}{3}y^3-\frac{1}{10}h^2y\right)\right]+v\left[\frac{1}{3}y^3-\frac{h^2}{4}y+\frac{h^3}{12}\right]\right\} \quad (11.30a)$$

$$\varepsilon_y = \frac{q}{2EI}\left\{-\left[\frac{1}{3}y^3-\frac{h^2}{4}y+\frac{h^3}{12}\right]-v\left[\left(\frac{l^2}{4}-x^2\right)y+\left(\frac{2}{3}y^3-\frac{1}{10}h^2y\right)\right]\right\} \quad (11.30b)$$

$$\gamma_{xy} = -\frac{q}{EI}(1+v)\left(\frac{h^2}{4}-y^2\right)x \quad (11.30c)$$

Integrating the first two of these equations, we have

$$u = \frac{q}{2EI}\left\{\left[\left(\frac{l^2}{4}x-\frac{x^3}{3}\right)y+\left(\frac{2}{3}y^3-\frac{1}{10}h^2y\right)x\right]+vx\left[\frac{1}{3}y^3-\frac{h^2}{4}y+\frac{h^3}{12}\right]\right\}+f(y) \quad (11.31a)$$

$$v = \frac{q}{2EI}\left\{-\left[\frac{y^4}{12}-\frac{h^2}{8}y^2+\frac{h^3}{12}y\right]-v\left[\left(\frac{l^2}{4}-x^2\right)\frac{y^2}{2}+\left(\frac{1}{6}y^4-\frac{h^2}{20}y^2\right)\right]\right\}+g(x) \quad (11.31b)$$

where f and g represent unknown functions of the coordinates y and x, respectively. Equation 11.30c then becomes

$$\frac{\partial u}{\partial y}+\frac{\partial v}{\partial x} = \frac{q}{2EI}\left\{\left[\left(\frac{l^2}{4}x-\frac{x^3}{3}\right)+\left(2y^2-\frac{h^2}{10}\right)x\right]\right.$$

$$+vx\left[y^2-\frac{h^2}{4}\right]\right\}+\frac{\partial f}{\partial y}$$

$$+\frac{q}{2EI}\left\{-v\left[(-2x)\frac{y^2}{2}\right]\right\}+\frac{\partial g}{\partial x}$$

$$=-\frac{q}{EI}(1+v)\left(\frac{h^2}{4}-y^2\right)x \quad (11.32)$$

Collecting terms, we obtain

$$\frac{\partial g}{\partial x}+\frac{q}{2EI}\left\{x\left[\frac{l^2}{4}+h^2\left(\frac{2}{5}+\frac{v}{4}\right)\right]-\frac{x^3}{3}\right\}=-\frac{\partial f}{\partial y} \quad (11.33)$$

Since the left-hand side is a function only of the variable x, while the right-hand side is a function only of the variable y, both must represent a constant C_1:

$$\frac{\partial g}{\partial x} = \frac{q}{2EI} \left\{ \frac{x^3}{3} - x \left[\frac{l^2}{4} + h^2 \left(\frac{2}{5} + \frac{v}{4} \right) \right] \right\} + C_1 \tag{11.34}$$

Integrating Equation 11.34, we find the function g:

$$g(x) = \frac{q}{2EI} \left\{ \frac{x^4}{12} - \frac{x^2}{2} \left[\frac{l^2}{4} + h^2 \left(\frac{2}{5} + \frac{v}{4} \right) \right] \right\} + C_1 x + C_2 \tag{11.35}$$

The constants C_1 and C_2 can be inferred from the vertical displacement in the center and from the conditions at the ends:

$$v(0,0) = \delta \tag{11.36a}$$

$$v\left(\pm \frac{l}{2}, 0 \right) = 0 \tag{11.36b}$$

Applying Equations 11.36 to Equations 11.31b and 11.35, we have

$$C_1 = 0 \tag{11.37a}$$

$$C_2 = \delta \tag{11.37b}$$

and, hence, the geometrical axis of the beam bends according to the following curve:

$$v(x,0) = \frac{q}{2EI} \left\{ \frac{x^4}{12} - \frac{x^2}{2} \left[\frac{l^2}{4} + h^2 \left(\frac{2}{5} + \frac{v}{4} \right) \right] \right\} + \delta \tag{11.38}$$

Since Equation 11.36b must hold, the vertical displacement in the center is

$$\delta = \frac{5}{384} \frac{q l^4}{EI} \left[1 + \frac{48}{25} \frac{h^2}{l^2} \left(1 + \frac{5}{8} v \right) \right] \tag{11.39}$$

It may be noted that the first term that contributes to Equation 11.39 coincides with the vertical displacement deriving from the elementary theory. The second term, which, on the other hand, represents its correction, diminishes with the increase in the ratio l/h. For low values of this ratio, it may be seen how the cross sections of the beam do not remain plane, much less orthogonal to the centroidal axis. The effect of shear, according to Saint-Venant's theory, can, on the other hand, be estimated by

$$\delta^T = \int_0^{l/2} \gamma_y dx = \int_0^{l/2} \frac{6}{5} \frac{T_y}{Gh} dx \tag{11.40}$$

Substituting the linear function of shear into Equation 11.40, we obtain

$$\delta^T = \frac{6}{5Gh} \int_0^{l/2} q\left(\frac{l}{2} - x\right) dx = \frac{1+\nu}{40}\left(\frac{h}{l}\right)^2 \frac{ql^4}{EI} \tag{11.41}$$

and hence

$$\delta = \delta^M + \delta^T = \frac{5}{384}\frac{ql^4}{EI}\left[1 + \frac{48}{25}\frac{h^2}{l^2}(1+\nu)\right] \tag{11.42}$$

It is interesting to note that, for $\nu = 0$, Equations 11.39 and 11.42 coincide, while, for $\nu > 0$, Equation 11.42 overestimates the exact deflection.

11.5 THICK-WALLED CYLINDER

The indefinite equations of equilibrium in cylindrical coordinates were presented for the solid of revolution in Section 3.11.

In the case of a plane stress condition, the components σ_z, τ_{rz}, $\tau_{\theta z}$ disappear, and the equations of equilibrium with regard to translation in the radial and circumferential directions (Figure 11.4) are obtained directly from Equation 3.134b:

$$\frac{\partial \sigma_r}{\partial r} + \frac{1}{r}\frac{\partial \tau_{r\vartheta}}{\partial \vartheta} + \frac{\sigma_r - \sigma_\vartheta}{r} + \mathcal{F}_r = 0 \tag{11.43a}$$

$$\frac{1}{r}\frac{\partial \sigma_\vartheta}{\partial \vartheta} + \frac{\partial \tau_{r\vartheta}}{\partial r} + \frac{2}{r}\tau_{r\vartheta} + \mathcal{F}_\vartheta = 0 \tag{11.43b}$$

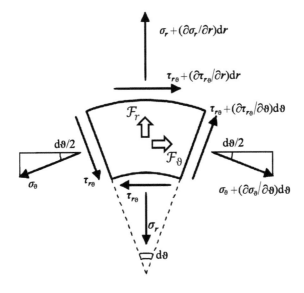

Figure 11.4

In the case where the body forces vanish, Equations 11.43 are identically satisfied by

$$\sigma_r = \frac{1}{r}\frac{\partial \Phi}{\partial r} + \frac{1}{r^2}\frac{\partial^2 \Phi}{\partial \vartheta^2} \tag{11.44a}$$

$$\sigma_\vartheta = \frac{\partial^2 \Phi}{\partial r^2} \tag{11.44b}$$

$$\tau_{r\vartheta} = \frac{1}{r^2}\frac{\partial \Phi}{\partial \vartheta} - \frac{1}{r}\frac{\partial^2 \Phi}{\partial r \partial \vartheta} = -\frac{\partial}{\partial r}\left(\frac{1}{r}\frac{\partial \Phi}{\partial \vartheta}\right) \tag{11.44c}$$

where Φ denotes the Airy stress function.

To obtain the biharmonic Equation 11.11 in polar coordinates, consider the coordinate transformation:

$$r^2 = x^2 + y^2 \tag{11.45a}$$

$$\vartheta = \arctan\frac{y}{x} \tag{11.45b}$$

from which there follows

$$\frac{\partial r}{\partial x} = \frac{x}{r} = \cos\vartheta \tag{11.46a}$$

$$\frac{\partial r}{\partial y} = \frac{y}{r} = \sin\vartheta \tag{11.46b}$$

$$\frac{\partial \vartheta}{\partial x} = -\frac{y}{r^2} = -\frac{\sin\vartheta}{r} \tag{11.46c}$$

$$\frac{\partial \vartheta}{\partial y} = \frac{x}{r^2} = \frac{\cos\vartheta}{r} \tag{11.46d}$$

The partial derivative of Φ with respect to x may be expressed, as is known, in the following way:

$$\frac{\partial \Phi}{\partial x} = \frac{\partial \Phi}{\partial r}\frac{\partial r}{\partial x} + \frac{\partial \Phi}{\partial \vartheta}\frac{\partial \vartheta}{\partial x}$$

$$= \cos\vartheta\frac{\partial \Phi}{\partial r} - \frac{\sin\vartheta}{r}\frac{\partial \Phi}{\partial \vartheta} \tag{11.47}$$

The second partial derivative is thus equal to

$$\frac{\partial^2 \Phi}{\partial x^2} = \left(\cos\vartheta \frac{\partial}{\partial r} - \frac{\sin\vartheta}{r}\frac{\partial}{\partial\vartheta} \right)\left(\cos\vartheta \frac{\partial\Phi}{\partial r} - \frac{\sin\vartheta}{r}\frac{\partial\Phi}{\partial\vartheta} \right)$$

$$= \cos^2\vartheta\frac{\partial^2\Phi}{\partial r^2} - \cos\vartheta\sin\vartheta\frac{\partial}{\partial r}\left(\frac{1}{r}\frac{\partial\Phi}{\partial\vartheta} \right)$$

$$-\frac{\sin\vartheta}{r}\frac{\partial}{\partial\vartheta}\left(\cos\vartheta\frac{\partial\Phi}{\partial r} \right) + \frac{\sin\vartheta}{r^2}\frac{\partial}{\partial\vartheta}\left(\sin\vartheta\frac{\partial\Phi}{\partial\vartheta} \right) \qquad (11.48)$$

Differentiating, we obtain

$$\frac{\partial^2\Phi}{\partial x^2} = \cos^2\vartheta\frac{\partial^2\Phi}{\partial r^2} + \sin^2\vartheta\left(\frac{1}{r}\frac{\partial\Phi}{\partial r} + \frac{1}{r^2}\frac{\partial^2\Phi}{\partial\vartheta^2} \right)$$

$$-2\sin\vartheta\cos\vartheta\frac{\partial}{\partial r}\left(\frac{1}{r}\frac{\partial\Phi}{\partial\vartheta} \right) \qquad (11.49a)$$

In like manner, we have

$$\frac{\partial^2\Phi}{\partial y^2} = \sin^2\vartheta\frac{\partial^2\Phi}{\partial r^2} + \cos^2\vartheta\left(\frac{1}{r}\frac{\partial\Phi}{\partial r} + \frac{1}{r^2}\frac{\partial^2\Phi}{\partial\vartheta^2} \right)$$

$$+2\sin\vartheta\cos\vartheta\frac{\partial}{\partial r}\left(\frac{1}{r}\frac{\partial\Phi}{\partial\vartheta} \right) \qquad (11.49b)$$

$$-\frac{\partial^2\Phi}{\partial x\partial y} = \sin\vartheta\cos\vartheta\left(\frac{1}{r}\frac{\partial\Phi}{\partial r} + \frac{1}{r^2}\frac{\partial^2\Phi}{\partial\vartheta^2} - \frac{\partial^2\Phi}{\partial r^2} \right)$$

$$-\left(\cos^2\vartheta - \sin^2\vartheta \right)\frac{\partial}{\partial r}\left(\frac{1}{r}\frac{\partial\Phi}{\partial\vartheta} \right) \qquad (11.49c)$$

Taking into account Equations 11.49a and b and Equations 11.44a and b, Equation 11.8c may be cast in the following form:

$$\left(\frac{\partial^2}{\partial x^2} + \frac{\partial^2}{\partial y^2} \right)\left(\sigma_x + \sigma_y \right)$$

$$= \left(\frac{\partial^2}{\partial r^2} + \frac{1}{r}\frac{\partial}{\partial r} + \frac{1}{r^2}\frac{\partial^2}{\partial\vartheta^2} \right)\left(\sigma_r + \sigma_\vartheta \right)$$

$$= \left(\frac{\partial^2}{\partial r^2} + \frac{1}{r}\frac{\partial}{\partial r} + \frac{1}{r^2}\frac{\partial^2}{\partial\vartheta^2} \right)^2\Phi = 0 \qquad (11.50)$$

the sum of the normal stresses, $\sigma_x + \sigma_y = \sigma_r + \sigma_\vartheta$, being constant and equal to the first stress invariant.

When there is polar symmetry, the Airy stress function depends only on the radial coordinate r, and the compatibility Equation 11.50 becomes

$$\left(\frac{d^2}{dr^2} + \frac{1}{r}\frac{d}{dr}\right)\left(\frac{d^2\Phi}{dr^2} + \frac{1}{r}\frac{d\Phi}{dr}\right)$$

$$= \frac{d^4\Phi}{dr^4} + \frac{2}{r}\frac{d^3\Phi}{dr^3} + \frac{1}{r^2}\frac{d^2\Phi}{dr^2} + \frac{1}{r^3}\frac{d\Phi}{dr} = 0 \qquad (11.51)$$

It is possible to verify that the complete integral of Equation 11.51 is

$$\Phi(r) = A\log r + Br^2\log r + Cr^2 + D \qquad (11.52)$$

The stress components are thus obtained from Equations 11.44:

$$\sigma_r = \frac{1}{r}\frac{d\Phi}{dr} = \frac{A}{r^2} + B(1 + 2\log r) + 2C \qquad (11.53a)$$

$$\sigma_\vartheta = \frac{d^2\Phi}{dr^2} = -\frac{A}{r^2} + B(3 + 2\log r) + 2C \qquad (11.53b)$$

$$\tau_{r\vartheta} = 0 \qquad (11.53c)$$

In the case where there is no hole at the origin, the only possible solution is that of uniform stress: $\sigma_r = \sigma_\vartheta = 2C$.

In the case of a thick-walled cylinder, subjected to uniform pressure both internally and externally (Figure 11.5), to guarantee single-valued tangential displacements, we have $B = 0$ with the boundary conditions:

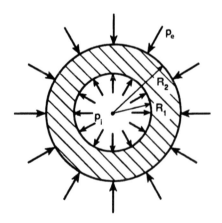

Figure 11.5

$$\sigma_r\left(r = R_1\right) = -p_i \tag{11.54a}$$

$$\sigma_r\left(r = R_2\right) = -p_e \tag{11.54b}$$

Imposing the conditions of Equations 11.54 on Equation 11.53a, we obtain two equations in the two unknowns A and C:

$$\frac{A}{R_1^2} + 2C = -p_i \tag{11.55a}$$

$$\frac{A}{R_2^2} + 2C = -p_e \tag{11.55b}$$

from which we find

$$A = \frac{R_1^2 R_2^2 \left(p_e - p_i\right)}{R_2^2 - R_1^2} \tag{11.56a}$$

$$2C = \frac{p_i R_1^2 - p_e R_2^2}{R_2^2 - R_1^2} \tag{11.56b}$$

Substituting these two constants into Equations 11.53, we have

$$\sigma_r = \frac{R_1^2 R_2^2 \left(p_e - p_i\right)}{R_2^2 - R_1^2} \frac{1}{r^2} + \frac{p_i R_1^2 - p_e R_2^2}{R_2^2 - R_1^2} \tag{11.57a}$$

$$\sigma_\vartheta = -\frac{R_1^2 R_2^2 \left(p_e - p_i\right)}{R_2^2 - R_1^2} \frac{1}{r^2} + \frac{p_i R_1^2 - p_e R_2^2}{R_2^2 - R_1^2} \tag{11.57b}$$

It is interesting to note that the sum $(\sigma_r + \sigma_\vartheta)$ is constant throughout the thickness of the cylinder. Consequently, the stresses σ_r and σ_ϑ produce a uniform dilation or contraction in the axial direction, so that the cross sections remain plane.

When the external pressure is zero, $p_e = 0$, and the cylinder is subjected to the internal pressure alone, Equations 11.57 become

$$\sigma_r = \frac{p_i R_1^2}{R_2^2 - R_1^2}\left(1 - \frac{R_2^2}{r^2}\right) \tag{11.58a}$$

$$\sigma_\vartheta = \frac{p_i R_1^2}{R_2^2 - R_1^2}\left(1 + \frac{R_2^2}{r^2}\right) \tag{11.58b}$$

These equations show that σ_r is always compressive and σ_ϑ is always tensile (Figure 11.6).

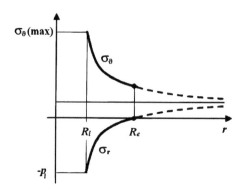

Figure 11.6

The latter is at its maximum on the internal surface of the cylinder where

$$\sigma_\vartheta (\max) = \frac{p_i \left(R_1^2 + R_2^2 \right)}{R_2^2 - R_1^2} \tag{11.59}$$

This tensile stress is greater than the pressure p_i for any value of the ratio R_2/R_1. In extreme cases, we have

$$\lim_{R_2/R_1 \to \infty} \sigma_\vartheta (\max) = p_i \tag{11.60a}$$

$$\lim_{R_2/R_1 \to 1} \sigma_\vartheta (\max) = \frac{p_i \left(2R^2 \right)}{t \left(2R \right)} = \frac{p_i R}{t} \tag{11.60b}$$

where R designates the mean radius and t the thickness, in the case of a thin cylinder. Equation 11.60b has already been obtained in Section 3.7, via another route.

11.6 CIRCULAR HOLE IN A PLATE SUBJECTED TO TENSION

Let us consider an infinite plate, subjected to a uniaxial tensile stress σ in the X direction (Figure 11.7). In the case where a circular hole of radius R is present in the plate, the distribution of the stresses is perturbed in the neighborhood of the hole itself. In the following, we intend to calculate the effect of **stress concentration**, that is, the amplification of the stresses on the edge of the hole.

Consider the portion of the plate that remains within the circumference of radius R', where $R' \gg R$. The stresses acting on this circumference are approximately the same as would result in the absence of the hole, and can therefore be deduced from Mohr's circle:

$$\sigma_r \left(r = R' \right) = \frac{1}{2} \sigma \left(1 + \cos 2\vartheta \right) \tag{11.61a}$$

$$\tau_{r\vartheta}(r = R') = -\frac{1}{2}\sigma \sin 2\vartheta \tag{11.61b}$$

The radial stress is made up of two parts: the first is constant and produces a stress field inside the ring, given by Equations 11.57:

$$\sigma_r = -\frac{\sigma}{2}\frac{R^2 R'^2}{R'^2 - R^2}\frac{1}{r^2} + \frac{\sigma}{2}\frac{R'^2}{R'^2 - R^2} \tag{11.62a}$$

$$\sigma_\vartheta = +\frac{\sigma}{2}\frac{R^2 R'^2}{R'^2 - R^2}\frac{1}{r^2} + \frac{\sigma}{2}\frac{R'^2}{R'^2 - R^2} \tag{11.62b}$$

In the limit whereby $R' \to \infty$, Equations 11.62 become

$$\sigma_r = \frac{\sigma}{2}\left(1 - \frac{R^2}{r^2}\right) \tag{11.63a}$$

$$\sigma_\vartheta = \frac{\sigma}{2}\left(1 + \frac{R^2}{r^2}\right) \tag{11.63b}$$

The second part of σ_r, 1/2 $\sigma\cos 2\vartheta$, together with the shearing stress, $-1/2\ \sigma\sin 2\vartheta$, produces a stress field that can be derived from an Airy stress function of the form:

$$\Phi = f(r)\cos 2\vartheta \tag{11.64}$$

Substituting this function into the compatibility Equation 11.50 we obtain the following total derivative differential equation:

$$\left(\frac{d^2}{dr^2} + \frac{1}{r}\frac{d}{dr} - \frac{4}{r^2}\right)^2 f = 0 \tag{11.65}$$

the complete integral of which is

$$f(r) = Ar^2 + Br^4 + \frac{C}{r^2} + D \tag{11.66}$$

The components of stress are thus found from Equations 11.44, 11.64, and 11.66:

$$\sigma_r = -\left(2A + \frac{6C}{r^4} + \frac{4D}{r^2}\right)\cos 2\vartheta \tag{11.67a}$$

$$\sigma_\vartheta = \left(2A + 12Br^2 + \frac{6C}{r^4}\right)\cos 2\vartheta \tag{11.67b}$$

$$\tau_{r\vartheta} = \left(2A + 6Br^2 - \frac{6C}{r^4} - \frac{2D}{r^2}\right)\sin 2\vartheta \qquad (11.67c)$$

The four constants of integration can be determined from the boundary conditions on the external circumference, that is, the two Equations 11.61, and from the boundary conditions on the internal circumference, respectively:

$$\sigma_r(R') = \frac{\sigma}{2}\cos 2\vartheta \qquad (11.68a)$$

$$\tau_{r\vartheta}(R') = -\frac{\sigma}{2}\sin 2\vartheta \qquad (11.68b)$$

$$\sigma_r(R) = 0 \qquad (11.68c)$$

$$\tau_{r\vartheta}(R) = 0 \qquad (11.68d)$$

From Equations 11.67 and 11.68, we obtain the system of equations:

$$2A + \frac{6C}{R'^4} + \frac{4D}{R'^2} = -\frac{\sigma}{2} \qquad (11.69a)$$

$$\qquad (11.69b)$$

$$2A + 6BR'^2 - \frac{6C}{R'^4} - \frac{2D}{R'^2} = -\frac{\sigma}{2}$$

$$2A + \frac{6C}{R^4} + \frac{4D}{R^2} = 0 \qquad (11.69c)$$

$$2A + 6BR^2 - \frac{6C}{R^4} - \frac{2D}{R^2} = 0 \qquad (11.69d)$$

For $R' \to \infty$, we have

$$A = -\frac{\sigma}{4}, \quad B = 0, \quad C = -\frac{\sigma}{4}R^4, \quad D = \frac{\sigma}{2}R^2 \qquad (11.70)$$

Substituting these values into Equations 11.67 and adding the contribution given by Equations 11.63, produced by the uniform stress σ/2, we finally arrive at

$$\sigma_r = \frac{\sigma}{2}\left(1 - \frac{R^2}{r^2}\right) + \frac{\sigma}{2}\left(1 + 3\frac{R^4}{r^4} - 4\frac{R^2}{r^2}\right)\cos 2\vartheta \qquad (11.71a)$$

$$\sigma_\vartheta = \frac{\sigma}{2}\left(1+\frac{R^2}{r^2}\right) - \frac{\sigma}{2}\left(1+3\frac{R^4}{r^4}\right)\cos 2\vartheta \qquad (11.71b)$$

$$\tau_{r\vartheta} = -\frac{\sigma}{2}\left(1-3\frac{R^4}{r^4}+2\frac{R^2}{r^2}\right)\sin 2\vartheta \qquad (11.71c)$$

This solution was obtained by Kirsch in 1898.

It may be verified how, for $r \to \infty$, the stress field, expressed by Equations 11.71, reproduces the conditions at infinity (Equation 11.61), while on the edge of the hole, $r = R$, we have

$$\sigma_\vartheta = \sigma(1 - 2\cos 2\vartheta) \qquad (11.72)$$

with $\sigma_r = \tau_{r\vartheta} = 0$.

The circumferential stress σ_ϑ is at its maximum for $\vartheta = \pi/2$ and $\vartheta = 3/2\pi$, that is, at the extremities of the diameter orthogonal to the direction of tension (Figure 11.7):

$$\sigma_\vartheta(\text{max}) = 3\sigma \qquad (11.73)$$

The so-called **stress concentration factor**, in the case of a circular hole, is therefore equal to 3, regardless of the radius of the hole.

On the other hand, the circumferential stress σ_ϑ is at its minimum for $\vartheta = 0$ and $\vartheta = \pi$, that is, at the extremities of the diameter collinear to the direction of tension (Figure 11.7):

$$\sigma_\vartheta(\text{min}) = -\sigma \qquad (11.74)$$

At these points, a compression is thus expected, as in all the other points for which $-1/6\pi < \vartheta < 1/6\pi$, and $5/6\pi < \vartheta < 7/6\pi$.

On the section of the plate perpendicular to the axis X and passing through the origin of the axes, we have

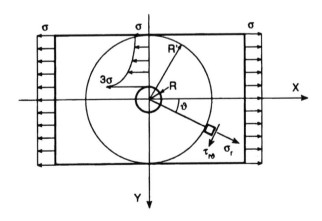

Figure 11.7

$$\sigma_\vartheta = \frac{\sigma}{2}\left(2 + \frac{R^2}{r^2} + 3\frac{R^4}{r^4}\right) \tag{11.75a}$$

$$\tau_{r\vartheta} = 0 \tag{11.75b}$$

From an examination of Equation 11.75a, the local character of stress concentration around the hole is evident. As r increases, the stress σ_ϑ tends rapidly to the value σ, as is shown by the diagram of Figure 11.7. At a distance from the edge of the hole equal to its diameter $2R$, the stress σ_ϑ is higher than the asymptotic value σ by 22%, whereas at a distance equal to twice the diameter it is higher only by 4%. Equations 11.71 are applicable also to a plate of finite width, provided that this width is not less than four times the diameter of the hole. In these cases, the error does not exceed 6%.

In cases of biaxial tensile and/or compressive loading, the stress field is obtained, by superposition, from Equations 11.71. When, for instance, there are two orthogonal tensile stresses of intensity σ, that is, a **uniform tensile condition** σ, the circumferential stress on the edge of the hole is also uniform and equal to 2σ. When, instead, there is a tensile stress σ in the X direction and a compressive stress $-\sigma$ in the Y direction, that is, in the case of **pure shear**, $\tau = \sigma$, the stress is at its maximum and equal to 4σ (tensile) for $\vartheta = \pi/2$ and $\vartheta = 3/2\pi$, whereas it is at its minimum and equal to -4σ (compressive) for $\vartheta = 0$ and $\vartheta = \pi$.

11.7 CONCENTRATED FORCE ACTING ON THE EDGE OF AN ELASTIC HALF-PLANE

Consider a vertical force acting on the horizontal boundary of an elastic half-plane (Figure 11.8a). Let the distribution of the force along the thickness of the plate be uniform (Figure 11.8b), and let P denote the load per unit thickness.

The solution of this elastic problem may be found from the Airy stress function:

$$\Phi = -\frac{P}{\pi} r\vartheta \sin\vartheta \tag{11.76}$$

Using Equations 11.44, we obtain the stresses

$$\sigma_r = -\frac{2P}{\pi}\frac{\cos\vartheta}{r} \tag{11.77a}$$

$$\sigma_\vartheta = \tau_{r\vartheta} = 0 \tag{11.77b}$$

which produce a field of radial compression. The boundary conditions are satisfied by Equation 11.77b, while the compression Equation 11.77a presents a singularity at the point of application of the force. The resultant of the radial forces that act on a semicylindrical surface of radius r (Figure 11.8b) is in equilibrium with the external force P. Summing up the vertical components $\sigma_r r d\vartheta \cos\vartheta$ of the elementary forces that act on each elementary portion $rd\vartheta$ of the surface, we have

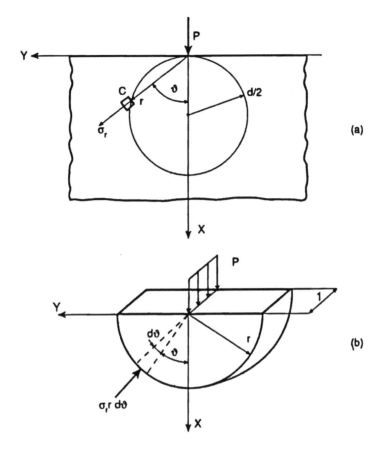

Figure 11.8

$$2\int_0^{\pi/2} \sigma_r r \cos\vartheta\, d\vartheta = -\frac{4P}{\pi}\int_0^{\pi/2} \cos^2\vartheta\, d\vartheta = -P \tag{11.78}$$

If we consider a circumference of diameter d, with its center on the X axis and tangent to the Y axis at the point of application of the force (Figure 11.8a), for each point C of this circumference we have $r = d\cos\vartheta$, and hence Equation 11.77a becomes

$$\sigma_r = -\frac{2P}{\pi d} \tag{11.79}$$

This means that the radial stress is the same at each point of the circumference, with the exception of the point of application of the force.

Rendering explicit the strain components (Equation 3.134a) corresponding to the cylindrical geometry, we have

$$\varepsilon_r = \frac{\partial u}{\partial r} = -\frac{2P}{\pi E}\frac{\cos\vartheta}{r} \tag{11.80a}$$

$$\varepsilon_\vartheta = \frac{u}{r} + \frac{1}{r}\frac{\partial v}{\partial \vartheta} = v\frac{2P}{\pi E}\frac{\cos\vartheta}{r} \tag{11.80b}$$

$$\gamma_{r\vartheta} = \frac{1}{r}\frac{\partial u}{\partial \vartheta} + \frac{\partial v}{\partial r} - \frac{v}{r} = 0 \tag{11.80c}$$

Integrating the first equation, we have

$$u = -\frac{2P}{\pi E}\cos\vartheta \log r + f(\vartheta) \tag{11.81}$$

where f is a function only of ϑ. Substituting into Equation 11.80b and integrating, we find

$$v = \frac{2vP}{\pi E}\sin\vartheta + \frac{2P}{\pi E}\log r \sin\vartheta - \int f(\vartheta)d\vartheta + g(r) \tag{11.82}$$

where g is a function only of r. Substituting Equations 11.81 and 11.82 into Equation 11.80c, we have, finally

$$f(\vartheta) = -\frac{(1-v)P}{\pi E}\vartheta\sin\vartheta + A\sin\vartheta + B\cos\vartheta \tag{11.83a}$$

$$g(r) = Cr \tag{11.83b}$$

where A, B, C are constants of integration.

From Equations 11.81 and 11.82, we obtain the field of displacements:

$$u = -\frac{2P}{\pi E}\cos\vartheta \log r - \frac{(1-v)P}{\pi E}\vartheta\sin\vartheta + A\sin\vartheta + B\cos\vartheta \tag{11.84a}$$

$$v = \frac{2vP}{\pi E}\sin\vartheta + \frac{2P}{\pi E}\log r \sin\vartheta - \frac{(1-v)P}{\pi E}\vartheta\cos\vartheta$$

$$+ \frac{(1-v)P}{\pi E}\sin\vartheta + A\cos\vartheta - B\sin\vartheta + Cr \tag{11.84b}$$

From symmetry, we have $v = 0$ for $\vartheta = 0$, and hence we must find $A = C = 0$. With these values of the integration constants, the vertical displacements of the points of the X axis are

$$u(\vartheta = 0) = -\frac{2P}{\pi E}\log r + B \tag{11.85}$$

To determine the constant B, it may be assumed that a point of the X axis at a distance d from the origin does not undergo a vertical displacement. From Equation 11.85, we obtain

$$B = \frac{2P}{\pi E} \log d \tag{11.86}$$

and hence

$$u(\vartheta = 0) = \frac{2P}{\pi E} \log \frac{d}{r} \tag{11.87}$$

As may be noted, the solution thus determined predicts an infinite displacement of the point of application of the force, and likewise a nonintegrable strain energy in the area around the same point:

$$\int_{-\pi/2}^{\pi/2} \int_{\varepsilon}^{R} \frac{\sigma_r^2}{2E} r\,dr\,d\vartheta = \frac{2P^2}{\pi^2 E} \int_{-\pi/2}^{\pi/2} \cos^2 \vartheta\,d\vartheta \int_{\varepsilon}^{R} \frac{dr}{r}$$

$$= \frac{P^2}{\pi E} \log \frac{R}{\varepsilon} \tag{11.88}$$

The limit for $\varepsilon \to 0$ of Equation 11.88 is infinite, but the same expression can be considered finite for $\varepsilon \neq 0$. This means that, if ideally we take away a portion of material around the point of application of the force, the incongruences pointed out can, in a sense, be removed. This portion of material is the one that actually undergoes plastic deformation and flows under the action of the concentrated force.

11.8 ANALYTICAL FUNCTIONS

A function Z is said to be **complex** when it is made up of two parts, one real, U, and the other imaginary, V:

$$Z = U + iV \tag{11.89}$$

Z is also referred to as the **dependent variable**, and, in general, each of its two parts, U and V, is a function of the **independent variable**:

$$z = x + iy \tag{11.90}$$

where x is the real part and y is the imaginary part of that complex variable. It may be stated that $Z(z) = U(z) + iV(z)$ is a **complex function of a complex variable**.

The complex function of a complex variable $Z(z)$ is said to be **analytical** at a point in the complex plane when its derivative is unique and does not depend on the direction of the increment.

Conditions are shown to exist on the components U and V that imply that the function Z is analytical. As is usually done in the case of real functions, let the derivative of the function Z be defined as the limit of the difference quotient:

$$Z' = \lim_{\Delta z \to 0} \frac{\Delta Z}{\Delta z} = \frac{dZ}{dz} = \frac{\dfrac{\partial U}{\partial x}dx + \dfrac{\partial U}{\partial y}dy + i\left(\dfrac{\partial V}{\partial x}dx + \dfrac{\partial V}{\partial y}dy\right)}{dx + idy} \tag{11.91}$$

If we multiply both numerator and denominator by $(dx-idy)$:

$$Z' = \frac{\left(\dfrac{\partial U}{\partial x}dx + \dfrac{\partial U}{\partial y}dy + i\dfrac{\partial V}{\partial x}dx + i\dfrac{\partial V}{\partial y}dy\right)(dx - idy)}{dx^2 + dy^2} \tag{11.92}$$

and designate as m the slope dy/dx, we obtain

$$Z' = \frac{\dfrac{\partial U}{\partial x} + m\left(\dfrac{\partial U}{\partial y} + \dfrac{\partial V}{\partial x}\right) + m^2\dfrac{\partial V}{\partial y} + i\left[\dfrac{\partial V}{\partial x} + m\left(\dfrac{\partial V}{\partial y} - \dfrac{\partial U}{\partial x}\right) - m^2\dfrac{\partial U}{\partial y}\right]}{1 + m^2} \tag{11.93}$$

The function Z is analytical and, consequently, its derivative is unique in z if and only if the following differential relations, referred to as the **Cauchy–Riemann conditions**, hold:

$$\frac{\partial U}{\partial x} = \frac{\partial V}{\partial y} \tag{11.94a}$$

$$\frac{\partial U}{\partial y} = -\frac{\partial V}{\partial x} \tag{11.94b}$$

If these conditions do hold, we have

$$Z' = \frac{\partial U}{\partial x} + i\frac{\partial V}{\partial x} \tag{11.95}$$

If a function Z is analytical, its derivative is also analytical and *vice versa*. To demonstrate this, it is sufficient to differentiate Equations 11.94 with respect to x:

$$\frac{\partial^2 U}{\partial x^2} = \frac{\partial^2 V}{\partial y \partial x} \tag{11.96a}$$

$$\frac{\partial^2 U}{\partial x \partial y} = -\frac{\partial^2 V}{\partial x^2} \tag{11.96b}$$

Taking into account Equation 11.95, Equations 11.96 can be transformed as follows:

$$\frac{\partial}{\partial x}\operatorname{Re} Z' = \frac{\partial}{\partial y}\operatorname{Im} Z' \tag{11.97a}$$

$$\frac{\partial}{\partial y} \operatorname{Re} Z' = -\frac{\partial}{\partial x} \operatorname{Im} Z'$$ (11.97b)

which represent the Cauchy–Riemann conditions for the derivative function.

If a function Z is analytical, its real and imaginary parts are harmonic functions. To demonstrate this, it is sufficient to derive Equations 11.94a and b with respect to x and y:

$$\frac{\partial^2 U}{\partial x^2} = \frac{\partial^2 V}{\partial x \partial y}$$ (11.98a)

$$\frac{\partial^2 U}{\partial y^2} = -\frac{\partial^2 V}{\partial x \partial y}$$ (11.98b)

and to apply the transitive law:

$$\frac{\partial^2 U}{\partial x^2} + \frac{\partial^2 U}{\partial y^2} = 0$$ (11.99)

Differentiating Equation 11.94a with respect to y, and Equation 11.94b with respect to x, we obtain, on the other hand,

$$\frac{\partial^2 U}{\partial x \partial y} = \frac{\partial^2 V}{\partial y^2}$$ (11.100a)

$$\frac{\partial^2 U}{\partial x \partial y} = -\frac{\partial^2 V}{\partial x^2}$$ (11.100b)

and hence

$$\frac{\partial^2 V}{\partial x^2} + \frac{\partial^2 V}{\partial y^2} = 0$$ (11.101)

Consider, for example, the function:

$$Z(z) = z = x + iy$$

Since

$$\frac{\partial U}{\partial x} = \frac{\partial V}{\partial y} = 1$$

and

$$\frac{\partial U}{\partial y} = -\frac{\partial V}{\partial x} = 0$$

the Cauchy–Riemann conditions are satisfied at each point in the complex plane, and hence the function z is always analytical.

On the other hand, the function:

$$Z(z) = \bar{z} = x - iy$$

which associates each point of the complex plane with its conjugate, is not analytical at any point in the complex plane. We have, in fact,

$$\frac{\partial U}{\partial x} = 1, \ \frac{\partial V}{\partial y} = -1$$

that is,

$$\frac{\partial U}{\partial x} \neq \frac{\partial V}{\partial y}$$

The function:

$$Z(z) = z^2 = (x+iy)(x+iy) = (x^2 - y^2) + 2ixy$$

is analytical because:

$$\frac{\partial U}{\partial x} = \frac{\partial V}{\partial y} = 2x$$

$$\frac{\partial U}{\partial y} = -\frac{\partial V}{\partial x} = -2y$$

More generally, all the polynomial functions are analytical.

The function:

$$Z(z) = \frac{1}{z} = \frac{x}{x^2 + y^2} - i\frac{y}{x^2 + y^2}$$

since it is verified that

$$\frac{\partial U}{\partial x} = \frac{\partial V}{\partial y} = \frac{y^2 - x^2}{\left(x^2 + y^2\right)^2}$$

$$\frac{\partial U}{\partial y} = -\frac{\partial V}{\partial x} = -\frac{2xy}{\left(x^2 + y^2\right)^2}$$

is analytical at all points in the complex plane, excluding the origin, where $1/z$ is not definite.
Ultimately, as regards the **square modulus** function:

$$Z(z) = |z|^2 = x^2 + y^2$$

because of

$$\frac{\partial U}{\partial x} = 2x, \ \frac{\partial V}{\partial y} = 0$$

$$\frac{\partial U}{\partial y} = 2y, \ \frac{\partial V}{\partial x} = 0$$

it is analytical only at the origin, while at the other points in the complex plane, it does not
satisfy the Cauchy–Riemann conditions.

11.9 KOLOSOFF–MUSKHELISHVILI METHOD

Using the properties of the analytical functions, presented in the foregoing section, Kolosoff
(1909) and subsequently Muskhelishvili (1933) developed a method for resolving plane elastic problems. According to this method, the Airy stress function becomes expressible as
follows:

$$\Phi = \operatorname{Re}(\bar{z}\psi + \chi) \tag{11.102}$$

where ψ and χ are two analytical functions, called **complex potentials**. It may be demonstrated that the biharmonic Equation 11.11 is satisfied by Equation 11.102.
Equation 11.102 may be transformed as follows:

$$\Phi = \operatorname{Re}\left(z\bar{z}\frac{\psi}{z} + \chi\right) \tag{11.103}$$

whence we have

$$\Phi = (x^2 + y^2)\operatorname{Re}\frac{\psi}{z} + \operatorname{Re}\chi \tag{11.104}$$

Since $1/z$, ψ, and χ are analytical functions, and the product of two analytical functions is
itself an analytical function, as can very easily be demonstrated, $\operatorname{Re}(\psi/z)$ and $\operatorname{Re}\chi$ are two
harmonic functions. The procedure, therefore, is to demonstrate that the product $(x^2+y^2)f$,
with f as a harmonic function, represents a biharmonic function.
Applying the Laplace operator, we obtain

$$\nabla^2\left[(x^2+y^2)f\right] = \frac{\partial}{\partial x}\left[\frac{\partial f}{\partial x}(x^2+y^2)+2xf\right] + \frac{\partial}{\partial y}\left[\frac{\partial f}{\partial y}(x^2+y^2)+2yf\right]$$

$$= (x^2+y^2)\nabla^2 f + 4x\frac{\partial f}{\partial x} + 4y\frac{\partial f}{\partial y} + 4f \tag{11.105}$$

and hence, if f is harmonic:

$$\nabla^2\left[\left(x^2+y^2\right)f\right]=4\left(x\frac{\partial f}{\partial x}+y\frac{\partial f}{\partial y}+f\right) \tag{11.106}$$

To demonstrate that the function is biharmonic, a second application of the Laplacian is necessary:

$$\nabla^2\nabla^2\left[\left(x^2+y^2\right)f\right]=4\left[\nabla^2\left(x\frac{\partial f}{\partial x}\right)+\nabla^2\left(y\frac{\partial f}{\partial y}\right)\right] \tag{11.107}$$

Performing the calculations, we have

$$\nabla^2\left(x\frac{\partial f}{\partial x}\right)=\frac{\partial}{\partial x}\left(\frac{\partial f}{\partial x}+x\frac{\partial^2 f}{\partial x^2}\right)+\frac{\partial}{\partial y}\left(x\frac{\partial^2 f}{\partial x\partial y}\right)$$

$$=2\frac{\partial^2 f}{\partial x^2}+x\frac{\partial}{\partial x}\left(\frac{\partial^2 f}{\partial x^2}+\frac{\partial^2 f}{\partial y^2}\right) \tag{11.108}$$

Since f is a harmonic function, it follows that

$$\nabla^2\left(x\frac{\partial f}{\partial x}\right)=2\frac{\partial^2 f}{\partial x^2} \tag{11.109a}$$

and likewise

$$\nabla^2\left(y\frac{\partial f}{\partial y}\right)=2\frac{\partial^2 f}{\partial y^2} \tag{11.109b}$$

Finally, substituting Equations 11.109 into Equation 11.107, we obtain

$$\nabla^4\left[\left(x^2+y^2\right)f\right]=8\nabla^2 f=0 \tag{11.110}$$

Since the sum of a complex function and its conjugate is equal to twice its real part:

$$f(z)+\bar{f}(z)=2\,\mathrm{Re}f(z) \tag{11.111}$$

from Equation 11.102 we find

$$2\Phi=\bar{z}\psi(z)+\chi(z)+z\bar{\psi}(z)+\bar{\chi}(z) \tag{11.112}$$

Applying to Equation 11.112 the following rules of differentiation of composite functions:

$$\frac{\partial f}{\partial x} = \frac{df}{dz}\frac{\partial z}{\partial x} = \frac{df}{dz} = f' \tag{11.113a}$$

$$\frac{\partial f}{\partial y} = \frac{df}{dz}\frac{\partial z}{\partial y} = i\frac{df}{dz} = if' \tag{11.113b}$$

$$\frac{\partial \bar{f}}{\partial x} = \frac{d\bar{f}}{d\bar{z}}\frac{\partial \bar{z}}{\partial x} = \frac{d\bar{f}}{d\bar{z}} = \bar{f}' \tag{11.113c}$$

$$\frac{\partial \bar{f}}{\partial y} = \frac{d\bar{f}}{d\bar{z}}\frac{\partial \bar{z}}{\partial y} = -i\frac{d\bar{f}}{d\bar{z}} = -i\bar{f}' \tag{11.113d}$$

we obtain

$$2\frac{\partial \Phi}{\partial x} = \psi(z) + \bar{z}\psi'(z) + \chi'(z) + \bar{\psi}(z) + z\bar{\psi}'(z) + \bar{\chi}'(z) \tag{11.114a}$$

$$2\frac{\partial \Phi}{\partial y} = -i\psi(z) + i\bar{z}\psi'(z) + i\chi'(z) + i\bar{\psi}(z) - iz\bar{\psi}'(z) - i\bar{\chi}'(z) \tag{11.114b}$$

Multiplying Equation 11.114b by the imaginary unit, we have

$$2i\frac{\partial \Phi}{\partial y} = \psi(z) + \bar{z}\psi'(z) - \chi'(z) - \bar{\psi}(z) + z\bar{\psi}'(z) + \bar{\chi}'(z) \tag{11.115}$$

The sum of Equations 11.114a and 11.115 yields

$$\frac{\partial \Phi}{\partial x} + i\frac{\partial \Phi}{\partial y} = \psi(z) + z\bar{\psi}'(z) + \bar{\chi}'(z) \tag{11.116}$$

The partial derivation with respect to x and to y of Equation 11.116 leads to the following equations:

$$\frac{\partial^2 \Phi}{\partial x^2} + i\frac{\partial^2 \Phi}{\partial x\partial y} = \psi'(z) + \bar{\psi}'(z) + z\bar{\psi}''(z) + \bar{\chi}''(z) \tag{11.117a}$$

$$\frac{\partial^2 \Phi}{\partial x\partial y} + i\frac{\partial^2 \Phi}{\partial y^2} = i\psi'(z) + i\bar{\psi}'(z) - iz\bar{\psi}''(z) - i\bar{\chi}''(z) \tag{11.117b}$$

Multiplying Equation 11.117b by the imaginary unit, we have

$$-\frac{\partial^2 \Phi}{\partial y^2} + i\frac{\partial^2 \Phi}{\partial x\partial y} = -\psi'(z) - \bar{\psi}'(z) + z\bar{\psi}''(z) + \bar{\chi}''(z) \tag{11.118}$$

The difference and the sum of Equations 11.117a and 11.118 give, respectively

$$\frac{\partial^2 \Phi}{\partial x^2} + \frac{\partial^2 \Phi}{\partial y^2} = 2\psi'(z) + 2\overline{\psi}'(z) = 4\operatorname{Re}\psi'(z) \tag{11.119a}$$

$$\frac{\partial^2 \Phi}{\partial x^2} - \frac{\partial^2 \Phi}{\partial y^2} + 2i\frac{\partial^2 \Phi}{\partial x \partial y} = 2\left[z\overline{\psi}''(z) + \overline{\chi}''(z)\right] \tag{11.119b}$$

Since the second partial derivatives of the Airy stress function represent the stress components, in agreement with Equation 11.9, Equations 11.119 can be written as follows:

$$\sigma_x + \sigma_y = 4\operatorname{Re}\psi'(z) \tag{11.120a}$$

$$\sigma_y - \sigma_x - 2i\tau_{xy} = 2\left[z\overline{\psi}''(z) + \overline{\chi}''(z)\right] \tag{11.120b}$$

or

$$\sigma_x + \sigma_y = 4\operatorname{Re}\psi'(z) \tag{11.121a}$$

$$\sigma_y - \sigma_x + 2i\tau_{xy} = 2\left[\overline{z}\psi''(z) + \chi''(z)\right] \tag{11.121b}$$

Equations 11.121 may be cast in an alternative form, which allows the individual stress components to be separated. Subtracting and adding them, we obtain, respectively:

$$2(\sigma_x - i\tau_{xy}) = 4\operatorname{Re}\psi'(z) - 2\left[\overline{z}\psi''(z) + \chi''(z)\right] \tag{11.122a}$$

$$2(\sigma_y + i\tau_{xy}) = 4\operatorname{Re}\psi'(z) + 2\left[\overline{z}\psi''(z) + \chi''(z)\right] \tag{11.122b}$$

while from Equation 11.121b, we find

$$\tau_{xy} = \operatorname{Im}\left[\overline{z}\psi''(z) + \chi''(z)\right] \tag{11.122c}$$

Isolating the real parts of Equations 11.122a and b, we obtain

$$\sigma_x = \operatorname{Re}\left[2\psi'(z)\right] - \operatorname{Re}\left[\chi''(z)\right] - x\operatorname{Re}\left[\psi''(z)\right] - y\operatorname{Im}\left[\psi''(z)\right] \tag{11.123a}$$

$$\sigma_y = \operatorname{Re}\left[2\psi'(z)\right] + \operatorname{Re}\left[\chi''(z)\right] + x\operatorname{Re}\left[\psi''(z)\right] + y\operatorname{Im}\left[\psi''(z)\right] \tag{11.123b}$$

while from Equation 11.122c, we have

$$\tau_{xy} = \operatorname{Im}\left[\chi''(z)\right] - y\operatorname{Re}\left[\psi''(z)\right] + x\operatorname{Im}\left[\psi''(z)\right] \tag{11.123c}$$

In the case of problems that are symmetrical with respect to the X axis, it must follow that

$$\tau_{xy}(x,0) = 0 \qquad\qquad\qquad\qquad (11.124)$$

and hence, from Equation 11.122c:

$$\text{Im}\left[\bar{z}\psi''(z) + \chi''(z)\right] = 0, \qquad \text{for } y = 0 \qquad\qquad (11.125)$$

On the real axis, since $z = \bar{z}$, the following relation also holds good:

$$\text{Im}\left[z\psi''(z) + \chi''(z)\right] = 0, \qquad \text{for } y = 0 \qquad\qquad (11.126)$$

If we extrapolate the foregoing condition to the entire complex plane, we can write

$$z\psi''(z) + \chi''(z) + B = 0, \qquad \forall z \in C \qquad\qquad (11.127)$$

where B is a real constant and C is the set of all the points of the complex plane. Note that B cannot be a real function of the complex variable z, since if it were, it would not obey the Cauchy–Riemann conditions of Equations 11.94 at each point of the complex plane.

If Equation 11.127 is identically zero over the entire complex plane, so must its real and imaginary parts likewise be zero:

$$x\,\text{Re}\left[\psi''(z)\right] - y\,\text{Im}\left[\psi''(z)\right] + \text{Re}\left[\chi''(z)\right] + B = 0 \qquad\qquad (11.128a)$$

$$x\,\text{Im}\left[\psi''(z)\right] + y\,\text{Re}\left[\psi''(z)\right] + \text{Im}\left[\chi''(z)\right] = 0 \qquad\qquad (11.128b)$$

Substituting Equations 11.128 into Equations 11.123, we obtain, finally:

$$\sigma_x = \text{Re}\left[2\psi'(z)\right] - y\,\text{Im}\left[2\psi''(z)\right] + B \qquad\qquad (11.129a)$$

$$\sigma_y = \text{Re}\left[2\psi'(z)\right] + y\,\text{Im}\left[2\psi''(z)\right] - B \qquad\qquad (11.129b)$$

$$\tau_{xy} = -y\,\text{Re}\left[2\psi''(z)\right] \qquad\qquad\qquad (11.129c)$$

The extrapolation of the condition of Equation 11.127 to the entire complex plane thus allows one of the two Muskhelishvili complex potentials to be eliminated. As shall be seen in the next chapter, the application of Equations 11.129 leads to a relatively simple solution of the problem of a plate in tension with a rectilinear crack. This solution was obtained by Westergaard in 1939 on the basis of Muskhelishvili's treatment, which had previously been published in Russian in 1933 and was subsequently republished in English in 1953. However, in his original publication, Westergaard used a different notation, $Z_I(z)$ instead of $2\psi'(z)$, and disregarded the real constant B. The latter erroneous assumption was pointed out by Sih in 1966.

11.10 ELLIPTICAL HOLE IN A PLATE SUBJECTED TO TENSION

Let us consider an infinite plate in a condition of uniaxial tension σ in a direction that forms an angle β with the X axis (Figure 11.9). Let this uniaxial condition be perturbed by an elliptical hole having its major axis along the X axis and its minor axis along the Y axis.

Let X^*Y^* be the Cartesian axes obtained from rotation of the XY axes by the angle β, such as to bring the X axis parallel to the tension σ (Figure 11.9).

We can obtain the equations of transformation:

$$\sigma_x^* = \sigma_x \cos^2\beta + \sigma_y \sin^2\beta + 2\tau_{xy}\sin\beta\,\cos\beta \tag{11.130a}$$

$$\sigma_y^* = \sigma_x \sin^2\beta + \sigma_y \cos^2\beta - 2\tau_{xy}\sin\beta\,\cos\beta \tag{11.130b}$$

$$\tau_{xy}^* = \left(\sigma_y - \sigma_x\right)\sin\beta\,\cos\beta + \tau_{xy}\left(\cos^2\beta - \sin^2\beta\right) \tag{11.130c}$$

which, via the well-known trigonometric formulas, become

$$\sigma_x^* = \frac{1}{2}\left(\sigma_x + \sigma_y\right) + \frac{1}{2}\left(\sigma_x - \sigma_y\right)\cos 2\beta + \tau_{xy}\sin 2\beta \tag{11.131a}$$

$$\sigma_y^* = \frac{1}{2}\left(\sigma_x + \sigma_y\right) - \frac{1}{2}\left(\sigma_x - \sigma_y\right)\cos 2\beta - \tau_{xy}\sin 2\beta \tag{11.131b}$$

$$\tau_{xy}^* = -\frac{1}{2}\left(\sigma_x - \sigma_y\right)\sin 2\beta + \tau_{xy}\cos 2\beta \tag{11.131c}$$

From Equations 11.131, the following relations are easily obtained:

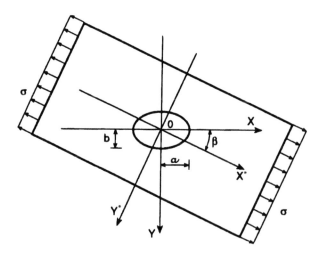

Figure 11.9

$$\sigma_x^* + \sigma_y^* = \sigma_x + \sigma_y \tag{11.132a}$$

$$\sigma_y^* - \sigma_x^* + 2i\tau_{xy}^* = e^{2i\beta}\left(\sigma_y - \sigma_x + 2i\tau_{xy}\right) \tag{11.132b}$$

Since at infinity we have

$$\sigma_x^* = \sigma, \quad \sigma_y^* = \tau_{xy}^* = 0 \tag{11.133}$$

Equations 11.132 yield

$$\sigma_x + \sigma_y = \sigma \tag{11.134a}$$

$$\sigma_y - \sigma_x + 2i\tau_{xy} = -\sigma e^{-2i\beta} \tag{11.134b}$$

and hence, via Equations 11.121, at infinity we have

$$4\operatorname{Re}\psi'(z) = \sigma \tag{11.135a}$$

$$2\left[\bar{z}\psi''(z) + \chi''(z)\right] = -\sigma e^{-2i\beta} \tag{11.135b}$$

The elliptical coordinates ξ, η, shown in Figure 11.10, are defined by the following relations:

$$z = c\cosh\zeta \tag{11.136a}$$

with

$$\zeta = \xi + i\eta \tag{11.136b}$$

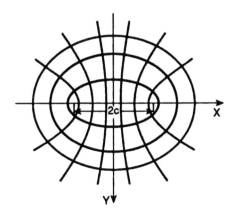

Figure 11.10

From Equations 11.136, we have

$$x = c \cosh \xi \cos \eta \tag{11.137a}$$

$$y = c \sinh \xi \sin \eta \tag{11.137b}$$

The coordinate ξ is constant and equal to ξ_0 on an ellipse of semiaxes $c \cosh \xi_0$ and $c \sinh \xi_0$, just as the coordinate η is constant and equal to η_0 on a hyperbola that has the same focuses $(\pm c, 0)$ as the ellipse. It is sufficient, in fact, to take into account the relations:

$$\cos^2 \eta + \sin^2 \eta = 1 \tag{11.138a}$$

$$\cosh^2 \xi - \sinh^2 \xi = 1 \tag{11.138b}$$

to obtain

$$\frac{x^2}{c^2 \cosh^2 \xi} + \frac{y^2}{c^2 \sinh^2 \xi} = 1 \tag{11.139a}$$

$$\frac{x^2}{c^2 \cos^2 \eta} - \frac{y^2}{c^2 \sin^2 \eta} = 1 \tag{11.139b}$$

Whereas, then, the semiaxes of the ellipse to which the point of **elliptical coordinates** (ξ_0, η_0) belongs are

$$a = c \cosh \xi_0, \quad b = c \sinh \xi_0 \tag{11.140}$$

η_0 represents the angular coordinate of the same point, on the basis of the scheme of Figure 11.11.

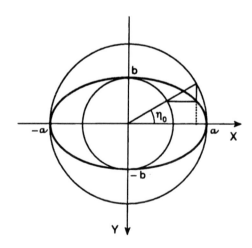

Figure 11.11

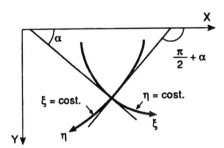

Figure 11.12

Since the ellipses $\xi=\xi_0$ and the hyperbolas $\eta=\eta_0$ are mutually orthogonal, it is possible to write a transformation analogous to Equations 11.132:

$$\sigma_\xi + \sigma_\eta = \sigma_x + \sigma_y \tag{11.141a}$$

$$\sigma_\eta - \sigma_\xi + 2i\tau_{\xi\eta} = e^{2i\alpha}\left(\sigma_y - \sigma_x + 2i\tau_{xy}\right) \tag{11.141b}$$

where:

σ_ξ and σ_η are the normal stresses on the curves $\xi=$ constant and $\eta=$ constant, respectively

$\tau_{\xi\eta}$ is the shearing stress along the same curves

α is the angle that the tangent to the curve $\eta=$ constant forms with the X axis (Figure 11.12)

On the edge of the elliptical hole of equation $\xi=\xi_0$, we have

$$\sigma_\xi = \tau_{\xi\eta} = 0 \tag{11.142}$$

whereby, subtracting Equation 11.141b from Equation 11.141a, we obtain

$$2\operatorname{Re}\psi'(z) - \left[\bar{z}\psi''(z) + \chi''(z)\right]e^{2i\alpha} = 0 \tag{11.143}$$

for $\xi=\xi_0$.

The boundary conditions at infinity, given by Equations 11.135, and on the edge of the hole, given by Equation 11.143, can be satisfied by the following complex potentials:

$$4\psi(z) = Ac\cosh\xi + Bc\sinh\zeta \tag{11.144a}$$

$$4\chi(z) = Cc^2\zeta + Dc^2\cosh 2\zeta + Ec^2\sinh 2\zeta \tag{11.144b}$$

where A, B, C, D, E are constants to be determined.

Substituting the foregoing forms given by Equations 11.144 in Equations 11.135, we obtain

$$\operatorname{Re}A + \operatorname{Re}B = \sigma \tag{11.145a}$$

$$2(D+E) = -\sigma e^{-2i\beta} \tag{11.145b}$$

just as, substituting the same equations into Equation 11.143, we have

$$\operatorname{cosech}\zeta\Big[(2A+B\operatorname{cotanh}\zeta)\sinh\overline{\zeta}$$

$$+\big(\overline{B}+B\operatorname{cosech}^2\zeta\big)\cosh\overline{\zeta}$$

$$+(C+2E)\operatorname{cosech}\zeta\operatorname{cotanh}\zeta$$

$$-4D\sinh\zeta-4E\cosh\zeta\Big]=0 \tag{11.146}$$

once having taken into account that

$$e^{2i\alpha}=\frac{\sinh\zeta}{\sinh\overline{\zeta}} \tag{11.147}$$

On the edge of the hole, we have $\xi=\xi_0$ and $\overline{\zeta}=2\xi_0-\zeta$. If this expression for $\overline{\zeta}$ is substituted into Equation 11.146, and the functions $\sinh(2\xi_0-\zeta)$ and $\cosh(2\xi_0-\zeta)$ are suitably expanded, the same equation becomes

$$(2A\sinh2\xi_0-2i\operatorname{Im}B\cosh2\xi_0-4E)\cosh\zeta$$

$$-(2A\cosh2\xi_0-2i\operatorname{Im}B\sinh2\xi_0+4D)\sinh\zeta$$

$$+(C+2E+B\cosh2\xi_0)\operatorname{cotanh}\zeta\operatorname{cosech}\zeta=0 \tag{11.148}$$

This equation is satisfied if the coefficients of $\cosh\zeta$, $\sinh\zeta$, and $\operatorname{cotanh}\zeta\operatorname{cosech}\zeta$ become zero. Therefore, we need to find three equations that, together with the two Equations 11.145, yield a system of five equations in the five unknowns A, B, C, D, E. The solution is given by the following expressions:

$$A=\sigma e^{2\xi_0}\cos2\beta \tag{11.149a}$$

$$B=\sigma\big(1-e^{2\xi_0+2i\beta}\big) \tag{11.149b}$$

$$C=-\sigma\big(\cosh2\xi_0-\cos2\beta\big) \tag{11.149c}$$

$$D=-\frac{1}{2}\sigma e^{2\xi_0}\cosh2\big(\xi_0+i\beta\big) \tag{11.149d}$$

$$E=\frac{1}{2}\sigma e^{2\xi_0}\sinh2\big(\xi_0+i\beta\big) \tag{11.149e}$$

The complex potentials in Equations 11.144 are given congruently by

$$4\psi(z) = \sigma c \left[e^{2\xi_0} \cos 2\beta \cosh \zeta + \left(1 - e^{2\xi_0 + 2i\beta}\right) \sinh \zeta \right]$$ (11.150a)

$$4\chi(z) = -\sigma c^2 \left[\left(\cosh 2\xi_0 - \cos 2\beta\right) \zeta \right.$$

$$\left. + \frac{1}{2} e^{2\xi_0} \cosh 2 \left(\zeta - \xi_0 - i\beta\right) \right]$$ (11.150b)

The normal stress σ_η along the contour of the hole may be obtained from Equation 11.141a, because there σ_ξ becomes zero:

$$\sigma_\eta = 4 \, \mathrm{Re} \, \psi'(z)$$ (11.151)

whence, via Equation 11.150a, we find

$$\sigma_\eta(\xi = \xi_0) = \sigma \frac{\sinh 2\xi_0 + \cos 2\beta - e^{2\xi_0} \cos 2(\beta - \eta)}{\cosh 2\xi_0 - \cos 2\eta}$$ (11.152)

When the stress σ is orthogonal to the major axis, that is, for $\beta = \pi/2$, Equation 11.152 becomes

$$\sigma_\eta(\xi = \xi_0) = \sigma e^{2\xi_0} \left[\frac{\sinh 2\xi_0 \left(1 + e^{-2\xi_0}\right)}{\cosh 2\xi_0 - \cos 2\eta} - 1 \right]$$ (11.153)

and the maximum value of σ_η is reached at the ends of the major axis $(\cos 2\eta = 1)$:

$$\sigma_\eta(\max) = \sigma \left(1 + 2 \frac{a}{b}\right)$$ (11.154)

This value tends to infinity for $a/b \to \infty$, that is, in the case where the ellipse becomes particularly eccentric; while it is equal to 3σ when $a = b$, that is, in the case of a circular hole. This latter result has already been obtained in Section 11.6. The minimum value of σ_η is instead $-\sigma$ and, as in the case of the circular hole, is reached at the ends of the axis collinear to the external load $(\cos 2\eta = -1)$.

When the stress σ is orthogonal to the minor axis, that is, for $\beta = 0$, the maximum value of σ_η is reached at the ends of the minor axis, and is equal to $\sigma[1 + 2(b/a)]$. This value tends to σ in the case of very eccentric ellipses. At the ends of the major axis, on the other hand, the stress is $-\sigma$ for any value of the ratio a/b.

The case of **uniform tension** σ can be considered as the combination of the two cases previously considered. The maximum stress $2\sigma(a/b)$ is therefore reached at the ends of the major axis, while the minimum stress $2\sigma(b/a)$ is reached at the ends of the minor axis.

The perturbing effect of the elliptical hole on a condition of **pure shear**, $\tau = \sigma$, parallel to the axes XY, is obtainable by superposition of the two cases, that of tension σ with $\beta = \pi/4$, and that of compression $-\sigma$ with $\beta = 3/4 \, \pi$:

$$\sigma_\eta \left(\xi = \xi_0 \right) = -2\sigma \frac{e^{2\xi_0} \sin 2\eta}{\cosh 2\xi_0 - \cos 2\eta} \tag{11.155}$$

This stress vanishes at the ends of both the axes and presents the extreme values:

$$\sigma_\eta \begin{pmatrix} \text{max} \\ \text{min} \end{pmatrix} = \pm \sigma \frac{(a+b)^2}{ab} \tag{11.156}$$

at the points for which $\tan \eta = \mp b/a$.

When the ellipse is very eccentric, the stresses, given by Equation 11.156, are very high, and the points where these are developed are very close to the ends of the major axis. When, instead, $a = b$, we find again the result for the circular hole, with a concentration factor equal to 4.

To summarize the conclusions both of Section 11.6 and the present section, Table 11.1 gives a complete presentation of the **stress concentration factors** for circular and elliptical holes.

Table 11.1 Stress concentration factors

Scheme	Hole shape	Loading condition	Stress concentration factor
	Circular	Uniaxial	3
	Circular	Uniform	2
	Circular	Pure shear	4
	Elliptical	Uniaxial along major axis	$1 + 2b/a$
	Elliptical	Uniaxial along minor axis	$1 + 2a/b$
	Elliptical	Uniform	$2a/b$
	Elliptical	Pure shear	$(a+b)^2/ab$

Chapter 12

Mechanics of fracture

12.1 INTRODUCTION

With the scientific advances of the last few decades in the field of **material mechanics**, it has been realized that the classical concept of **strength**, understood as force per unit surface causing fracture, is in need of revision, especially in cases where particularly large or particularly small structures are involved. That is, the **strength** of the material must be compared against another characteristic, the **toughness** of the material, to define, via the **dimension** of the structure, the **ductility** or the **brittleness** of the structure itself. Two intrinsic characteristics of the material, plus a geometrical characteristic of the structure, are, in fact, the minimum basis for being able to predict the type of structural response. The **fracture energy**, $\mathcal{G}_{IC}$, is one of the parameters capable of measuring the toughness of the material. Through it, we will describe how the structural response to uniaxial tension varies as $\mathcal{G}_{IC}$ and/or the length of a bar longitudinally subjected to tension varies. In that case, a tendency emerges toward ductile behavior in the case of short lengths of the bar and, on the other hand, a tendency toward brittle behavior (**snap-back**) in the case of greater bar lengths. This tendency will also be encountered in the case of two- and three-dimensional solids, in such a way as to associate ductile behavior with relatively small solids, and brittle behavior with relatively large solids. Just as in structures acted on prevalently by a compressive force (Figure 7.11), there is a transition from **plastic collapse** to **buckling instability** as **slenderness** increases, so in structures acted on prevalently by a tensile force, there is a transition from **plastic collapse** to **brittle fracture** as the **size scale** increases.

Two extreme cases of these properties are shown in Figure 12.1. The first (Figure 12.1a) depicts one of the hundreds of **Liberty ships** that, in the years of the Second World War, split into two parts, with extremely brittle fractures and without the slightest evidence forewarning of such an eventuality. What caused profound astonishment in the technicians who first looked into those accidents was, on the one hand, the extremely low stresses present in the hull at the moment of failure and, on the other, the contrast between the extreme brittleness of the failure and the considerable ductility shown in the laboratory by specimens of the same steel.

The second case (Figure 12.1b) depicts a microscopic filament of glass, used for fiber reinforcement of polymer materials. The filament is elastically bent with a large curvature, so that it undergoes a regime of large strains and stresses as much as two orders of magnitude greater than the tensile strength of glass, as measured in the laboratory with specimens of macroscopic dimensions.

The two cases just examined illustrate starkly and unequivocally how both strength and ductility are functions of the size scale, such as to lead to brittleness and low strength in enormous steel structures as well as to ductility and high strength in microscopic structures

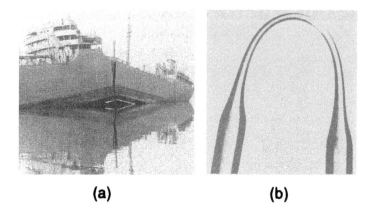

(a)   **(b)**

Figure 12.1

made of glass. On the other hand, it is well known that, at the size scale of the laboratory, steel is a particularly ductile material and glass a particularly brittle one.

It is not necessary, however, to consider extreme cases to realize how ductility is not a characteristic of the material, but rather a characteristic of the entire structure. Even at laboratory scale, the **ductile–brittle transition**, with an increase in the size of the specimen, has been brought to light (Figure 12.2). If the material and the geometrical shape are kept unvaried, increasing the size scale leads to a distinct transition toward a brittle type of behavior accompanied by a sudden drop in the loading capacity and a rapid crack propagation that is, in fact, found for all materials, whether they be metal, polymer, ceramic, or cement. On the other hand, with specimens of relatively modest dimensions, a ductile behavior with slow crack propagation is encountered. In the case, for instance, of three-point bending, it is possible to witness the formation of a plastic hinge in the center and the impossibility of separating the specimen into two distinct parts by applying a simple monotonic loading (Figure 12.2).

In this chapter, after a brief reference to the by now classic **Griffith's energy criterion** (1920), the major physicomathematical theories that, between 1920 and 1950, paved the way to modern-day fracture mechanics, will be presented. These are

1. **Westergaard's method** (1939), or the **method of complex potentials**, which, in fact, is a simplification of Muskhelishvili's method (1933), already referred to in Section 11.9
2. **Williams' method** (1952), or the **series expansion method**, which proves to be a more general approach

Both these fundamental methods lead to the determination of the power of the stress singularity that is produced at the tip of a crack, and thus to the definition of the **stress-intensity factor.**

In addition to the problem of the opening of a symmetrically loaded crack (Mode I), the problem of the same crack loaded skew-symmetrically by in-plane shear forces (Mode II) will also be addressed. In this more general context, the more widely known branching criteria will be proposed.

Subsequently, the concept of fracture energy will be taken up and directly correlated to the critical value of the stress-intensity factor (Irwin 1957). The plastic zone (or process zone) that always develops at the tip of each real crack, will be considered; the amplitude of the zone will be estimated, and an analytical and numerical model will be proposed that is

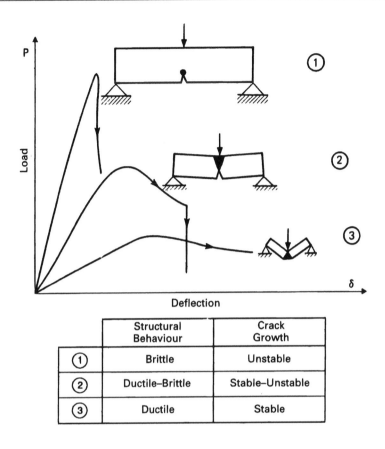

	Structural Behaviour	Crack Growth
①	Brittle	Unstable
②	Ductile–Brittle	Stable–Unstable
③	Ductile	Stable

Figure 12.2

able to provide a continuous description of the ductile–brittle size scale transition referred to previously.

Ultimately, the bridging actions of fibers or reinforcements will be estimated, together with the stability of crack propagation.

12.2 GRIFFITH'S ENERGY CRITERION

Flaws in materials are often to be considered as the major causes of the onset of brittle fractures. The effects of **stress concentration** in the vicinity of imperfections or irregularities have been well known for a long time. In 1898, Kirsch had already provided a solution to the problem of an infinite plate with a circular hole subjected to tension. As was shown in Section 11.6, the maximum stress on the edge of the hole is three times as great as that applied externally (Figure 12.3a). This means that the strength of a plate of dimensions much greater than the hole present in it is reduced to one-third of that of the integer plate, regardless of the size of the hole. Thus, there is a compromise situation between the amount of material removed and the curvature of the hole. At the limit, even an infinitesimal hole causes a concentration factor equal to 3, even though the amount of material removed is practically zero. The radius of curvature of the hole is, in fact, very small in this case and creates conditions of particular severity.

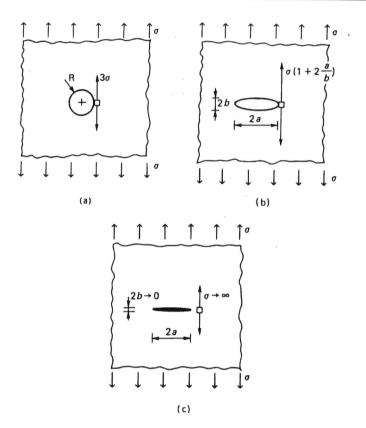

Figure 12.3

Inglis (1913) extended investigations into stress concentration to the more general case of elliptical hole (Figure 12.3b). As has been shown in Section 11.10, the maximum stress on the edge of the hole with the major axis orthogonal to the external force is, in this case, multiplied by the factor $[1+2(a/b)]$. The strength of the plate with an elliptical hole thus comes to depend solely on the ratio between the semiaxes. The stress concentration factor increases with the increase in the eccentricity of the ellipse. For $a/b \to \infty$, that is, when the ellipse is very eccentric, the concentration factor tends to infinity. This model does not, therefore, prove useful for describing the critical condition of a crack of length $2a$ and initial width $2b$ tending to zero (Figure 12.3c). In fact, very small external stresses suffice to exceed the tensile strength σ_u at the tip of the crack. Instead, in actual practice, cracked solids can even stand up to considerable stresses.

In 1920, Griffith, a young aeronautical engineer engaged in the study of glass fiber–reinforced materials, thus felt the need to introduce energy considerations, and not only stress considerations, into the analysis of the fracture propagation phenomenon. He showed that the elastic strain energy W_e released by a uniformly extended plate of unit thickness, when a crack of length $2a$ is formed and the displacements at infinity are maintained as constant, is equal to the energy contained in the circle of radius a before the crack originates (Figure 12.4):

$$W_e = \pi a^2 \frac{\sigma^2}{E} \qquad (12.1)$$

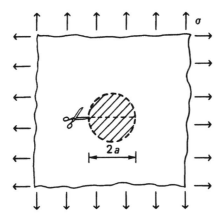

Figure 12.4

where E is the elastic modulus of the material.

On the other hand, to create a crack of length $2a$ requires a **surface energy** equal to

$$W_s = 4a\gamma \tag{12.2}$$

where γ is the surface energy of the material, a thermodynamic quantity that is usually considered for liquids rather than for solids.

Griffith supposed that, for a preexisting crack of length $2a$ to extend, the elastic energy released in a virtual extension must be greater than or equal to that required by the new portion of free surface that is created:

$$\frac{dW_e}{da} \geq \frac{dW_s}{da} \tag{12.3}$$

The condition of instability is therefore the following:

$$2\pi a \frac{\sigma^2}{E} \geq 4\gamma \tag{12.4}$$

The foregoing inequality is valid both for the applied stress σ and for the half-length a of the crack. The pairs of values σ and a that fall beneath the curve of Figure 12.5 constitute stable cases, whereas the pairs that fall above it concern unstable cases. Rendering Equation 12.4 explicit with respect to the applied stress, we obtain, in fact

$$\sigma \geq \sqrt{\frac{2\gamma E}{\pi a}} \tag{12.5}$$

Twice the value of the surface energy is usually termed **fracture energy**, $\mathcal{G}_{IC}$, so that Equation 12.5 takes the form

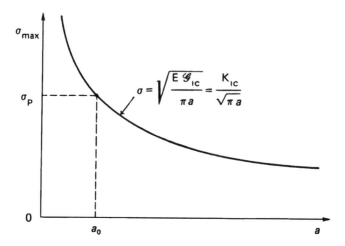

Figure 12.5

$$\sigma \geq \sqrt{\frac{\mathcal{G}_{IC}E}{\pi a}} \tag{12.6}$$

The curve of Figure 12.5, which indicates instability stress as a function of the half-length of the crack, presents two asymptotes. The horizontal asymptotic branch represents the decrease of plate strength with the increase in the crack length. When this length tends to infinity, the strength of the plate consistently tends to zero. The vertical asymptotic branch represents the increase of plate strength with the decrease in crack length. For $a \to 0$, the strength tends to infinity. This result is not, on the other hand, consistent with the assumed intrinsic finite strength σ_P of the material of which the plate is made. A similar problem has already been met with in our discussion of the limits of validity of Euler's formula in Section 7.4 (Figure 7.11). In that case, a limit slenderness was defined, beneath which the compressive yielding of the beam precedes its buckling instability. In the present case, analogously, it is possible to define a length $2a_0$ of the crack, beneath which tensile yielding of the entire plate precedes the unstable propagation of the crack:

$$a_0 = \frac{1}{\pi}\frac{\mathcal{G}_{IC}E}{\sigma_P^2} \tag{12.7}$$

The length $2a_0$ represents the equivalent dimension of the microcracks and of the flaws pre-existing in the material of which the plate is made.

Equation 12.7 offers an explanation of the fact that glass filaments show a strength as much as two orders of magnitude greater than that found using macroscopic test specimens. The fiber cross section has, in fact, a diameter that is considerably less than the size of the flaws that are found in macroscopic test specimens. An increase of two orders of magnitude in the apparent strength σ_P derives, in fact, from the reduction in the characteristic length $2a_0$ by a factor of 10^4. In this perspective, the tensile strength ceases to represent a characteristic of the material and becomes a function of the length $2a_0$ of the preexisting microcracks.

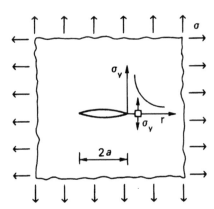

Figure 12.6

12.3 WESTERGAARD'S METHOD

In this section, we shall show how Westergaard (1939), taking up the treatment of the subject by Muskhelishvili (1933), identified the power of the singularity that the stresses present at the tips of a crack. It is known that if discontinuities or slits are considered instead of elliptical holes (Figure 12.3c), the stress that develops in the area around the tip tends to infinity as the distance from the tip itself tends to zero (Figure 12.6). At this point, the stress field is not defined and, in actual practice, around the tip there forms a zone of plastic deformation, albeit small, even in the most brittle materials.

It is possible to demonstrate that each function of the form

$$\Phi = U_1 + xU_2 + yU_3 \tag{12.8}$$

where U_1, U_2, U_3 are harmonic real functions, satisfies the Airy Equation 11.11. Obviously

$$\nabla^4\left(U_1\right) = \nabla^2\left(\nabla^2 U_1\right) = \nabla^2\left(0\right) = 0 \tag{12.9}$$

On the other hand, we have

$$\nabla^2\left(xU_2\right) = \left(\frac{\partial^2}{\partial x^2} + \frac{\partial^2}{\partial y^2}\right)\left(xU_2\right) \tag{12.10}$$

with

$$\frac{\partial^2}{\partial x^2}\left(xU_2\right) = \frac{\partial}{\partial x}\left[\frac{\partial}{\partial x}\left(xU_2\right)\right] = \frac{\partial}{\partial x}\left[U_2 + x\frac{\partial U_2}{\partial x}\right] = 2\frac{\partial U_2}{\partial x} + x\frac{\partial^2 U_2}{\partial x^2} \tag{12.11a}$$

and

$$\frac{\partial^2}{\partial y^2}\left(xU_2\right) = x\frac{\partial^2 U_2}{\partial y^2} \tag{12.11b}$$

Equation 12.10, on the basis of Equations 12.11, becomes

$$\nabla^2 (xU_2) = 2\frac{\partial U_2}{\partial x} + x\nabla^2 U_2 \tag{12.12}$$

and, since the function U_2 is harmonic by hypothesis:

$$\nabla^2 (xU_2) = 2\frac{\partial U_2}{\partial x} \tag{12.13}$$

Further applying the Laplace operator to Equation 12.13 and reversing the order of this with the partial derivative, we obtain, finally

$$\nabla^4 (xU_2) = \nabla^2 \left(2\frac{\partial U_2}{\partial x}\right) = 2\frac{\partial}{\partial x}\left(\nabla^2 U_2\right) = 0 \tag{12.14a}$$

Likewise, it is also possible to demonstrate that

$$\nabla^4 (yU_3) = 0 \tag{12.14b}$$

Consider an analytical function $\overline{\overline{Z}}(z)$, and its subsequent derivatives, themselves also analytical:

$$\frac{d\overline{\overline{Z}}}{dz} = \overline{Z}, \quad \frac{d\overline{Z}}{dz} = Z, \quad \frac{dZ}{dz} = Z' \tag{12.15}$$

where the overbars represent Westergaard's original notation and have nothing to do with the symbol indicating the conjugate of a complex number. The rules of differentiation (Equation 11.113) in this case become

$$\frac{\partial \overline{\overline{Z}}}{\partial x} = \frac{d\overline{\overline{Z}}}{dz}\frac{\partial z}{\partial x} = \overline{Z} \tag{12.16a}$$

$$\frac{\partial \overline{\overline{Z}}}{\partial y} = \frac{d\overline{\overline{Z}}}{dz}\frac{\partial z}{\partial y} = i\overline{Z} \tag{12.16b}$$

and likewise for the subsequent derivatives.

From Equations 12.16, there follow the rules of differentiation of the real part and the imaginary part of an analytical function:

$$\frac{\partial}{\partial x}\operatorname{Re}\overline{\overline{Z}} = \operatorname{Re}\frac{\partial \overline{\overline{Z}}}{\partial x} = \operatorname{Re}\overline{Z} \tag{12.17a}$$

$$\frac{\partial}{\partial y}\operatorname{Re}\overline{\overline{Z}} = \operatorname{Re}\frac{\partial \overline{\overline{Z}}}{\partial y} = -\operatorname{Im}\overline{Z} \tag{12.17b}$$

$$\frac{\partial}{\partial x} \operatorname{Im}\overline{\overline{Z}} = \operatorname{Im}\frac{\partial \overline{\overline{Z}}}{\partial x} = \operatorname{Im}\overline{Z} \tag{12.17c}$$

$$\frac{\partial}{\partial y} \operatorname{Im}\overline{\overline{Z}} = \operatorname{Im}\frac{\partial \overline{\overline{Z}}}{\partial y} = \operatorname{Re}\overline{Z} \tag{12.17d}$$

Notice that, by the transitive law, from the two pairs of Equations 12.17a and d and Equations 12.17b and c, we find again the Cauchy–Riemann conditions of Equations 11.94.

The first of Westergaard's hypotheses concerns the Airy stress function, which is written in the form:

$$\Phi_I = \operatorname{Re}\overline{\overline{Z}}_I + y\operatorname{Im}\overline{Z}_I + \frac{1}{2}B\left(y^2 - x^2\right) \tag{12.18}$$

where subscript I indicates a symmetrical situation with respect to the X axis. The real and imaginary parts of an analytical function are, in fact, harmonic functions, just as it is easy to verify that the function $1/2\ B(y^2 - x^2)$, with B as a real constant, is also harmonic.

The stress components are obtained by double derivation of the Airy stress function. Applying the rules of derivation (Equations 12.17), we have

$$\frac{\partial \Phi_I}{\partial x} = \operatorname{Re}\overline{Z}_I + y\operatorname{Im}Z_I - Bx \tag{12.19a}$$

$$\frac{\partial \Phi_I}{\partial y} = y\operatorname{Re}Z_I + By \tag{12.19b}$$

and hence, via Equations 11.9:

$$\sigma_x = \frac{\partial^2 \Phi_I}{\partial y^2} = \operatorname{Re}Z_I - y\operatorname{Im}Z_I' + B \tag{12.20a}$$

$$\sigma_y = \frac{\partial^2 \Phi_I}{\partial x^2} = \operatorname{Re}Z_I + y\operatorname{Im}Z_I' - B \tag{12.20b}$$

$$\tau_{xy} = -\frac{\partial^2 \Phi_I}{\partial x \partial y} = -y\operatorname{Re}Z_I' \tag{12.20c}$$

Note that, as had already been anticipated in the foregoing chapter, Equations 12.20 coincide with Equations 11.129, once the following substitution of complex potential is made:

$$Z_I(z) = 2\psi'(z) \tag{12.21}$$

Now consider a rectilinear crack of length $2a$ along the X axis between $-a$ and $+a$ (Figure 12.7). The boundary conditions expressing the absence of stresses on the faces of the crack are

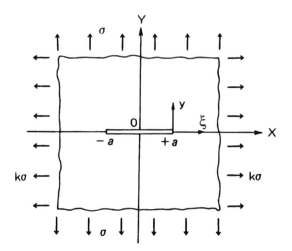

Figure 12.7

$$\sigma_y(x,0) = \tau_{xy}(x,0) = 0, \quad \text{for} - a < x < a \tag{12.22}$$

The second of Westergaard's hypotheses concerns the complex potential, which is written in the form

$$Z_I = \frac{g(z)}{\left[(z+a)(z-a)\right]^{1/2}} + B, \quad \forall z \in C \tag{12.23}$$

where $g(z)$ is a real function if it is defined on the real axis and B is the real constant previously introduced in Equation 12.18. Equation 12.23 satisfies, via Equations 12.20b and c, the boundary conditions of Equation 12.22. On the basis of Equation 12.20a, we then find on the faces of the crack:

$$\sigma_x(x,0) = 2B, \quad \text{for} - a < x < a \tag{12.24}$$

Carrying out in Equation 12.23 the substitution of the variable:

$$z = \zeta + a \tag{12.25}$$

and, therefore, considering a reference system centered in the right-hand tip of the crack (Figure 12.7), we obtain

$$Z_I = \frac{g(\zeta+a)/(\zeta+2a)^{1/2}}{\zeta^{1/2}} + B \tag{12.26}$$

In the area surrounding the right-hand tip of the crack, Equation 12.26 can be approximated by

$$Z_I = \frac{g(a)/\sqrt{2a}}{\zeta^{1/2}} + B \qquad (12.27)$$

If we set

$$\frac{g(a)}{\sqrt{a}} = \frac{K_I}{\sqrt{\pi}} \qquad (12.28)$$

finally, we obtain

$$Z_I = \frac{K_I}{\sqrt{2\pi\zeta}} + B \qquad (12.29)$$

The real constant K_I represents the so-called **stress-intensity factor**.

To be able to understand the physical meaning of the factor K_I, it is necessary to introduce the polar coordinates into the study of the stresses given by Equations 12.20. From complex analysis it is known that

$$\zeta = re^{i\vartheta} = r(\cos\vartheta + i\sin\vartheta) \qquad (12.30a)$$

$$\zeta^{-1/2} = r^{-1/2}e^{-(1/2)i\vartheta} = r^{-1/2}\left(\cos\frac{\vartheta}{2} - i\sin\frac{\vartheta}{2}\right) \qquad (12.30b)$$

$$\zeta^{-3/2} = r^{-3/2}e^{-(3/2)i\vartheta} = r^{-3/2}\left(\cos\frac{3}{2}\vartheta - i\sin\frac{3}{2}\vartheta\right) \qquad (12.30c)$$

$$y = r\sin\vartheta = 2r\sin\frac{\vartheta}{2}\cos\frac{\vartheta}{2} \qquad (12.30d)$$

The complex potential of Equation 12.29 can thus be expressed as follows:

$$Z_I = \frac{K_I}{\sqrt{2\pi r}}\left(\cos\frac{\vartheta}{2} - i\sin\frac{\vartheta}{2}\right) + B \qquad (12.31)$$

having taken care to introduce also the constant B, which, at noninfinitesimal distances from the origin, cannot be considered negligible.

The derivative of the complex potential of Equation 12.29 may itself also be expressed in polar coordinates:

$$Z_I' = \frac{K_I}{\sqrt{2\pi}}\left(-\frac{1}{2}\right)\zeta^{-3/2} = -\frac{K_I}{2\sqrt{2\pi}\left(r^{3/2}\right)}\left(\cos\frac{3}{2}\vartheta - i\sin\frac{3}{2}\vartheta\right) \qquad (12.32)$$

Substituting Equations 12.30d, 12.31, and 12.32 into Equations 12.20, we obtain the stress field that is valid in the crack tip vicinity:

$$\sigma_x = \frac{K_I}{\sqrt{2\pi r}}\cos\frac{\vartheta}{2} - 2r\sin\frac{\vartheta}{2}\cos\frac{\vartheta}{2}\frac{K_I}{2\sqrt{2\pi}\left(r^{3/2}\right)}\sin\frac{3}{2}\vartheta + 2B \tag{12.33a}$$

$$\sigma_y = \frac{K_I}{\sqrt{2\pi r}}\cos\frac{\vartheta}{2} + 2r\sin\frac{\vartheta}{2}\cos\frac{\vartheta}{2}\frac{K_I}{2\sqrt{2\pi}\left(r^{3/2}\right)}\sin\frac{3}{2}\vartheta \tag{12.33b}$$

$$\tau_{xy} = -2r\sin\frac{\vartheta}{2}\cos\frac{\vartheta}{2}\left(-\frac{K_I}{2\sqrt{2\pi}\left(r^{3/2}\right)}\cos\frac{3}{2}\vartheta\right) \tag{12.33c}$$

Gathering common factors, we have, finally:

$$\sigma_x = \frac{K_I}{\sqrt{2\pi r}}\cos\frac{\vartheta}{2}\left(1 - \sin\frac{\vartheta}{2}\sin\frac{3}{2}\vartheta\right) + 2B \tag{12.34a}$$

$$\sigma_y = \frac{K_I}{\sqrt{2\pi r}}\cos\frac{\vartheta}{2}\left(1 + \sin\frac{\vartheta}{2}\sin\frac{3}{2}\vartheta\right) \tag{12.34b}$$

$$\tau_{xy} = \frac{K_I}{\sqrt{2\pi r}}\sin\frac{\vartheta}{2}\cos\frac{\vartheta}{2}\cos\frac{3}{2}\vartheta \tag{12.34c}$$

In regard to the stress components of Equations 12.34, the following observations may be made:

1. All three stress components in Equations 12.34 present the singularity $r^{-1/2}$ at the tip of the crack. The power $-1/2$ of this singularity depends only on the boundary conditions on the faces of the crack, and not on the conditions at infinity.
2. The angular profile of the stress field itself also depends on the boundary conditions on the faces of the crack, and not on the conditions at infinity.
3. The stress field in the crack tip vicinity is uniquely defined by the factor K_I, which is, on the other hand, a function of the conditions at infinity, or, in the case of plates of finite dimensions, a function of the conditions imposed on the external contour.
4. The physical dimensions of K_I are somewhat unusual: $[F][L]^{-3/2}$. It is precisely these dimensions that are the substantial cause of the size effects, both in fracture mechanics and, indirectly, in the strength of materials.

The third of Westergaard's hypotheses concerns the function $g(z)$, present in Equation 12.23 of the complex potential, and is related to the conditions at infinity, which have so far been disregarded. Let it be assumed that the stress condition at infinity presents the principal directions parallel to the XY coordinate axes, with principal stresses equal, respectively, to $k\sigma$ and σ, k being a real constant (Figure 12.7). Setting

$$g(z) = \sigma z \tag{12.35}$$

and hence

$$Z_I = \frac{\sigma z}{\left[(z+a)(z-a)\right]^{1/2}} + B, \quad \forall z \in C \tag{12.36}$$

the aforementioned conditions at infinity remain satisfied. From Equations 12.20, we have, in fact:

$$\lim_{z \to \infty} \sigma_x = \sigma + 2B \tag{12.37a}$$

$$\lim_{z \to \infty} \sigma_y = \sigma \tag{12.37b}$$

$$\lim_{z \to \infty} \tau_{xy} = 0 \tag{12.37c}$$

and the limit of Equation 12.37a yields the value $k\sigma$ for

$$B = \frac{1}{2}\sigma(k-1) \tag{12.38}$$

From Equations 12.28 and 12.35, we obtain the expression of the **stress-intensity factor**:

$$K_I = \sigma\sqrt{\pi a} \tag{12.39}$$

which turns out to depend on the stress at infinity orthogonal to the crack and on the half-length of the crack. The normal stress at infinity parallel to the crack does not enter into Equation 12.39, since it does not affect the stress intensification.

Moreover, from Equations 12.24 and 12.38, we derive the value of the stress σ_x on the faces of the crack:

$$\sigma_x(x,0) = \sigma(k-1), \quad \text{for} -a < x < a \tag{12.40}$$

Whereas the radial and angular variation of the stress field around the crack tip is independent of the specific geometry under examination and is described by Equations 12.34, the information on the geometry and on the external boundary conditions (loads and constraints) is summed up in the factor K_I. In the case, for instance, of a plate of finite width $2h$ with a centered crack of length $2a$, loaded at infinity with a stress σ orthogonal to the crack (Figure 12.8), we have

$$K_I = \sigma\sqrt{\pi a}\left(\sec\frac{\pi a}{2h}\right)^{1/2} \tag{12.41}$$

For $h/a \to \infty$, the foregoing expression tends to Equation 12.39.

In the case of a three-point bending specimen, with an edge crack of length a in the center, we have (Figure 12.9)

$$K_I = \frac{Pl}{th^{3/2}}f\left(\frac{a}{h}\right) \tag{12.42a}$$

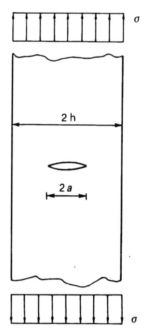

Figure 12.8

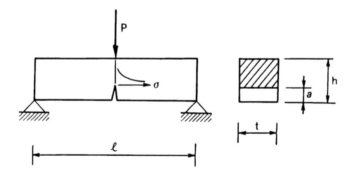

Figure 12.9

with

$$f\left(\frac{a}{h}\right) = 2.9\left(\frac{a}{h}\right)^{1/2} - 4.6\left(\frac{a}{h}\right)^{3/2}$$

$$+ 21.8\left(\frac{a}{h}\right)^{5/2} - 37.6\left(\frac{a}{h}\right)^{7/2} + 38.7\left(\frac{a}{h}\right)^{9/2} \tag{12.42b}$$

where:
- h is the depth
- t is the thickness
- l is the length of the plate
- P is the external force

As regards the elastic **crack opening displacement** (COD), this may be found from the stress field, via the dilation:

$$\varepsilon_y = \frac{\partial \upsilon}{\partial y} = \frac{1}{E}\left(\sigma_y - \nu\sigma_x\right) \tag{12.43}$$

in the case of a plane stress condition. From Equations 12.20a and b we have

$$\upsilon = \int \varepsilon_y \, dy$$

$$= \frac{1}{E}\int \left(\operatorname{Re} Z_I + y\operatorname{Im} Z'_I - B\right)dy - \frac{\nu}{E}\int \left(\operatorname{Re} Z_I - y\operatorname{Im} Z'_I + B\right)dy \tag{12.44}$$

It is easy to verify that the derivative of the following expression coincides with the integrand of Equation 12.44:

$$\upsilon = \frac{2}{E}\operatorname{Im} \bar{Z}_I - \frac{1+\nu}{E}y\operatorname{Re} Z_I - \frac{1+\nu}{E}By \tag{12.45}$$

From Equation 12.29 of the complex potential, we obtain by integration:

$$\bar{Z}_I = \frac{K_I}{\sqrt{2\pi}}2\zeta^{1/2} + B\zeta + C \tag{12.46}$$

and, in polar coordinates:

$$\bar{Z}_I = \frac{2K_I}{\sqrt{2\pi}}r^{1/2}\left(\cos\frac{\vartheta}{2} + i\sin\frac{\vartheta}{2}\right) + Br\left(\cos\vartheta + i\sin\vartheta\right) + C \tag{12.47}$$

The displacements in the Y direction of the points belonging to the upper face of the crack are, therefore:

$$\upsilon(\vartheta = \pi) = 2\left(\frac{2}{\pi}\right)^{1/2}\frac{K_I}{E}r^{1/2} \tag{12.48a}$$

whereas, of course, the points belonging to the lower face present opposite values:

$$\upsilon(\vartheta = -\pi) = -2\left(\frac{2}{\pi}\right)^{1/2}\frac{K_I}{E}r^{1/2} \tag{12.48b}$$

The relative displacement of crack opening in the vicinity of the tip is thus

$$\text{COD} = \upsilon(\pi) - \upsilon(-\pi) = 4\left(\frac{2}{\pi}\right)^{1/2}\frac{K_I}{E}r^{1/2} \tag{12.49}$$

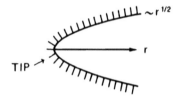

Figure 12.10

The COD is directly proportional to the factor K_I (which, in turn, is always directly proportional to the external load) and inversely proportional to the elastic modulus E. It varies according to a parabolic law along the crack itself, presenting, consistently, a null value in the tip (Figure 12.10). It is interesting to observe how the deformed configuration of the crack reveals a blunting with vertical tangent at the tip. The shearing strain is, in fact, tending to infinity as the shearing stress.

12.4 MODE II AND MIXED MODES

Westergaard's treatment may also concern Mode II, that is, those cases in which the crack undergoes skew-symmetrical loadings with respect to the X axis (Figure 12.11). It will be shown that, as with Mode I (symmetrical loadings), the stress field around the tip of the crack has also with Mode II a radial variation $r^{-1/2}$, with a singularity at the tip with a similar power of $-1/2$, and that the angular variation does not depend on the geometry or on the boundary conditions.

For the in-plane skew-symmetrical cases (Mode II), it is possible to choose an Airy stress function of the following form:

$$\Phi_{II} = -y \operatorname{Re} \bar{Z}_{II} \tag{12.50}$$

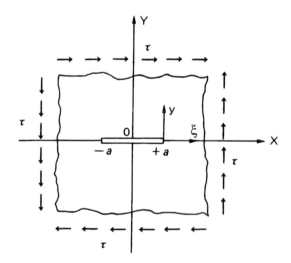

Figure 12.11

The stresses are obtained with a double derivation of Equation 12.50:

$$\sigma_x = 2\operatorname{Im} Z_{II} + y\operatorname{Re} Z'_{II} \tag{12.51a}$$

$$\sigma_y = -y\operatorname{Re} Z'_{II} \tag{12.51b}$$

$$\tau_{xy} = \operatorname{Re} Z_{II} - y\operatorname{Im} Z'_{II} \tag{12.51c}$$

The boundary conditions on the faces of the crack are still represented by Equations 12.22, and are satisfied by a potential of the form

$$Z_{II} = \frac{f(z)}{\left[(z+a)(z-a)\right]^{1/2}}, \quad \forall z \in C \tag{12.52}$$

with f as a real function.

The substitution of Equation 12.25, in the vicinity of the right-hand tip of the crack, yields

$$Z_{II} = \frac{K_{II}}{\sqrt{2\pi\zeta}} \tag{12.53}$$

where

$$K_{II} = f(a)\sqrt{\frac{\pi}{a}} \tag{12.54}$$

is the **second stress-intensity factor**.

Differentiating Equation 12.53 and expressing Equations 12.51 in polar coordinates, we obtain

$$\sigma_x = -\frac{K_{II}}{\sqrt{2\pi r}}\sin\frac{\vartheta}{2}\left(2+\cos\frac{\vartheta}{2}\cos\frac{3}{2}\vartheta\right) \tag{12.55a}$$

$$\sigma_y = \frac{K_{II}}{\sqrt{2\pi r}}\cos\frac{\vartheta}{2}\sin\frac{\vartheta}{2}\cos\frac{3}{2}\vartheta \tag{12.55b}$$

$$\tau_{xy} = \frac{K_{II}}{\sqrt{2\pi r}}\cos\frac{\vartheta}{2}\left(1-\sin\frac{\vartheta}{2}\sin\frac{3}{2}\vartheta\right) \tag{12.55c}$$

The stress field given by Equations 12.55 holds in the crack tip vicinity and is independent of the skew-symmetrical conditions at infinity, except for the factor K_{II}, which is, instead, a function thereof. In the particular condition of pure shear at infinity, parallel to the XY axes, with tension $\sigma = \tau$ at $45°$ and compression $\sigma = -\tau$ at $-45°$ (Figure 12.11), let us assume the function:

$$f(z) = \tau z \qquad (12.56)$$

so that Equation 12.52 becomes

$$Z_{II} = \frac{\tau z}{\left[(z+a)(z-a)\right]^{1/2}}, \quad \forall z \in C \qquad (12.57)$$

This equation satisfies the conditions at infinity. In fact, via Equations 12.51, we have

$$\lim_{z \to \infty} \sigma_x = 0 \qquad (12.58a)$$

$$\lim_{z \to \infty} \sigma_y = 0 \qquad (12.58b)$$

$$\lim_{z \to \infty} \tau_{xy} = \tau \qquad (12.58c)$$

Finally, from Equations 12.54 and 12.56, we find the expression for the stress-intensity factor:

$$K_{II} = \tau \sqrt{\pi a} \qquad (12.59)$$

which turns out to be analogous to Equation 12.39.

Having separately resolved the symmetrical problem with Equations 12.34, and the skew-symmetrical problem with Equations 12.55, it is possible to demonstrate that each generic problem presents a solution, asymptotically valid in the crack tip vicinity, which is the sum of the two elementary modes, that is, Modes I and II:

$$\{\sigma\} = \frac{K_I}{\sqrt{2\pi r}} \{\Theta_I(\vartheta)\} + \frac{K_{II}}{\sqrt{2\pi r}} \{\Theta_{II}(\vartheta)\} \qquad (12.60)$$

Therefore, it is sufficient to know the expressions of the factors K_I and K_{II} to define univocally the stress field around the tip of the crack. In the particular case of a stress condition set at infinity (Figure 12.12), with the principal directions inclined with respect to the crack, the principle of superposition furnishes with Equation 12.60, with K_I and K_{II} given by Equations 12.39 and 12.59, respectively.

So far, we have considered the two elementary modes of loading the crack corresponding to plane stress or plane strain conditions: Mode I, or the **opening mode**, which is symmetrical with respect to the crack (Figure 12.13a); and Mode II, or the **sliding mode**, which is skew-symmetrical with respect to the X axis (Figure 12.13b). There also exists a third elementary mode, corresponding to three-dimensional conditions: Mode III, or the **tearing mode**, which is skew-symmetrical with respect to the XZ plane. This mode is characteristic of the tearing of a sheet of paper (Figure 12.13c). These three modes represent all the modes that exist for subjecting a crack to stress, in the sense that, around a point belonging to the front of a skewed crack that is, in turn, contained in a three-dimensional solid

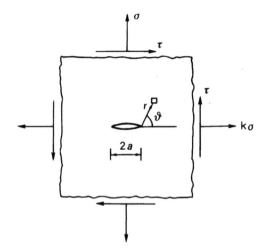

Figure 12.12

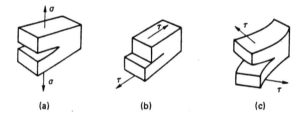

(a)   (b)   (c)

Figure 12.13

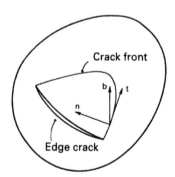

Figure 12.14

(Figure 12.14), the stress field, consisting of the five stress components, σ_n, σ_b, τ_{nb}, τ_{nt}, τ_{bt}, can be expressed as follows:

$$\{\sigma\}_{5\times1} = (2\pi r)^{-1/2} \left[F(\vartheta,\varphi) \right]_{5\times3} \{K\}_{3\times1} \tag{12.61}$$

where r is the radial distance from the point of the crack front, and

$$\{K\} = \begin{bmatrix} K_I \\ K_{II} \\ K_{III} \end{bmatrix} \tag{12.62}$$

is the vector of the stress-intensity factors for that same point. $\left[F(\vartheta,\varphi)\right]$ is a (5×3) matrix that represents the angular profile of the asymptotic field, as a function of the latitude ϑ and the longitude φ in the local reference system tnb, consisting of the tangent, the normal, and the binormal to the crack front.

12.5 WILLIAMS' METHOD

The problem of the determination of the stress field around the vertex of a **reentrant corner** was tackled and solved by Williams in 1952. Five years later, the same author extrapolated it to the limit case of an edge-crack, and so confirmed the stress singularity $r^{-1/2}$, already identified by Muskhelishvili and Westergaard. Williams' method is also known as the **series expansion method**, because the Airy stress function is expanded, as we shall see, in a series of functions. Very important and pioneering contributions on the same subject had been published by Wieghardt (1907) and Brahtz (1933).

Consider a plane sector of elastic material with angular amplitude 2α, and a polar reference system centered at the vertex of this sector (Figure 12.15). Let r and ϑ be the radial and the angular coordinates, respectively, the angle ϑ being considered positive if it is counterclockwise and zero if the position vector is in the direction of the bisector inside the elastic sector. Let it be assumed that a function series expansion may be carried out on the Airy stress function, and that each term of the series may be separated into a power-law radial function with an *a priori* unknown exponent (**eigenvalue**) and an *a priori* unknown angular function (**eigenfunction**):

$$\Phi(r,\vartheta) = \sum_n r^{\lambda_n+1} f_n(\vartheta) \tag{12.63}$$

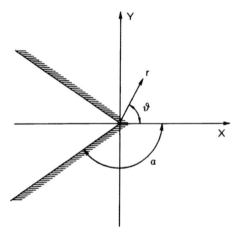

Figure 12.15

Whereas the exponents $(\lambda_n + 1)$ and, hence, the power of the stress singularity, will be identified on the basis of the boundary conditions on the free edges of the sector, the functions f_n can be fully defined only on the basis of the loading conditions at infinity.

Applying Equations 11.44 to the series expansion in Equation 12.63, we obtain the stresses:

$$\sigma_r = \sum_n r^{\lambda_n - 1} \left[f_n''(\vartheta) + (\lambda_n + 1) f_n(\vartheta) \right] \tag{12.64a}$$

$$\sigma_\vartheta = \sum_n r^{\lambda_n - 1} \left[\lambda_n (\lambda_n + 1) f_n(\vartheta) \right] \tag{12.64b}$$

$$\tau_{r\vartheta} = \sum_n r^{\lambda_n - 1} \lambda_n f_n'(\vartheta) \tag{12.64c}$$

where the prime indicates derivation with respect to ϑ. The equation of congruence (Equation 11.50) can therefore be evaluated by summing the following terms:

$$\frac{1}{r^2} \frac{\partial^2}{\partial \vartheta^2} (\sigma_r + \sigma_\vartheta) = \sum_n r^{\lambda_n - 3} \left[f_n^{IV} + (\lambda_n + 1)^2 f_n'' \right] \tag{12.65a}$$

$$\frac{1}{r} \frac{\partial}{\partial r} (\sigma_r + \sigma_\vartheta) = \sum_n r^{\lambda_n - 3} (\lambda_n - 1) \left[f_n'' + (\lambda_n + 1)^2 f_n \right] \tag{12.65b}$$

$$\frac{\partial^2}{\partial r^2} (\sigma_r + \sigma_\vartheta) = \sum_n r^{\lambda_n - 3} (\lambda_n - 1)(\lambda_n - 2) \left[f_n'' + (\lambda_n + 1)^2 f_n \right] \tag{12.65c}$$

Gathering common factors, we finally obtain

$$\nabla^2 (\sigma_r + \sigma_\vartheta) = \sum_n r^{\lambda_n - 3} \left\{ (\lambda_n - 1)^2 \left[f_n'' + (\lambda_n + 1)^2 f_n \right] \right.$$

$$\left. + \left[f_n^{IV} + (\lambda_n + 1)^2 f_n'' \right] \right\} = 0 \tag{12.66}$$

Equation 12.66 is identically satisfied by equating to zero the expression within the braces, which contains only angular functions:

$$f_n^{IV} + \left[(\lambda_n - 1)^2 + (\lambda_n + 1)^2 \right] f_n'' + \left[(\lambda_n - 1)^2 (\lambda_n + 1)^2 \right] f_n = 0 \tag{12.67}$$

The fourth-order differential Equation 12.67, and hence the congruence, are identically satisfied by the following trigonometric form:

$$f_n(\vartheta) = A_n \cos(\lambda_n + 1)\vartheta + B_n \cos(\lambda_n - 1)\vartheta$$

$$+ C_n \sin(\lambda_n + 1)\vartheta + D_n \sin(\lambda_n - 1)\vartheta \tag{12.68}$$

While the first two terms of Equation 12.68 represent the symmetrical solution (Mode I), the two remaining terms represent the skew-symmetrical solution (Mode II).

The boundary conditions on the edges of the elastic sector express the fact that the circumferential stress (and hence the one normal to the edge) and the shearing stress become zero:

$$\sigma_\vartheta(\pm\alpha) = 0 \tag{12.69a}$$

$$\tau_{r\vartheta}(\pm\alpha) = 0 \tag{12.69b}$$

for any radius $r > 0$.

From Equations 12.64b and c, we obtain

$$f_n(\pm\alpha) = 0 \tag{12.70a}$$

$$f_n'(\pm\alpha) = 0 \tag{12.70b}$$

Using Equation 12.68, Equations 12.70 become

$$A_n \cos(\lambda_n + 1)\alpha + B_n \cos(\lambda_n - 1)\alpha \pm C_n \sin(\lambda_n + 1)\alpha$$
$$\pm D_n \sin(\lambda_n - 1)\alpha = 0 \tag{12.71a}$$

$$\pm A_n(\lambda_n + 1)\sin(\lambda_n + 1)\alpha \pm B_n(\lambda_n - 1)\sin(\lambda_n - 1)\alpha$$
$$+ C_n(\lambda_n + 1)\cos(\lambda_n + 1)\alpha + D_n(\lambda_n - 1)\cos(\lambda_n - 1)\alpha = 0 \tag{12.71b}$$

These two equations can be separated, so as to obtain two systems of homogeneous linear algebraic equations in the unknowns A_n, B_n and C_n, D_n, respectively:

$$A_n \cos(\lambda_n + 1)\alpha + B_n \cos(\lambda_n - 1)\alpha = 0 \tag{12.72a}$$

$$A_n(\lambda_n + 1)\sin(\lambda_n + 1)\alpha + B_n(\lambda_n - 1)\sin(\lambda_n - 1)\alpha = 0 \tag{12.72b}$$

$$C_n \sin(\lambda_n + 1)\alpha + D_n \sin(\lambda_n - 1)\alpha = 0 \tag{12.72c}$$

$$C_n(\lambda_n + 1)\cos(\lambda_n + 1)\alpha + D_n(\lambda_n - 1)\cos(\lambda_n - 1)\alpha = 0 \tag{12.72d}$$

The first two equations correspond to the symmetrical problems (Mode I), whereas the last two correspond to the skew-symmetrical problems (Mode II). To obtain solutions different from the trivial one, the determinants of the coefficients of the two systems must become zero. The unknowns A_n and B_n will therefore be defined but for one factor:

$$(\lambda_n - 1)\sin(\lambda_n - 1)\alpha\cos(\lambda_n + 1)\alpha$$
$$- (\lambda_n + 1)\cos(\lambda_n - 1)\alpha\sin(\lambda_n + 1)\alpha = 0 \tag{12.73a}$$

just as the unknowns C_n and D_n will be defined but for one factor:

$$(\lambda_n + 1)\sin(\lambda_n - 1)\alpha \cos(\lambda_n + 1)\alpha$$

$$-(\lambda_n - 1)\cos(\lambda_n - 1)\alpha \sin(\lambda_n + 1)\alpha = 0 \tag{12.73b}$$

Note that, in the case of A_1 and B_1, the aforementioned proportionality factor coincides with the stress-intensity factor K_I, just as it coincides with K_{II} in the case of C_1 and D_1.

From Equation 12.73a, and taking into account the well-known trigonometric relations:

$$\sin x \cos y - \cos x \sin y = \sin(x - y) \tag{12.74a}$$

$$\sin x \cos y + \cos x \sin y = \sin(x + y) \tag{12.74b}$$

we find the condition

$$-\lambda_n \sin 2\alpha = \sin 2\lambda_n \alpha \tag{12.75a}$$

Likewise, from Equation 12.73b we have

$$+\lambda_n \sin 2\alpha = \sin 2\lambda_n \alpha \tag{12.75b}$$

Equations 12.75 are the eigenvalue equations for the elastic sector problem, from which the exponents $(\lambda_n + 1)$ of the series expansion of Equation 12.63 are obtainable. More precisely, from Equation 12.75a we obtain the eigenvalues of the symmetrical problem, while from Equation 12.75b we obtain the eigenvalues of the skew-symmetrical problem.

The terms of the series expansion of Equations 12.64 are finite or infinitesimal for $r \to 0^+$, should the corresponding eigenvalues satisfy the inequality

$$\lambda_n \geq 1 \tag{12.76}$$

On the other hand, the strain energy contained in an infinitesimal circular area of radius R around the vertex of the sector is infinite if

$$\lambda_n \leq 0 \tag{12.77}$$

Thus, we have

$$W(R) \propto \int_0^R r^{2(\lambda_n - 1)} r \, dr \tag{12.78}$$

and the integral is divergent for $(2\lambda_n - 1) \leq -1$. Hence, for the analysis of the dominant singularity of the stress field at the vertex of the sector, the only eigenvalues of interest are those contained in the interval

$$0 < \lambda_n < 1 \tag{12.79}$$

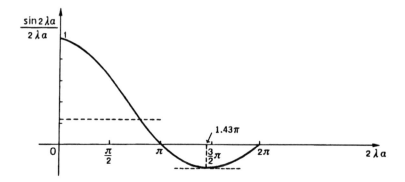

Figure 12.16

The eigenvalue Equations 12.75 can be written in the following form:

$$\frac{\sin 2\lambda_n \alpha}{2\lambda_n \alpha} = \mp \frac{\sin 2\alpha}{2\alpha} \qquad (12.80)$$

with $0 \leq 2\alpha \leq 2\pi$.

From the graphical viewpoint, Equations 12.80 may be resolved rather neatly by intersecting the oscillating function $y = \sin 2\lambda\alpha/(2\lambda\alpha)$ with the horizontal straight lines $y = \mp\sin 2\alpha/(2\alpha)$. In this way, four principal cases can be distinguished (Figure 12.16).

1. $0 \leq 2\alpha \leq \pi$ (convex angle or wedge). The first eigenvalue of the symmetrical problem, λ_I, does not exist or, otherwise, $\lambda_I \geq 1$. The first eigenvalue of the skew-symmetrical problem is $\lambda_{II} = 1$. Consequently, there is no stress singularity in the case where the elastic sector is convex.
2. $\pi < 2\alpha \leq 1.43\pi$ (obtuse concave angle). The first eigenvalue of the symmetrical problem is $\lambda_I < 1$, while we again have $\lambda_{II} = 1$. There is, therefore, only one symmetrical stress singularity.
3. $1.43\pi < 2\alpha < 2\pi$ (acute concave angle). The first eigenvalues, whether of the symmetrical problem or of the skew-symmetrical one, are both less than 1: $\lambda_I < 1$, $\lambda_{II} < 1$. We thus have both the symmetrical and the skew-symmetrical stress singularities, albeit the symmetrical one is of a higher order.
4. $2\alpha = 2\pi$ (null angle or crack). In this case, we have $\lambda_I = \lambda_{II} = 1/2$. Only in the case of a null angle (as well as that of a flat angle) is the first eigenvalue followed by a numerable infinity of other eigenvalues:

$$\lambda_{In} = \lambda_{IIn} = \frac{1}{2}, 1, \frac{3}{2}, 2, \ldots \qquad (12.81)$$

or equivalently

$$\lambda_n = \frac{n}{2}, \qquad n = \text{natural number} \qquad (12.82)$$

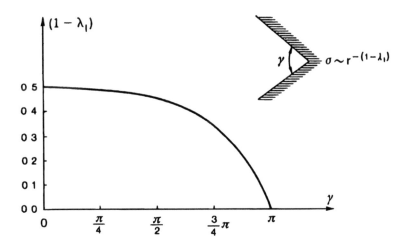

Figure 12.17

The power of the symmetrical stress singularity is represented in Figure 12.17 as a function of the V-notch angle γ. For $\gamma = 0$, the notch becomes a crack and, in fact, we find again the classical singularity $r^{-1/2}$. As γ increases, there is a transition, which is very slow up to $\gamma \simeq \pi/2$, but subsequently, between $\pi/2$ and π, undergoes a rapid acceleration. Obviously, when $\gamma = \pi$, the notch disappears, just as the singularity of the stress field vanishes. When, instead, the reentrant corner angle is a right angle, $\gamma = \pi/2$, we have the power $(1 - \lambda_I) \simeq 0.45$.

12.6 RELATION BETWEEN ENERGY AND STRESS TREATMENTS: IRWIN'S THEOREM

Griffith's criterion, discussed in Section 12.2, represents the first energy criterion of fracture mechanics. In the years that followed, between 1920 and 1950, the efforts of the research workers were all directed, as has been seen in the foregoing sections, toward defining the singular stress field around the tip of the crack. It was only in 1957 that Irwin made a direct correlation of the two different treatments: the energy approach of Griffith and the stress approach of Muskhelishvili, Westergaard, and Williams.

As regards a more general energy criterion than that of Griffith, which had reference to a particular geometry (infinite plate with rectilinear crack, subjected to a loading condition uniform at infinity) and to a **deformation-controlled** loading process, we shall see how the concept of **total potential energy** allows a criterion to be defined that is independent of specimen geometry and control exercised over the loading process.

Let us consider an **imposed-force** loading process on a plate with an initial crack of length $2a$ (Figure 12.18a). For a certain critical value of the force F, let the crack be assumed to extend for the length $2da$ (Figure 12.18b), so as to produce an increment of compliance dC and, hence, an incremental displacement of each end of the plate equal to (Figure 12.18c)

$$d\delta = FdC \qquad (12.83)$$

The variation in the total potential energy due to the infinitesimal propagation of the crack is

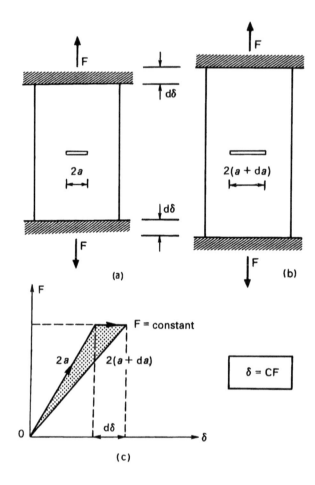

Figure 12.18

$$dW = dL - 2Fd\delta \tag{12.84}$$

where dL denotes the variation in elastic strain energy and the second term represents the variation in the potential energy of the external loads. By virtue of Clapeyron's theorem and evaluating graphically the shaded area of the triangle in Figure 12.18c, we have

$$dL = 2\left(\frac{1}{2}Fd\delta\right) \tag{12.85}$$

and, hence, applying Equation 12.83:

$$dL = F^2dC \tag{12.86}$$

In conclusion, we therefore obtain a decrease in total potential energy:

$$dW = -F^2dC \tag{12.87}$$

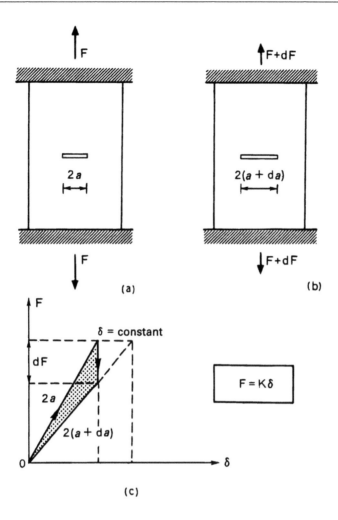

Figure 12.19

Let us now consider an **imposed-displacement** loading process on the previously considered plate (Figure 12.19a). For a certain critical value of the displacement δ, let us assume that the crack is extended by the length $2\mathrm{d}a$ (Figure 12.19b), so as to produce a decrement of stiffness $\mathrm{d}K$ and, hence, a decrement of the external force equal to (Figure 12.19c)

$$\mathrm{d}F = \delta\mathrm{d}K \tag{12.88}$$

The variation in the total potential energy due to the infinitesimal propagation of the crack, in this second case, is equal to the variation in elastic strain energy since, by hypothesis, the external loads do not perform incremental work:

$$\mathrm{d}W = \mathrm{d}L \tag{12.89}$$

By virtue of Clapeyron's theorem and evaluating graphically the shaded area of the triangle in Figure 12.19c, we have

$$dW = 2\left(\frac{1}{2}\delta dF\right) \tag{12.90}$$

and hence, applying Equation 12.88:

$$dW = \delta^2 dK \tag{12.91}$$

Since stiffness is the inverse of compliance, we have

$$dK = d\left(\frac{1}{C}\right) = -\frac{1}{C^2}dC \tag{12.92}$$

Substituting Equation 12.92 into Equation 12.91, we obtain

$$dW = -\frac{\delta^2}{C^2}dC \tag{12.93}$$

and finally, since $\delta/C = F$, we once again obtain Equation 12.87.

Differential calculus basically shows how the difference between the areas of the two shaded triangles of Figures 12.18c and 12.19c constitutes an infinitesimal of an order higher than that of the areas of the triangles themselves. We have, therefore, demonstrated how the total potential energy always diminishes by the same amount F^2dC, following an infinitesimal extension of the crack, regardless of the control exercised over the loading process.

By virtue of the principle of conservation of energy, the following balance between the variation in the total potential energy and the fracture energy must hold:

$$dW + 4\gamma da = 0 \tag{12.94}$$

where $\gamma = \mathcal{G}_{IC}/2$ is the surface energy, that is, the energy necessary for breaking the chemical and atomic bonds connecting two unit and contiguous surfaces of matter. Equation 12.94 represents the more general formulation of Griffith's criterion, expressed by Equation 12.3.

Considering also **virtual**, and not only real, propagations of the crack, the concept of **strain energy release rate** is defined as

$$dW + \mathcal{G}_I dA = 0 \tag{12.95}$$

where dA represents the incremental fracture area. The parameter $\mathcal{G}_I$ is thus defined as the total potential energy released per unit increment in the fracture area:

$$\mathcal{G}_I = -\frac{dW}{dA} \tag{12.96}$$

and is a positive quantity, dW always representing a decrement.

Brittle crack propagation occurs **really** when $\mathcal{G}_I$ reaches its critical value:

$$\mathcal{G}_I = \mathcal{G}_{IC} \tag{12.97a}$$

Since the stress field in the crack tip vicinity is univocally defined by the factor K_I, it is, on the other hand, legitimate to assume that the unstable propagation of the crack occurs when it attains its critical value:

$$K_I = K_{IC} \tag{12.97b}$$

It is thus evident how the two fracture criteria, concerning energy (Equation 12.97a) and stress (Equation 12.97b), have completely different origins. The two critical values, $\mathcal{G}_{IC}$ of the variation in total potential energy (fracture energy), and K_{IC} of the stress-intensity factor, are not, however, independent, but linked by a fundamental relation that will be described in the following.

A first simple way of arriving at the relation that links $\mathcal{G}_{IC}$ and K_{IC} is that of considering the case of the infinite cracked plate, loaded at infinity by a uniform stress condition. According to Griffith, the condition of instability is given by Equation 12.6, while according to Westergaard and taking into account Equation 12.39, it is

$$\sigma \geq \frac{K_{IC}}{\sqrt{\pi a}} \tag{12.98}$$

Since the conditions expressed by Equation 12.6 and Equation 12.98 concern the same physical problem and both of them present the half-length a of the crack raised to the power $-1/2$, we immediately obtain

$$K_{IC} = \sqrt{\mathcal{G}_{IC} E} \tag{12.99}$$

It may be noted that K_{IC} and $\mathcal{G}_{IC}$ are related via the elastic modulus E of the material. In Chapter 11, it was demonstrated that the plane stress fields do not depend on E. There thus follows the independence of the factor K_I from the elastic modulus E, as well as from Poisson's ratio ν. If, instead, one reasons in energy terms, and hence in terms of fracture energy, the influence of E emerges clearly.

Equation 12.99 concerns the critical values of the stress-intensity factor and the strain energy release rate. It can, however, also be extended to the generic values of these two parameters, using a fundamental demonstration due to Irwin.

Consider a cracked plate, subjected to a plane stress condition and to displacements imposed on its external boundary (**fixed grip condition**). Let a be the length of the crack and Δa the extension of the segment of the X axis on which the stresses σ_y are assumed to be known (Figure 12.20a). Consider, then, a virtual extension of the crack, so that it presents the incremented length $a + \Delta a$ (Figure 12.20b). Let υ be the transverse displacements of the faces of the crack in this new configuration, which we take as being known on the same segment of extension Δa.

If we assume that the extension Δa is so small that Westergaard's asymptotic stress and displacement fields hold, and if we apply Clapeyron's theorem to the phenomenon of crack reclosure (from Scheme (b) to Scheme (a) in Figure 12.20), we have the following variation in total potential energy:

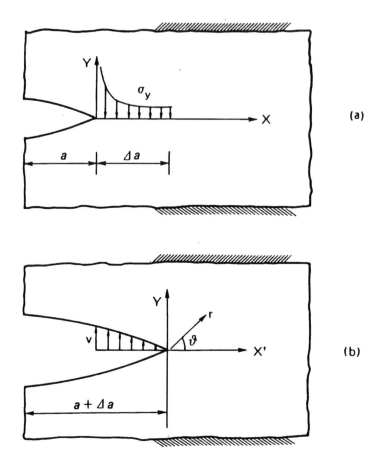

Figure 12.20

$$\Delta W = 2\int_0^{\Delta a} \frac{1}{2}\sigma_y \upsilon \, dr \tag{12.100}$$

with

$$\sigma_y = \sigma_y \left(\vartheta = 0 \right) = \frac{K_I}{\left[2\pi \left(\Delta a - r \right) \right]^{1/2}} \tag{12.101}$$

from Equation 12.34b and

$$\upsilon = \upsilon \left(\vartheta = \pi \right) = 2 \left(\frac{2}{\pi} \right)^{1/2} \frac{K_I}{E} r^{1/2} \tag{12.102}$$

from Equation 12.48a.

Substituting Equations 12.101 and 12.102 into the integral of Equation 12.100, we have

$$\Delta W = \frac{2}{\pi}\frac{K_I^2}{E}\int_0^{\Delta a}\left(\frac{r}{\Delta a - r}\right)^{1/2}dr \qquad (12.103)$$

and evaluating the integral gives

$$\Delta W = \frac{K_I^2}{E}\Delta a \qquad (12.104)$$

On the other hand, from Equation 12.95 we have

$$\Delta W = \mathcal{G}_I \Delta a \qquad (12.105)$$

omitting the negative algebraic sign, since the process of crack reclosure is exactly the opposite of the one so far considered.

The comparison between Equations 12.104 and 12.105 gives, finally, the generalization of Equation 12.99:

$$\mathcal{G}_I = \frac{K_I^2}{E} \qquad (12.106a)$$

which applies in cases where a **plane stress condition** holds.

For **plane strain conditions**, it is not difficult to demonstrate, via a revision of Equation 12.43, that the following relation holds instead:

$$\mathcal{G}_I = \frac{K_I^2}{E}\left(1 - v^2\right) \qquad (12.106b)$$

A check on the coherence of Equations 12.106 is afforded by dimensional analysis:

$$[\mathcal{G}_I] = \frac{[F]^2[L]^{-3}}{[F][L]^{-2}} = [F][L]^{-1} \qquad (12.107)$$

The physical dimension of fracture energy corresponds, in fact, to that of work per unit area, or force per unit length.

In the case of the **mixed mode** (Mode I + Mode II) condition, it is possible to extrapolate the foregoing reasoning:

$$\Delta W = 2\int_0^{\Delta a}\frac{1}{2}\sigma_y v\,dr + 2\int_0^{\Delta a}\frac{1}{2}\tau_{xy}u\,dr \qquad (12.108a)$$

$$\Delta W = \mathcal{G}\,\Delta a \qquad (12.108b)$$

Performing the calculations, we obtain

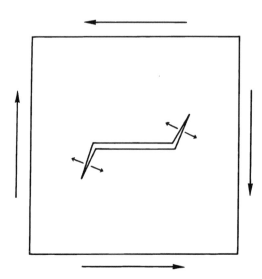

Figure 12.21

$$\mathcal{G} = \frac{K_I^2}{E} + \frac{K_{II}^2}{E} \qquad (12.109)$$

In this case, $\mathcal{G}$ represents the variation in total potential energy per **virtual extension** of the crack. In fact, when subjected to a mixed mode loading, a crack does not extend collinearly to itself (**self-similar propagation**). In reality, as we shall see in the next section, it branches (Figure 12.21).

12.7 CRACK BRANCHING CRITERION IN MIXED MODE CONDITION

As already mentioned in the previous section, Griffith's energy criterion applies consistently only to the case of collinear crack propagation, that is, in the case of Mode I. It cannot be conveniently applied to situations where the crack branches out and changes direction, once subjected to biaxial load conditions. These conditions produce a superposition of Mode I and Mode II, which is conventionally termed **mixed mode**. The procedure will then be to determine all the pairs of values K_I and K_{II} that cause the critical condition around the crack tip and, hence, crack branching.

The first branching criterion, chronologically speaking, is that of **maximum circumferential** stress, proposed by Erdogan and Sih in 1963. It is based on the hypothesis that the crack extends starting from its tip, in the direction normal to that of maximum circumferential stress σ_ϑ. Since the stresses around the crack tip are expressible as products of a radial function by an angular function, this direction does not depend on the radius r of the circumference on which the maximum of the stress σ_ϑ is evaluated.

Translating Equations 12.34 and 12.55 into polar coordinates, and summing up the corresponding results, we obtain

$$\sigma_r = \frac{1}{(2\pi r)^{1/2}} \cos\frac{\vartheta}{2}\left[K_I\left(1 + \sin^2\frac{\vartheta}{2}\right) + K_{II}\left(\frac{3}{2}\sin\vartheta - 2\tan\frac{\vartheta}{2}\right)\right] \qquad (12.110a)$$

$$\sigma_\vartheta = \frac{1}{(2\pi r)^{1/2}} \cos\frac{\vartheta}{2}\left[K_I \cos^2\frac{\vartheta}{2} - \frac{3}{2}K_{II}\sin\vartheta\right] \tag{12.110b}$$

$$\tau_{r\vartheta} = \frac{1}{2(2\pi r)^{1/2}} \cos\frac{\vartheta}{2}\left[K_I \sin\vartheta + K_{II}(3\cos\vartheta - 1)\right] \tag{12.110c}$$

The branching angle ϑ is obtained from the condition of stationarity:

$$\frac{\partial\sigma_\vartheta}{\partial\vartheta} = -\frac{3}{4(2\pi r)^{1/2}}\left[K_I \sin\vartheta + K_{II}(3\cos\vartheta - 1)\right]\cos\frac{\vartheta}{2}$$

$$= -\frac{3}{2}\tau_{r\vartheta} = 0 \tag{12.111}$$

which can be satisfied by setting $\cos\vartheta/2 = 0$, corresponding to the crack surface condition of zero shearing stress ($\tau_{r\vartheta} = 0$) for $\vartheta = \pm\pi$, or

$$K_I \sin\vartheta + K_{II}(3\cos\vartheta - 1) = 0 \tag{12.112}$$

which yields the branching angle of the crack.

For a crack of length $2a$, subjected to a generic biaxial stress condition at infinity (Figure 12.22), the stress-intensity factors are

$$K_I = \sigma_\beta\sqrt{\pi a} \tag{12.113a}$$

$$K_{II} = \tau_\beta\sqrt{\pi a} \tag{12.113b}$$

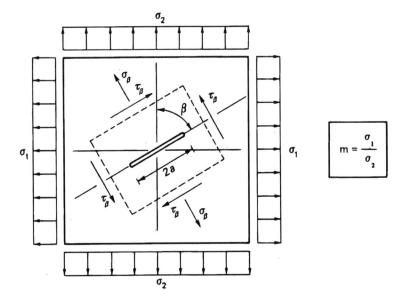

Figure 12.22

where σ_β and τ_β are, respectively, the normal stress and the shearing stress with respect to the crack line, acting at infinity. The usual Mohr relations lead to the following expressions:

$$K_I = \left(\frac{\sigma_1 + \sigma_2}{2} + \frac{\sigma_1 - \sigma_2}{2} \cos 2\beta \right) \sqrt{\pi a} \qquad (12.114a)$$

$$K_{II} = \left(\frac{\sigma_1 - \sigma_2}{2} \sin 2\beta \right) \sqrt{\pi a} \qquad (12.114b)$$

where σ_1, σ_2 are the principal stresses at infinity and β is the angle of inclination of the crack (Figure 12.22). If we denote by m the ratio σ_1/σ_2, Equations 12.114 can be recast in the following form:

$$K_I = \sigma_2 \sqrt{\pi a} \left[m + (1 - m) \sin^2 \beta \right] \qquad (12.115a)$$

$$K_{II} = \sigma_2 \sqrt{\pi a} (1 - m) \sin \beta \cos \beta \qquad (12.115b)$$

Equations 12.112 and 12.115 lead to a condition that relates the branching angle ϑ to the angle of inclination β:

$$\left[m + (1 - m) \sin^2 \beta \right] \sin \vartheta + \left[\frac{1}{2} (1 - m) \sin 2\beta \right] (3 \cos \vartheta - 1) = 0 \qquad (12.116)$$

Equation 12.116 is equivalent to the following:

$$2(1 - m) \sin 2\beta \left(\tan \frac{\vartheta}{2} \right)^2$$

$$-2 \left[m + (1 - m) \sin^2 \beta \right] \left(\tan \frac{\vartheta}{2} \right) - (1 - m) \sin 2\beta = 0 \qquad (12.117)$$

The solution is represented in Figure 12.23 for various ratios m.

If $m = 1$ (uniform stress at infinity), we always have $\vartheta = 0$, and the extension of the crack is collinear by symmetry. On the other hand, if $m = 0$ (uniaxial stress at infinity), a discontinuity occurs for $\beta = 0$. In fact

$$\vartheta(m = 0, \beta = 0) = 0 \qquad (12.118a)$$

by symmetry, whereas instead

$$\lim_{\beta \to 0^+} \vartheta(m = 0, \beta) \approx 70° \qquad (12.118b)$$

Then, if m is small but nonzero, the discontinuity disappears and is replaced by a rapid variation, represented by a very steep branch in Figure 12.23. From a mathematical

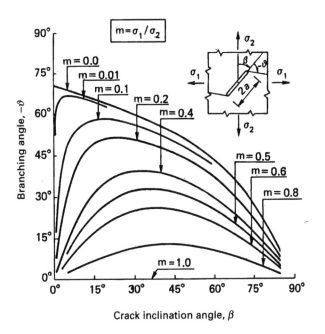

Figure 12.23

standpoint, this is a case of nonuniform convergence of the function $\vartheta(m,\beta)$ in $\beta=0$ for $m \to 0^+$.

Whereas Equation 12.111 defines the direction of maximum circumferential stress, the biaxial critical condition may be obtained from the comparison with the simple Mode I case:

$$\sqrt{2\pi r}\,\sigma_\vartheta = K_{IC} \tag{12.119}$$

Introducing the nondimensional factors:

$$K_I^* = K_I/K_{IC}, \qquad K_{II}^* = K_{II}/K_{IC} \tag{12.120}$$

the branching conditions given by Equation 12.111 and Equation 12.119 may be expressed as follows:

$$K_I^* \sin\vartheta + K_{II}^* \left(3\cos\vartheta - 1\right) = 0 \tag{12.121a}$$

$$K_I^* \cos^2\frac{\vartheta}{2} - \frac{3}{2}K_{II}^* \sin\vartheta = \frac{1}{\cos\left(\vartheta/2\right)} \tag{12.121b}$$

As the angle ϑ varies, all the points of the critical domain are thus defined in parametric form. These points are symmetrical with respect to the axis K_I^* and valid only in the half-plane $K_I^* \geq 0$ (Figure 12.24).

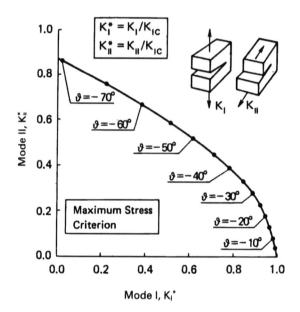

Figure 12.24

12.8 PLASTIC ZONE AT THE CRACK TIP

The stress components around the tip of a real crack present a radial variation $r^{-1/2}$ only beyond a certain distance from that singular point. For smaller distances, plastic phenomena occur, which means that the stresses are, in fact, smaller than those theoretically expected. In this way, a plastic zone is created around the tip of the crack, and the more extensive the greater the ductility of the material. To a first approximation, since, in front of the crack tip (Figure 12.6):

$$\sigma_y = \frac{K_I}{\sqrt{2\pi r}} \tag{12.122}$$

it follows that the radius r_P of the plastic zone can be estimated from (Figure 12.25)

$$\sigma_P = \frac{K_I}{\sqrt{2\pi r_P}} \tag{12.123}$$

where σ_P is the yield stress of the material. At the moment of crack propagation, we thus have the following estimation:

$$r_{PC} = \frac{1}{2\pi} \frac{K_{IC}^2}{\sigma_P^2} \tag{12.124}$$

As will be better understood in the following, the ratio K_{IC}/σ_P therefore represents a measure of the material's **ductility**.

In actual fact, as Irwin observed in 1960, Equation 12.124 provides only the order of magnitude of the plastic radius. A more accurate evaluation can be achieved by considering

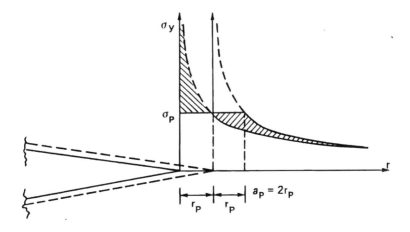

Figure 12.25

the redistribution of stresses, both elastic and plastic, that develops ahead of the crack. In other words, the singular stress distribution of Figure 12.25 is to be translated along the axis r, so that the integral of elastic and plastic stresses is equal to the integral of the aforesaid distribution. From the graphical viewpoint, therefore, the hatched areas of Figure 12.25 must be equal.

The integral of the singular stress distribution between the crack tip and the plastic radius r_P is

$$\int_0^{r_P} \frac{K_I}{\sqrt{2\pi r}} \, dr = \left(\frac{2}{\pi}\right)^{1/2} K_I r_P^{1/2} \tag{12.125}$$

Finding K_I from Equation 12.123 and inserting it into Equation 12.125, we obtain

$$\int_0^{r_P} \frac{K_I}{\sqrt{2\pi r}} \, dr = 2\sigma_P r_P \tag{12.126}$$

From Equation 12.126, we deduce that the left-hand hatched area (Figure 12.25) is equal to that of the rectangle of sides σ_P, r_P. Also the right-hand hatched area, obtained with a translation equal to r_P, is equal to that of the rectangle, since both of them are complementary of the same area. Finally, we obtain the following extension of the plastic zone at the moment of crack propagation, according to Irwin's evaluation:

$$a_{PC} = 2r_{PC} \tag{12.127}$$

or, considering Equation 12.124:

$$a_{PC} = \frac{1}{\pi} \frac{K_{IC}^2}{\sigma_P^2} \tag{12.128}$$

Equation 12.128 coincides with Equation 12.7; that is, the size of the microcrack characteristic for the material corresponds to the size of the plastic zone at crack propagation.

A fracture may be defined as **brittle** when the plastic zone is much smaller than the initial crack and the solid containing it:

$$a_{PC} \ll a \tag{12.129a}$$

$$a_{PC} \ll h \tag{12.129b}$$

where h denotes a characteristic dimension of the cracked solid under examination.

From Equations 12.129, and taking into account Equation 12.128, we obtain the following limitations in nondimensional form:

$$\frac{K_{IC}}{\sigma_P \sqrt{a}} \ll \sqrt{\pi} \tag{12.130a}$$

$$\frac{K_{IC}}{\sigma_P \sqrt{h}} \ll \sqrt{\pi} \tag{12.130b}$$

While Equation 12.130a may be obtained trivially from the condition

$$\sigma \ll \sigma_P \tag{12.131}$$

once account is taken of Equation 12.39 and the graph of Figure 12.5, Equation 12.130b is highly significant for structural purposes, and will be taken up again in the next section.

A different evaluation of the extension of the plastic zone is due to Dugdale (1960) and is based on the simulation of the plastic stresses via a uniform distribution of forces directly applied to the faces of a **fictitious crack**, longer than the real one (Figure 12.26). The condition to be applied is that making the total stress-intensity factor zero:

$$K_I(\sigma) + K_I(\sigma_P) = 0 \tag{12.132}$$

where the first term is that of the stresses applied at infinity, while the second term represents the stress intensity due to the restraining stresses σ_P, applied orthogonally to the faces of the crack at distances from the tips less than or equal to a_P (Figure 12.26).

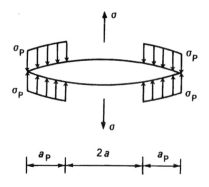

Figure 12.26

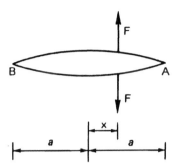

Figure 12.27

Since two concentrated forces F, applied orthogonally to the faces of the crack at a distance x from the center, cause, at the two tips—nearer and further, respectively—the following stress-intensity factors (Figure 12.27):

$$K_I(A) = \frac{F}{\sqrt{\pi a}} \left(\frac{a+x}{a-x} \right)^{1/2} \tag{12.133a}$$

$$K_I(B) = \frac{F}{\sqrt{\pi a}} \left(\frac{a-x}{a+x} \right)^{1/2} \tag{12.133b}$$

where $2a$ is the length of the crack, integrating the effects of the plastic stresses σ_P (Figure 12.26), we have

$$-K_I(\sigma_P) = \frac{\sigma_P}{\left[\pi(a+a_P) \right]^{1/2}} \int_a^{a+a_P} \left(\frac{(a+a_P)+x}{(a+a_P)-x} \right)^{1/2} + \left(\frac{(a+a_P)-x}{(a+a_P)+x} \right)^{1/2} dx \tag{12.134}$$

Evaluating the integral, we obtain

$$-K_I(\sigma_P) = 2\sigma_P \left(\frac{a+a_P}{\pi} \right)^{1/2} \arccos\left(\frac{a}{a+a_P} \right) \tag{12.135}$$

while the factor corresponding to the external load σ equals

$$K_I(\sigma) = \sigma\sqrt{\pi(a+a_P)} \tag{12.136}$$

Substituting Equations 12.135 and 12.136 into Equation 12.132, we have

$$\frac{a}{a+a_P} = \cos\frac{\pi\sigma}{2\sigma_P} \tag{12.137}$$

The limit cases of zero external stress, or external stress equal to the yield strength σ_P, consistently produce, according to Dugdale's model, a null plastic zone ($a_P = 0$), or a general yielding ($a_P \to \infty$), respectively.

Neglecting the terms of a higher order in the series expansion of the cosine, Equation 12.137 is transformed as follows:

$$\frac{a}{a + a_P} = 1 - \frac{1}{2}\left(\frac{\pi\sigma}{2\sigma_P}\right)^2 \tag{12.138}$$

from which we obtain

$$\frac{a_P}{a + a_P} = \frac{\pi^2\sigma^2}{8\sigma_P^2} \tag{12.139}$$

Substituting Equation 12.136 into the foregoing equation, we find, finally

$$a_P = \frac{\pi}{8}\frac{K_I^2}{\sigma_P^2} \tag{12.140}$$

The extension of the plastic zone at the moment of crack propagation, according to Dugdale's evaluation, is, therefore

$$a_{PC} = \frac{\pi}{8}\frac{K_{IC}^2}{\sigma_P^2} \tag{12.141}$$

Also in this case, in estimating a_{PC}, the material's ductility ratio K_{IC}/σ_P is present.

The comparison between the plastic extensions according to Irwin (Equation 12.128) and Dugdale (Equation 12.141) shows how the two models, albeit notably different, lead to estimations that closely resemble one another. The plastic extension, as given by Dugdale, is greater than that given by Irwin by only 23%:

$$\frac{a_{PC}(\text{Dugdale})}{a_{PC}(\text{Irwin})} = \frac{\pi^2}{8} \simeq 1.23 \tag{12.142}$$

12.9 SIZE EFFECTS AND DUCTILE–BRITTLE TRANSITION

A first size effect has already been considered in Section 12.2, which is that corresponding to the length of the crack. As emerges clearly from Figure 12.5, for crack half-lengths greater than a_0, where a_0 is a characteristic length given by Equation 12.7, the collapse due to brittle crack propagation precedes the plastic collapse of the plate. For $a < a_0$, the plastic collapse of the plate instead precedes the collapse due to brittle crack propagation. Recalling the fundamental Equation 12.99 that links $\mathcal{G}_{IC}$ and K_{IC}, the characteristic length, given by Equation 12.7, can also be expressed as a function of K_{IC}:

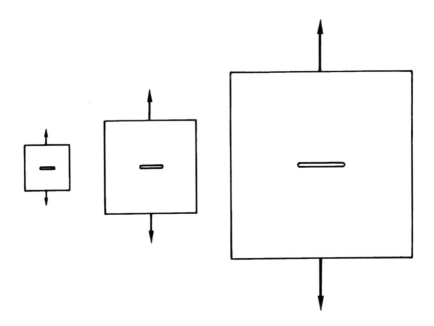

Figure 12.28

$$a_0 = \frac{1}{\pi} \frac{K_{IC}^2}{\sigma_P^2} \qquad\qquad (12.143)$$

Equation 12.143 coincides with Equation 12.128.

The limitations defined by the inequalities in Equations 12.130 thus take the form

$$a \gg a_0 \qquad\qquad (12.144a)$$

$$h \gg a_0 \qquad\qquad (12.144b)$$

A second dimensional effect, which derives directly from the one just considered, is that which corresponds to the dimensions of the cracked body, once constant ratios are assumed between crack length and characteristic dimensions of the body. In the case of the geometrically similar plates of Figure 12.28, the collapse stress will only be a function of the half-length of the crack, when the plates are sufficiently large to allow the effects of the free edge to be neglected:

$$\sigma = \frac{K_{IC}}{\sqrt{\pi a}}, \ \text{ for } a \geq a_0 \qquad\qquad (12.145a)$$

$$\sigma = \sigma_P, \quad \text{ for } a < a_0 \qquad\qquad (12.145b)$$

Since, on the other hand, by virtue of the supposed geometrical similitude, the half-length a is proportional to the characteristic dimension h of the plate:

$$a = \xi h \tag{12.146}$$

where ξ is the relative crack length, Equations 12.145 can be recast in the form

$$\sigma = \frac{K_{IC}}{\sqrt{\pi \xi h}}, \quad \text{for } h \geq \frac{a_0}{\xi} \tag{12.147a}$$

$$\sigma = \sigma_P, \quad \text{for } h < \frac{a_0}{\xi} \tag{12.147b}$$

There thus exists a dimension of the plate, $h_0 = a_0/\xi$, below which the plastic collapse of the plate precedes the brittle propagation of the crack. This dimension depends not only on the geometrical shape of the plate and of the crack, but also on the ductility K_{IC}/σ_P of the material from which the plate is made.

From the simple example just dealt with, the absence of **physical similitude** in the tensile collapse of geometrically similar bodies is immediately inferred once the existence of a crack of a length proportional to the dimension of the body is assumed (Figure 12.28). As has already been observed, it is not possible to state the same in the case of an elliptical hole, for which the stress concentration factor depends on the ratio between the semiaxes, and not on their absolute dimensions.

The hypothesis of negligibility of edge effects can, on the other hand, be removed without vitiating the important conclusions displayed earlier; indeed, they are enriched with fresh insights. Let us consider a plate of finite width $2h$, with a crack of length $2a$, $0 < a/h < 1$, loaded at infinity by a stress σ orthogonal to the crack (Figure 12.8). Since the stress-intensity factor is given by Equation 12.41, the brittle propagation of the crack occurs for

$$\sigma = \frac{K_{IC}}{\sqrt{\pi a}} \left(\cos \frac{\pi a}{2h} \right)^{1/2} \tag{12.148}$$

or, in nondimensional form

$$\frac{\sigma}{\sigma_P} = \frac{K_{IC}}{\sigma_P \sqrt{2h}} \left(\frac{\cos\left(\dfrac{\pi a}{2h} \right)}{\dfrac{\pi a}{2h}} \right)^{1/2} \tag{12.149}$$

Denoting by

$$s = \frac{K_{IC}}{\sigma_P \sqrt{2h}} \tag{12.150}$$

the so-called **brittleness number**, we obtain

$$\frac{\sigma}{\sigma_P} = s \left(\frac{\cos\left(\dfrac{\pi a}{2h}\right)}{\dfrac{\pi a}{2h}} \right)^{1/2} \tag{12.151}$$

On the other hand, the plastic limit analysis carried out on the net section complementary to the crack, which is referred to as the **ligament**, provides a second collapse condition different from Equation 12.151:

$$\frac{\sigma}{\sigma_P} = 1 - \frac{a}{h} \tag{12.152}$$

The diagrams of Equations 12.151 and 12.152 are presented in Figure 12.29 as functions of the relative crack length a/h. While the first of these equations gives a family of curves related to the nondimensional number s, the second is represented by a single curve (thick line). When $s \leq 0.54$, it may be noted that plastic collapse precedes brittle crack propagation, both for sufficiently short cracks and sufficiently long ones. While the first tendency is by now familiar, starting from Section 12.2 onward, the second represents a new, nonintuitive development. It is basically due to the unlikelihood of a singular stress distribution developing in the cases where there is an excessively reduced ligament. As the number s increases, the interval of a/h for which brittle propagation of the crack precedes plastic collapse contracts until it vanishes for $s = s_0 \approx 0.54$, the value for which the corresponding fracture curve is tangential to the curve of plastic collapse. For $s \geq 0.54$, plastic collapse precedes brittle

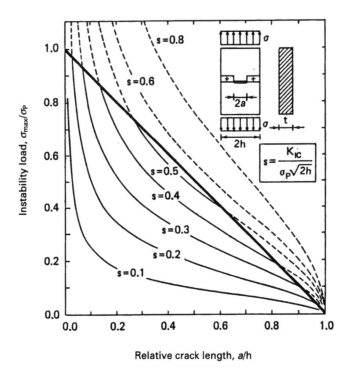

Figure 12.29

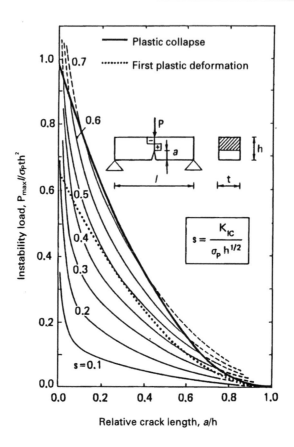

Figure 12.30

crack propagation for any relative crack length; there exists no point of intersection between the fracture curve and the plastic collapse curve. Consequently, the condition expressed by Equation 12.130b is reconfirmed following another path. This condition means that brittle types of collapse tend to occur with low material toughness, high yielding stress, and/or large structural sizes. It is not the individual values of K_{IC}, σ_P, and h that are responsible for the nature of the collapse mechanism, but rather only their function s (cf. Equation 12.150).

Also in the case of three-point bending of a plate (Figure 12.30), it is possible to arrive at the same conclusions. Recalling Equation 12.42a of the factor K_I at the moment of potential collapse due to brittle crack propagation, we have

$$K_{IC} = \frac{P_{max}l}{th^{3/2}} f\left(\frac{a}{h}\right)$$

(12.153)

from which, in nondimensional form

$$\frac{P_{max}l}{\sigma_P th^2} = \frac{s}{f\left(\dfrac{a}{h}\right)}$$

(12.154)

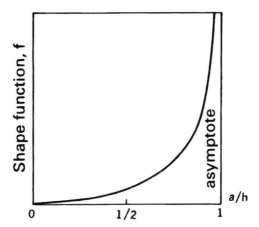

Figure 12.31

where

$$s = \frac{K_{IC}}{\sigma_P h^{1/2}} \tag{12.155}$$

denotes the brittleness number that considers the plate depth h as the characteristic dimension, and f is the shape function of Equation 12.42b, which, like the function $[(\pi a/2h)\sec(\pi a/2h)]^{1/2}$ of Equation 12.41, vanishes for $a/h = 0$ and tends to infinity for $a/h \to 1^{-}$ (Figure 12.31).

On the other hand, the force P that potentially produces plastic collapse can be held to be the one that generates a plastic hinge at the ligament:

$$\frac{1}{4} P_{\text{max}} l = \sigma_P t \frac{(h-a)^2}{4} \tag{12.156}$$

from which follows, in nondimensional form

$$\frac{P_{\text{max}} l}{\sigma_P t h^2} = \left(1 - \frac{a}{h}\right)^2 \tag{12.157}$$

The diagrams of Equations 12.154 and 12.157 are presented in Figure 12.30. For this structural geometry, the brittleness number that marks the transition from ductile collapse to brittle collapse is $s_0 \approx 0.75$. For this value, the fracture curve is tangential to the plastic collapse curve.

If, therefore, the ductility of a material is measurable via the ratio K_{IC}/σ_P, for the ductility of a structure to be defined, it is necessary that a dimension of that structure also be entered into the equation. The brittleness number s provided by Equation 12.155 is certainly the most synthetic way of describing the degree of ductility of a structure. Plastic limit analysis, therefore, represents a reliable method of calculation only in the cases where the brittleness number of the structure being examined is not excessively low.

Table 12.1 Strength and toughness

	Strength σ_P (MN/m²)	Toughness K_{IC} (MN/m³/²)	Brittleness σ_P/K_{IC} (m⁻¹/²)
Concrete	3.57	1.96	1.8
Aluminum	500	100	5
Plexiglass	33	5.5	6
Glass	170	0.25	680

Table 12.1 gives indicative values of tensile strength σ_P and fracture toughness K_{IC} for some materials. The ratio σ_P/K_{IC} then provides a measure (in meters to the power of –1/2) of the brittleness of the material. Glass proves to be, by far, the most brittle, while concrete unexpectedly proves to be the most ductile. This ductility in the case of concrete cannot be put down to a hardening behavior of the material, but to its softening behavior. On the other hand, in view of the dimension h in the brittleness number s, it is, at this point, easy to understand how glass can prove ductile for small structural dimensions and steel prove brittle for large ones (Figure 12.1). Basically, then, the size effects are caused by the different physical dimensions of **strength** and **toughness**.

Finally, notice how, via Equation 12.143 concerning the characteristic length of the micro-flaws, it is possible to give the brittleness number the following alternative form

$$s = \left(\pi \frac{a_0}{h} \right)^{1/2} \tag{12.158}$$

12.10 COHESIVE CRACK MODEL AND SNAP-BACK INSTABILITY

One way of describing the behavior of materials in a consistent manner is that of using a pair of constitutive laws:

1. A **stress–strain** relation that describes the elastic and hardening behavior of the uncracked material up to the maximum stress σ_u, unloadings included (Figure 12.32a)
2. A **stress–COD** relation that describes the softening behavior of the cracked material up to the critical opening w_c, beyond which the interaction between the crack faces becomes zero (Figure 12.32b)

The double constitutive law represented in Figure 12.32 is proposed for brittle materials having an elastic-softening behavior, where the energy is dissipated exclusively on the crack surface. In the more general case of material presenting an elastic-hardening-softening behavior (Figure 12.32), the energy is dissipated both in the volume of the uncracked material and on the surface of the crack. The energy J_V dissipated in the unit volume is equal to the hatched area in Figure 12.32a, while the energy J_S dissipated over the unit surface is equal to the hatched area in Figure 12.32b. If a bar is subjected to tension, it cracks and eventually breaks into two parts, the total dissipated energy being given by the sum of the energy dissipated in the volume of the bar plus that dissipated over the surface of the crack:

$$\text{energy dissipated} = J_V \times \text{area} \times l + J_S \times \text{area} \tag{12.159}$$

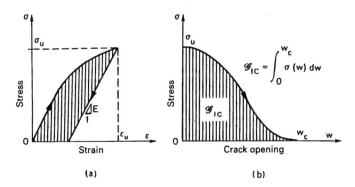

Figure 12.32

from which we find that the energy dissipated per unit surface of the crack, J_C, depends on the length of the bar:

$$J_C = J_V l + J_S \tag{12.160}$$

Only in the case of elastic-softening material, for which $J_V = 0$, does the so-called integral J_C not depend on the dimension of the bar:

$$J_C = J_S = \mathcal{G}_{IC} \tag{12.161}$$

In the case of steels, the constitutive law is generally of the more complex type, that is, elastic-hardening-softening, and consequently it is particularly difficult to find models to describe their behavior, as any model must account for two different mechanisms of dissipation, on the surface and in the volume. However, in the case of concrete, rocks, and ceramic materials, the simpler elastic-softening law is able to describe the actual behavior very consistently.

The so-called **cohesive crack model** is analogous to **Dugdale's model** of Figure 12.26, but with the difference that, in the former, the distribution of the cohesive forces is not uniform, but decreases as the crack opening increases, following a softening law like that of Figure 12.32b (Hillerborg and co-workers 1976). The zone ahead of the real crack tip appears damaged and presents microcracks. It represents a portion of the developing macrocrack, still, however, partially sutured by inclusions, aggregates, or fibers (Figure 12.33a). This zone, in which nonlinear and dissipative phenomena of a microscopic nature occur, is termed the **process zone** (or **plastic zone**). If it is sufficiently small compared with the real crack, then the concepts of linear elastic fracture mechanics (LEFM) are fully applicable. On the other hand, if the extension of the process zone is comparable with that of the real crack, then it must be appropriately modeled (Figure 12.33b). The tip of the **cohesive** (or **fictitious**) **crack** coincides with the tip of the process zone, in which the opening w is still equal to zero and the restraining stress is equal to the tensile strength σ_u. The tip of the **real crack** is found, instead, at the critical crack opening w_c, for which the interaction vanishes according to the cohesive law. In the intermediate points of the process zone, the pairs σ–w are given by the diagram in Figure 12.32b. A previous analogous crack model able to simulate microcracks in atomic lattices was proposed by Barenblatt (1962).

Figure 12.34a represents the structural response, in terms of load versus deflection curve, of a three-point-bending concrete slab, as the relative depth a/h of the initial crack varies, for $\mathcal{G}_{IC} = 0.05$ kg/cm and $h = 15$ cm. The response predicted by the cohesive crack model is

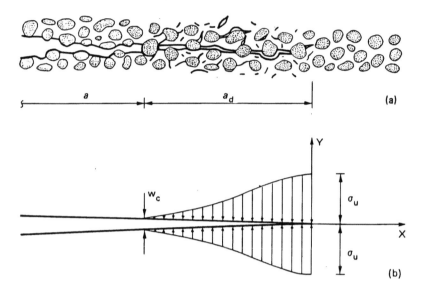

Figure 12.33

always of a softening type, and, as can be noted, with the increase in a/h, a decrease in stiffness and in loading capacity is found, together with an increase in ductility. The tail of the P–δ response proves insensitive to the length of the initial crack.

Figure 12.34b represents the structural response of the same slab for a lower fracture energy, $\mathcal{G}_{IC}=0.01$ kg/cm. The trends are the same as in the previous case, but the responses all appear more brittle, and especially the one corresponding to the initially uncracked slab $(a/h=0.0)$, which shows a marked phenomenon of snap-back. Snap-back disappears for $a/h \geq 0.25$. This type of instability has already been encountered in Section 7.9, where mention was made of the snap-through and snap-back instabilities of vertically loaded shallow domes and axially loaded cylindrical shells.

Figure 12.35 gives the diagrams of the load P as a function of the **crack mouth opening displacement** (CMOD). Whereas, with the tougher material, the crack starts to open before the maximum load is reached, with the more brittle material, the onset of crack opening corresponds exactly to the point of maximum load, and the crack continues to open in a monotonic way as the load diminishes in the softening stage. From this diagram, it is possible to understand how control is necessary via crack opening to detect the snap-back branch BC experimentally (Figure 12.34b), and not via deflection, which proves not to be a monotonic function of time or crack length.

The notable embrittlement of the structural response, which in Figure 12.34 is produced by a drop in fracture energy $\mathcal{G}_{IC}$ of the material, can equivalently be generated by a dilation of the size scale h. Figure 12.36 gives the structural responses of the previously considered slab for $\mathcal{G}_{IC}=0.05$ kg/cm and four different sizes: (a) $h=10$ cm, (b) $h=20$ cm, (c) $h=40$ cm, and (d) $h=80$ cm. Whereas with $h=10$ cm, the response is of a softening type for each depth of the initial crack, with $h=20$ cm there is a practically vertical drop in the loading capacity for the initially uncracked slab; with $h=40$ cm, a clear instance of snap-back is recorded, which, with $h=80$ cm, turns into a very sharply pointed cusp.

The embrittlement of the structural response, produced both by the decrease in fracture energy $\mathcal{G}_{IC}$ and the increase in strength σ_u and/or the size h, can be described in a unitary and synthetic manner via the variation in the following dimensionless number:

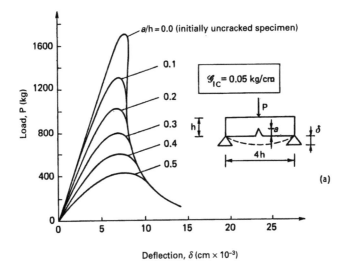

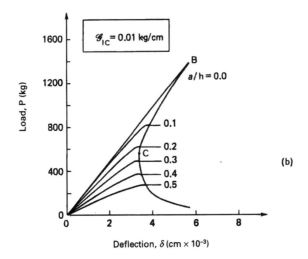

Figure 12.34

$$s_E = \frac{\mathcal{G}_{IC}}{\sigma_u h} \qquad\qquad (12.162)$$

Whereas the brittleness number of Equation 12.155 is of a stress type, the brittleness number of Equation 12.162 is of an energy type. Taking into account the fundamental Equation 12.99 that links K_{IC} and $\mathcal{G}_{IC}$, it is possible to demonstrate that, between the two brittleness numbers referred to, there is the following relation:

$$s_E = \varepsilon_u s^2 \qquad\qquad (12.163)$$

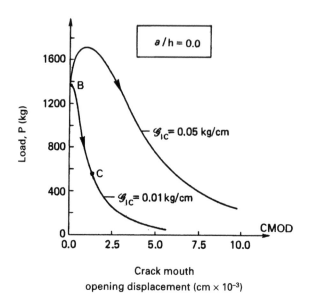

Figure 12.35

where $\varepsilon_u = \sigma_u/E$ represents the ultimate tensile dilation. It may be shown that there is a perfect physical similarity in the failure behavior, when two of the three pure numbers s, s_E, ε_u are equal.

Figure 12.37a gives the load versus deflection response in nondimensional form, for $a/h = 0.1$, $\varepsilon_u = 0.87 \times 10^{-4}$, $v = 0.1$, $l = 4h$, as the brittleness number s_E varies. It is clearly evident that, as s_E varies through three orders of magnitude, the shape of the nondimensional curve changes totally, from ductile to brittle. For $s_E \leq 12.45 \times 10^{-5}$, the softening branch acquires, at least for a portion, a positive slope, and hence the phenomenon of snap-back is seen to occur.

The area bounded by each individual curve of Figure 12.37a and the horizontal axis represents the product between fracture energy $\mathcal{G}_{IC}$ and the initial area of the ligament $(h-a)$ t. The areas under the nondimensional P–δ curves are therefore proportional to the respective brittleness numbers s_E. This simple result is made possible by the hypothesis that the energy dissipation occurs exclusively on the fracture surface and not in the volume of the slab. Figures 12.37b and c illustrate the cases $a/h = 0.3$ and 0.5, respectively, which show a greater ductility.

The maximum load deriving from the cohesive crack model can be compared with the load of brittle crack propagation, expressed by Equation 12.154. The values of their ratio are presented in the graphs of Figure 12.38 as functions of the inverse of the brittleness number s_E. This ratio can be regarded as the ratio between **fictitious toughness** and **actual toughness**. Fictitious toughness is always lower than actual toughness, because, for high values of s_E, plastic collapse tends to precede potential collapse due to brittle crack propagation. It is evident that, for $s_E \to 0$, the results of the cohesive crack model tend to converge with those deriving from LEFM and that, consequently, the phenomenon of snap-back tends to represent the classic Griffith's instability. The cohesive crack model thus manifests its ability to describe the ductile–brittle transition, the existence of which has already been revealed in the foregoing sections.

Another demonstration of the fact that the cohesive crack model tends to approach LEFM asymptotically is provided by the diagrams in Figure 12.39, which give the relative depth

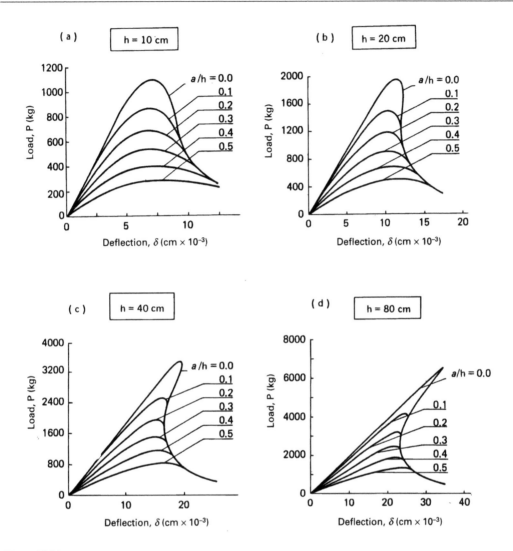

Figure 12.36

of the fictitious crack at maximum load as a function of the inverse of the number s_E. For $s_E \to 0$, this depth tends to that of the initial crack and hence, at the point of snap-back instability, there is an absence of the process zone, or a completely brittle-type fracture. On the other hand, for $s_E \to \infty$, the process zone at maximum load invades the entire ligament.

The cohesive crack model algorithm is based on the following assumptions:

1. The cohesive fracture zone (process zone) begins to develop when the maximum principal stress achieves the ultimate tensile strength σ_u.
2. The material in the process zone is partially damaged but is still able to transfer stress. Such a stress is dependent on the COD, w.

The closing stresses acting on the crack surfaces can be replaced by nodal forces. The intensity of these forces depends on the opening of the fictitious crack w, according to the

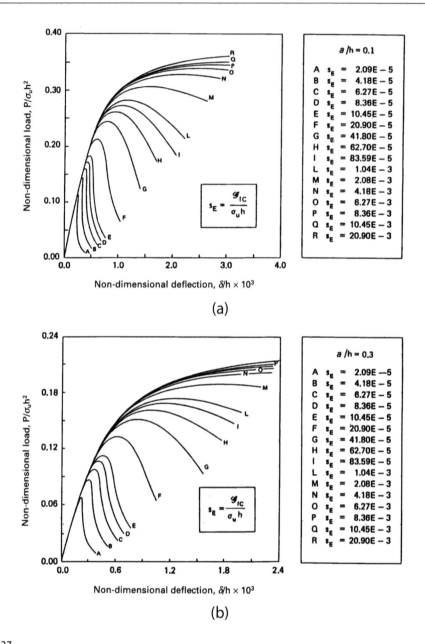

Figure 12.37

$\sigma - w$ cohesive law of the material. When the tensile strength σ_u is achieved at the fictitious crack tip, the top node is opened and a cohesive force starts acting across the crack, while the fictitious tip moves to the next node.

With reference to a three-point-bending test (TPBT) geometry (Figure 12.40), the nodes are distributed along the potential fracture line. The coefficients of influence in terms of node openings and central beam deflection are computed by a finite element analysis, where the fictitious structure is subjected to $(n+1)$ different loading conditions. Consider the TPBT in Figure 12.40a with the initial crack of length a_0 whose tip is at node k. The CODs at the n fracture nodes may be expressed as follows:

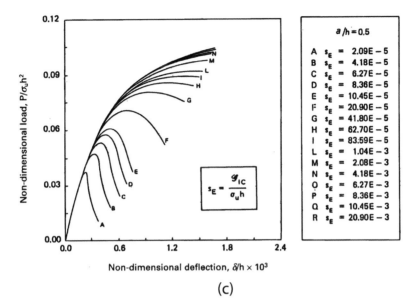

$$s_E = \frac{\mathcal{G}_{IC}}{\sigma_u h}$$

	$a/h = 0.5$	
A	s_E =	2.09E − 5
B	s_E =	4.18E − 5
C	s_E =	6.27E − 5
D	s_E =	8.36E − 5
E	s_E =	10.45E − 5
F	s_E =	20.90E − 5
G	s_E =	41.80E − 5
H	s_E =	62.70E − 5
I	s_E =	83.59E − 5
L	s_E =	1.04E − 3
M	s_E =	2.08E − 3
N	s_E =	4.18E − 3
O	s_E =	6.27E − 3
P	s_E =	8.36E − 3
Q	s_E =	10.45E − 3
R	s_E =	20.90E − 3

(c)

Figure 12.37 (Continued)

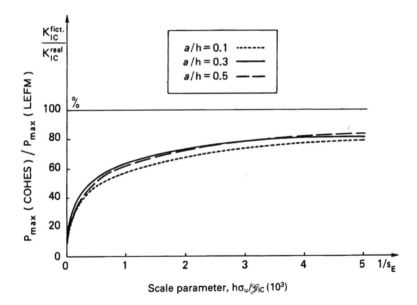

Figure 12.38

$$\{w\} = [H]\{F\} + \{C\}P \tag{12.164}$$

where:
  $\{w\}$ is the vector of the crack openings
  $[H]$ is the matrix of the coefficients of influence ($F_i = 1$)
  $\{F\}$ is the vector of the closing forces

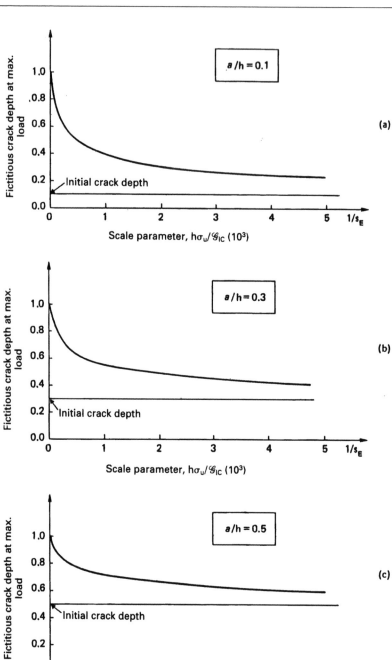

Figure 12.39

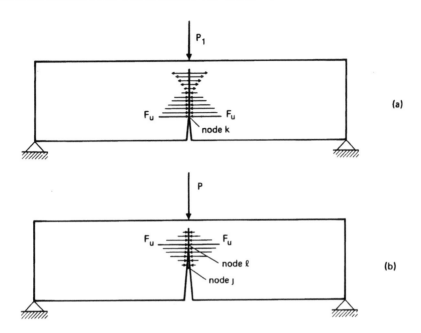

Figure 12.40

{C}   is the vector of the coefficients of influence $(P=1)$

P is the external load

When the process zone is absent (Figure 12.40a), we have a linear algebraic system of $(2n)$ equations in the $(2n)$ unknowns $\{w\}$ and $\{F\}$:

$$F_i = 0, \qquad \text{for } i = 1,2,\ldots,(k-1) \tag{12.165a}$$

$$w_i = 0, \qquad \text{for } i = k,(k+1),\ldots,n \tag{12.165b}$$

On the other hand, when the process zone is present between nodes j and l (Figure 12.40b), we have a linear algebraic system of $(2n+1)$ equations in the $(2n+1)$ unknowns $\{w\}$, $\{F\}$, and P:

$$F_i = 0, \qquad\qquad \text{for } i = 1,2,\ldots,(j-1) \tag{12.166a}$$

$$F_i = F_u\left(1 - \frac{w_i}{w_c}\right), \qquad \text{for } i = j,(j+1),\ldots,l \tag{12.166b}$$

$$w_i = 0, \qquad\qquad \text{for } i = l,(l+1),\ldots,n \tag{12.166c}$$

The beam deflection can be computed as

$$\delta = \{C\}^{\mathrm{T}}\{F\} + D_P P \tag{12.167}$$

where D_p is the deflection for $P=1$.

When the cohesive zone is missing ($l=j=k$), the routine computes the load P_1 producing the ultimate strength nodal force F_u at the initial crack tip (node k) and the corresponding midspan deflection. When the cohesive zone is between nodes k and $k+1$, the routine computes the load P_2 producing F_u at the second fictitious crack tip (node $k+1$). Then, the length of the fictitious crack is incremented by one finite element. The real crack depth, the external load, and the deflection are obtained using an iterative procedure. The routine stops when untying the node n and, consequently, with the determination of the last couple of values P_n and δ_n.

12.11 ECCENTRIC COMPRESSION ON A CRACKED BEAM: OPENING VERSUS CLOSING OF THE CRACK

Let us take the dimensionless crack depth $\xi=a/b$ as the damage parameter, and the stress-intensity factor K_I as the loading parameter (Figure 12.41).

The bending moment M produces a stress-intensity factor expressed as

$$K_I^{(M)} = \frac{M}{tb^{3/2}} Y_M(\xi) \tag{12.168a}$$

with

$$Y_M(\xi) = 6\left(1.99\xi^{1/2} - 2.47\xi^{3/2} + 12.97\xi^{5/2} - 23.17\xi^{7/2} + 24.8\xi^{9/2}\right) \tag{12.168b}$$

Similarly, a tensile axial force F produces

$$K_I^{(F)} = \frac{F}{tb^{1/2}} Y_F(\xi) \tag{12.169a}$$

with

$$Y_F(\xi) = 1.99\xi^{1/2} - 0.41\xi^{3/2} + 18.70\xi^{5/2} - 38.50\xi^{7/2} + 53.86\xi^{9/2} \tag{12.169b}$$

When the axial force is compressive and the bending moment tends to open the crack, the total stress-intensity factor can be determined by applying the superposition principle:

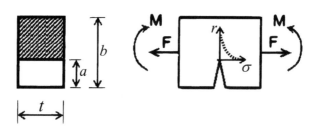

Figure 12.41

$$K_I = K_I^{(M)} - K_I^{(F)} = \frac{F}{tb^{1/2}}\left[\frac{e}{b}Y_M(\xi) - Y_F(\xi)\right] \tag{12.170}$$

where e stands for the eccentricity of the equivalent axial force with respect to the centroid of the cross-sectional area.

From the critical condition $K_I = K_{IC}$, it is possible to determine the dimensionless crack extension axial force as a function of crack depth ξ and relative eccentricity of the load, e/b:

$$\overline{F}_C = \frac{F_C}{tb^{1/2}K_{IC}} = \frac{1}{\dfrac{e}{b}Y_M(\xi) - Y_F(\xi)} \tag{12.171}$$

The curves in Figure 12.42 graphically represent this expression and show how, when the eccentricity e/b is fixed, the fracturing process reaches a condition of stability only after demonstrating an unstable condition. If the load F is unable to follow the decreasing unstable branch of the e/b=constant curve in a strain-softening unloading process, the fracturing process shows a catastrophic behavior and the representative point advances horizontally until it meets the e/b=constant curve again on the stable branch (**snap-through**). On the other hand, the possibility of load relaxation and a less catastrophic fracturing behavior depends on the structural geometry and the material characteristics, and is affected in particular by the degree of redundancy and the structural size.

It is also important to consider that, for any relative crack depth ξ, there is a relative eccentricity below which the crack tends, at least partially, to close. From the closing condition $K_I = 0$, we obtain

$$K_I = \frac{F}{tb^{1/2}}\left[\frac{e}{b}Y_M(\xi) - Y_F(\xi)\right] = 0 \tag{12.172}$$

from which there follows

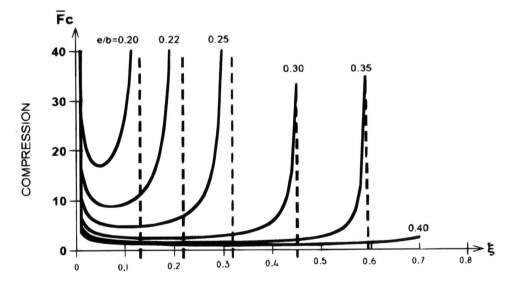

Figure 12.42

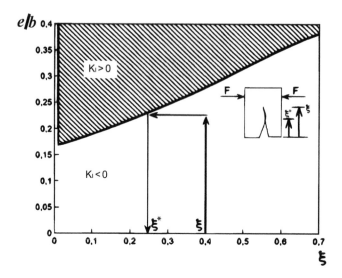

Figure 12.43

$$\frac{e}{b} = \frac{Y_F(\xi)}{Y_M(\xi)} \tag{12.173}$$

Equation 12.173 is graphically represented in Figure 12.43. The points below the curve represent the crack and loading conditions whereby $K_I < 0$. The value ξ^* represents the depth of the partially open crack.

12.12 STABILITY OF FRACTURING PROCESS IN REINFORCED CONCRETE BEAMS: THE BRIDGED CRACK MODEL

Let the cracked concrete beam element in Figure 12.44 be subjected to a bending moment M and to an eccentric axial force F, due to the statically indeterminate reaction of the reinforcement. Bending moment M^* and axial force F^* induce stress-intensity factors at the crack tip given, respectively, by Equations 12.168a and 12.169a.

On the other hand, M^* and F^* produce local rotations equal to, respectively

$$\varphi^{(M)} = \lambda_{MM} M^* \tag{12.174a}$$

$$\varphi^{(F)} = \lambda_{MF} F^* \tag{12.174b}$$

where

$$\lambda_{MM} = \frac{2}{b^2 tE} \int_0^\xi Y_M^2(\xi) d\xi \tag{12.175a}$$

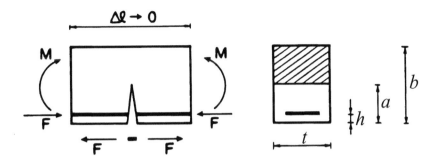

Figure 12.44

$$\lambda_{MF} = \frac{2}{btE} \int_0^{\xi} Y_M(\xi) Y_F(\xi) \, d\xi \tag{12.175b}$$

Up to the moment of steel yielding or slippage, the local rotation in the cracked cross section is equal to zero:

$$\varphi = \varphi^{(M)} + \varphi^{(F)} = 0 \tag{12.176}$$

Equation 12.176 is the congruence condition giving the unknown force F. Recalling that (Figure 12.44)

$$M^* = M - F(b/2 - h) \tag{12.177a}$$

$$F^* = -F \tag{12.177b}$$

Equations 12.174 and 12.176 provide

$$\frac{Fb}{M} = \frac{1}{(0.5 - h/b) + r(\xi)} \tag{12.178}$$

where

$$r(\xi) = \frac{\displaystyle\int_0^{\xi} Y_M(\xi) Y_F(\xi) \, d\xi}{\displaystyle\int_0^{\xi} Y_M^2(\xi) \, d\xi} \tag{12.179}$$

The statically indeterminate reaction of the reinforcement against the relative crack depth for $h/b = 1/10, 1/20$, is reported in Figure 12.45.

If a perfectly plastic behavior of the reinforcement is considered (yielding or slippage), from Equation 12.178 the moment of plastic flow for the reinforcement turns out to be

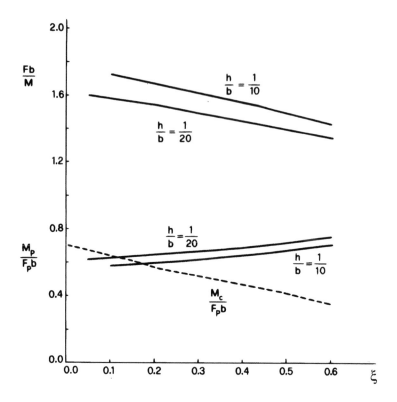

Figure 12.45

$$M_P = F_P b \left[(0.5 - h/b) + r(\xi) \right] \tag{12.180}$$

Such a moment against the relative crack depth, for $h/b = 1/10$, $1/20$, is reported in Figure 12.45. A limited increase in the moment of reinforcement plastic flow M_P occurs by increasing the crack depth ξ.

However, it should be observed that if concrete presents a low crushing strength f_c and/or steel a high yield strength f_y and/or a large area A_s, the crushing of concrete can precede the plastic flow of reinforcement.

If M_c is the bending moment of concrete crushing and a hypothesis of linear stress variation through the ligament holds, it follows that

$$\frac{M_c}{F_P b} = \frac{f_c}{f_y \dfrac{A_s}{A}} \frac{(1 - \xi)\left(2 + \xi - 3\dfrac{h}{b}\right)}{6} \tag{12.181}$$

The dashed line in the diagram in Figure 12.45 represents Equation 12.181 for $f_c = 19.62$ MPa; $f_y = 353.16$ MPa; $A_s/A = 0.024$; and $h/b = 1/10$. It can be observed that, although values very

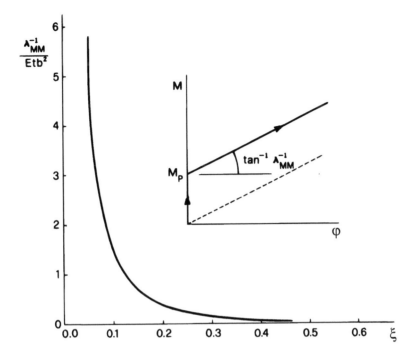

Figure 12.46

favorable to concrete crushing collapse have been chosen, such collapse, in fact, comes before the steel plastic flow only for sufficiently high values of the crack depth ($\xi > 0.175$).

The mechanical behavior of the cracked reinforced concrete beam section is rigid until the bending moment M_P is exceeded; that is, $\varphi = 0$ for $M \leq M_P$. On the other hand, for $M > M_P$ the $M-\varphi$ diagram becomes linear hardening:

$$\varphi = \lambda_{MM}\left[M - F_P\left(\frac{b}{2} - h\right)\right] - \lambda_{MF}F_P \qquad (12.182)$$

The $M-\varphi$ diagram is represented in Figure 12.46. This diagram expresses the equivalence of the beam section with a rigid–linear hardening spring. It is interesting to observe that the hardening line is parallel to the $M-\varphi$ diagram relating to the same cracked beam section without reinforcement (dashed line). The hardening coefficient λ_{MM} against the relative crack depth ξ is reported in Figure 12.46. By increasing the crack depth, the hardening line becomes more and more inclined, until producing a rigid–perfectly plastic behavior. On the other hand, for $\xi \to 0$, the hardening line becomes nearly vertical.

Therefore, it is possible to conclude that, by increasing ξ, the moment of steel plastic flow lightly increases (Figure 12.45), while the slope of the hardening line decreases (Figure 12.46). Some $M-\varphi$ diagrams for $h/b = 1/20$ are reported in Figure 12.47, with ξ varying between 0.05 and 0.50. The moment of steel plastic flow increases very little by increasing ξ, whereas the slope of the hardening line decreases sharply. The envelope of the hardening lines, by varying the crack depth ξ, represents the hardening response when the crack can grow in a stable manner.

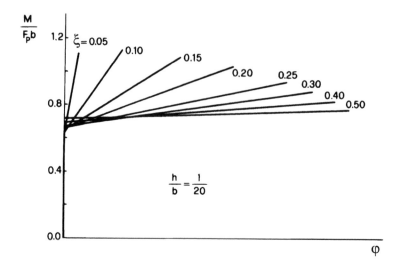

Figure 12.47

After the plastic flow of reinforcement, the stress-intensity factor is given by the superposition principle:

$$K_I = K_I^{(M)} + K_I^{(F)} \tag{12.183}$$

Recalling Equations 12.174 and considering the equivalent loads:

$$M^* = M - F_P(b/2 - h) \tag{12.184a}$$

$$F^* = -F_P \tag{12.184b}$$

the global stress-intensity factor is

$$K_I = \frac{Y_M(\xi)}{b^{3/2}t}\left[M - F_P\left(\frac{b}{2} - h\right)\right] - \frac{F_P}{b^{1/2}t}Y_F(\xi) \tag{12.185}$$

The moment of crack propagation is then

$$\frac{M_F}{K_{IC}b^{3/2}t} = \frac{1}{Y_M(\xi)} + N_P\left[\frac{Y_F(\xi)}{Y_M(\xi)} + \frac{1}{2} - \frac{h}{b}\right] \tag{12.186}$$

with

$$N_P = \frac{f_y b^{1/2}}{K_{IC}}\frac{A_s}{A} \tag{12.187}$$

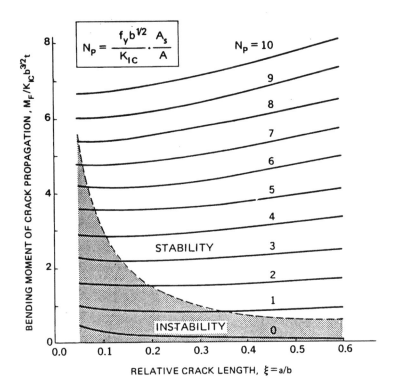

Figure 12.48

while the rotation at the crack propagation is

$$\varphi_F = \lambda_{MM}(M_F - M_P)$$ (12.188)

The crack propagation moment is plotted in Figure 12.48 as a function of the crack depth ξ and varying the brittleness number N_P. For low N_P values, that is, for low reinforced beams, the fracture moment decreases while the crack extends, and a typical phenomenon of unstable fracture occurs. For $N_P \geq 0.7$, a stable branch follows the unstable one; whereas, for $N_P \geq 8.5$, only the stable branch remains. The locus of the minima is represented by a dashed line in Figure 12.48. In the upper zone, the fracture process is stable, whereas it is unstable in the lower one.

Rigid behavior $(0 \leq M \leq M_P)$ is followed by linear hardening $(M_P < M < M_F)$. The latter stops when crack propagation occurs. If the fracture process is unstable, diagram M–φ presents a discontinuity and drops from M_F to $F_P b$ with a negative jump. In fact, in this case, a complete and instantaneous disconnection of concrete occurs. The new moment $F_P b$ can be estimated according to the scheme in Figure 12.49.

The nonlinear descending law:

$$M = F_P(b-h)\cos(\varphi/2)$$ (12.189)

can thus be approximated by the perfectly-plastic one:

$$M = F_P b$$ (12.190)

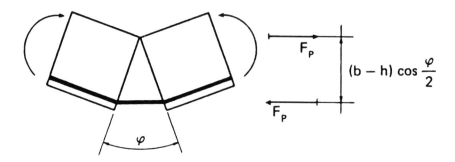

Figure 12.49

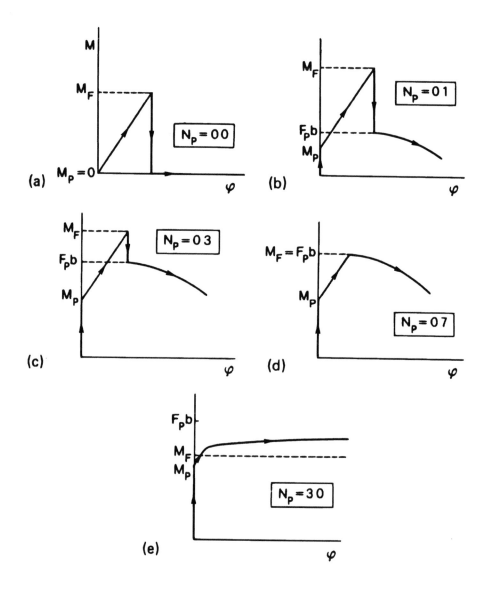

Figure 12.50

It is not difficult to demonstrate that, according to Equations 12.186, 12.187, and 12.190, the brittleness number N_P represents the ratio of the ultimate moment $F_p b$ to the fracture moment M_F.

On the other hand, if the fracture process is stable, diagram $M-\varphi$ does not present any discontinuity but rather an asymptote (see the envelope in Figure 12.47).

In Figure 12.50, the moment-rotation diagrams are reported for $\xi_0 = 0.1$ and five different values of brittleness number N_P. Once the cross section sizes and the material properties have been defined, they represent five different steel areas. For $N_P \leq 0.7$, it is $F_p b < M_F$ and, therefore, a discontinuity appears in the $M-\varphi$ diagram (Figures 12.50a through c). On the other hand, for $N_P \leq 0.7$, the curves in Figure 12.48 lie completely in the unstable zone. It is possible to conclude that, by increasing the steel percentage A_s/A, the concrete fracturing process becomes stable.

In Figure 12.51, the load versus midspan deflection diagrams are reported for a reinforced concrete beam subjected to TPBT, while varying the percentage of steel reinforcement with values of $\rho = 0.00\%$ (no reinforcement), 0.08%, 0.26%, and 0.65%. The same diagrams show the experimental and numerical results (cohesive crack model and reinforcement effect) obtained for a beam with a span of 0.60 m, a thickness of 0.15 m, and a depth of 0.10 m. Such curves evidence a transition from an overall softening response to a hardening response by increasing the steel percentage.

In the diagram where $\rho = 0.08\%$ (very low percentage of reinforcement), the contributions of the reinforcement do not involve any appreciable increase in the load-bearing

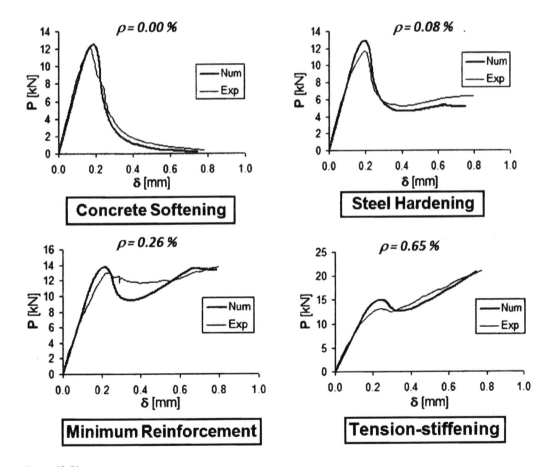

Figure 12.51

capacity with regard to the plain cross section. Therefore, the phenomenon of crack propagation remains unstable. On the other hand, when $\rho = 0.26\%$ or 0.65%, the benefit of the reinforcement is manifest, and the load-bearing capacity increases. The behavior changes from unstable to stable, and the failure does not turn out to be brittle if it is strain controlled. However, a snap-through instability is predicted if the process is load controlled.

In the first case, we have a simple **softening** behavior in flexure; in the second, we have a **plastic behavior in the ultimate stage**; in the third, the condition of **minimum reinforcement** is achieved; and, finally, the fourth represents a **tension-stiffening** behavior.

References

Albrecht, P., Namini, A., and Bosch, H. (1992). Finite element-based flutter analysis of cable-suspended bridges, *Journal of Structural Engineering* (ASCE), 118: 1509–1526.

Baldacci, R., Ceradini, G., and Giangreco, E. (1971). *Dinamica e Stabilità*, CISIA-Tamburini, Milan.

Baldacci, R., Ceradini, G., and Giangreco, E. (1974). *Plasticità*, CISIA-Tamburini, Milan.

Barenblatt, G.I. (1962). The mathematical theory of equilibrium cracks in brittle fracture, *Advances in Applied Mechanics*, 7: 55–129.

Bathe, K.J. and Wilson, E.L. (1976). *Numerical Methods in Finite Element Analysis*, Prentice-Hall, Englewood Cliffs, NJ.

Bažant, Z.P. and Cedolin, L. (1991). *Stability of Structures: Elastic, Inelastic, Fracture, and Damage Theories*, Oxford University Press, New York.

Belluzzi, O. (1941–1968). *Scienza delle Costruzioni*, Vols. 1–4, Zanichelli, Bologna.

Benvenuto, E. (1991). *An Introduction to the History of Structural Mechanics*, Springer-Verlag, New York.

Bolotin, V.V. (1963). *Nonconservative Problems of the Theory of Elastic Stability*, Pergamon, London.

Boresi, A.P. and Chong, K.P. (1987). *Elasticity in Engineering Mechanics*, Elsevier, New York.

Boscotrecase, L. and Di Tommaso, A. (1976). *Statica Applicata alle Costruzioni*, Patron, Bologna.

Brahtz, J.H.A. (1933). Stress distribution in a reentrant corner, *Transactions of the American Society of Mechanical Engineers* (ASME), 55: 31–37.

Bushnell, D. (1985). *Computerized Buckling Analysis of Shells*, Martinus Nijhoff, Dordrecht.

Capurso, M. (1971). *Lezioni di Scienza delle Costruzioni*, Pitagora, Bologna.

Capurso, M. (1981). Sul calcolo dei sistemi spaziali di controventamento, *Giornale del Genio Civile*, Fasc. I, II, III: 523–538.

Carpinteri, A. (1982). Notch sensitivity in fracture testing of aggregative materials, *Engineering Fracture Mechanics*, 16: 467–481.

Carpinteri, A. (1984). Stability of fracturing process in RC beams, *Journal of Structural Engineering* (ASCE), 110: 544–558.

Carpinteri, A. (1986). *Mechanical Damage and Crack Growth in Concrete: Plastic Collapse to Brittle Fracture*, Martinus Nijhoff, Dordrecht.

Carpinteri, A. (1989). Cusp catastrophe interpretation of fracture instability, *Journal of the Mechanics and Physics of Solids*, 37: 567–582.

Carpinteri, A. (1992a). *Meccanica dei Materiali e della Frattura*, Pitagora, Bologna.

Carpinteri, A. (1992b). *Scienza delle Costruzioni*, Vols. 1, 2, Pitagora, Bologna.

Carpinteri, A. (1997a). *Calcolo Automatico delle Strutture*, Pitagora, Bologna.

Carpinteri, A. (1997b). *Structural Mechanics: A Unified Approach*, Chapman & Hall, London.

Carpinteri, A. (1998a). *Analisi Non-lineare delle Strutture*, Pitagora, Bologna.

Carpinteri, A. (1998b). *Dinamica delle Strutture*, Pitagora, Bologna.

Carpinteri, A. (2014). *Structural Mechanics Fundamentals*, CRC Press, Boca Raton, FL.

Carpinteri, A. and Carpinteri, An. (1985). Lateral loading distribution between the elements of a three-dimensional civil structure, *Computers and Structures*, 21: 563–580.

Carpinteri, A. and Cornetti, P. (1997). Lastre a doppia curvatura, in Carpinteri, A., *Calcolo Automatico delle Strutture*, Pitagora, Bologna, 159–215.

Carpinteri, A., Lacidogna, G., and Accornero, F. (2015). Evolution of the fracturing process in masonry arches, *Journal of Structural Engineering* (ASCE), 141: 1–10.

Carpinteri, A., Lacidogna, G., and Puzzi, S. (2010). A global approach for three-dimensional analysis of tall buildings, *The Structural Design of Tall and Special Buildings*, 19: 518–536.

Carpinteri, A. and Paggi, M. (2013). A theoretical approach to the interaction between buckling and resonance instabilities, *Journal of Engineering Mathematics*, 78: 19–35.

Carpinteri, A., Paggi, M., and Zavarise, G. (2009). Cusp catastrophe interpretation of the stick-slip behaviour of rough surfaces, *Computer Modeling in Engineering and Sciences*, 1521: 1–23.

Chen, W.F. and Duan, L. (2000). *Bridge Engineering Handbook*, CRC Press, Boca Raton, FL.

Clough, R.W. and Penzien, J. (1975). *Dynamics of Structures*, McGraw-Hill, New York.

Colonnetti, G. (1941). *Scienza delle Costruzioni*, Einaudi, Turin.

Corradi, L. (1978). *Instabilità delle Strutture*, Clup, Milan.

Di Pasquale, S. (1975). *Scienza delle Costruzioni: Introduzione alla Progettazione Strutturale*, Tamburini, Milan.

Di Tommaso, A. (1981–1993). *Fondamenti di Scienza delle Costruzioni*, Vols. 1, 2, Patron, Bologna.

Drucker, D.C. (1951). A more fundamental approach to plastic stress-strain relations, *Proceedings of the 1st National Congress of Applied Mechanics*, I-487.

Drucker, D.C., Greenberg, H.J., and Prager, W. (1951). The safety factor of an elastic-plastic body in plane strain, *Journal of Applied Mechanics* (ASME), 18: 371–378.

Dugdale, D.S. (1960). Yielding of steel sheets containing slits, *Journal of the Mechanics and Physics of Solids*, 8: 100–104.

Erdogan, F., Sih, G.C. (1963). On the crack extension in plates under plane loading and transverse shear, *Journal of Basic Engineering*, 85: 519–525.

Fung, Y.C. (1965). *Foundations of Solid Mechanics*, Prentice-Hall, Englewood Cliffs, NJ.

Gavarini, C. (1978). *Dinamica delle Strutture*, Esa, Rome.

Griffith, A.A. (1921). The phenomena of rupture in solids, *Philosophical Transaction of the Royal Society*, A221: 163–198.

Gurtin, M.E. (1981). *An Introduction to Continuum Mechanics*, Academic Press, New York.

Hillerborg, A., Modéer, M., Petersson, P.E. (1976). Analysis of crack formation and crack growth in concrete by means of fracture mechanics and finite elements, *Cement and Concrete Research*, 6: 773–781.

Inglis, C.E. (1913). Stresses in a plate due to the presence of cracks and sharp corners, *Transaction of the Royal Institution of Naval Architects*, 60: 219–241.

Irwin, G.R. (1957). Analysis of stresses and strains near the end of a crack traversing a plate, *Journal of Applied Mechanics* (ASME), 24: 361–364.

Irwin, G.R. (1960). Plastic zone near a crack and fracture toughness, *Proceedings of the 7th Sagamore Conference*, IV-63.

Isaacson, E. and Keller, H.B. (1966). *Analysis of Numerical Methods*, Wiley, New York.

Johansen, K.W. (1962). *Yield-Line Theory*, Cement and Concrete Association, London.

Jones, R.M. (1975). *Mechanics of Composite Materials*, McGraw-Hill, New York.

Levi-Civita, T. and Amaldi, U. (1938). *Compendio di Meccanica Razionale*, Zanichelli, Bologna.

Malvern, M. (1969). *Introduction to the Mechanics of a Continuous Medium*, Prentice-Hall, Englewood Cliffs, NJ.

Massonnet, C. and Save, M. (1977). *Calcolo a Rottura delle Strutture*, Zanichelli, Bologna.

Muskhelishvili, N.I. (1933). Some basic problems of the mathematical theory of elasticity (in Russian), *Proceedings of the USSR Academy of Sciences*, Leningrad; (1953) 4th edn (in English), Noordhoff, Groningen.

Neal, B.G. (1977). *The Plastic Methods of Structural Analysis*, Chapman & Hall, London.

Novozhilov, V.V. (1961). *Theory of Elasticity*, Pergamon Press, Oxford.

Novozhilov, V.V. (1970). *Thin Shell Theory*, Noordhoff, Groningen.

Paz, M. (1980). *Structural Dynamics: Theory and Computation*, Van Nostrand, London.

Pignataro, M., Rizzi, N., and Luongo, A. (1991). *Stability, Bifurcation and Postcritical Behaviour of Elastic Structures*, Elsevier, Amsterdam.

Scanlan, R.H. and Simiu, E. (1987). *Wind Effects on Structures: Fundamentals and Applications to Design*, Wiley, New York.

Sih, G.C. (1966). On the Westergaard method of crack analysis, *International Journal of Fracture Mechanics*, 2: 628–631.

Stafford Smith, B., and Coull, A. (1991). *Tall Building Structures: Analysis and Design*, Wiley, New York.

Taranath, B.S. (1988). *Structural Analysis and Design of Tall Buildings*, McGraw-Hill, New York.

Timoshenko, S.P. (1928). *Vibration Problems in Engineering*, Van Nostrand, London.

Timoshenko, S.P. (1940). *Strength of Materials*, Van Nostrand, New York.

Timoshenko, S.P. (1945). Theory of bending, torsion and buckling of thin-walled members of open cross section, *Journal of the Franklin Institute*, 239: 201–361.

Timoshenko, S.P. and Goodier, J.N. (1970). *Theory of Elasticity*, McGraw-Hill, Tokyo.

Vlasov, V. (1961). *Thin-Walled Elastic Beams* (Israeli Program for Scientific Translation), US Science Foundation, Washington, DC.

Viola, E. (2010). *Teoria delle Strutture*, Pitagora, Bologna.

Virgin, N.L. (2007). *Vibration of Axially Loaded Structures*, Cambridge University Press, New York.

Westergaard, H.M. (1939). Bearing pressures and cracks, *Journal of Applied Mechanics* (ASME), 6: 49–53.

Wieghardt, K. (1907). Über das Spalten und Zerreißen elastischer Körper, *Zeitschrift für Mathematik und Physik*, 55: 60–103.

Williams, M.L. (1952). Stress singularities resulting from various boundary conditions in angular corners of plate in extension, *Journal of Applied Mechanics* (ASME), 19: 526–528.

Zienkiewicz, O.C. (1971). *The Finite Element Method in Engineering Science*, McGraw-Hill, London.

Appendix I: Static–kinematic duality for the shells of revolution—Reasons for the absence of the terms sin α/r from the kinematic matrix operator

As clearly shown by Equations 3.51 and 3.52, for the shells of revolution, the static and kinematic matrix operators are not exactly the transpose of the other but for the change of sign in the algebraic terms, as occurs for deformable three-dimensional solids, beams, and plates. From a mathematical point of view, this peculiarity derives from the application of Green's theorem to a surface element whose area cannot be expressed simply as the product of two differentials. In this appendix, we will prove this fundamental result, that is, the static–kinematic duality for the shells of revolution loaded symmetrically, where actually the infinitesimal element area is given by $r\mathrm{d}\vartheta\mathrm{d}s$.

In more detail, we will show how the kinematic equations can be derived from the static equations by means of the principle of virtual work. Let us consider a finite portion of a shell of revolution (S being its surface). Without losing generality, we can assume the portion to be bounded by two parallels and two meridians (together forming its contour C). The principle of virtual work states that external virtual work equals internal virtual work, that is to say

$$\int_S \left(p_s u + qw + m_s\varphi_s\right) r\,\mathrm{d}\vartheta\,\mathrm{d}s + \oint_C \left(N_s u + T_s w + M_s\varphi_s\right) r\,\mathrm{d}\vartheta$$

$$= \int_S \left(N_s\varepsilon_s + N_\vartheta\varepsilon_\vartheta + T_s\gamma_s + M_s\chi_s + M_\vartheta\chi_\vartheta\right) r\,\mathrm{d}\vartheta\,\mathrm{d}s \tag{I.1}$$

provided that the static system is equilibrated and the kinematic system is congruent.

By means of the static Equations 3.52, the external distributed forces can be substituted in the first integral at the left-hand side, which thus reads

$$\int_S \left[\left(-\frac{\mathrm{d}N_s}{\mathrm{d}s} - \frac{\sin\alpha}{r}N_s + \frac{\sin\alpha}{r}N_\vartheta - \frac{T_s}{R_1}\right)u + \left(\frac{N_s}{R_1} + \frac{N_\vartheta}{R_2} - \frac{\mathrm{d}T_s}{\mathrm{d}s} - \frac{\sin\alpha}{r}T_s\right)w\right.$$

$$\left. + \left(T_s - \frac{\mathrm{d}M_s}{\mathrm{d}s} - \frac{\sin\alpha}{r}M_s + \frac{\sin\alpha}{r}M_\vartheta\right)\varphi_s\right] r\,\mathrm{d}\vartheta\,\mathrm{d}s \tag{I.2}$$

The integrand function contains three derivatives. To these terms, we apply Green's theorem. The first one reads

$$-\int_S \frac{dN_s}{ds} u\, r\, d\vartheta\, ds = \int_S N_s \frac{d(u\, r)}{ds} d\vartheta\, ds - \oint_C N_s u\, r\, d\vartheta$$

$$= \int_S N_s \left(\frac{du}{ds} + \frac{\sin\alpha}{r} u \right) r\, d\vartheta\, ds - \oint_C N_s u\, r\, d\vartheta \tag{I.3a}$$

Analogously, for the other two terms, we have

$$-\int_S \frac{dT_s}{ds} w\, r\, d\vartheta\, ds = \int_S T_s \left(\frac{dw}{ds} + \frac{\sin\alpha}{r} w \right) r\, d\vartheta\, ds - \oint_C T_s w\, r\, d\vartheta \tag{I.3b}$$

$$-\int_S \frac{dM_s}{ds} \varphi_s\, r\, d\vartheta\, ds = \int_S M_s \left(\frac{d\varphi_s}{ds} + \frac{\sin\alpha}{r} \varphi_s \right) r\, d\vartheta\, ds - \oint_C M_s \varphi_s\, r\, d\vartheta \tag{I.3c}$$

Upon substitution of Equations I.3a through c into Equation I.2 and then into Equation I.1, we observe that several terms cancel each other, particularly the line integrals along the contour C. Thus, the principle of virtual work in Equation I.1 now reads

$$\int_S \left[N_s \left(\frac{du}{ds} + \frac{w}{R_1} \right) + N_\vartheta \left(\frac{\sin\alpha}{r} u + \frac{w}{R_2} \right) + T_s \left(-\frac{u}{R_1} + \frac{dw}{ds} + \varphi_s \right) + M_s \left(\frac{d\varphi_s}{ds} \right) \right.$$

$$\left. + M_\vartheta \left(\frac{\sin\alpha}{r} \varphi_s \right) \right] r\, d\vartheta\, ds$$

$$= \int_S (N_s \varepsilon_s + N_\vartheta \varepsilon_\vartheta + T_s \gamma_s + M_s \chi_s + M_\vartheta \chi_\vartheta) r\, d\vartheta\, ds \tag{I.4}$$

For the arbitrariness of the integration domain, the two integrand functions must be equal. Furthermore, because of the arbitrariness of the static system, the integrand functions equal each other if and only if

$$\varepsilon_s = \frac{du}{ds} + \frac{w}{R_1} \tag{I.5a}$$

$$\varepsilon_\vartheta = \frac{\sin\alpha}{r} u + \frac{w}{R_2} \tag{I.5b}$$

$$\gamma_s = -\frac{u}{R_1} + \frac{dw}{ds} + \varphi_s \tag{I.5c}$$

$$\chi_s = \frac{d\varphi_s}{ds} \tag{I.5d}$$

$$\chi_\vartheta = \frac{\sin\alpha}{r} \varphi_s \tag{I.5e}$$

Equations I.5 represent the kinematic equations for the shells of revolution, which, in matrix form, are given by Equation 3.51. This demonstration highlights how the static–kinematic duality is still valid, although three of the sin α/r terms are absent from the kinematic matrix operator. It can be easily extended to the general case of shells with double curvature.

Appendix II: Application of the finite element method to diffusion problems

The phenomenon of **heat conduction** in solid bodies can be described by a set of quantities and of equations that correspond exactly to those introduced in Chapter 3 for studying elastic plates and shells. These equations can, moreover, be discretized in the form described in Chapter 4.

Let us take as the principal unknown the **temperature** T. This is a scalar quantity and corresponds to the displacement vector in the elastic case. On the other hand, the quantity that corresponds to the deformation characteristics vector is the **temperature gradient:**

$$\{\operatorname{grad} T\}_{(3\times1)} = [\partial]_{(3\times1)} T_{(1\times1)} \tag{II.1}$$

which is a three-component vector.

The kinematic operator $[\partial]$, in this case, is the following vector operator:

$$[\partial] = \begin{bmatrix} \dfrac{\partial}{\partial x} \\[2ex] \dfrac{\partial}{\partial y} \\[2ex] \dfrac{\partial}{\partial z} \end{bmatrix} \tag{II.2}$$

The quantity corresponding to the static characteristics vector is the **heat flux**, which is a vector with three components that, according to the corresponding **coefficients of thermal conductivity,** are proportional to the respective components of the temperature gradient. Expressed in formulas, we have the following constitutive equation:

$$\begin{bmatrix} q_x \\ q_y \\ q_z \end{bmatrix} = -\begin{bmatrix} k_x & 0 & 0 \\ 0 & k_y & 0 \\ 0 & 0 & k_z \end{bmatrix} \begin{bmatrix} \dfrac{\partial T}{\partial x} \\[2ex] \dfrac{\partial T}{\partial y} \\[2ex] \dfrac{\partial T}{\partial z} \end{bmatrix} \tag{II.3}$$

or, in compact form:

$$\{q\} = - [k] \{\text{grad } T\}$$
$$\begin{array}{ccc} (3\times1) & (3\times3) & (3\times1) \end{array}$$
$$\tag{II.4}$$

The energy balance for the infinitesimal element in the **steady-state regime** yields, on the other hand, the following scalar equation:

$$\text{div}\{q\} = \dot{Q} \tag{II.5}$$

where $\dot{Q}$ denotes the **power generation per unit volume**, and the divergence operator can be represented in matrix form thus:

$$\text{div} = [\partial]^T \atop (1\times3) \tag{II.6}$$

where $[\partial]$ is the differential operator in Equation II.2.

Finally, combining the static Equation II.5 with the constitutive Equation II.4 and the kinematic Equation II.1, we obtain the operator equation:

$$\left(\underset{(1\times3)\ (3\times3)\ (3\times1)}{[\partial]^T [k] [\partial]} \right) T + \dot{Q} = 0 \tag{II.7}$$

which has the same form as the Lamé equation for elastic problems. At this point it is clear how, whereas the gradient and the divergence correspond, respectively, to the kinematic and static operators, the coefficients of thermal conductivity correspond to the stiffness, the power generation to the external body force, and the temperature to the displacement.

Since the energy balance in the **transient regime** yields, instead of Equation II.5, the following equation:

$$\text{div}\{q\} + c\frac{\partial T}{\partial t} = \dot{Q} \tag{II.8}$$

where c is the **thermal capacity** of the material, and t represents the time, Equation II.7 can be generalized in this regime as follows:

$$\left([\partial]^T [k][\partial] \right) T + \dot{Q} = c\frac{\partial T}{\partial t} \tag{II.9}$$

As regards the boundary conditions, these may be of two kinds, as in the elastic problems. In fact, on the boundary, it is possible to assign the **temperature** or the **normal heat flux**:

$$T = T_0, \quad \forall P \in S_T \tag{II.10a}$$

$$\left(\underset{(1\times3)\ (3\times3)\ (3\times1)}{[\mathcal{N}]^T [k] [\partial]} \right) T = -q_n, \quad \forall P \in S_q \tag{II.10b}$$

The matrix $[\mathcal{N}]$ represents the normal unit vector on the external surface:

$$[\mathcal{N}] = \begin{bmatrix} n_x \\ n_y \\ n_z \end{bmatrix} \tag{II.11}$$

and corresponds to Equation II.2 in the spirit of Green's theorem.

The finite element method can therefore be applied to discretize heat conduction problems and, more generally, all **diffusion problems** that are governed by equations altogether analogous to those introduced hitherto. Referring to Table 4.1, for such problems we have $g = 1$, $d = 3$. This means that the principal unknown is scalar, while the characteristic vector represents a flux that is, in isotropic cases ($k_x = k_y = k_z = k$), proportional to the gradient of the scalar unknown.

Examples of diffusion problems of applicational importance are **infiltration of fluids in porous media,** for which the scalar is the **fluid pressure,** while the constant k represents the **permeability** of the medium; **electrical conduction,** for which the scalar is the **potential,** while the constant k represents the **electrical conductivity.**

Finally, it should be noted that, in the case of a thermally isotropic material, Equation II.9 simplifies to the well-known form:

$$k\nabla^2 T + \dot{Q} = c\frac{\partial T}{\partial t} \tag{II.12}$$

Appendix III: Initial strains and residual stresses

Initial strains $\{\varepsilon_0\}$ may be due to nonmechanical internal causes, such as temperature variations, shrinkage, and phase transformations. The stresses will result from the difference between the actual and the initial strains. On the other hand, at the outset of analysis, the body could be stressed by some known field of initial **residual stresses** $\{\sigma_0\}$, due, for instance, to imposed displacements, constraint settlements, assemblage defects, and welding effects. These stresses must simply be added onto the general definition.

If a general elastic behavior is assumed, the relationship between stresses and strains will be linear and of the form

$$\{\sigma\} = [H](\{\varepsilon\} - \{\varepsilon_0\}) + \{\sigma_0\} \tag{III.1}$$

where $[H]$ is the Hessian of the elastic potential energy. Substituting the constitutive Equation III.1 and the kinematic equation into the static equation, we obtain

$$\left([\partial]^T [H][\partial]\right)\{\eta\} - [\partial]^T [H]\{\varepsilon_0\} + [\partial]^T \{\sigma_0\} + \{\mathcal{F}\} = \{0\} \tag{III.2}$$

or, in compact form

$$[\mathcal{L}]\{\eta\} = -\{\mathcal{F}\} - \{\mathcal{F}_{\sigma_0}\} + \{\mathcal{F}_{\varepsilon_0}\} \tag{III.3}$$

where $\{\mathcal{F}_{\sigma_0}\}$ and $-\{\mathcal{F}_{\varepsilon_0}\}$ represent equivalent body forces, due to residual stresses and initial strains, respectively.

If the effects of residual stresses and initial strains are merged with the body forces, the principle of minimum total potential energy may be demonstrated as in Section 4.3, and likewise the finite element method may be defined in the same manner as in Sections 4.3 and 4.4. In particular, the vectors of the equivalent nodal forces, according to Equation 4.44, are

$$\{F_e\} = \int_{V_e} [\eta_e]^T \{\mathcal{F}\} dV \tag{III.4a}$$

$$\{F_e^{\sigma_0}\} = \int_{V_e} [B_e]^T \{\sigma_0\} dV \tag{III.4b}$$

$$\left\{ F_e^{\varepsilon_0} \right\} = -\int_{V_e} \left[B_e \right]^{\mathrm{T}} \left[H \right] \{ \varepsilon_0 \} \, \mathrm{d}V \qquad\qquad\qquad\qquad\qquad \text{(III.4c)}$$

where the matrix $[B_e]$ is given in Equation 4.39. After the expansion and assemblage procedures, the finite element Equation 4.55 may be generalized as follows:

$$[K]\{\delta\} = \{F\} + \left\{ F^{\sigma_0} \right\} + \left\{ F^{\varepsilon_0} \right\} \qquad\qquad\qquad\qquad\qquad \text{(III.5)}$$

Appendix IV: Plane elasticity with couple stresses

In the classical theory of elasticity by Cauchy, it is assumed that the action of the material on one side of an elementary surface on the material on the other side of the same surface is equivalent to a force. In the **couple-stress theory**, introduced by Cosserat, the interaction is assumed to be equivalent to a force and a couple. The **couple stresses** are taken to be moments per unit area, just as the body couples are moments per unit volume.

For the plane problem, the indefinite equations of equilibrium in the case of a medium that can support couple stresses are (Figure IV.1)

$$
\begin{bmatrix}
\dfrac{\partial}{\partial x} & 0 & 0 & \dfrac{\partial}{\partial y} & 0 & 0 \\[2ex]
0 & \dfrac{\partial}{\partial y} & \dfrac{\partial}{\partial x} & 0 & 0 & 0 \\[2ex]
0 & 0 & +1 & -1 & \dfrac{\partial}{\partial x} & \dfrac{\partial}{\partial y}
\end{bmatrix}
\begin{bmatrix}
\sigma_x \\ \sigma_y \\ \tau_{xy} \\ \tau_{yx} \\ m_{xz} \\ m_{yz}
\end{bmatrix}
+
\begin{bmatrix}
\mathcal{F}_x \\ \mathcal{F}_y \\ \mathcal{M}_z
\end{bmatrix}
=
\begin{bmatrix}
0 \\ 0 \\ 0
\end{bmatrix}
\tag{IV.1}
$$

Accordingly, for nonconstant couple stresses ($\partial m_{xz}/\partial x \neq 0$, $\partial m_{yz}/\partial y \neq 0$), the shearing stresses are not necessarily equal (i.e., $\tau_{xy} \neq \tau_{yx}$).

In the definition of the deformation characteristics, it is possible to follow a heuristic procedure and consider the adjoint of the static operator:

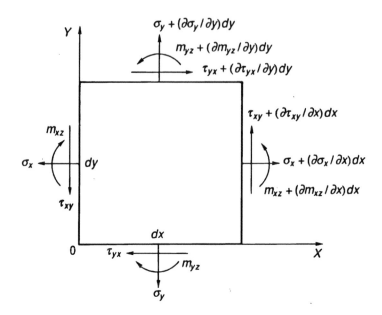

Figure IV.1

$$
\begin{bmatrix} \varepsilon_x \\[4pt] \varepsilon_y \\[4pt] \gamma_{xy} \\[4pt] \gamma_{yx} \\[4pt] \chi_{xz} \\[4pt] \chi_{yz} \end{bmatrix} = \begin{bmatrix} \dfrac{\partial}{\partial x} & 0 & 0 \\[8pt] 0 & \dfrac{\partial}{\partial y} & 0 \\[8pt] 0 & \dfrac{\partial}{\partial x} & -1 \\[8pt] \dfrac{\partial}{\partial y} & 0 & +1 \\[8pt] 0 & 0 & \dfrac{\partial}{\partial x} \\[8pt] 0 & 0 & \dfrac{\partial}{\partial y} \end{bmatrix} \begin{bmatrix} u \\[4pt] \upsilon \\[4pt] \varphi_z \end{bmatrix} \qquad (IV.2)
$$

Observe that, just as rotational equilibrium is contemplated in the static Equations IV.1, so in the kinematic Equations IV.2, the dual generalized displacement also appears, namely, the rotation φ_z. Consistent with this, in addition to the dilations, two shearing strains and two curvatures form the vector of the deformation characteristics. It is evident how the structure of Equations IV.1 and IV.2 is a combination of the equations for the elastic solid and those for the elastic beam.

The constitutive equations may be written as follows:

$$
\begin{bmatrix} \sigma_x \\ \sigma_y \\ \tau_{xy} \\ \tau_{yx} \\ m_{xz} \\ m_{yz} \end{bmatrix} = \begin{bmatrix} A & B & 0 & 0 & 0 & 0 \\ B & A & 0 & 0 & 0 & 0 \\ 0 & 0 & G & 0 & 0 & 0 \\ 0 & 0 & 0 & G & 0 & 0 \\ 0 & 0 & 0 & 0 & C & 0 \\ 0 & 0 & 0 & 0 & 0 & C \end{bmatrix} \begin{bmatrix} \varepsilon_x \\ \varepsilon_y \\ \gamma_{xy} \\ \gamma_{yx} \\ \chi_{xz} \\ \chi_{yz} \end{bmatrix} \tag{IV.3}
$$

with

$$
A = \frac{E}{1-v^2}, \quad B = \frac{Ev}{1-v^2}, \quad G = \frac{E}{2(1+v)}
$$

where E is Young's modulus and v is the Poisson ratio.

The three elements of the elasticity matrix, A, B, G, have the dimensions of stress, whereas C, that is, the bending modulus, has the dimensions of a force. By considering the ratio $C/G \sim l^2$, we can define a length scale l, which represents the scale at which local rotations φ_z take place. Above this scale, rotations may be neglected as in the classical theory of elasticity.

Appendix V: Shape functions

V.I RECTANGULAR FINITE ELEMENTS: LAGRANGE FAMILY

A recursive and relatively simple method for generating shape functions of any order is that of multiplying appropriate polynomials that tend to zero at the mesh nodes. Let us take, for instance, the element of Figure V.1, where a series of nodes, both internal and boundary, is located on a regular grid. Suppose we have to define the shape function of the node indicated by the double circle. Of course, the shape function sought will be given by the product of a fifth-order polynomial in ξ, having a value of unity in the second column of nodes and zero in all the others, and a fourth-order polynomial in η, having a value of unity in the first row of nodes and zero in all the others.

Polynomials in one variable that present the foregoing properties are termed **Lagrange polynomials**. For the nodes of abscissa ξ_i, we have the following polynomial:

$$L_i^{(n)} = \frac{(\xi - \xi_1)(\xi - \xi_2)\ldots(\xi - \xi_{i-1})(\xi - \xi_{i+1})\ldots(\xi - \xi_n)}{(\xi_i - \xi_1)(\xi_i - \xi_2)\ldots(\xi_i - \xi_{i-1})(\xi_i - \xi_{i+1})\ldots(\xi_i - \xi_n)} \tag{V.1}$$

The shape function of the node of coordinates (ξ_i, η_j) is thus given by the product

$$N_{ij}(\xi, \eta) = L_i^{(n)}(\xi) L_j^{(m)}(\eta) \tag{V.2}$$

where n and m represent the number of subdivisions in each direction.

Figure V.2 shows some elements of this unlimited family. The shape functions of element (b), which presents an internal node, are shown in Figure 4.5. Notwithstanding the ease with which such shape functions may be generated, their use is limited, on account of the high number of internal nodes and the poor ability of higher-order polynomials to approximate the solutions. The next section will outline a method of obviating such drawbacks.

V.2 RECTANGULAR FINITE ELEMENTS: SERENDIPITY FAMILY

Consider the elements of Figure V.3, which present the nodal points, spaced at equal intervals, only on the boundary sides. In the case of Element (a), Equation V.1 yields

$$N_{11}(\xi, \eta) = \frac{1}{4}(1 - \xi)(1 + \eta) \tag{V.3a}$$

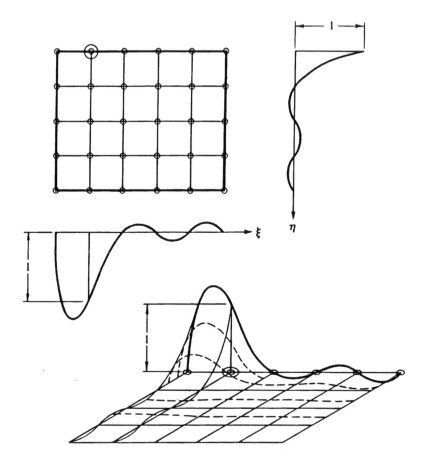

Figure V.1

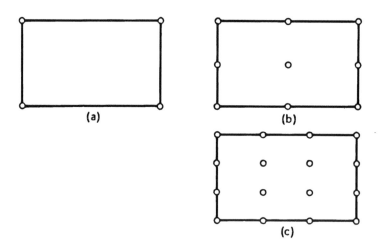

Figure V.2

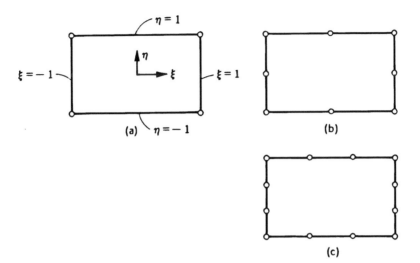

Figure V.3

$$N_{12}(\xi,\eta) = \frac{1}{4}(1+\xi)(1+\eta) \tag{V.3b}$$

$$N_{21}(\xi,\eta) = \frac{1}{4}(1-\xi)(1-\eta) \tag{V.3c}$$

$$N_{22}(\xi,\eta) = \frac{1}{4}(1+\xi)(1-\eta) \tag{V.3d}$$

In the case of element (b), which presents intermediate nodes on the sides, we have

$$N_{ij}(\xi,\eta) = \frac{1}{4}(1+\xi_0)(1+\eta_0)(\xi_0+\eta_0-1) \tag{V.4a}$$

at the corner nodes, and

$$N_{ij}(\xi,\eta) = \frac{1}{2}(1-\xi^2)(1+\eta_0), \quad \text{for } \xi_i = 0 \tag{V.4b}$$

$$N_{ij}(\xi,\eta) = \frac{1}{2}(1+\xi_0)(1-\eta^2), \quad \text{for } \eta_j = 0 \tag{V.4c}$$

at the mid-side nodes, having introduced the new coordinates:

$$\xi_0 = \xi\xi_i, \ \eta_0 = \eta\eta_j \tag{V.5}$$

It is possible to verify that in the case of the element of Figure V.3c, we have

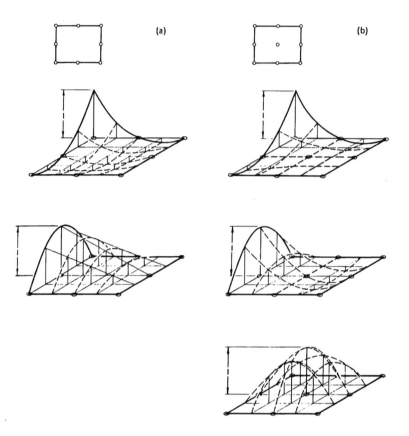

Figure V.4

$$N_{ij}(\xi,\eta) = \frac{1}{32}(1+\xi_0)(1+\eta_0)\left[-10+9\left(\xi^2+\eta^2\right)\right] \qquad (V.6a)$$

at the corner nodes, and

$$N_{ij}(\xi,\eta) = \frac{9}{32}(1+\xi_0)\left(1-\eta^2\right)(1+9\eta_0), \quad \text{for } \xi_i = \pm 1, \ \eta_j = \pm\frac{1}{3} \qquad (V.6b)$$

$$N_{ij}(\xi,\eta) = \frac{9}{32}\left(1-\xi^2\right)(1+\eta_0)(1+9\xi_0), \quad \text{for } \xi_i = \pm\frac{1}{3}, \ \eta_j = \pm 1 \qquad (V.6c)$$

at the mid-side nodes.

It is interesting to note how the Lagrangian and serendipity finite elements are identical only in their linear form (Figures V.2a and V.3a), whereas they differ, by the existence or otherwise of the central node, in their quadratic forms (Figures V.2b and V.3b). The corresponding shape functions are shown in Figure V.4a (serendipity element) and Figure V.4b (Lagrangian element).

V.3 TRIANGULAR FINITE ELEMENTS

It is well known how the triangular shape is the simplest and most appropriate for con-
structing meshes on complex-shape plane structural elements. Whereas Cartesian coordinates constitute the most natural choice for the rectangular element, the most appropriate
choice for the triangular element is represented by **area coordinates** (Figure V.5):

$$L_1 = \frac{\text{Area } P23}{\text{Area } 123} \tag{V.7a}$$

$$L_2 = \frac{\text{Area } 1P3}{\text{Area } 123} \tag{V.7b}$$

$$L_3 = \frac{\text{Area } 12P}{\text{Area } 123} \tag{V.7c}$$

For the triangular element of Figure V.6a, the shape functions are given simply by the area
coordinates:

$$N_1 = L_1, \; N_2 = L_2, \; N_3 = L_3 \tag{V.8}$$

For the triangular element of Figure V.6b, which also presents mid-side nodes, we have

$$N_1 = (2L_1 - 1)L_1 \tag{V.9a}$$

$$N_2 = (2L_2 - 1)L_2 \tag{V.9b}$$

$$N_3 = (2L_3 - 1)L_3 \tag{V.9c}$$

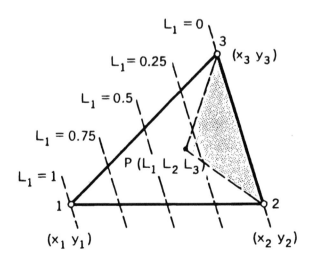

Figure V.5

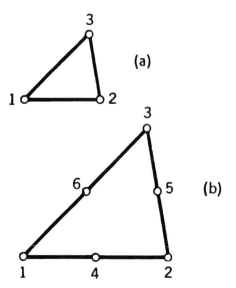

Figure V.6

at the corner nodes, and

$$N_4 = 4L_1L_2 \tag{V.9d}$$

$$N_5 = 4L_2L_3 \tag{V.9e}$$

$$N_6 = 4L_1L_3 \tag{V.9f}$$

at the mid-side nodes.

The area coordinates are, on the other hand, linked to the Cartesian coordinates by the following relations:

$$x = L_1x_1 + L_2x_2 + L_3x_3 \tag{V.10a}$$

$$y = L_1y_1 + L_2y_2 + L_3y_3 \tag{V.10b}$$

$$1 = L_1 + L_2 + L_3 \tag{V.10c}$$

V.4 THREE-DIMENSIONAL FINITE ELEMENTS

In the case of three-dimensional finite elements, shape functions altogether similar to the foregoing ones may be generated by simply adding a dimension.

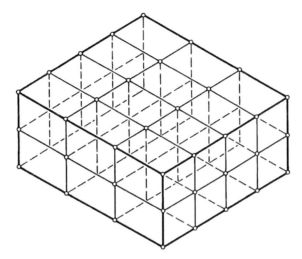

Figure V.7

In the case of **rectangular prisms**, the Lagrange family of functions (Figure V.7) is generated from the product of three polynomials. Extending the notation of Equation V.2, we have

$$N_{ijk}(\xi,\eta,\zeta) = L_i^{(n)}(\xi) L_j^{(m)}(\eta) L_k^{(l)}(\zeta) \tag{V.11}$$

On the other hand, the family of elements shown in Figure V.8 is altogether analogous to that of Figure V.3 (serendipity). For eight-node linear elements, we have (Figure V.8a)

$$N_{ijk}(\xi,\eta,\zeta) = \frac{1}{8}(1+\xi_0)(1+\eta_0)(1+\zeta_0) \tag{V.12}$$

whereas for 20-node quadratic elements, we have (Figure V.8b)

$$N_{ijk}(\xi,\eta,\zeta) = \frac{1}{8}(1+\xi_0)(1+\eta_0)(1+\zeta_0)(\xi_0+\eta_0+\zeta_0-2) \tag{V.13a}$$

at the corner nodes, and

$$N_{ijk}(\xi,\eta,\zeta) = \frac{1}{4}(1-\xi^2)(1+\eta_0)(1+\zeta_0), \quad \text{for}\,\xi_i = 0,\ \eta_j = \pm1,\ \zeta_k = \pm1 \tag{V.13b}$$

at four typical mid-side nodes.

Finally, for **tetrahedral elements** (Figure V.9), the properties are similar to those of triangular plane elements. Introducing volume coordinates analogous to those of Equations V.7:

$$L_1 = \frac{\text{Volume}\,P234}{\text{Volume}\,1234}, \text{etc.} \tag{V.14}$$

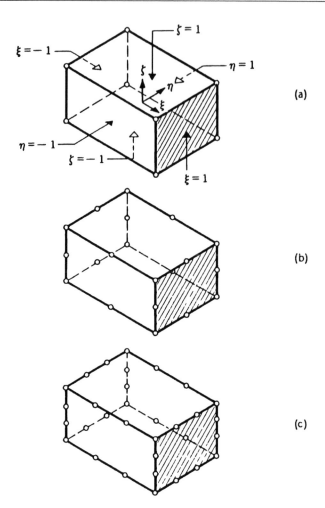

Figure V.8

we obtain linear shape functions coinciding with the corresponding volume coordinates (Figure V.9a), while the quadratic shape functions (Figure V.9b) for the 10-node tetrahedron are

$$N_i = (2L_i - 1)L_i, \qquad i = 1, 2, 3, 4 \tag{V.15}$$

for the corner nodes, and

$$N_5 = 4L_1L_2 \tag{V.16a}$$

$$N_6 = 4L_1L_3 \tag{V.16b}$$

$$N_7 = 4L_1L_4 \tag{V.16c}$$

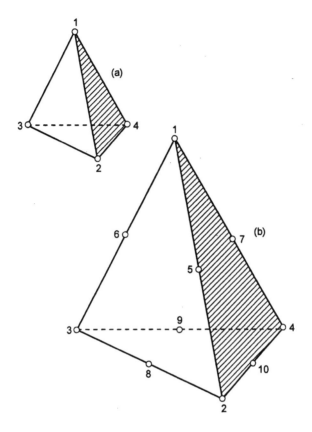

Figure V.9

$$N_8 = 4L_2L_3 \qquad\qquad\qquad\qquad\qquad\qquad\qquad\text{(V.16d)}$$

$$N_9 = 4L_3L_4 \qquad\qquad\qquad\qquad\qquad\qquad\qquad\text{(V.16e)}$$

$$N_{10} = 4L_2L_4 \qquad\qquad\qquad\qquad\qquad\qquad\qquad\text{(V.16f)}$$

for the mid-side nodes.

Appendix VI: Rotating circular disk

Equation 3.136 can be written in the form

$$\frac{d}{dr}(r\sigma_r) - \sigma_\vartheta + \rho\omega^2 r^2 = 0 \tag{VI.1}$$

where:
  $\rho\omega^2 r$ is the inertial force
  ρ is the material density
  ω is the angular velocity of the rotating circular disk

From Equation 3.134a, the strain components in the case of symmetry are

$$\varepsilon_r = \frac{du}{dr} \tag{VI.2a}$$

$$\varepsilon_\vartheta = \frac{u}{r} \tag{VI.2b}$$

The constitutive elastic equations then are

$$\sigma_r = \frac{E}{1-v^2}\left(\frac{du}{dr} + v\frac{u}{r}\right) \tag{VI.3a}$$

$$\sigma_\vartheta = \frac{E}{1-v^2}\left(\frac{u}{r} + v\frac{du}{dr}\right) \tag{VI.3b}$$

When the stresses in Equations VI.3 are substituted in Equation VI.1, we find that u must satisfy

$$r^2\frac{d^2u}{dr^2} + r\frac{du}{dr} - u = -\frac{1-v^2}{E}\rho\omega^2 r^3 \tag{VI.4}$$

The general solution of this equation is

$$u = \frac{1}{E}\left[(1-v)Cr - (1+v)C_1\frac{1}{r} - \frac{1-v^2}{8}\rho\omega^2 r^3\right] \tag{VI.5}$$

where C and C_1 are arbitrary constants. The corresponding stress components are now found from Equations VI.3:

$$\sigma_r = C + C_1 \frac{1}{r^2} - \frac{3+v}{8} \rho \omega^2 r^2 \qquad\qquad \text{(VI.6a)}$$

$$\sigma_\vartheta = C - C_1 \frac{1}{r^2} - \frac{1+3v}{8} \rho \omega^2 r^2 \qquad\qquad \text{(VI.6b)}$$

The integration constants C and C_1 are determined from the boundary conditions.

For a **solid disk**, we must take $C_1 = 0$ to have $u = 0$ at the center. The constant C is determined from the condition at the periphery ($r = b$) of the disk:

$$\sigma_r (r = b) = C - \frac{3+v}{8} \rho \omega^2 b^2 = 0 \qquad\qquad \text{(VI.7)}$$

from which

$$C = \frac{3+v}{8} \rho \omega^2 b^2 \qquad\qquad \text{(VI.8)}$$

The stress components of Equations VI.6 then take the following form:

$$\sigma_r = \frac{3+v}{8} \rho \omega^2 \left(b^2 - r^2 \right) \qquad\qquad \text{(VI.9a)}$$

$$\sigma_\vartheta = \frac{3+v}{8} \rho \omega^2 b^2 - \frac{1+3v}{8} \rho \omega^2 r^2 \qquad\qquad \text{(VI.9b)}$$

These stresses are greatest at the center of the disk:

$$\sigma_r (\max) = \sigma_\vartheta (\max) = \frac{3+v}{8} \rho \omega^2 b^2 \qquad\qquad \text{(VI.10)}$$

In the case of a **disk with a circular hole** of radius a at the center, the constants of integration in Equations VI.6 are obtained from the conditions at the inner and outer boundaries:

$$\sigma_r (r = a) = \sigma_r (r = b) = 0 \qquad\qquad \text{(VI.11)}$$

The calculation gives

$$C = \frac{3+v}{8} \rho \omega^2 \left(a^2 + b^2 \right) \qquad\qquad \text{(VI.12a)}$$

$$C_1 = \frac{3+v}{8} \rho \omega^2 a^2 b^2 \qquad\qquad \text{(VI.12b)}$$

Substituting in Equations VI.6, we have

$$\sigma_r = \frac{3+\nu}{8}\rho\omega^2\left(a^2 + b^2 - \frac{a^2b^2}{r^2} - r^2\right) \tag{VI.13a}$$

$$\sigma_\vartheta = \frac{3+\nu}{8}\rho\omega^2\left(a^2 + b^2 - \frac{a^2b^2}{r^2} - \frac{1+3\nu}{3+\nu}r^2\right) \tag{VI.13b}$$

We find the maximum radial stress at $r = \sqrt{ab}$,

$$\sigma_r(\max) = \frac{3+\nu}{8}\rho\omega^2(b-a)^2 \tag{VI.14a}$$

and the maximum circumferential stress at the inner boundary:

$$\sigma_\vartheta(\max) = \frac{3+\nu}{4}\rho\omega^2\left(b^2 + \frac{1-\nu}{3+\nu}a^2\right) \tag{VI.14b}$$

The latter is larger than σ_r (max).

When the radius a of the hole approaches zero, the maximum circumferential stress approaches a value twice as great as that for a solid disk in Equation VI.10; that is, by making a small circular hole at the center of a rotating disk, we double the maximum stress. This is a phenomenon of stress concentration similar to that discussed in Section 11.6.

Appendix VII: Thermal stress in a circular disk

The thermal stresses σ_r and σ_ϑ satisfy Equation 3.136 with $\mathcal{F}_r = 0$ in the case of a circular disk with a temperature distribution symmetrical about the center. The strain is due partly to stress and partly to thermal expansion:

$$\varepsilon_r = \frac{1}{E}\left(\sigma_r - \nu\sigma_\vartheta\right) + \alpha T \tag{VII.1a}$$

$$\varepsilon_\vartheta = \frac{1}{E}\left(\sigma_\vartheta - \nu\sigma_r\right) + \alpha T \tag{VII.1b}$$

Solving Equations VII.1 for σ_r, σ_ϑ, we find

$$\sigma_r = \frac{E}{1-\nu^2}\left[\varepsilon_r + \nu\varepsilon_\vartheta - (1+\nu)\alpha T\right] \tag{VII.2a}$$

$$\sigma_\vartheta = \frac{E}{1-\nu^2}\left[\varepsilon_\vartheta + \nu\varepsilon_r - (1+\nu)\alpha T\right] \tag{VII.2b}$$

Equation 3.136 then becomes

$$r\frac{d}{dr}\left(\varepsilon_r + \nu\varepsilon_\vartheta\right) + (1-\nu)\left(\varepsilon_r - \varepsilon_\vartheta\right) = (1+\nu)\alpha r\frac{dT}{dr} \tag{VII.3}$$

Substituting Equations VI.2 in Equation VII.3, we obtain

$$\frac{d^2u}{dr^2} + \frac{1}{r}\frac{du}{dr} - \frac{u}{r^2} = (1+\nu)\alpha\frac{dT}{dr} \tag{VII.4}$$

which may be written as

$$\frac{d}{dr}\left[\frac{1}{r}\frac{d(ur)}{dr}\right] = (1+\nu)\alpha\frac{dT}{dr} \tag{VII.5}$$

Integration of this equation yields

$$u = (1+v)\alpha\frac{1}{r}\int_a^r Tr\,dr + C_1 r + \frac{C_2}{r} \tag{VII.6}$$

where a is the inner radius for a disk with a hole, or zero for a solid disk.

The stress components are found by using the solution of Equation VII.6 in Equations VI.2 and substituting the results in Equations VII.2:

$$\sigma_r = -\alpha E\frac{1}{r^2}\int_a^r Tr\,dr + \frac{E}{1-v^2}\left[C_1(1+v) - C_2(1-v)\frac{1}{r^2}\right] \tag{VII.7a}$$

$$\sigma_\vartheta = \alpha E\frac{1}{r^2}\int_a^r Tr\,dr - \alpha ET + \frac{E}{1-v^2}\left[C_1(1+v) + C_2(1-v)\frac{1}{r^2}\right] \tag{VII.7b}$$

The constants C_1 and C_2 are determined by the boundary conditions.

For a **solid disk** $(a=0)$, the constant C_2 must be equal to zero in order that u may be zero at the center, since

$$\lim_{r\to 0}\frac{1}{r}\int_0^r Tr\,dr = \lim_{r\to 0}\left(\frac{1}{2}T_0 r\right) = 0 \tag{VII.8}$$

T_0 being the temperature at the center.

The boundary condition $\sigma_r\,(r=b)=0$ gives

$$C_1 = (1-v)\frac{\alpha}{b^2}\int_0^b Tr\,dr \tag{VII.9}$$

so that the final expressions for the stresses are, consequently,

$$\sigma_r = \alpha E\left(\frac{1}{b^2}\int_0^b Tr\,dr - \frac{1}{r^2}\int_0^r Tr\,dr\right) \tag{VII.10a}$$

$$\sigma_\vartheta = \alpha E\left(-T + \frac{1}{b^2}\int_0^b Tr\,dr + \frac{1}{r^2}\int_0^r Tr\,dr\right) \tag{VII.10b}$$

These give finite values at the center, since

$$\lim_{r\to 0}\frac{1}{r^2}\int_0^r Tr\,dr = \frac{1}{2}T_0 \tag{VII.11}$$

Index

Aerodynamic damping, 288
Aerodynamic moment, 279
Aeroelastic flutter, 283–291
Aeroelastic instability, 278
Airy stress function, 435
Alternating plastic deformation, 378–379
Angular congruence, 17, 22, 38
Assemblage, 124
Axial forces, 37, 58

Beam
    boundary conditions for single, 165–172
        cantilever, 167–168
        clamped–hinged, 172
        double clamped, 170–171
        rope in tension, 168–169
        simply supported, 165–167
        unconstrained, 169–170
    continuous, on three/more supports,
        173–175
    eccentric compression on cracked, 482–484
    finite element formulation for, 266–270
    flexural oscillations of, 261–264
    framed, systems, 223–227
    frames with nonorthogonal, 40–44
    modal analysis of deflected, 162–165
    oscillations and lateral torsional buckling of
        deep, 264–266
    rectilinear elastic, 214–223
    stability of fracturing process in reinforced
        concrete, 484–492
    Vlasov's theory of thin-walled open-section,
        in torsion, 301–309
Beam systems
    automatic computation of, 58–65
    with axial skew-symmetry, 9–13
    with axial symmetry, 4–9
    dynamics of, 180–182
    incremental plastic analysis of, 338–351
    loaded proportionally
        by concentrated forces, 357–362
        by distributed forces, 362–371
    nonproportionally loaded, 371–375
    parallel-arranged, 54–58

with polar skew-symmetry, 14–16
with polar symmetry, 13–14
Betti's reciprocal theorem, 53, 117, 119
Bimoment, 302
Bridged crack model, 484–492
Bridge flutter analysis, 284
Buckling, dynamics and, 253–291
    discrete systems with one/two degrees of
        freedom, 254–260
    finite element formulation, 266–270
    flexural oscillations of beams, 261–264
    flutter, 283–291
    galloping, 280–283
    influence of dead loads on natural
        frequencies, 254
    nonconservative loading and flutter, 270–277
    oscillations and lateral torsional, 264–266
    torsional divergence, 278–280
    wind effects on suspension/cable-stayed
        bridges, 277–278
Buckling instability, 203–251
    discrete mechanical systems
        with one degree of freedom, 204–206
        with two/more degrees of freedom,
            206–214
    framed beam systems, 223–227
    lateral torsional buckling, 231–233
    plates subjected to compression, 234–238
    rectilinear elastic beams, 214–223
    rings and cylindrical shells, 227–231
    shallow arches and shells subjected to
        vertical loading, 239–244
    trussed vaults and domes, 244–251

Cable-stayed bridges, 277–278
Cantilever beam, 167–168
Capurso's method, 309–313
Cauchy–Riemann conditions, 412–415
Circular disk
    rotating, 521–523
    thermal stress in, 525–526
Circular plates, 97–103
Clamped–hinged beam, 172
Clapeyron's theorem, 118, 452, 454

Classical flutter, 286
CMOD, *see* Crack mouth opening displacement
  (CMOD)
COD, *see* Crack opening displacement (COD)
Coefficient of distribution, 51
Cohesive crack model, 472–482
Collapse, mechanism of, 359, 364, 372
Combining mechanisms, method of, 359
Complex potentials, method of, *see*
  Westergaard's method
Compressive stresses, 396
Continuous elastic systems, dynamics of,
  161–201
  beam on three/more supports, 173–175
  boundary conditions for single beam, 165–172
    cantilever, 167–168
    clamped–hinged, 172
    double clamped, 170–171
    rope in tension, 168–169
    simply supported, 165–167
    unconstrained, 169–170
  dynamics of
    beam systems, 180–182
    elastic solids with linear viscous damping,
      200–201
    shells and three-dimensional elastic solids,
      195–199
  forced oscillations of shear-type multistory
    frames, 182–189
  method of approximation of Rayleigh–Ritz,
    175–180
  modal analysis of deflected beams, 162–165
  vibrating membranes, 189–193
  vibrating plates, 193–195
Convexity, of plastic limit surface, 351–354
Coordinate transformation, 199
Coupled flutter, *see* Classical flutter
Couple stresses, 507–509
Crack branching criterion, 458–462
Crack mouth opening displacement (CMOD),
  474
Crack opening displacement (COD), 441–442
Crack propagation, 489
Curvilinear coordinate, 89
Cylindrical shells, 104–106

D'Alembert's principle, 274
Damped free vibrations, 132–137
Damping-driven flutter, *see* Uncoupled flutter
Deflected circular plates, 379–383
Deflected rectangular plates, 383–388
Differential motion equation, 192
Diffusion problems, 501–503
Direct Fourier transform, 148
Discrete mechanical systems
  with one degree of freedom, 204–206
  with two/more degrees of freedom, 206–214
Discrete systems, dynamics of, 129–160

elastic–perfectly plastic spring, 152–154
free vibrations, 129–137
  damped, 132–137
  undamped, 131–132
general dynamic loading, 146–149
harmonic loading and resonance, 137–141
  systems with viscous damping, 138–141
  undamped systems, 137–138
impulsive loading, 143–146
linear elastic systems, 154–156
nonlinear elastic systems, 149–151
periodic loading, 142–143
Rayleigh ratio, 156–157
Stodola–Vianello method, 158–160
Displacements, method of, 49–73
  beam systems
    automatic computation of, 58–65
    parallel-arranged, 54–58
  parallel-arranged bar systems, 49–54
  plane frames, 67–69
  plane grids, 69–71
  plane trusses, 65–67
  space trusses and frames, 71–73
Double clamped beam, 170–171
Double-layer grid, 245
Ductile–brittle transition, 466–472
Ductility, 471
Dugdale's model, 466

Eigenfunctions, 161, 263
Eigenvalue, 156, 214, 258, 260, 262, 449–450
Eigenvector, 158–160, 182, 197, 199, 208–209
Elastic deformed configuration, 100, 103
Elastic equilibrium, 203
Elastic-hardening-softening behavior, 472–473
Elastic–perfectly plastic spring, 152–154
Elastic–plastic flexure, 332–338
Euler method, 272, 276
Euler's critical load, 219
Euler's hyperbola, 220

Finite element method, 113–127, 181, 266–270
  application of
    to diffusion problems, 501–503
    principle of virtual work, 121–126
  kinematic boundary conditions, 127
  principle of minimum total potential energy,
    116–118
  Ritz–Galerkin method, 118–121
  single-degree-of-freedom system, 113–115
Fixed-joint nodes, 58
Flexural oscillations, of beams, 261–264
Flexural rigidity, 266
Flutter load, 274
Flutter speed, and frequency, 284
Fracture, mechanics of, 427–492
  bridged crack model, 484–492
  cohesive crack model, 472–482

crack branching criterion in mixed mode
        condition, 458–462
eccentric compression on cracked beam,
        482–484
Griffith's energy criterion, 429–433
Irwin's theorem, 451–458
mode II and mixed modes, 442–446
plastic zone at crack tip, 462–466
relation between energy and stress
        treatments, 451–458
size effects and ductile–brittle transition,
        466–472
snap-back instability, 472–482
stability of, in reinforced concrete beams,
        484–492
Westergaard's method, 433–442
Williams' method, 446–451
Framed beam systems, 223–227
Free vibrations, 129–137
    damped, 132–137
    undamped, 131–132
Fundamental frequency, and mode, 156

Galloping, 280–283
General dynamic loading, 146–149
Geometric stiffness matrix, 266–269
Golden Gate Bridge, 288, 289
Green's theorem, 497
Griffith's energy criterion, 428, 429–433

Harmonic loading, and resonance, 137–141
    systems with viscous damping, 138–141
    undamped systems, 137–138
Heat conduction, 501, 503
High-rise structures, 293–327
    Capurso's method, 309–313
    diagonalization of Vlasov's equations,
        314–316
    dynamic analysis of tall buildings, 316–319
    numerical example, 319–327
    thin-walled open-section beams in torsion,
        301–309
    vertical cantilevers
        lateral loading distribution between thin-
            walled open-section, 309–313
        parallel-arranged system of, 294–301
    Vlasov's theory, 301–309
Hinge-nodes, 58
Huygens–Steiner theorem, 310

Impulsive loading, 143–146
Incremental plastic analysis, 331, 338–351
Incremental plastic deformation, 351–354,
        357, 382
Initial strains, 505–506
Initial stress matrix, 267
Interpolation method, 115
Inverse Fourier transform, 148

Irwin's theorem, 451–458
Isoparametric finite element method, 120

John Hancock Center, 294, 295

Kinematic equations, 84, 87, 497
Kinematic matrix operator, 497–499
Kinematic theorem, 356
Kirchhoff kinematic hypothesis, 75, 77
Kolosoff–Muskhelishvili method, 415–419

Lagrange polynomials, 511
Lamé's equation, 116
Lateral torsional buckling, 231–233
Law of normality, 351–354
Linear elastic fracture mechanics (LEFM), 473
Linear elastic systems, 154–156
Load, of plastic collapse, 331
Load-geometric matrix, see Initial stress matrix
Load vs. midspan deflection diagrams, 491
Long-duration loadings, 145
Long-span structures, 253–291
    discrete systems with one/two degrees of
        freedom, 254–260
    finite element formulation, 266–270
    flexural oscillations of beams, 261–264
    flutter, 283–291
    galloping, 280–283
    influence of dead loads on natural
        frequencies, 254
    nonconservative loading and flutter, 270–277
    oscillations and lateral torsional buckling,
        264–266
    torsional divergence, 278–280
    wind effects on suspension/cable-stayed
        bridges, 277–278
Lower bound theorem, see Kinematic theorem

Membranes
    and thin shells of revolution, 93–97
    vibrating, 189–193
Meridians, 89
Meusnier's theorem, 89
Minimum total potential energy, principle of,
        115, 116–118
Mixed theorem, 356
Modal analysis, 156
Mode II, and mixed modes, 442–446
Moment diagram, 19, 21
Moment-rotation diagrams, 491
Motion dependent force, 278
Motion independent force, 278

Neal and Symonds' method, 359, 361
Nonconservative loading, and flutter,
        270–277
Nonlinear elastic systems, 149–151
Nonlinear step-by-step analysis, 245

Oscillations, and lateral torsional buckling, 264–266
Oscillatory instability/flutter, 274

Parallel-arranged bar systems, 49–54
Parallel-arranged beam systems, 54–58
Parallels, 89
Periodic loading, 142–143
Plane elasticity, 507–509
Plane frames, 1–47, 67–69
    beam systems
        with axial skew-symmetry, 9–13
        with axial symmetry, 4–9
        with polar skew-symmetry, 14–16
        with polar symmetry, 13–14
    loaded out of their own plane, 44–47
    with nonorthogonal beams, 40–44
    rotating-node, 17–27
    thermal loads and imposed displacements, 35–40
    translating-node, 27–35
Plane grids, 69–71
Plane stress, and plane strain conditions, 389–426
    analytical functions, 411–415
    circular hole in plate subjected to tension, 404–408
    concentrated force acting on edge of elastic half-plane, 408–411
    in cylindrical/or prismatic solids, 392–394
    deep beam, 394–399
    elliptical hole in plate subjected to tension, 420–426
    Kolosoff–Muskhelishvili method, 415–419
    thick-walled cylinder, 399–404
    in thin plates, 389–392
Plane trusses, 65–67
Plastic collapse, 468, 469–470, 471
Plastic deformation, 332–333, 335, 352
Plasticity, theory of, 329–388
    beam systems
        loaded proportionally by concentrated forces, 357–362
        loaded proportionally by distributed forces, 362–371
        nonproportionally loaded, 371–375
    convexity of plastic limit surface, 351–354
    cyclic loading and shake-down, 375–379
    deflected circular plates, 379–383
    deflected rectangular plates, 383–388
    elastic–plastic flexure, 332–338
    incremental plastic analysis of beam systems, 338–351
    law of normality of incremental plastic deformation, 351–354
    theorems of plastic limit analysis, 354–357
        addition of material, 356–357
        kinematic, 356

lower bound, 356
maximum dissipated energy, 354–355
mixed theorem, 356
static, 355–356
upper bound, 355–356
Plastic limit analysis, theorems of, 331, 354–357
    addition of material, 356–357
    kinematic, 356
    lower bound, 356
    maximum dissipated energy, 354–355
    mixed theorem, 356
    static, 355–356
    upper bound, 355–356
Plastic zone, see Process zone
Plate, 75–111
    circular, 97–103
    deflected circular, 379–383
    deflected rectangular, 383–388
    finite element formulation for, 266–270
    in flexure, 75–82
    subjected to compression, 234–238
    subjected to tension
        circular hole in, 404–408
        elliptical hole in, 420–426
    vibrating, 193–195
Potential well, 114
Power method, 158
Prandtl's formula, 233
Process zone, 473

Quasi-static loading, 145

Rayleigh ratio, 156–157
Rayleigh–Ritz approximation method, 175–180
Rectangular finite elements, 511–514
    Lagrange family, 511
    serendipity family, 511–514
Rectilinear elastic beams, 214–223
Redundant moments, 27
Residual stresses, 505–506
Rings, and cylindrical shells, 227–231
Ritz–Galerkin method, 118–121, 237
Rotating-node frames, 1, 17–27, 67, 68

Saint-Venant's theory, 302, 315, 396, 398
Sears Tower (now Willis Tower), 294, 295
Serendipity finite elements, 511–514
Series expansion method, see Williams' method
Shallow arches, and shells, 239–244
Shape functions, 511–519
    rectangular finite elements, 511–514
        Lagrange family, 511
        serendipity family, 511–514
    three-dimensional finite elements, 516–519
    triangular finite elements, 515–516
Shear diagram, 19, 21, 33, 37
Shearing forces, 58
Shearing stresses, 396

Shear-type frames, *see* Translating-node frames
Shear-type multistory frames, 182–189
Shear walls, 301
Shell of revolution
    membranes and thin, 93–97
    nonsymmetrically loaded, 88–90
    symmetrically loaded, 91–92
Shells, 75–111
    cylindrical, 104–106
        pressurized vessels with bottoms, 106–110
        subjected to external pressure, 227–231
    with double curvature, 84–88
    dynamics of, and three-dimensional elastic
        solids, 195–199
    finite element formulation for, 266–270
    of revolution
        membranes and thin, 93–97
        nonsymmetrically loaded, 88–90
        symmetrically loaded, 91–92
    subjected to vertical loading, 239–244
Short-duration impulsive loadings, 145, 146
Simply supported beam, 165–167
Single-degree-of-freedom system, 113–115
Single-layer grid, 245
Snap-back instability, 472–482
Snap-through, phenomena of
    case of progressive, 244–251
    interaction between buckling and, 239–244
Sophie Germain's equation, 82–84
Space trusses, and frames, 71–73
Splines, 120
Statically indeterminate beam systems, 49–73
    automatic computation of, 58–65
    parallel-arranged, 54–58
    parallel-arranged bar systems, 49–54
    plane frames, 67–69
    plane grids, 69–71
    plane trusses, 65–67
    space trusses and frames, 71–73
Static equations, 497
Static–kinematic duality, 497–499
Static theorem, 355–356
Steady-state response, 138, 139
Stiffness-driven flutter, *see* Classical flutter
Stiffness matrix, 59, 64, 124, 126, 195, 196, 210

Stodola–Vianello method, 158–160
Stress–COD relation, 472
Stress concentration, 430
Stress–strain relation, 472
Suspension bridges, 277–278

Tacoma Narrows bridge, 284, 288
Thin-walled open-section beams, in torsion,
    301–309
Three-dimensional bodies, of revolution,
    110–111
Three-dimensional finite elements, 516–519
Three-point-bending test (TPBT), 478
Torsional divergence, 278–280
TPBT, *see* Three-point-bending test (TPBT)
Transient response, 138
Translating-node frames, 1, 27–35, 67, 68
Tresca's criterion, 354
Triangular finite elements, 515–516
Trussed vaults, and domes, 244–251
Truss structure, 2, 3, 27, 31, 37, 40

Unconstrained beam, 169–170
Uncoupled flutter, 286
Undamped free vibrations, 131–132
Undamped systems, 137–138
Upper bound theorem, *see* Static theorem

Variational formulation, 118
Vertical cantilevers, 294–301
    lateral loading distribution between thin-
        walled open-section, 309–313
    parallel-arranged system of, 294–301
Virtual work, principle of, 38, 121–126, 361,
    364, 386, 497
Vlasov's equations, 314–316
Vlasov's theory, 301–309

Warping, 302
Westergaard, H. M., 419
Westergaard's method, 428, 433–442
Williams' method, 428, 446–451
Willis Tower (formerly Sears Tower), 294, 295
Wind effects, on suspension/cable-stayed
    bridges, 277–278

Milton Keynes UK
Ingram Content Group UK Ltd.
UKHW052026071024
449327UK00027B/2441